INTRODUCTION TO NONDESTRUCTIVE TESTING

INTRODUCTION TO NONDESTRUCTIVE TESTING

A Training Guide

Second Edition

PAUL E. MIX

A JOHN WILEY & SONS, INC., PUBLICATION

Copyright © 2005 by John Wiley & Sons, Inc. All rights reserved

Published by John Wiley & Sons, Inc., Hoboken, New Jersey
Published simultaneously in Canada

No part of this publication may be reproduced, stored in a retrieval system, or transmitted in any form or by any means, electronic, mechanical, photocopying, recording, scanning, or otherwise, except as permitted under Section 107 or 108 of the 1976 United States Copyright Act, without either the prior written permission of the Publisher, or authorization through payment of the appropriate per-copy fee to the Copyright Clearance Center, Inc., 222 Rosewood Drive, Danvers, MA 01923, (978) 750-8400, fax (978) 750-4470, or on the web at www.copyright.com. Requests to the Publisher for permission should be addressed to the Permissions Department, John Wiley & Sons, Inc., 111 River Street, Hoboken, NJ 07030, (201) 748-6011, fax (201) 748-6008, or online at http://www.wiley.com/go/permission.

Limit of Liability/Disclaimer of Warranty: While the publisher and author have used their best efforts in preparing this book, they make no representations or warranties with respect to the accuracy or completeness of the contents of this book and specifically disclaim any implied warranties of merchantability or fitness for a particular purpose. No warranty may be created or extended by sales representatives or written sales materials. The advice and strategies contained herein may not be suitable for your situation. You should consult with a professional where appropriate. Neither the publisher nor author shall be liable for any loss of profit or any other commercial damages, including but not limited to special, incidental, consequential, or other damages.

For general information on our other products and services or for technical support, please contact our Customer Care Department within the United States at (800) 762-2974, outside the United States at (317) 572-3993 or fax (317) 572-4002.

Wiley also publishes its books in a variety of electronic formats. Some content that appears in print may not be available in electronic formats. For more information about Wiley products, visit our web site at www.wiley.com.

Library of Congress Cataloging-in-Publication Data:
Mix, Paul E.
    Introduction to nondestructive testing : a training guide / Paul E. Mix.—2nd ed.
       p. cm.
    Includes index.
    ISBN-13 978-0-471-42029-3 (cloth)
    ISBN-10 0-471-42029-8 (cloth)
  1. Nondestructive testing. I. Title.
    TA417. 2. M59 2005
    620. 1′127–dc22
2004020584

This book is dedicated to the memory of Robert C. McMaster (1913–1985) and Friedrich Foerster (1908–1999), who were pioneers in the field of nondestructive testing and shared a common interest in electromagnetic testing. As a young engineer working for DuPont, under contract to the U.S. Atomic Energy Commission, the ASNT's Nondestructive Testing Handbook *and CRC's* Handbook of Chemistry and Physics *quickly became indispensable reference sources.*

CONTENTS

Preface   xxv

1 Introduction   1

    1.1 Digital Technology, 1
    1.2 Smaller Is Better, 2
    1.3 Medical Marvels, 5
    1.4 Improving Shuttle Safety, 6
    1.5 Airport Security, 9
    1.6 Process Control, 9
    1.7 Instrument Synchronization with PXI, 10
    1.8 PCI vs. PXI, 11
    1.9 60,000-Mile-High Elevator, 11
    1.10 Proliferation of Information, 12

2 Acoustic Emissions   15

    2.1 Principles and Theory, 15
    2.2 Signal Propagation, 16
    2.3 Physical Considerations, 16
    2.4 The AE Process Chain, 17
    2.5 Time Considerations, 18
    2.6 AE Parameters, 18
    2.7 The AE Measurement Chain, 20
        2.7.1 Coupling Agents, 21
        2.7.2 AE Sensors, 21

- 2.7.3 Sensor Attachment, 22
- 2.7.4 Sensor to Preamplifier Cable, 22
- 2.7.5 AE Preamplifier, 23
- 2.7.6 Preamplifier to System Cable, 23
- 2.8 Vallen AMSY-5 High-Speed AE System, 24
  - 2.8.1 Frequency Filter, 24
  - 2.8.2 The A/D Converter, 25
  - 2.8.3 Feature Extraction, 25
  - 2.8.4 Transient Recorder, 25
  - 2.8.5 Data Buffer, 26
  - 2.8.6 Personal Computer and Software, 26
  - 2.8.7 Sensor Coupling Test (Autocalibration), 26
- 2.9 Location Calculation and Clustering, 27
  - 2.9.1 Location Calculation Based on Time Differences, 27
  - 2.9.2 Clustering, 29
  - 2.9.3 Sample Analysis Screen, 30
  - 2.9.4 Visualization of Measurement Results, 32
- 2.10 Advantages and Limitations of AE Testing, 32
  - 2.10.1 Advantages, 32
  - 2.10.2 Advantages of Using Operating Medium (Gas or Liquid), 32
  - 2.10.3 Advantages Compared to Other NDT Methods, 33
  - 2.10.4 Limitations, 33
  - 2.10.5 Location Errors, 33
- 2.11 AMSY-5 Main Features, 34
- 2.12 AE Transducers, 34
- 2.13 Kistler Piezotron® Acoustic Emission Sensors and Couplers, 35
- 2.14 AE Sensor Construction, 35
- 2.15 Summary of AE Sensor Features, 36
- 2.16 Technical Specifications—8152B2 Sensor, 36
- 2.17 AE Coupler Features, 36
- 2.18 Technical Specifications—5125B Coupler, 38
  - 2.18.1 Input, 38
  - 2.18.2 Output, 38
- 2.19 Acoustic Emission Technology, 38
- 2.20 AE Applications, 39
- 2.21 AE Theory, 39
- 2.22 Applications, 41
  - 2.22.1 Behavior of Materials—Metals, Ceramics, Composites, Rocks, Concrete, 41
  - 2.22.2 Nondestructive Testing During Manufacturing Processes, 41
  - 2.22.3 Monitoring Structures, 41
  - 2.22.4 Special Applications, 41

CONTENTS ix

- 2.23 Advanced Equipment, 42
  - 2.23.1 PCI-2 AE Unit, 42
  - 2.23.2 Key Features, 42
  - 2.23.3 PCI-8, 16-Bit, 8-Channel AE Unit, 43
  - 2.23.4 MicroSAMOS™, Budget, Compact AE System, 44
  - 2.23.5 DiSP Systems, 45
  - 2.23.6 PCI/DSP-4 Card, 45
  - 2.23.7 Features of PCI/DSP-4 System Board, 47
  - 2.23.8 PCI/DSP-4 Board Operation and Functions, 47
  - 2.23.9 DiSP System Block Diagram, 49
  - 2.23.10 Other Company Products, 50
- 2.24 Codes, Standards, Practices, Guidelines, and Societies, 50
  - 2.24.1 Sheer Numbers, 52
  - 2.24.2 Terminology, 52
  - 2.24.3 Common Term Definitions, 52
  - 2.24.4 General Principles, 53
  - 2.24.5 Measurement Techniques and Calibration, 53
  - 2.24.6 Areas of Opportunity, 53
- 2.25 Application and Product-Specific Procedures, 54
- 2.26 Impact-Echo Method, 54
  - 2.26.1 Background, 54
  - 2.26.2 Finite Element Code, 55
  - 2.26.3 Ball Bearing–Generated Stress, 55
  - 2.26.4 Impact-Echo Transducer Development, 56
  - 2.26.5 Frequency Domain Analysis, 56
  - 2.26.6 Theory of Operations, 56
  - 2.26.7 Propagation of Waves, 57
  - 2.26.8 Impact-Echo Instrumentation, 59
    - 2.26.8.1 System Components, 59
    - 2.26.8.2 Heavy-Duty Carrying Case, 60
    - 2.26.8.3 Computer Recommendations, 60
- 2.27 Technical Specifications, 61
  - 2.27.1 Hand-Held Transducer Unit, 61
  - 2.27.2 A/D Data Acquisition System, 62
  - 2.27.3 Windows-Based Software, 63
- 2.28 Applications, 64

3 Electromagnetic Testing Method   65

- 3.1 Eddy Current Theory, 66
  - 3.1.1 Surface Mounted Coils, 66
  - 3.1.2 Encircling Coils, 71
- 3.2 Magnetic Flux Leakage Theory, 73
- 3.3 Eddy Current Sensing Probes, 79

3.4 Flux Leakage Sensing Probes, 83
   3.4.1 Induction Coils, 83
   3.4.2 Hall Effect Sensors, 84
3.5 Factors Affecting Flux Leakage, 87
3.6 Signal-to-Noise Ratio, 88
3.7 Test Frequency, 88
3.8 Magnetization for Flux Leakage Testing, 90
3.9 Coupling, 95
3.10 Eddy Current Techniques, 95
3.11 Instrument Design Considerations, 96
3.12 UniWest US-454 EddyView™, 98
   3.12.1 E-Lab Model US-450, 101
   3.12.2 ETC-2000 Scanner, 102
3.13 Institut Dr. Foerster, 103
3.14 Magnetic Flux Leakage Testing, 106
3.15 Applications, 108
   3.15.1 General Eddy Current Applications, 108
   3.15.2 Specific Eddy Current Applications, 110
   3.15.3 General Flux Leakage Applications, 111
   3.15.4 Specific Leakage Flux Applications, 111
3.16 Use of Computers, 112
3.17 Barkhausen Noise/Micromagnetic Testing, 112
   3.17.1 Introduction, 112
3.18 Early Applications, 113
3.19 Principles of Measurement, 114
3.20 Equipment, 115
3.21 Technical Specifications, 117
3.22 Calibration and Testing, 117
3.23 Current Applications, 120
   3.23.1 Applications in Aircraft/Automotive/Marine Industries, 120
3.24 General Applications, 121
   3.24.1 Pipe/Tubing/Sheet/Plate Manufacturing, 121
3.25 Electromechanical Acoustic Transducers (EMATs), 121
   3.25.1 EMATs Advantages Over Piezoelectric Transducers, 122
3.26 Basic Theory of Operation, 122
3.27 Recent Applications and Developments, 123
3.28 Alternating Current Field Measurement (ACFM) Method, 124
3.29 ACFM Principles of Operation, 125
   3.29.1 Bx and Bz Components, 126
   3.29.2 Butterfly Plot, 127
3.30 Probe Design, 127
3.31 Applications, 128

4 Laser Testing Methods 131

- 4.1 Introduction, 131
- 4.2 Disadvantages, 132
- 4.3 Main Advantages, 132
- 4.4 Laser Theory, 132
- 4.5 Laser Safety, 133
- 4.6 Laser Classification, 133
- 4.7 Training, 134
- 4.8 Profilometry Methods, 134
  - 4.8.1 Stylus Profilometry, 135
  - 4.8.2 Optical Profilometry, 135
  - 4.8.3 White Light Interferometry, 135
- 4.9 Basic TV Holography/ESPI Interferometry, 136
  - 4.9.1 Single Laser Operation, 136
  - 4.9.2 Camera Operation, 136
  - 4.9.3 Applications, 137
  - 4.9.4 Thermal Stresses, 139
  - 4.9.5 Quantitative Aspects of Mechanical Stress, 140
  - 4.9.6 Qualitative Aspects, 141
- 4.10 Nanometric Profiling Measurements, 141
  - 4.10.1 Introduction, 141
  - 4.10.2 Autofocus Principle, 142
  - 4.10.3 Specifications, 142
    - 4.10.3.1 Sensor, 142
    - 4.10.3.2 Camera, 142
- 4.11 Conoscopic Holography, 143
  - 4.11.1 Theory, 143
  - 4.11.2 Specifications, 143
- 4.12 Confocal Measurement, 144
  - 4.12.1 Specifications, 144
    - 4.12.1.1 Sensor, 144
    - 4.12.1.2 Camera, 144
- 4.13 NanoSurf Confocal Microscopy, 145
  - 4.13.1 Introduction, 145
  - 4.13.2 Standard Components, 146
  - 4.13.3 Options, 146
- 4.14 3D Confocal Microscopy, 146
  - 4.14.1 Principle of Operation, 146
  - 4.14.2 Advantages, 146
  - 4.14.3 Specifications, 146
- 4.15 Nanometric Profiling Applications, 147
- 4.16 Scanning Laser Profilometry, 147
  - 4.16.1 Optical Principle, 148
  - 4.16.2 Probes, 149

- 4.16.3 3D Profiler, 149
- 4.16.4 LP-2000™ Control Unit, 150
- 4.17 Laser-Scanned Penetrant Inspection (LSPI™-Patent Pending), 152
  - 4.17.1 Applications, 154
- 4.18 Advanced Techniques, 154
- 4.19 Natural and External Excitation, 154
- 4.20 Strain/Stress Measurement, 155
  - 4.20.1 Theory of Operation, 155
  - 4.20.2 Technical Data, 156
- 4.21 Longer Range 3D Speckle Interferometry System, 157
  - 4.21.1 Technical Data, 158
  - 4.21.2 Hardware and Software Options, 158
  - 4.21.3 Applications for 3D-ESPI Systems, 158
  - 4.21.4 Technical Data, 160
- 4.22 Nondestructive Testing (NDT), 161
- 4.23 Shearography, 161
  - 4.23.1 Principle of Laser Shearography, 161
  - 4.23.2 Compact Shearography System, 162
  - 4.23.3 Technical Data, 163
- 4.24 Portable Shearography System, 164
  - 4.24.1 Technical Data, 164
  - 4.24.2 Other Applications, 165
- 4.25 Feltmetal Inspection System, 166
  - 4.25.1 Setup and Technique, 166
  - 4.25.2 Technical Data, 166
- 4.26 Optional Applications, 168
- 4.27 Optical Inspection Systems, 168
  - 4.27.1 ARAMIS, 168
  - 4.27.2 Industry-Specific Applications, 170
  - 4.27.3 Measuring Procedure, 170
  - 4.27.4 Measurement Results, 170
  - 4.27.5 Measurement Advantages, 170
  - 4.27.6 Comparison of ESPI and 3D Image Correlation, 171
  - 4.27.7 ARAMIS HR Specifications, 172
- 4.28 ARGUS, 172
- 4.29 TRITOP, 174
  - 4.29.1 Photogrammetric Offline System, 174
  - 4.29.2 Measurable Object Size, 174
  - 4.29.3 Digital Photogrammetry Functionality, 174
- 4.30 System Advantages, 175
- 4.31 Portable Measuring System Technique, 175
- 4.32 Dynamic TRITOP, 177
- 4.33 Other Laser Methods, 177
  - 4.33.1 Measurement of Hot Spots in Metal/Semiconductor Field-Effect Transistors, 177

5 Leak Testing Methods 179

- 5.1 Introduction, 179
- 5.2 Fundamentals, 180
- 5.3 Ultrasonic Leak Testing, 180
  - 5.3.1 Ultrasonic Leak Detectors, 180
- 5.4 Bubble Leak Testing, 183
- 5.5 Dye Penetrant Leak Testing, 183
- 5.6 Pressure Change Leak Testing, 183
- 5.7 Helium Mass Spectrometer Leak Testing, 184
- 5.8 Mass Spectrometer Leak Detector, 184
- 5.9 MSLD Subsystems, 184
  - 5.9.1 Spectrometer Tube, 185
  - 5.9.2 Vacuum Systems, 186
- 5.10 Vacuum System Configurations, 186
  - 5.10.1 Conventional (Direct) Flow, 186
  - 5.10.2 Contraflow (Reverse) Flow, 187
  - 5.10.3 Midstage Flow, 188
  - 5.10.4 Multiple Flow, 188
- 5.11 Electronics, 189
  - 5.11.1 I/O Functions, 190
- 5.12 Methods of Leak Detection, 190
- 5.13 Vacuum Testing Method (Outside-In), 191
  - 5.13.1 Locating Leaks, 191
  - 5.13.2 Measuring Leaks, 192
- 5.14 Pressure Test Method (Inside-Out), 192
  - 5.14.1 Locating Leaks, 192
  - 5.14.2 Measuring Leaks, 193
- 5.15 Accumulation Testing Method, 194
- 5.16 Vacuum Systems, 194
- 5.17 Pressurized Systems, 195
- 5.18 MSLD Configurations, 196
  - 5.18.1 "Wet" or "Dry" Pumps, 196
  - 5.18.2 Cabinet or Workstation Models, 196
  - 5.18.3 Portable Units, 197
  - 5.18.4 Component or Integratable Units, 197
- 5.19 Calibration, 197
  - 5.19.1 Calibrated Leaks, 198
- 5.20 Radioisotope Tracer Leak Testing, 198
- 5.21 Bubble Leak Testing, 199
  - 5.21.1 Leak Detector Solution, 199
  - 5.21.2 Vacuum Box Bubble Tracer Leak Testing, 199
  - 5.21.3 Pressure Bubble Leak Testing, 200
  - 5.21.4 Indications, 201
- 5.22 Pressure Change Leak Testing, 202

        5.22.1   Principles, 202
        5.22.2   Terminology, 202
        5.22.3   Equipment, 203
        5.22.4   Pressurizing Gases, 204
  5.23 Pressure Change Measurement Testing, 205
        5.23.1   Reference System Technique, 207
  5.24 Leakage Rate and Flow Measurement Testing, 207
  5.25 Nuclear Reactor Systems, 208
  5.26 Halogen Diode Leak Testing, 209
        5.26.1   Principles, 209
        5.26.2   Terminology, 210
        5.26.3   Gases and Equipment, 210
        5.26.4   Calibration, 210
  5.27 "Sniffer" Techniques, 212
        5.27.1   Equipment Operation and Servicing, 212
        5.27.2   Normal Operation, 212
  5.28 VIC MSLD Leak Detectors, 213
  5.29 MSLD Subsystems, 216
        5.29.1   Spectrometer Tube, 217
        5.29.2   Vacuum System, 218
  5.30 Operating Sequence (MS-40 and MS-40 Dry), 219
  5.31 Calibration Sequence (MS-40 and MS-40 Dry), 220

6 Liquid Penetrant Tests   221

  6.1   Introduction, 221
  6.2   Processing, 222
  6.3   Test Methods, 224
        6.3.1   Water Washable Fluorescent Penetrant Process, 224
        6.3.2   Post-Emulsification Fluorescent Process, 226
        6.3.3   Reverse Fluorescent Dye Penetrant Process, 227
        6.3.4   Visible Dye Penetrant Process, 227
        6.3.5   Water Emulsifiable Visible Dye Penetrant Process, 228
        6.3.6   Water Washable Visible Dye Penetrant Process, 228
        6.3.7   Post-Emulsifiable Visible Dye Penetrant Process, 229
        6.3.8   Solvent Clean Visible Dye Penetrant Process, 229
  6.4   Advantages and Disadvantages of Various Methods, 230
  6.5   Test Equipment, 231
  6.6   Penetrant Materials, 236
  6.7   System Comparisons, 238
  6.8   Applications, 239
  6.9   Measurement of UV and Visible Light, 242
  6.10 Automatic and Semiautomatic Penetrant Testing Methods, 245

7 Magnetic Particle Testing — 247

- 7.1 Magnetic Principles, 247
- 7.2 Magnets and Magnetic Fields, 249
- 7.3 Discontinuities and Defects, 252
- 7.4 Induced Magnetic Fields, 254
- 7.5 Circular and Longitudinal Fields, 257
- 7.6 Selection of Magnetizing Method, 262
- 7.7 Commercial Equipment, 263
- 7.8 Wet and Dry Particle Inspection, 264
- 7.9 MT Improvements, 267
  - 7.9.1 Remote Magnetic Particle Inspection, 269
  - 7.9.2 Probe Power, 269
  - 7.9.3 Lightweight UV Lamps, 270
  - 7.9.4 Dual Light (UV/Visible and Visible) Particle Indications, 270
- 7.10 Applications, 270
- 7.11 Residual Fields and Demagnetization, 273
- 7.12 Magnetic Flux Strips, 275
- 7.13 Hall Effect Gaussmeter, 276
- 7.14 The Hysteresis Curve, 277
- 7.15 Selection of Equipment, 280
- 7.16 Advantages and Disadvantages of the Method, 285
- 7.17 Magnetic Rubber Inspection, 285
  - 7.17.1 Introduction, 285
  - 7.17.2 Inspection Principles, 285
  - 7.17.3 Advantages of MRI, 286
  - 7.17.4 Formulations, 287
- 7.18 Underwater MRI, 288
  - 7.18.1 Technique, 288
  - 7.18.2 Disadvantages, 288
- 7.19 Magnetic Penetrameters, 289
- 7.20 Automatic and Semiautomatic Inspection, 289
- 7.21 Magwerks Integrated System Tracking Technology, 290
  - 7.21.1 Basic Operation, 290
    - 7.21.1.1 Basic Operation—Automatic Mode, 291
    - 7.21.1.2 Applications, 295
- 7.22 Discontinuities and Their Appearances, 296
- 7.23 Nonrelevant Indications, 297

8 Neutron Radiographic Testing — 301

- 8.1 Introduction, 301
- 8.2 Physical Principles, 303
- 8.3 Neutron Radiation Sources, 304

- 8.4 Neutron Activation Analysis, 304
- 8.5 Ward Center TRIGA Reactor, 307
- 8.6 Radiation Hazards and Personal Protection, 309
- 8.7 Radiation Detection Imaging, 311
  - 8.7.1 Conversion Screens, 312
  - 8.7.2 Indirect Transfer Method, 312
  - 8.7.3 Direct Transfer Method, 312
  - 8.7.4 Fluorescent Screens, 313
- 8.8 Electronic Imaging, 313
- 8.9 Nonimaging Detectors, 313
- 8.10 Neutron Radiographic Process, 313
- 8.11 Interpretation of Results, 315
- 8.12 Other Neutron Source Applications, 316
- 8.13 Neutron Level Gauges, 320
- 8.14 Californium-252 Sources, 321
- 8.15 Neutron Radioscopic Systems, 321
  - 8.15.1 Introduction, 321
  - 8.15.2 Neutron Imaging System Components, 322
  - 8.15.3 Online Inspection Systems, 323
  - 8.15.4 Characteristics of Aluminum Corrosion, 323
  - 8.15.5 Thermal Neutron Inspection System Requirements, 324
  - 8.15.6 Conclusions, 324

9 Radiographic Testing Method 325

- 9.1 Industrial Radiography, 325
  - 9.1.1 Personnel Monitoring, 325
  - 9.1.2 Selected Definitions, 326
  - 9.1.3 Survey Instruments, 327
  - 9.1.4 Leak Testing of Sealed Sources, 329
  - 9.1.5 Survey Reports, 331
- 9.2 Work Practices, 331
- 9.3 Time—Distance—Shielding—Containment, 332
- 9.4 Regulatory Requirements, 335
- 9.5 Exposure Devices, 335
- 9.6 State and Federal Regulations, 337
- 9.7 Basic Radiographic Physics, 338
  - 9.7.1 Introduction—Isotope Production, 338
- 9.8 Fundamental Properties of Matter, 339
- 9.9 Radioactive Materials, 340
  - 9.9.1 Stability and Decay, 341
  - 9.9.2 Activity, 341
  - 9.9.3 Half-Life, 342
- 9.10 Types of Radiation, 343
- 9.11 Interaction of Radiation with Matter, 346

- 9.12 Biological Effects, 348
- 9.13 Radiation Detection, 352
  - 9.13.1 Survey Instruments, 354
- 9.14 Radiation Sources, 356
  - 9.14.1 Isotope Sources, 356
- 9.15 Portable Linear Accelerators, 359
- 9.16 Special Radiographic Techniques, 360
- 9.17 Standard Radiographic Techniques, 361
  - 9.17.1 Introduction, 361
  - 9.17.2 Basic Principles, 363
  - 9.17.3 Screens, 364
  - 9.17.4 Film Composition, 365
- 9.18 The Radiograph, 365
  - 9.18.1 Image Quality, 370
  - 9.18.2 Film Handling, Loading, and Processing, 374
  - 9.18.3 High-Intensity Illuminators, 376
- 9.19 Fluoroscopy Techniques, 377
- 9.20 Flat Panel Digital Imaging Systems, 378
- 9.21 Flat Panel Systems vs. Fuji Dynamix CR Imaging System, 379
  - 9.21.1 Resolution, 379
  - 9.21.2 Ghost Images, 380
  - 9.21.3 Image Lag, 380
  - 9.21.4 Dark Current Noise, 381
  - 9.21.5 Portability, 381
  - 9.21.6 Temperature Sensitivity, 381
  - 9.21.7 Flexibility, 381
  - 9.21.8 Fragility, 381
  - 9.21.9 Advantages, 381
- 9.22 Industrial Computed Tomography, 382
  - 9.22.1 Scan Procedure, 382
  - 9.22.2 Applications of Industrial Computed Tomography, 383
  - 9.22.3 CT System Components, 384
- 9.23 Automatic Defect Recognition, 387
  - 9.23.1 Imaging Improvements, 387
  - 9.23.2 LDA Design and Operation, 389
  - 9.23.3 ADR Techniques, 389
  - 9.23.4 Neural Network Artificial Intelligence (AI), 390
  - 9.23.5 Rule Base Using Specific Algorithms, 392
    - 9.23.5.1 Operating Sequence, 392
  - 9.23.6 ADR Advances of a PC Platform Over Proprietary Hardware, 392
  - 9.23.7 ADR Techniques, 392
  - 9.23.8 SADR, 392
  - 9.23.9 Conclusions, 393
- 9.24 The Digitome® Process, 393

9.24.1 Examination Concept, 394
9.24.2 Digital Flat Panel Detector, 395
9.24.3 Image Acquisition, 396
9.24.4 Flaw Location and Measurement, 396
9.24.5 Other Applications, 396
9.25 Manufacturing Processes and Discontinuities, 397
9.26 Other Isotope Applications, 397
9.26.1 Electron Capture Detection, 397
9.26.2 Moisture Gauging, 397
9.26.3 Bone Density, 400
9.26.4 Gamma and Beta Thickness Gauging, 401
9.26.5 Gamma and Beta Backscatter Thickness Gauging, 401
9.26.6 Gamma Level Gauging, 402
9.26.7 Gamma Density Measurement, 402
9.26.8 Point Level Switch, 404
9.26.8.1 Features and Benefits, 405
9.26.9 Oil Well Logging, 405

10 Thermal/Infrared Testing Method 407

10.1 Basic Modes of Heat Transfer, 407
10.2 The Nature of Heat Flow, 408
10.2.1 Exothermic and Endothermic Reactions, 408
10.2.1.1 Exothermic Reactions, 408
10.2.1.2 Endothermic Reactions, 409
10.3 Temperature Measurement, 409
10.4 Common Temperature Measurements, 410
10.4.1 Melting Point Indicators, 410
10.5 Color Change Thermometry, 411
10.5.1 Irreversible Color Change Indicators, 411
10.5.2 Thermochromic Liquid Crystal Indicators, 413
10.5.3 Liquid in Glass Thermometers, 415
10.6 Temperature Sensors with External Readouts, 416
10.6.1 Thermocouple Sensors, 416
10.6.2 Special Thermocouple Products, 418
10.6.3 Resistance Temperature Devices (RTDs), 418
10.6.3.1 RTD Sensing Elements and Typical Temperature Ranges, 418
10.6.4 Resistance Temperature Elements (RTEs), 420
10.7 Infrared Imaging Energy, 420
10.8 Heat and Light Concepts, 421
10.9 Pyrometers, 422
10.9.1 Error Correction, 422
10.9.2 Principles of Operation, 423
10.9.2.1 Narrow-Band Optical Pyrometers, 423
10.9.2.2 Broad-Band Optical Pyrometers, 424

CONTENTS

- 10.9.3 Design and Operations of Optical Pyrometers, 426
- 10.9.4 Applications for Broad-Band Optical Pyrometers, 427
- 10.9.5 Installation of Optical Pyrometers, 427
- 10.10 Infrared Imaging Systems, 427
  - 10.10.1 Blackbody Calibration Sources, 427
- 10.11 Spacial Resolution Concepts, 428
  - 10.11.1 FOV, IFOV, MIFOV, and GIFOV, 428
  - 10.11.2 Angular Resolving Power, 428
  - 10.11.3 Error Potential in Radiant Measurements, 429
- 10.12 Infrared Testing Method, 429
  - 10.12.1 Preventive and Predictive Maintenance Programs, 429
  - 10.12.2 Electrical PdM Applications, 429
  - 10.12.3 Mechanical PdM Applications, 430
- 10.13 High-Performance Thermal Imager for Predictive Maintenance, 430
  - 10.13.1 Predictive Maintenance Program, 431
  - 10.13.2 Specifications, 432
    - 10.13.2.1 Thermal, 432
    - 10.13.2.2 Controls, 433
    - 10.13.2.3 Optional Features, 433
    - 10.13.2.4 Other, 433
- 10.14 High-Performance Radiometric IR System, 433
  - 10.14.1 Introduction, 433
  - 10.14.2 Applications, 434
  - 10.14.3 Theory of Operation, 434
  - 10.14.4 Operating Technique, 436
  - 10.14.5 Typical Specifications, 438
- 10.15 Mikron Instrument Company, Inc., 439
- 10.16 Mikron 7200V Thermal Imager and Visible Light Camera, 440
  - 10.16.1 General Features, 440
  - 10.16.2 Technical Data, 440
    - 10.16.2.1 Performance, 440
    - 10.16.2.2 Presentation, 441
    - 10.16.2.3 Measurement, 441
    - 10.16.2.4 Interface, 442
- 10.17 High-Speed IR Line Cameras, 442
  - 10.17.1 General Information—MikroLine Series 2128, 442
  - 10.17.2 High-Speed Temperature Measurement of Tires, 442
    - 10.17.2.1 Camera Specifications, 443
- 10.18 Other Thermal Testing Methods, 444
  - 10.18.1 Fourier Transform Infrared Spectrometer, 444
    - 10.18.1.1 DLATGS Pyroelectric Detectors, 447
    - 10.18.1.2 FTIR Evaluation of Hard Disk Fluororesin Coating, 447
    - 10.18.1.3 Measurement of Film Thickness on a Silicon Wafer, 448

10.18.2 Advanced Mercury Analyzer, 448
    10.18.2.1 Introduction, 448
    10.18.2.2 Theory of Operation, 449
    10.18.2.3 Software, 450
10.18.3 Identification of Materials, 450
    10.18.3.1 Thermoelectric Alloy Sorting, 450
    10.18.3.2 Applications, 453
10.18.4 Advantages and Disadvantages, 454
    10.18.4.1 Advantages, 454
    10.18.4.2 Disadvantages, 456

11 Ultrasonic Testing   457

11.1 Introduction, 457
11.2 Definition of Acoustic Parameters of a Transducer, 458
11.3 Noncontacting Ultrasonic Testing, 458
    11.3.1 NCU Transducers, 460
    11.3.2 Instant Picture Analysis System, 463
    11.3.3 Limitations, 465
    11.3.4 Bioterrorism, 466
11.4 Ultrasonic Pulsers/Receivers, 466
11.5 Multilayer Ultrasonic Thickness Gauge, 470
11.6 Conventional Ultrasound, 471
    11.6.1 Flaw Detection, 473
    11.6.2 Frequency, 474
    11.6.3 Ultrasonic Wave Propagation, 476
    11.6.4 Acoustic Impedance, 477
    11.6.5 Reflection and Refraction, 478
    11.6.6 Diffraction, Dispersion, and Attenuation, 481
    11.6.7 Fresnel and Fraunhofer Fields, 482
    11.6.8 Generation of Ultrasonic Waves, 483
    11.6.9 Search Unit Construction, 484
    11.6.10 Test Methods, 489
11.7 Ultrasonic Testing Equipment, 498
    11.7.1 Equipment Operation, 507
    11.7.2 Flaw Transducers, 509
        11.7.2.1 Instrument Features, 509
        11.7.2.2 Ultrasonic Specifications, 510
        11.7.2.3 Physical Description and Power Supply, 510
    11.7.3 Testing Procedures, 512
        11.7.3.1 Variables Affecting Results, 517
11.8 Time-of-Flight Diffraction (TOFD), 519

12 Vibration Analysis Method 521

- 12.1 Introduction, 521
- 12.2 Principles/Theory, 522
    - 12.2.1 Modes of Vibration, 522
    - 12.2.2 Resonance, 523
    - 12.2.3 Degrees of Freedom, 524
- 12.3 Sources of Vibration, 524
- 12.4 Noise Analysis, 525
- 12.5 Stress Analysis, 525
- 12.6 Modal Analysis, 526
- 12.7 Vibration Analysis/Troubleshooting, 527
    - 12.7.1 Rotating Equipment Analysis, 527
    - 12.7.2 Order Analysis, 527
- 12.8 Transfer Functions, 528
- 12.9 Predictive Maintenance, 528
- 12.10 Failure Analysis, 529
- 12.11 Impact Testing and Frequency Response, 529
- 12.12 Pass and Fail Testing, 530
- 12.13 Correction Methods, 530
    - 12.13.1 Alignment and Balance, 530
    - 12.13.2 Beat Frequency, 530
    - 12.13.3 Vibration Damping, 532
    - 12.13.4 Dynamic Absorber/Increasing Mass, 534
    - 12.13.5 Looseness/Nonlinear Mechanical Systems, 536
    - 12.13.6 Isolation Treatments, 536
    - 12.13.7 Speed Change, 540
    - 12.13.8 Stiffening, 540
- 12.14 Machine Diagnosis, 541
- 12.15 Sensors, 543
    - 12.15.1 Strain Gauges, 543
    - 12.15.2 Accelerometers, 544
    - 12.15.3 Velocity Sensors, 545
    - 12.15.4 Displacement Sensors, 545
- 12.16 Rolling Element Bearing Failures, 547
- 12.17 Bearing Vibration/Noise, 548
- 12.18 Blowers and Fans, 550
- 12.19 Vibrotest 60 Version 4, 550
- 12.20 Signal Conditioning, 555
    - 12.20.1 Acoustic Filters, 555
- 12.21 Equipment Response to Environmental Factors, 555
    - 12.21.1 Temperature/Humidity, 555
- 12.22 Data Presentation, 555
    - 12.22.1 Acceleration, Velocity, and Displacement, 555
    - 12.22.2 Fast Fourier Transform (FFT)/Time Waveform, 556

12.22.3 Cepstrum Analysis, 557
12.22.4 Nyquist Frequency/Plot, 557
12.22.5 Orbit, Lissajous, X-Y, and Hysteresis Plots, 559
12.23 Online Monitoring, 560
12.23.1 Trend Analysis, 560
12.24 Portable Noise and Vibration Analysis System, 560
12.24.1 Typical Applications, 562
12.24.2 System Requirements, 562
12.25 Laser Methods, 562
12.25.1 Theory of Operation, 563
12.25.2 Applications, 565
12.25.3 Specifications, 566
12.26 TEC's Aviation Products, 567
12.26.1 Analyzer Plus Model 1700, 567
12.26.1.1 Flexible System, 568
12.26.1.2 User Friendly, 568
12.26.1.3 Expandability, 568
12.26.1.4 Quality Commitment, 568
12.26.1.5 Engine Fan Balancing Application, 569
12.26.1.6 Technical Specifications, 569
12.26.2 ProBalancer Analyzer 2020, 570
12.26.2.1 Software Features, 571
12.26.2.2 Technical Specifications, 572
12.26.3 Viper 4040, 572
12.26.3.1 Automated Track and Balancing, 572
12.26.3.2 Vibration Analysis, 574
12.26.3.3 Acoustic Analysis, 574
12.26.3.4 Technical Specifications, 574

13 Visual and Optical Testing 575

13.1 Fundamentals, 575
13.2 Principles and Theory of Visual Testing, 576
13.3 Selection of Correct Visual Technique, 576
13.4 Equipment, 578
13.4.1 Borescopes, 578
13.4.2 Jet Engine Inspection, 581
13.4.3 Nuclear Applications, 582
13.4.4 Other Applications, 584
13.5 Fiberscopes and Videoscopes, 584
13.5.1 Applications, 585
13.6 SnakeEye™ Diagnostic Tool, 587
13.7 Industrial Videoscopes, 589
13.7.1 Equipment and Features, 589
13.7.2 Instrument Setup, 590

CONTENTS   xxiii

        13.7.3 3D Viewing, 592
        13.7.4 Applications, 592
        13.7.5 Working Tools, 592
  13.8 Projection Microscopes, 593
        13.8.1 Leica FS4000 Forensic Comparison Microscope, 596
  13.9 The Long-Distance Microscope, 600
        13.9.1 New Developments, 600
        13.9.2 Model K-2 Long-Distance Microscope, 601
                13.9.2.1 Numerical Aperture (NA), 604
                13.9.2.2 Care and Cleaning, 605
        13.9.3 InfiniVar CFM-2 Video Inspection Microscope, 605
        13.9.4 Accordion™ Machine Vision, 607
        13.9.5 InFocus Microscope Enhancement System, 607
                13.9.5.1 Spherical Aberrations, 607
                13.9.5.2 InFocus Corrections, 608
                13.9.5.3 Applications, 608
  13.10 InfiniMax™ Long-Distance Microscope, 611
  13.11 Remote Visual Inspection, 611
        13.11.1 Industries—Applications, 614
        13.11.2 Camera Head Options, 616
        13.11.3 Camera Pan and Tilt Features, 617
        13.11.4 Hand-Held Controller, 618
        13.11.5 Camera Control Unit, 619
        13.11.6 Hand-Held Controller Details, 620
        13.11.7 Applications, 622
  13.12 Robotic Crawler Units, 623
        13.12.1 Control Unit, 623
        13.12.2 Cable Reels, 623
        13.12.3 Crawler and Camera Options, 624
        13.12.4 Applications, 624
  13.13 Pipe and Vessel Inspections/Metal Joining Processes, 626
  13.14 Ocean Optics Photometers, 629
        13.14.1 Optical Resolution, 630
        13.14.2 System Sensitivity, 633
        13.14.3 Specifications, 634
        13.14.4 Applications, 636

14 Overview of Recommended Practice No. SNT-TC-1A, 2001 Edition   **639**

  14.1 Purpose, 639
        14.1.1 Personnel Qualification and Certification in Nondestructive Testing 639
  14.2 NDT Levels of Qualification, 640

14.3 Recommended NDT Level III Education, Training, and Experience, 640
14.4 Written Practice, 641
14.5 Charts, 641
14.6 Recommended Training Courses, 641
    14.6.1 Acoustic Emissions Testing Method, 641
    14.6.2 Electromagnetic Testing Method, 643
    14.6.3 Laser Testing Methods—Holography/Shearography, 644
    14.6.4 Laser Testing Methods—Profilometry, 646
    14.6.5 Leak Testing Methods, 646
    14.6.6 Liquid Penetrant Testing Methods, 648
    14.6.7 Magnetic Particle Testing Method, 648
    14.6.8 Neutron Radiographic Testing Method, 649
    14.6.9 Radiographic Testing Method, 651
    14.6.10 Thermal/Infrared Testing Method, 653
    14.6.11 Ultrasonic Testing Method, 654
    14.6.12 Vibration Analysis Method, 655
    14.6.13 Visual Testing Method, 656
    14.6.14 Appendix, 657
        14.6.14.1 Example Questions, 657
        14.6.14.2 Answers to Example Questions, 658
    14.6.15 A Dynamic Document, 658
    14.6.16 Special Disclaimer, 659

Appendix 1: Bibliography of Credits   **661**

Appendix 2: Company Contributors   **667**

Index   **671**

PREFACE

This book has been written to provide a single volume with basic background information needed to help students and *nondestructive testing (NDT)* personnel qualify for Levels I, II, and III certification in the NDT methods of their choice, in accordance with the American Society for Nondestructive Testing (ASNT) Recommended Practice No. SNT-TC-1A (2001 edition). It is also recommended that students and NDT personnel become thoroughly familiar with the most current recommended practice as well. The contents of this book tend to follow the general outline of the recommended standard practice. Questions and answers for each NDT method are not included in this volume, but are available in separate volumes elsewhere. Acronyms are spelled out when first used and a complete list of commonly used NDT acronyms may be available from the American Society for Nondestructive Testing.

In the case of Level III certification, the book is meant to supplement the required hands-on laboratory and field training. A detailed discussion of standard procedures and practices is beyond the scope of this book. There are six major organizations concerned with codes, standards, practices, and guidelines:

1. ISO—International Organization for Standardization
2. CEN—European Standardization Committee
3. ASTM—American Society for Testing and Materials
4. EWGAE—European Working Group on Acoustic Emission
5. AFNOR—French Standardization Society
6. DGZIP—German Society for Nondestructive Testing

The ASTM is a leading international organization concerned with codes, standards, practices, and guidelines. Because of the proliferation of organizations and standards, it is very difficult to get universal approval of proposed standards, practices, or guidelines. This is a continuing problem and one that will not be solved soon.

College and university instructors should find the text especially useful in working with private industry and NDT employers who want to set up on-site NDT training and certification programs. At the same time, coverage of the text is broad enough in scope to be suitable for use as a classroom text and general source of information on NDT.

The author hopes that this text will interest a new generation of young people in considering careers in nondestructive testing or *nondestructive evaluation (NDE)*, the term currently favored by some practitioners. The ASNT's mission is to create a safer world by promoting the profession and technologies of nondestructive testing. The mission of an engineer is to create a safer world though the ethical application of his or her specific knowledge and discipline.

PAUL E. MIX, PE, EE
Austin, Texas

1

INTRODUCTION

1.1 DIGITAL TECHNOLOGY

Digital technology is virtually sweeping the nondestructive testing industry as well as affecting every aspect of American life. We have digital television, digital cameras and video recording systems, digital telecommunications, digital global positioning, digital satellite radios, digital appliances, and personal digital computers with high-speed memory and capacity that were unheard of 20 years ago. Some optical and digital electronic gadgets just recently introduced include small hand-held 10× binoculars with 8× digital cameras, shirt-pocket MP3 players, and real two-way wrist radios with a range of $1\frac{1}{2}$ miles. Dick Tracy had to wait a long time for that one. Can personal identification chips with global tracking be far behind? Standard "Walkie-Talkie" range has reached up to 10 miles.

High-speed computers with high-capacity memory and high-speed data transfer can provide real-time evaluation and control in many nondestructive testing applications. Huge amounts of information can be stored for later review and analysis when desired. AMD beat Intel in the race to be the first on the market to introduce a new 64-bit microprocessor chip. Have we reached the ultimate in computer memory, speed, and data transfer capacity? No, the future still lies ahead.

At the same time, wireless technology is advancing by leaps and bounds. Most people seem to have cells phones rather than those old Alexander

---

Introduction to Nondestructive Testing: A Training Guide, Second Edition, by Paul E. Mix
Copyright © 2005 John Wiley & Sons, Inc.

Graham Bell telephones with wires. The advantages of wireless for industrial automation and the process control industries include worker and work station mobility and the elimination of thousands of miles of expensive conduit and cable. However, there are still many concerns regarding system design and potential signal transmission problems such as signal interference, signal hacking, sudden signal loss and retries, RF interferences, and multipath fading from unwanted reflections. However, it is probably safe to say that we can look forward to continued miniaturization and improvements in all forms of wireless technology.

City parks in Austin, Texas currently offer wireless Internet access; however, there are still some concerns that Bluetooth technology can be compromised. Bluetooth is a universal radio interface in the 2.45 GHz ISM frequency band designed to function on a worldwide basis. A Bluetooth system consists of a radio unit, link controller, link manager, and software. Spectrum spreading facilitates optional operation at power levels up to 100 mW worldwide. This is accomplished by frequency hopping in 79 hops displaced by 1 MHz, from 2.402 GHz to 2.480 Hz. The maximum frequency hopping rate is 1600 hops/sec. Bluetooth devices must be able to recognize each other and load the appropriate software to use the higher-level abilities each device supports and existing protocols.

Notebook PC computers can be used for remote networking using Bluetooth telephone systems, Bluetooth phones, cellular phones and notebooks for conference calls, speakerphone applications, business card exchange and calendar sychronization. Bluetooth technology is an operating system that is independent of any specific operating system. Advantages of Bluetooth technology are:

- Data exchange; signals penetrate solid objects.
- Remote networking and maximum mobility.
- Omnidirectional with synchronous voice channels.

The main disadvantage is that signals can be monitored by a snooping device from any direction or hidden location. Encryption with authenticity check is possible using a challenge-response protocol utilizing a secret key or password. Both devices must share the same secret key. The technology is suitable for many industrial data-sharing applications.

Will wireless RF ID tags help scientists track mad cows from country to country and state to state? Only if cattle ranchers and farmers all over the world are forced to comply with this requirement and that isn't very likely, is it?

1.2 SMALLER IS BETTER

Virtually all sensors, whether they are laser, infrared, acoustic, ultrasonic, or eddy current, have benefited from the high-tech explosion as well. Generally, high-resolution sensors have become smaller, more sensitive, and more robust. For flaw detection, many sensors can be focused more sharply as parts are scanned at faster rates, resulting in high-speed, high-resolution flaw detection. Eddy current and ultrasonic transducer arrays have greatly increased the single pass surface area scanned, while decreasing scanning times.

Piezo-composite ultrasonic transducers have greatly increased the sensitivity and range of ultrasonic transducers while reducing noise. In some applications noncontacting ultrasonic probes with perfect air/gas (compressed fiber) impedance matching can compete with laser profiling applications and other methods for the detection of minute surface defects. Noncontacting sensors also have some advantages in medical applications.

Micro-electromechanical systems (MEMS) have been around for about 20 years and are increasingly important to many manufacturing industries including semiconductor, automotive, electrical, mechanical, chemical, medical, aerospace, and defense. A very rapid growth of MEMS is expected over the next decade.

Small, sensitive airbag accelerometers help protect us in our automobiles and miniature flow valves provide beautiful letter-quality ink jet printing. Other MEMS developments include:

- Micropressure and acceleration sensors for restricted spaces
- Microelectronic components such as capacitors, inductors, and filters
- Micromechanical components such as valves and particle filters

National security applications for MEMS include nonproliferation, counterterrorism, land mine, chemical and biological warfare, and WMD stockpiling detection. Spin-off applications, which benefit mankind, include biomedical diagnostics, food and water safety, and industrial process and environmental monitoring.

The design, fabrication, testing, and inspection of microcomponents and assemblies challenge engineers and designers because the software, tooling, mechanics, size and shape, fluidity, damping, and electrostatic effects encountered in the microcomponent world are considerably different from those associated with the more conventional macrocomponent world.

While MEMS may still be considered in its youth, the birth of nanotechnology has progressed to at least that of a preschooler. Nanotechnology is now widely recognized by the government and various technical groups. New products are being developed and evaluated by many sectors. Nanotechnology has been defined as the manipulation or self-assembly of individual atoms, molecules, or molecular clusters into structures having dimensions in the 10 to 100 nanometer range to create new materials and devices with new or vastly different properties. Scientists believe the ability to move and combine individual atoms and molecules will revolutionize the production of virtually every human-made object and usher in a new high-tech revolution.

DOE nanotechnology accomplishments include:

- Addition of aluminum oxide nanoparticles that converts aluminum metal into a material with wear resistance equal to that of the best bearing steel
- Novel optical properties of semiconducting nanocrystals that are used to label and track molecular processes in living cells
- Nanoscaled layered materials that can yield a fourfold increase in the performance of permanent magnets
- Layered quantum well structures to produce highly efficient, low-power light sources and photovoltaic cells
- Novel chemical properties of nanocrystals that show promise to speed the breakdown of toxic wastes
- Meso-porous inorganic hosts with self-assembled organic microlayers that are used to trap and remove heavy metal from the environment

Unlike one old science-fiction thriller, nanobots may not be able to cure a young man's cancer, phenomenally increase his personal endurance and strength, and protect him against all harmful outside elements by stimulating the growth of gills in his neck, growing eyes in the back of his head, and developing an alligator skin for him, but it can make structural elements smaller, stronger, lighter, and safer. In turn, nanotechnology can make larger structures and all forms of transportation safer for us mere mortals.

Benoy George Thomas, in an article for *PCQuest* (September 2003, p. 174), mentions that scientists Robert A. Freitas and Christopher J. Phoenix claim that someday nanobots may change the very essence of life by replacing the blood currently coursing through our arteries and veins with over 500 trillion oxygen- and nutrient-carrying nanobots. In this scenario, the nanobots would duplicate just about every function of blood, but do it more efficiently.

The bloodstream would be made up of respirocytes each consisting of 18 billion precisely aligned structural atoms. Each respirocyte would have an onboard computer, power plant, and molecular pumps and storage hulls to transport molecules of oxygen and carbon dioxide. These nanobots would be a thousand times more efficient than the red blood cells (RBCs) they replace. If it sounds too good to be true, then it probably is.

While nanotechnology has been heralded as the driving force for America's next industrial revolution, extreme care must be exercised along the way. At present, the hazards and risks associated with nanoparticles are poorly defined. Toxicologists at Southern University in Dallas have discovered that C60 buckyballs (nanoparticles) in modest concentrations can kill water fleas (a source of food for newly hatched fish) and cause damaging biochemical reactions in the brains of largemouth bass fingerlings. Preliminary studies also indicated that similar problems were observed when nanoparticles were inhaled by animals. Therefore, the toxicology effects of nanoparticles must be considered for all phases of work in this field.

1.3 MEDICAL MARVELS

While doctors and scientists can't yet make a fantastic voyage in a MEMS or nanosubmarine through human arteries and blood vessels, they can virtually examine every artery and cavity in the human body. Doctors can go through the groin to open partially plugged carotid arteries leading to the brain, remove small blood clots from the brain using a small corkscrew-shaped device at the end of a microcatheter, or even correct small aneurisms in the brain. And, doctors can even fuse vertebrae disks by going through an incision in the front of the throat. There ought to be easier ways to get to some of these places.

With the new SilverHawk procedure, developed by Dr. John Simpson, leg arteries with 85% plaque blockage can be restored to normal flow and the arterial wall plaque can be saved for additional medical studies. Figure 1.1 shows the SilverHawk tool. The composition of the removed arterial plaque is then studied by heart doctors to help determine if early warning signs of heart attacks and strokes can be developed for otherwise normally healthy

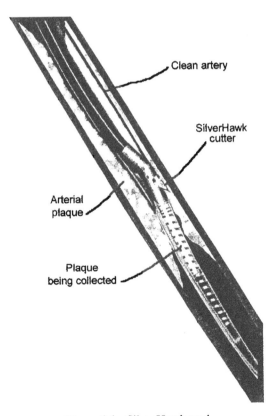

Figure 1.1 SilverHawk tool.

patients. Best of all, these new catheter procedures are highly reliable and relatively inexpensive compared to surgical procedures.

Verging on what might seem science fiction to some, heart doctors now have the ability to give patients with totally plugged heart arteries angio-genesis therapy (AGT), a modified gene therapy cocktail injected in heart arteries to encourage the growth of natural heart artery bypasses. This is an ongoing investigational study that probably will be continued for several years.

Preparation for AGT starts with the patient on the procedure table. A staff member informs the patient about the procedure, gives him oxygen, and places a full-face mask on him and leaves the room. When all staff members and doctor return wearing complete operating attire, rubber gloves, and full-face masks, patients may think they have just slipped into the Twilight Zone. However, if the procedure is successful, substantial improvements in health may be noted.

Enhanced External Counterpulsation (EECP®) therapy is one technique that is truly noninvasive. The goal of this therapy is to stimulate the formation of natural bypasses around narrowed or blocked arteries in the legs and heart.

The EECP system compresses the lower legs, upper legs, and lower buttock to increase blood flow toward the heart. The heart rate is monitored and each pressure wave is timed to increase blood flow to the heart when the heart is relaxing. When the heart pumps, the pressure is released until the heart relaxes again.

The goal of this therapy is to stimulate the growth of collateral blood vessels both aiding normal blood circulation and relieving chronic angina, which has proved unresponsive to other medical therapy. When successful, EECP can eliminate or reduce nitrate use and provide improved ability for patients to exercise more.

While these medical marvels are not nonintrusive for the most part, microsurgery and gene therapy are not very destructive in nature either; they owe much of their success to scientists and engineers working closely with the medical community to help prolong and extend the quality of human life. Once again, it proves there are no limits to imagination and innovation.

1.4 IMPROVING SHUTTLE SAFETY

Primary reaction control system (PRCS) thrusters are a critical part of the power and guidance systems of space shuttle orbiters. A space shuttle orbiter has 38 PRCS thrusters to help power and position the vehicle for maneuvers in space, including reentry and establishing earth orbit. However, minor flaws in the ceramic lining of a thruster, such as a chip or crack, can cripple the operations of an orbiter in space and jeopardize a mission. In the past, these thrusters had to be detached and visually inspected in great detail at one of two NASA facilities—the White Sands facility or the Kennedy Space Center—before and after each mission.

In 2002, James Doyle, president of Laser Techniques, Inc., successfully demonstrated that a miniature, high-performance laser could locate and map hidden thruster features smaller than the head of a pin, to an accuracy of 0.0003 inch. Figure 1.2 shows James Doyle near the rear of a space shuttle with three vertical PRCS thrusters pictured. His initial development work led to the issuance of a NASA contract to build a full-scale, portable in-situ thruster mapping system.

A cutaway view of a thruster shows the laser inspection system and related mechanical actuator in place and ready for inspection in Figure 1.3. The mechanical actuator for the sensor carrier arm and module are retracted when the assembly is placed in the thruster. The thruster interface unit helps center and align the assembly. When a vacuum is pulled on the vacuum locking device, special o-rings lock the assembly in place, readying it for inspection. The sensor carrier arm can be extended, retracted, and rotated. The sensor, which is held by the carrier arm, also rotates about the axis of the thruster and has a tilt mechanism for contour following.

The high-performance laser sensor is shown in Figure 1.4 and compared in size to a 2002 penny. It is important to note that the sensor is used to inspect and map the inner thruster surface area starting about 0.5 inch from the injector face to about 1.5 inches downstream of the thruster throat. Most ceramic coating defects are upstream of the thruster throat and very difficult to eval-

Figure 1.2 James Doyle, president of Laser Techniques Co., at back of space shuttle near three vertical thrusters.

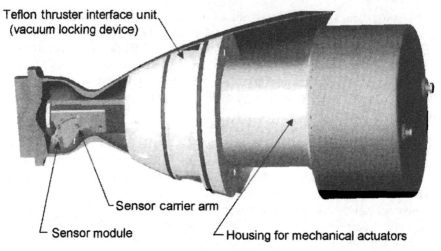

Figure 1.3 Cutaway view of thruster. Courtesy of Laser Techniques Co.

Figure 1.4 Comparison of penny to high-performance laser sensor. Courtesy of Laser Techniques Co.

uate visually. With the scanning laser system, this area of the thruster can be quickly inspected and mapped, providing technicians with accurate 3D data for evaluating the ceramic surface condition of the thrusters.

The portable laser scanner system has been sent to the White Sands test facility in New Mexico where it will be used in thruster life-testing projects and routine thruster overhaul and refurbishment programs.

Figure 1.5 Gilardoni airport security system for luggage inspection showing typical suitcase luggage content (normally in color). Courtesy of Gilardoni Scientific Industry, Italy.

At the ASNT 13th Annual Research Symposium, keynote speaker Bob DeVries reviewed his NDE team's investigative work following the Columbia shuttle disaster that centered on a piece of external tank foam that struck the leading edge of the space shuttle during its launch. After each NASA impact test on the thermal protection system, including the leading edge of the shuttle, nondestructive evaluations were made on the reinforced carbon-carbon components that were impacted. As a result of Mr. DeVries' team efforts, future improvements can be made in material design and structure.

1.5 AIRPORT SECURITY

Gilardoni of Italy fabricates high-tech X-ray inspection systems for increased airport security. These systems are capable of identifying the contents of passenger luggage and parcels as shown in Figure 1.5. Note that a knife and dynamite stick appear in the photo along with other suspicious items. With this equipment, the radiation dose received by luggage contents is so low that ordinary camera film may be packed with the luggage and exposed to the X-rays with no adverse affects. The X-ray inspection system features sophisticated electronics and advanced solid-state detectors. The double-energy X-ray scanning feature of this system provides for selective detection of explosive devices, and therefore this system is capable of improving the security of prisons, banks, hotels, exhibitions, and other locations.

1.6 PROCESS CONTROL

The three key variables for proper process control are flow, pressure, and temperature. Chemical and nuclear reactors must precisely control these variables to prevent fires, explosions, or even reactor core meltdown. These catastrophic events can result in injury or death to operating personnel. These disasters can also result in toxic fume releases to the atmosphere and contamination of the water supplies, thereby endangering the general population. World health depends on protecting the environment.

In the chemical and petrochemical industries, thermocouples, resistant bulb thermometers, contacting and noncontacting pyrometers, and infrared scanners are used to measure temperature, determine heat flux patterns, and evaluate temperature excursions.

From a mechanical standpoint, pumps, fans, motors, and compressors are subject to damage and possible failure when bearing wear and excessive vibration of rotating parts occur. Accelerometers and lasers may be used to analyze and determine the extent of vibrations. In some cases, simple sonic devices can also detect bearing noise and provide a relative indication of bearing wear.

The ultimate goal of the process control industry is to shut down the process or isolate defective equipment and repair it before catastrophic failure occurs. Other NDT methods commonly employed by process control industries include:

- Neutron or gamma radiation gauges for noncontacting process flow, level, or density
- IR analyzers for measuring moisture in anhydrous gases and some nonaqueous liquids
- Ultrasonic transmission to detect pipe and vessel thinning caused by corrosion
- Alloy analyzers for material and part identification
- Hand-held sonic detectors for overhead gas pipe leaks
- Visual inspection in combination with other methods to determine corrosion

1.7 INSTRUMENT SYCHRONIZATION WITH PXI

As electronic instruments become more complex, the need to synchronize multiple instruments for testing and characterization of various devices becomes more important. PXI is a modular-based instrumentation platform designed for measurement and automation applications. PXI is an acronym for *PCI eXtensions for Instrumentation*. PXI systems incorporate the PCI bus in rugged modular Eurocard mechanical packaging systems with electrical and software features that provide complete systems for test and measurement, data acquisition, and manufacturing applications.

PXI modular instrumentation (Figure 1.6) adds a dedicated system reference clock, PXI trigger bus, star trigger bus, and slot-to-slot local bus to provide

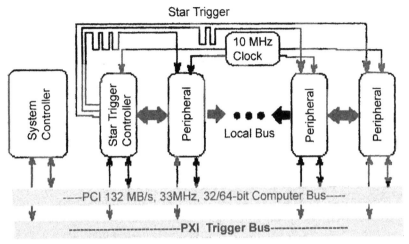

Figure 1.6 Instrument synchronization with PXI.

advanced timing, synchronization, and side-band communication while maintaining PCI bus advantages. System functions are as follows:

- The PXI backplane provides a common 10 MHz reference clock for sychronization of multiple modules.
- PXI defines eight trigger bus lines for sychronization and communication between modules.
- The star trigger bus provides an independent trigger for each slot oriented in a star configuration from slot 2 in any PXI chassis.
- The PXI local bus is daisy-chained to each peripheral slot with its adjacent peripheral slots. Each local bus is 13 lines wide and can handle analog signals as high as 42 V between cards or provide a high-speed side-band communications path that does not affect PCI bandwidth.

1.8 PCI VS. PXI

PXI offers the same performance features as PCI. PXI and CompactPCI systems can have up to seven peripheral slots per bus segment, whereas most desktop PCI systems have only three. Otherwise, all PCI features apply to PXI and CompactPCI.

1.9 60,000-MILE-HIGH SPACE ELEVATOR

While the ancient ones thought they could build a stairway to heaven, they never achieved their dream. Today modern man believes he can build a 60,000-

mile-high elevator in the next 10 to 20 years. If so, the space elevator envisioned by science fiction may soon become a reality, capable of greatly reducing the cost of putting modern satellites into space.

The expected success of a space elevator is based on the 1991 development of nanotubes made of cylindrical carbon molecules that are many times stronger than the strongest steel. Why blast satellites into outer space when you can give them a 60,000-mile-high elevator ride? To date, the most promising idea has been conceived by scientist Bradley Edwards, who has proposed a single, thinner-than-paper, 3-foot-wide ribbon, stretching from earth to the 60,000-mile goal.

As described in local newspapers, a spacecraft would first be launched to lower a narrow ribbon to earth where it would be tied down to a base station. Then the spacecraft would move outward, releasing more ribbon and establishing its position in space. Automated mechanical climbers would then be sent up the ribbon to widen and reinforce it. It is estimated that construction of the 3-foot-wide ribbon could be completed in about 30 months.

Following ribbon construction, as many as eight mechanical tractors could be used to each lift 13 tons of cargo into space. The tractors could be clamped on both sides of the ribbon by tank-like treads and powered by earth-bound lasers. When the construction of the elevator and its upper platform has been completed, satellites could be launched from the platform.

Many adverse factors could affect the building and life expectancy of the space elevator. Some of these factors include falling or orbiting space debris, radiation fields, storms, and terrorist attacks. Science fiction writer Arthur C. Clarke hopes to live long enough to see the construction of his real-world space elevator. However, if we concentrate on putting people on Mars, orbiting satellites and space shuttle missions may become fond memories of the past.

1.10 PROLIFERATION OF INFORMATION

The proliferation of NDT information on the Internet is analogous to the proliferation of standards and practices for the technical societies, as discussed in Chapter 2, Acoustic Emissions.

There appears to be an abundance of information on most company websites engaged in the manufacture and distribution of NDT products. Many of these sites adequately cover company history, their complete product line, applications, downloadable product information sheets, and in some cases, complete instruction manuals. However, simple theory of operation and system block diagrams showing how the equipment works or functions are often missing. Some educational websites that offer NDT training help by providing this basic information.

At times the NTD technology seems to be advancing so rapidly that, before the information can be recorded, newer, better equipment, and software packages are being developed and are "just around the corner." Unfortunately, sometimes it takes a very long time to get around the corner. When this happens, it's time to shoot the writer and publish the work. Perhaps company publications representatives could help their companies, customers, general public, and potential authors alike by releasing new information as quickly as possible, when company confidentiality is no longer needed.

2

ACOUSTIC EMISSIONS

2.1 PRINCIPLES AND THEORY

The *acoustic emission (AE)* testing method is a unique nondestructive testing (NDT) method where the material being inspected generates signals that warn of impending failure. Acoustic emission testing is based on the fact that solid materials emit sonic or ultrasonic acoustic emissions when they are mechanically or thermally stressed to the point where deformation or fracturing occurs. During plastic deformation, dislocations move through the material's crystal lattice structure producing low-amplitude AE signals, which can be measured only over short distances under laboratory conditions. The AE test method detects, locates, identifies, and displays flaw data for the stressed object the moment the flaw is created. Therefore, flaws can not be retested by the AE method. In contrast, ultrasonic testing detects and characterizes flaws after they have been created. Almost all materials produce acoustic emissions when they are stressed beyond their normal design ranges to final failure.

It has been said that the first practical use of AE occurred in about 6500 BC as pottery makers listened to the cracking sounds made by clay pots that had been allowed to cool too quickly. By experience the potters learned that cracked pots were structurally defective and would fail prematurely. However, the father of modern AE testing was Josef Kaiser of Germany.

In 1950, Kaiser published his Ph.D. thesis, which was the first comprehensive investigation of acoustic emissions. He made two important discoveries;

Introduction to Nondestructive Testing: A Training Guide, Second Edition, by Paul E. Mix
Copyright © 2005 John Wiley & Sons, Inc.

the first was that material emits minute pulses of elastic energy when placed under stress. His second discovery stated that once a given load was applied and the acoustic emission from that noise had ceased, no further emission would occur until the previous stress level was exceeded, even if the load was removed and later reapplied. This so-called "Kaiser effect" can be time dependent for materials with elastic aging. The principle is used in present-day AE proof testing of fiberglass and metallic pressure vessels.

Cracking in structures such as aircraft wings, pipes, circuit boards, and industrial storage tanks generates acoustic emissions. They are also generated by deformation and crack propagation in pipes, pressure vessels, and weldments. Other sources of AE are nugget formation and overwelding in spot-welding operations, leaks in steam valves and traps, and bearing failure in pumps, motors, and compressors.

2.2 SIGNAL PROPAGATION

A short, transient AE event is produced by a very fast release of elastic energy for a specific dislocation movement. This local dislocation is the source of the elastic wave that propagates in all directions and cannot be stopped. It is similar to a microscopic earthquake with its epicenter at the defect. On flat surfaces, the wave propagates as concentric circles around the source and multiple installed sensors can detect it. Like a pebble thrown in water, the amplitude of the concentric waves is attenuated with increasing time and distance. The maximum distance where an AE event can be detected depends on material properties, geometry of the test object, its content and environment, etc. On flat or cylindrical surfaces, AE events can be detected at several meters, which is a great advantage for this method. Figure 2.1 shows multiple sensor monitoring and how sound is propagated from the defect location.

2.3 PHYSICAL CONSIDERATIONS

With tank bottom testing, waves are analyzed that propagate through the liquid from the source to the tank wall. Without liquid in the tank, the waves would be greatly attenuated. The same analogy applies to liquid-filled pipelines where the maximum distance AE events can be detected is longer than in gas-filled tubes because the AE signal is attenuated less in the liquid volume than in thin-walled tubing.

AE waves reach the sensors with certain delays that depend on the position of the AE source. The position of the source can be calculated using the different arrival times. This is known as "location calculation." Because of high

THE AE PROCESS CHAIN

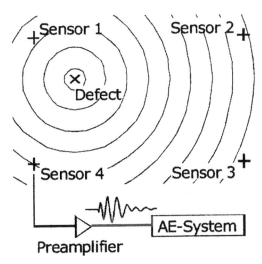

Figure 2.1 The propagation of sound. Courtesy of Vallen-Systeme, GmbH.

speed and large memories, modern PCs can calculate the location of the AE emissions in real time and display results immediately.

2.4 THE AE PROCESS CHAIN

AE testing can be thought of as a process chain consisting of the following steps:

1. *Test object and application of load:* produce mechanical tensions
2. *Source mechanisms:* releasing elastic energy
3. *Wave propagation:* from the source to the sensor
4. *Sensors:* converting a mechanical wave into an electrical AE signal
5. *Acquisition of measurement data:* converting the electrical AE signal into an electronic data set
6. *Display of measurement data:* plotting the recorded data into diagrams
7. *Evaluation of the display:* from diagrams to a safety-relevant interpretation

As shown in the process chain diagram of Figure 2.2, mechanical stress has to be produced within the test object, which is usually done by applying external forces, such as pressuring a tank or heating its contents. Material properties and environmental conditions influence the start of the release of elastic energy (i.e., the start of crack formation).

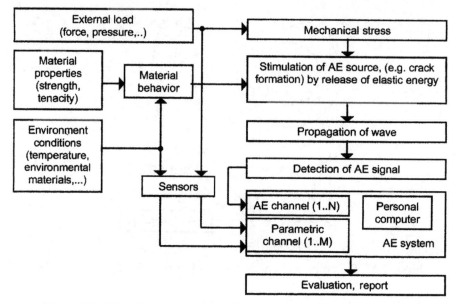

Figure 2.2 The AE process chain. Courtesy of Vallen-Systeme, GmbH.

2.5 TIME CONSIDERATIONS

There are two types of AE signals, namely transient and continuous. With transient AE signals, also called bursts, the starting and ending points of the transient signal are clearly differentiated from background noise. With continuous AE signals, amplitude and frequency variations can be seen but the signal never ends thereby appearing continuous. In this chapter we assume that fractures and crack growth produce burst-type signals and friction, process, and electrical background noise produce unwanted noise. In most cases, bandpass filters are used to minimize background noise while enhancing the AE signal.

AE systems convert AE bursts into compact data sets and eliminate background noise by filtering and establishing detection thresholds. Positive and negative threshold limits are set by the user to determine the starting point, which is the first threshold crossing or arrival time of detected AE burst signal. This information is used for the location calculation.

2.6 AE PARAMETERS

A single-burst waveform with highlighted AE parameters is shown in Figure 2.3. Note that the threshold detection points are at ±0.05 mV on the y-axis and the peak signal amplitude is about 0.14 mV. The risetime is the time from the

AE PARAMETERS

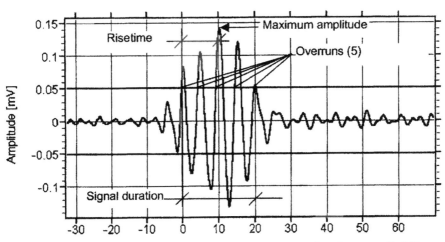

Figure 2.3 Features of transient signals. Courtesy of Vallen-Systeme, GmbH.

first threshold crossing until the signal reaches its maximum amplitude. There are five peaks (overruns) above the threshold limit. The first threshold crossing is at 0 on the time or x-axis and the signal duration is from 0 to 20 time units, typically microseconds (μs).

The waveform shown consists of many individual points or "samples." They correspond to many single measurements at constant time intervals. Digital systems sample the AE signal every 100 ns or 10 million times a second. The unit of time axis is typically specified in μs, where every 10 μs contains 100 samples. Therefore, a wave packet of 100 μs as shown consists of more than 1000 samples, which illustrates the large amount of memory required for analysis of a single burst.

In most cases, hundreds or even thousands of bursts are recorded for statistic evaluation. The AE parameters listed below can be evaluated statistically.

- Arrival time (absolute time of first threshold crossing)
- Peak amplitude
- Risetime (time interval between first threshold crossing and peak amplitude)
- Signal duration (time interval between first and last threshold crossing)
- Number of threshold crossings (counts) of the threshold of one polarity
- Energy integral of squared (or absolute) amplitude over time of signal duration
- RMS (root mean square) of the continuous background noise (before the burst)

AE bursts are produced by both defects of interest and background noise, which exceed the low threshold. Therefore, it is important to determine the characteristics that distinguish the wanted from the unwanted AE bursts.

Peak amplitude is one of the most important burst features. Crack signals show medium to high amplitudes for durations of some 10 µs, depending on the test object's properties. In most cases, bursts with less than three threshold crossings and durations less than 3 µs are regarded as unwanted noise signals. Very short signals usually indicate electrical noise peaks if they arrive at all channels at the same time. Additional logical filtering can be done in this case. External parameters such as pressure and temperature can also be measured, charted, and used as references for the measured AE data.

2.7 THE AE MEASUREMENT CHAIN

Figure 2.4 shows the schematic of an AE measurement chain.

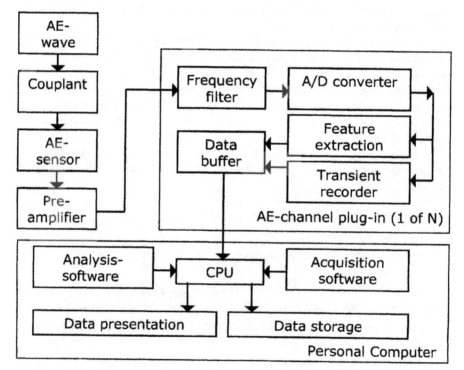

Figure 2.4 The AE measurement chain. Courtesy of Vallen-Systeme, GmbH.

2.7.1 Coupling Agents

Coupling agents are crucial to the quality of sensor coupling. They provide good acoustic contact between the sensor and the surface of the test object. Care must be taken to assure that the appropriate couplant is selected so that it is acceptable at the normal test temperature and does not corrode the test object's surface. A variety of silicone greases are available for industrial, aerospace, and other applications. See Chapter 11 on the ultrasonic testing method.

Generally, the coupling layer should be made as thin as possible by firmly pressing the sensor against the test object's surface. After attaching the sensors with an appropriate hold-down device, the quality of the coupling must be verified (pencil lead break, automatic coupling test). If test results are not satisfactory, the coupling procedure must be repeated.

2.7.2 AE Sensors

Piezoelectric sensors convert mechanical waves into electrical AE signals. They are hardy and more sensitive than other sensor techniques such as capacitive, electrodynamic, or laser-optical.

When testing metal vessels for integrity, frequencies between 100 and 300 kHz are often used. For this frequency range, the sensors have a resonance of about 150 kHz and cover the range of 150 to 300 kHz with a variation of sensitivity of about 6 dB. The resonant frequency indirectly determines the spatial range of the sensor. High-frequency sound attenuates faster, so it has a shorter detection distance. Background noises coming from longer distances typically consist of frequency components below 100 kHz, so they have only a small influence on the measurement chain tuned to 100 to 300 kHz.

For testing tank bottoms, sensors with a high sensitivity down to 25 kHz are required because these signals must run for long distances. Therefore it is very important to find and eliminate the lower-frequency potential noise sources.

Sensors often contain built-in preamlifiers and are attached to magnetic test objects using magnetic holding devices. The amplified AE signal is transmitted to the AE system via a signal cable. Typically, the 28 VDC power supply for the preamplifiers is fed through the signal cable and can have a length of several hundred meters.

The sensitivity of piezoelectric sensors can be as high as 1000 V/μm. In other words, a displacement of 0.1 picometer (pm) generates a 100 μVpk signal that can be easily distinguished from electrical noise, which is about 10 μVpk. For comparison purposes, the atomic radii of the elements are in the range of 150 pm, with manganese at 112 pm and lead at 175 pm. Therefore, displacements of 1/100 of an atomic radius can produce well-defined AE signals!

Figure 2.5 shows a number of Fuji Ceramics AE sensors used by Vallen-Systeme, GmbH and other AE equipment manufacturers. Differential sensors

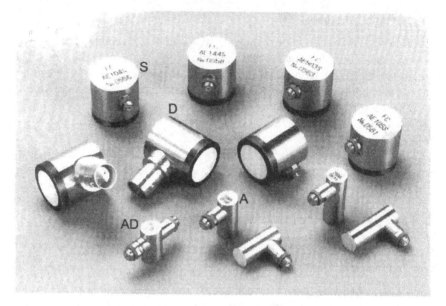

Sensor types—S: Single, A: Compact single, D: Differential, and AD: Compact differential

Figure 2.5 A selection of Fuji Ceramics AE transducers typically used by Vallen-Systeme, GmbH.

are typically used when there is a degradation of the AE signal by electrostatic or electromagnetic interferences.

2.7.3 Sensor Attachment

When testing magnetic materials, sensors are usually attached to the test object using magnetic holders as shown in Figure 2.6. When nonmagnetic objects are tested, elastic ties, tape, clamps, and glue may be used for sensor hold-down. When attaching sensors to the test objects, care must be taken so that unwanted noise is not introduced.

2.7.4 Sensor to Preamplifier Cable (Not Required for Sensors with Integrated Preamplifier)

This cable connects the sensor with the preamplifier and should not be longer than 1.2 m because of the capacitance load on the sensor. This cable is very sensitive and thin because of the miniature sensor connectors. The cable must not be bent sharply or strained and a tensile load should never be applied to the connectors. The cable may conduct unwanted acoustic noise through the

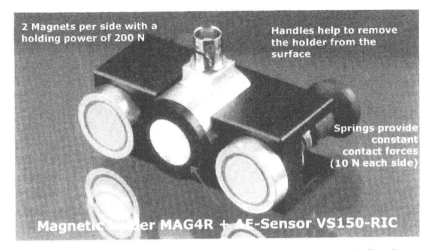

Figure 2.6 Vallen magnetic sensor holder with sensor. Courtesy of Vallen-Systeme, GmbH.

connector to the piezoelectric sensor. Because of these limitations, sensors with built-in preamplifiers are becoming more popular.

2.7.5 AE Preamplifier

The AE preamplifier can be a separate device or be integrated into the sensor. It amplifies the AE signal and drives the cable between the sensor and AE system. The preamplifier should have low input noise to distinguish the smallest sensor signals from electronic noise. It should also have a large dynamic range to process high-amplitude signals without saturation. The preamplifier should operate over a wide temperature range for applications near low-temperature vessels as well as above the transition temperature from brittle to ductile behavior. The preamplifier supply voltage (28 VDC) is sent via the signal cable. Preamplifiers have optional frequency filter capacity and provisions that permit a calibration pulse to be routed through the sensor.

2.7.6 Preamplifier to System Cable

RG 58 C/U cable with 50 ohm BNC connectors at both ends can be supplied for lengths to several hundred meters. Cables typically transmit the AE signal, DC power supply voltage, and calibration pulse for the sensor coupling test.

2.8 VALLEN AMSY-5 HIGH-SPEED AE SYSTEM

The Vallen AMSY-5 high-speed AE system is shown in Figure 2.7 along with its companion notebook PC. This system is the newest basic workhorse of the Vallen Systeme, GmbH. The PC can easily be separated from the system for report work, maintenance, upgrading to latest technology, etc. The AMSY-5 system shown here has an M16 master unit for up to 16 AE channels. Systems with up to 254 synchronized AE channels can be provided.

2.8.1 Frequency Filter

The frequency filter inside the channel is used to eliminate noise sources and matches the measurement chain to the requirements of the application. The 20 to 100 kHz filter is used for tank bottom tests where leakage and corrosion may be problems. The 100 to 300 kHz filter is used for integrity studies of metallic components. Filters with frequencies above 300 kHz can be used

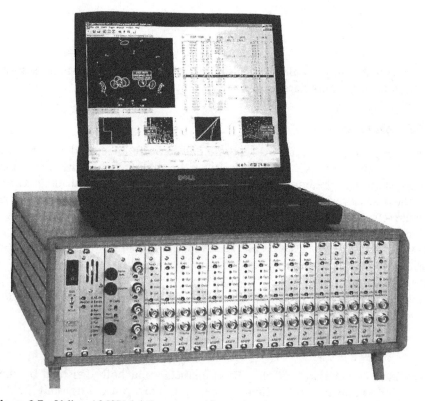

Figure 2.7 Vallen AMSY-5 AE system with notebook computer. Courtesy of Vallen-Systeme, GmbH.

where there are short distances between sensors. The manufacturer should be contacted for specific recommendations.

2.8.2 The A/D Converter

A/D converters are used to digitize the filtered AE signals. A large dynamic range is required for the converter to handle very strong AE bursts from nearby sources, which can be 1000 times greater than AE bursts located at long distances. The A/D converter must be able to handle the weak long-distance AE signals as well as the strong AE signals from nearby sources without saturating. A/D converters must also convert data quickly without extensive calculations, to prevent losing signals during times of high AE activity.

2.8.3 Feature Extraction

A continuous sampling rate of 10 MHz or 10 million measurement values per second per channel is used for feature extraction, which is processed in real time. This high-speed processing is possible through the use of field programmable gate arrays (FPGAs). These special ICs have thousands of processing elements that can be linked by PC software. For special processing setups used with pipeline structures, about 420 million instructions per second (420 MIPS) are performed. This corresponds to 42 instructions per second at 10 MHz. The results of the feature extraction are features such as maximum amplitude, rise-time, counts, energy, etc., for every hit in each channel. A typical example of feature extraction is shown in Figure 2.8.

2.8.4 Transient Recorder

Optional transient recorders provide a way to store large amounts of information about complex AE waves for further research and development in

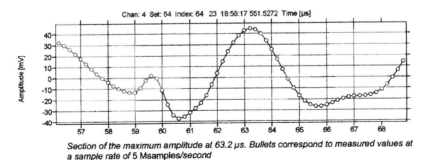

Section of the maximum amplitude at 63.2 μs. Bullets correspond to measured values at a sample rate of 5 Msamples/second

Figure 2.8 AE data set showing measured values. Courtesy of Vallen-Systeme, GmbH.

future applications. The transient recorder is an optional addition to the channel plug-in.

2.8.5 Data Buffer

The data buffer temporarily stores data when the CPU is busy with other tasks and cannot accept more data. Buffers are necessary because even today's most powerful PCs are not made for strict real-time processing. External buffers are very important for systems using standard Windows® operating systems such as the AMSY-5. With the new AMSY-5, the interconnection of multiple system units is user-friendly due to extensive self-test and self-configuration features.

2.8.6 Personal Computer and Software

Modern AE systems use PCs with menu-driven parameter input and system control. Software online help-screens provide quick access to information that explains the use of software features. The data acquisition file contains all the features for all the bursts of all sensors as well as the external parameters such as pressure and temperature. If the complete waveform is to be stored another file is created.

During the AE test the measured data are analyzed online and displayed so that the operator may immediately recognize the possible development of defects within the test object. He has the ability to halt the load increase, such as pressure, in order to minimize damage and protect personnel, the environment, and the test object.

PC tasks are:

- Data acquisition and storage—provide storage for all data.
- Provide data analysis—online and offline.
- Enable logical filtering to determine defect plausibility.
- Provide location calculation and clustering information.
- Provide statistical information for analysis.
- Display results—graphically and numerically.
- Provide self-test of system hardware.
- Initiate sensor coupling test and record sensor frequency response.

2.8.7 Sensor Coupling Test (Autocalibration)

One channel transmits an electrical test pulse to the connected sensor. The sensor emits a mechanical wave that is detected by neighboring sensors as shown in Figure 2.9. After three test pulses, the next sensor becomes the pulse emitter. The plausibility of the received amplitude enables the operator to draw conclusions about the quality of the coupling. Figure 2.10 shows sensor

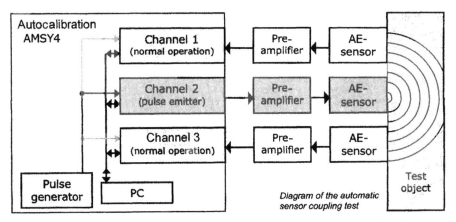

Figure 2.9 Automatic calibration of AMSY-4 system. Courtesy of Vallen-Systeme, GmbH.

	1	2	3	4	5	6
1	88	84	80	85	69	77
2	86	88	94	94	94	94
3	80	94	88	94	86	94
4	82	94	93	88	92	92
5	69	93	86	91	88	84
6	73	94	94	90	85	88

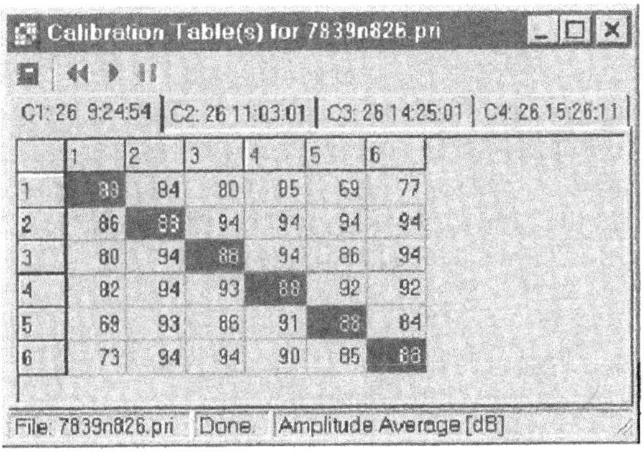

Figure 2.10 Vallen sensor calibration table. Courtesy of Vallen-Systeme, GmbH.

coupling test results for six channels in table form. Readings are in decibels (dB).

2.9 LOCATION CALCULATION AND CLUSTERING

2.9.1 Location Calculation Based on Time Differences

The determination of the source location for each event is an essential element of AE testing. The difference distance between a defect source and other sensors equals the *arrival time difference* times the *sound velocity*. The location calculation is based on the arrival time differences of the AE signal

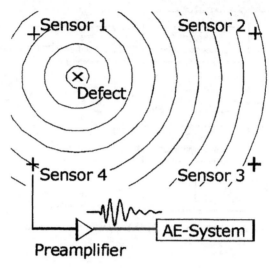

Figure 2.11 The Principle of Location. Courtesy of Vallen-Systeme, GmbH.

propagating from its source to different sensors as shown in Figure 2.11 where the AE wave is propagating in concentric circles from its source and arrives at different sensors at different times. This time delay is proportional to the distance between the sensor and the source. In the reference figure, the wave reaches sensor 1 first, then 4, 2, and 3 in that order.

When a pencil lead is broken on an acrylic glass plate, these same four sensors display the AE waveforms shown in Figure 2.12. Note that zero of the time axis marks the AE arrival time of the burst at sensor 1 (top section of the figure). The arrival time differences between channel 1 and channels 2, 3, and 4 can be read at the time axis of the waveform in the next three sections of the figure.

A hyperbola defines the case where points have a constant difference between their distances to two fixed points on a circular conic surface. Figure 2.13 shows three hyperbolae, each representing all points within the calculated distance difference to two sensors. At the point of intersection of the three hyperbolae, the three distance differences are equivalent to the measured time differences. Therefore, this is the location of the unknown source or defect being analyzed. As shown in this example, the arrival time at three sensors is needed to find the point of intersection. If an AE event arrives at only two sensors, there is only one set of sensors with one hyperbola and the planar location of the source cannot be determined. Hyperbola diagrams are used to check the plausibility of certain selected location results. This calculation is based on an inverse method, which in addition to the location results provides a measure of the plausibility of the location calculation when more than three sensors have been hit by the same event.

LOCATION CALCULATION AND CLUSTERING

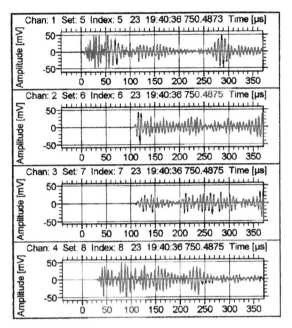

Figure 2.12 Four sensor test signals. Courtesy of Vallen-Systeme, GmbH.

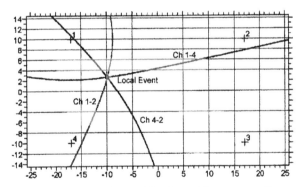

Figure 2.13 Hyperbola intersection with three hit sensors. Courtesy of Vallen-Systeme, GmbH.

2.9.2 Clustering

Most of the time, the results of a location calculation are plotted on a point diagram without hyperbolae. However when a 5-cm-thick glass foam plate is compressed, a huge number of events are recorded as shown in Figure 2.14 by the cluster chart on the left and planar diagram on the right. In this case,

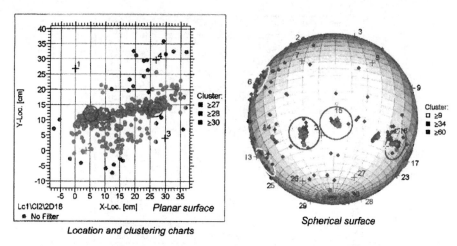

Figure 2.14 Clustering shown on clustering chart and spherical surfaces. Courtesy of Vallen-Systeme, GmbH.

overlapping hits cannot be seen and AE technology uses clustering to count, clarify, and mark the areas of high location density.

Clustering is a mathematical way to determine the point density within a certain area, marking those areas of high point density by colored rectangles or circles. The left diagram shows three clusters indicating three areas of high point density. The clusters appear as darker circles in this illustration. The spherical calculation provides a 3D rotating display of the sphere shown on the right. Spherical representation is required for spherical liquid gas and natural gas tanks.

Clustering always indicates a source, which repeatedly emits acoustic signals. Sometimes these sources are easily identified as abrasion marks or annexes, etc. If the source cannot be identified, a real defect must be assumed. Therefore, it is important to pay attention to the activity and intensity of a cluster chart. In Figure 2.15 the horizontal axis represents the cluster ID. The vertical axis on the left represents amplitude and the vertical axis on the right represents the number of events per cluster indicated by the stepped line. The diagram clearly shows the operator during the test when location clustering occurs and when a high numbers of events with relevant amplitude are forming. For example, the figure clearly shows that cluster number 11 contains 105 events with amplitudes of 39 to 67 dB.

2.9.3 Sample Analysis Screen

Figure 2.16 shows a sample analysis screen as the operator sees it during the test or during a post-test analysis. Some of the diagrams that require an explanation have already been discussed. The small information boxes at the right

LOCATION CALCULATION AND CLUSTERING

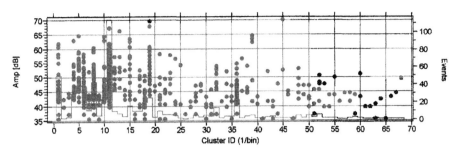

Figure 2.15 Amplitude and number of events for various clusters. Courtesy of Vallen-Systeme, GmbH.

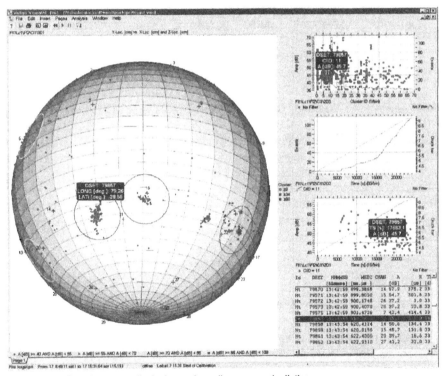

Sample screen displaying various diagrams and a listing

Figure 2.16. Actual AE instrument display with data. Courtesy of Vallen-Systeme, GmbH.

appear as soon as an AE event has been selected by a double-click, which allows the operator to easily refer to any event within various types of display.

2.9.4 Visualization of Measurement Results

The AE system software provides a clear, well-structured display of the following AE data:

- Numerical display
- Cumulative or differential diagrams such as total number of events, total number of bursts, total energy vs. time or pressure
- Pressure vs. time
- Distribution of amplitudes or other parameters
- Location diagrams including a picture of the test object and a cluster display
- Waveform diagrams in time or frequency domain

2.10 ADVANTAGES AND LIMITATIONS OF AE TESTING

Numerous test objects such as spherical natural or liquid gas tanks and petrochemical reactors have been successfully tested with the AE testing method. AE testing complements hydrotesting and pneumatic testing.

AE monitoring during pneumatic pressure testing can be used to assure a safe test. AE testing provides an early warning system for developing or growing defects. With AE monitoring, the pressure test can be stopped before failure occurs.

2.10.1 Advantages

- Pneumatic pressure testing can minimize potential overdesign of tank-support systems.
- AE monitoring during pressurization provides in-depth information beyond the statements—failure, leakage, or noticeable deformation during test.
- In many cases the pressure system can be tested within standard operating conditions and acceptable temperature ranges.

2.10.2 Advantages of Using Operating Medium (Gas or Liquid)

- Minimizes downtime caused by removal of the tank working medium if this medium is used for pressurization.
- Minimizes expensive drying operations following tank de-inventorying (as required with hydrotesting).

ADVANTAGES AND LIMITATIONS OF AE TESTING

- Minimizes de-inventorying water and subsequent decontamination expenses.
- Minimizes the problem of catalysts becoming unstable in chemical reactors during de-inventorying operations.
- AE testing can detect corrosion and minimize or eliminate the need for internal inspection.

2.10.3 Advantages Compared to Other NDT Methods

- AE monitors the dynamic reaction of the test object passively, without intervention, when the load is applied.
- In most cases, the AE method can detect sources at several meters distance from the sensor.
- AE provides 100% pressurized wall monitoring.
- AE provides real-time monitoring of the growth of unknown defects at a given load, even remotely by data transmission.
- It monitors a structure under all operating conditions.

2.10.4 Limitations

- Defects that do not grow or move cannot be detected.
- The only defects that can be detected without exceeding the highest preceding load are defects that are already active at the actual load level and endangering the component.
- Evaluation criteria do not exist in the form of commonly accessible data; the rating of AE results is dependent on the knowledge and experience of the service provider.
- AE testing is sensitive to process noise exceeding the detection threshold. Raising the threshold level decreases test sensitivity. Above a certain noise level, AE testing is no longer efficient.

2.10.5 Location Errors

Despite the improvements in AE testing, the AE tester must still be aware of influences causing location errors and be able to accurately rate them. Some factors affecting location accuracy are:

- A different wave mode than the assumed one determines the arrival time.
- A wave takes a different propagation path than assumed by the algorithm.
- Multiple waves that overlap at the sensor.
- Sources emit signals in such quick succession that there is not enough time for the signals in the structure to decay. Therefore they do not represent a new hit.

One big problem is that based on the origin of the AE signal. Inexperienced operators may find it difficult to determine whether the AE burst was produced by a real defect, and if so, the severity of the crack growth or flaw. In many cases, experience is the best teacher. Therefore, whenever possible, inexperienced operators should work alongside experienced operators who can help direct the work and interpret the results.

2.11 AMSY-5 MAIN FEATURES

- *Top AE speed:* stores over 30,000 AE hits/s, full data set, filtered, time sorted, sustained, to HDD.
- *Top TR-speed:* over 2.5 Mbytes/sec waveform data, filtered, time sorted, sustained, to HDD.
- *Top PC flexibility:* PC can be easily separated for report work, maintenance, upgrading to latest technology, etc.
- *Top software:* acquisition and analysis software for Windows® 2000 and Windows® XP. Analysis software is compatible with AMS3 and AMS4 systems, familiar to most previous users.
- *Top configuration flexibility:* Any number of AE channels up to 264 can be used by connecting up to four AMSY-5 units to one PCI interface board (ASyC), with up to four ASyC per PC.
- *Top economics:* Thousands of reliable AMSY-4 AE channels (shipped worldwide) can be easily moved to AMSY-5 units of latest technology.

Note: For additional information see the manufacturer's instruction manuals and booklet *AE Testing—Fundamental, Equipment, Applications*, by Dipl. Ing. Hartmut Vallen, Vallen-Systeme, GmbH, Icking (Munich), Germany, April 2002.

2.12 AE TRANSDUCERS

AE transducers are state-of-the-art piezoelectric sensors operating at frequencies from 20 kilohertz (kHz) to 1.5 megahertz (MHz); some sensors are capable of reliable operation at temperatures as high as 550° centigrade (C). For special applications, wideband sensors are available with flat frequency response from 100 kHz to 1 MHz. Recent innovations in transducer design include the incorporation of low-noise high-impedance inputs, 40 decibel (dB) preamplifiers, and noise filters. These integral preamp sensors, with resonant frequencies of 20, 60, 150, 300, and 500 kHz, are capable of driving 1000 ft of signal cable. A 200 kHz airborne sensor is used for remote leak monitoring applications.

2.13 KISTLER PIEZOTRON® ACOUSTIC EMISSION SENSORS AND COUPLERS

The Kistler Instrument Corporation manufactures small, rugged AE sensors and signal processing couplers for easy mounting near the sources of acoustic emissions. Acoustic emissions typically occur during plastic deformation, martensitic steel phase conversions, crack formation and propagation, and frictional processes such as rubbing between two surfaces or materials.

2.14 AE SENSOR CONSTRUCTION

The Kistler 8152B sensors (Figure 2.17) consist of a sensor case, piezoelectric measuring element, and integral impedance converter. The measuring element is mounted on a thin steel diaphragm. The diaphragm has a slightly protruding coupling surface so that it can be mounted with a precise force for accuracy and reproducibility. The measuring element is acoustically isolated from its case and other AE interference by design. Kistler AE sensors have a very high sensitivity to surface (Rayleigh) and longitudinal waves over a wide frequency range. A miniature impedance converter is built into the AE sensor so that a low-impedance voltage signal output is obtained. This greatly reduces unwanted noise pickup.

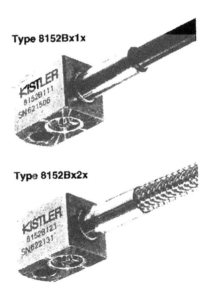

Figure 2.17 Kistler Piezotron® sensors. Courtesy of Kistler Instrument Corp.

2.15 SUMMARY OF AE SENSOR FEATURES

- High sensitivity and wide frequency response
- Inherent high-bandpass characteristics
- Inherent low sensitivity to electric and magnetic noise
- Robust design suitable for industrial use
- Ground isolation, which prevents ground loop currents
- Designs conforming to Canadian Electrical (CE) safety standards

2.16 TECHNICAL SPECIFICATIONS—8152B2 SENSOR

Frequency range (±10 dB): 100 to 900 kHz
Sensitivity*: 48 dB$_{ref\ 1V/(m/s)}$ or 250 V(m/s)
Operating temperature: −40 to +60°C
Mass: 29 g
Ground isolation: >1 MΩ
Overload shock, 0.5 ms pulse or 0.2 ms pulse: 2000 g
Overload vibration: ±1000 g
Supply, constant current: 3 to 6 mA
Supply, voltage (coupler): 5 to 36 VDC
Output voltage (full scale): ±2 V
Output voltage bias: 2.5 VDC
Output current: 2 mA
Output impedance: 10 Ω
Case material: inox/stainless steel
Standard cable construction: 2 m Viton hose, steel braided

Optional Feature

Cable length: cable lengths up to 5 m

2.17 AE COUPLER FEATURES

The Kistler 5125B coupler (Figure 2.18) has an integrated RMS converter and limit switch specifically designed for processing high-frequency AE signals. A jumper is used to select a gain of 1 or 10, which in turn amplifies the AE signal by a factor of 10 or 100. The amplifier has two selectable series-connected high-pass or low-pass filters. Typically, the two filters can be set up to provide

---

* Average sensitivity for surface waves (Rayleigh waves).

Figure 2.18 Kistler AE Piezotron® coupler type 5125B. Courtesy of Kistler Instrument Corp.

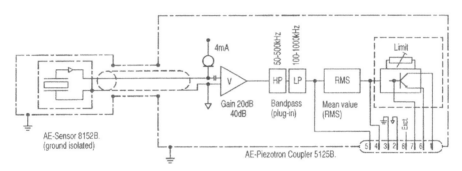

Figure 2.19 Kistler AE sensor and piezotron® coupler arrangement. Courtesy of Kistler Instrument, Corp.

a bandpass such as 100 kHz (high-pass filter) to 900 kHz (low-pass filter). A simplified operational diagram for the coupler is shown in Figure 2.19. The integrated time constant is also modular and selectable. A round 8-pin connector provides access to two analog outputs, the AE filtered output and RMS output, and one digital limit switch output. The RMS switching level can be monitored at the limit switch output (pin 7) with a DVM or oscilloscope. The manufacturer's grounding and mounting instructions must be rigidly adhered to for optimized performance.

The filtered output from the sensor has the effect of attenuating the AE signals above and below the desired frequency range. When the unwanted frequencies are removed, bandpass frequencies are enhanced. The RMS converter integrates the filtered high-frequency signal for the selected time constant to produce a refined high-frequency envelope. Finally, the limit switch provides simple limiting of the RMS signal—as soon as the RMS signal exceeds the preset limit, the optocoupler cuts off. In this manner, the limit switch removes baseline noise, transmitting only the high-frequency peaks above the preset limit.

2.18 TECHNICAL SPECIFICATIONS—5125B COUPLER

2.18.1 Input

Supply current for sensor: about 4.3 mA
Input voltage range: 0 to 8 V
Amplitude signal: $1.6\,V_{pp}$
Signal processing gain 1×: 10 V
Signal processing gain 10×: 100 V

2.18.2 Output

Frequency range (without filter): −5% at 15 to 1000 kHz
Frequency range (without filter): −3 dB at 5 to 1700 kHz
Standard low-pass filter: 2-pole Butterworth: −40 dB/dec
Standard low-pass filter: cut-off frequency (−3 dB): 1000 kHz
Voltage: 0 to ±5 V
Current: 0 to ±5 mA
Output resistance: 50 Ω
Zero offsets: <±20 mV
Noise: $<10\,mV_{pp}$
Tolerance: 5%

2.19 ACOUSTIC EMISSION TECHNOLOGY

The MISTRAS Holdings Group trades under the name of Physical Acoustics around the globe; Physical Acoustics Corporation (PAC) is a world leader in acoustic emission technology.

Acoustic emission (AE) testing is a powerful method for examining the behavior of materials deforming under stress. AE can be defined as a transient static wave generated by the rapid release of energy within a material. These materials emit sounds that warn of impending structural damage. AE sounds can be heard as cracks propagate and are arrested. Weaker sounds associated with plastic deformation take place at the tips of cracks prior to crack propagation.

Small-scale damage is detectable long before failure, so AE detection can be used as a nondestructive testing (NDT) technique to find defects during structural proof tests and plant operation. AE also offers unique opportunities for materials research and development in the laboratory. AE equipment is also used by production and quality control (QC) groups for testing, weld monitoring, and leak detection.

Physical Acoustics Corporation (PAC) designs, manufactures, and markets AE instruments and systems for the monitoring and nondestructive testing of the structural integrity and general quality of a variety of materials. The company also sells equipment for field testing services and contract research.

2.20 AE APPLICATIONS

AE is used by numerous industries for NDT. These industries include refineries, pipelines, power generation (including nuclear), structural, and aircraft. AE testing and evaluation is also used by offshore oil platforms and paper mills. Structures frequently tested include bridges, tunnels, towers, tanks, pipes, cranes, and heavy industrial equipment.

Most of PAC's products are computerized flaw detector systems and sensors. AE sensors can be used for proof testing or online monitoring of full-sized structures with multichannel detection and location capabilities. PAC also markets a series of laboratory instruments used for analyzing a variety of materials. New, fully integrated AE testing systems such as the PCI/DSP-4 card and DiSP provide real-time parallel processing of high-speed AE signals using digital signal processing (DiSP), application-specific integrated circuits (ASICs), and microprocessors. For field testing with mobile service units, PAC provides personnel, computerized systems, laboratory gear, special transducers, and other equipment needed to perform on-site tests.

2.21 AE THEORY

AE was originally conceived as an NDT method for locating flaws as they occurred in pressure vessels. Today AE encompasses a much wider scope and can be applied to all types of process monitoring as well as real-time flaw detection and structural integrity studies. Pressure, temperature, vacuum, mechanical tension, or compression are the most commonly applied stresses. Figure 2.20 illustrates that an acoustic noise source, such as a propagating crack, generates sound waves that radiate outward in all directions. AE sensors, typically piezoelectric transducers, can be placed anywhere on the object within range of AE energy waves. By using three transducers and triangulation techniques, the location of the flaw can be detected.

AE sensors are low-noise detectors that operate in the ultrasonic frequency range of 10 kHz to 2 MHz. Physical motions as small as 1×10^{-12} can be detected. AE sensors can hear the breaking of a single grain of metal or a single fiber in a fiber-reinforced composite material. AE sensors can also

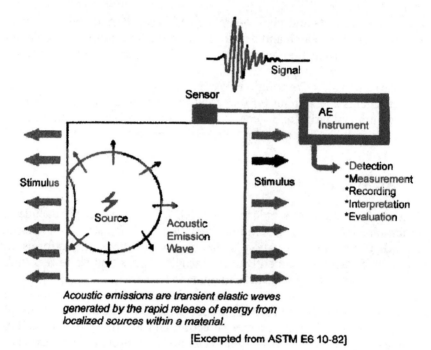

Figure 2.20 AE signal generation, transmission, and detection. Courtesy of Physical Acoustics Corp.

detect the sound of a tiny bubble of gas from a pinhole leak as it arrives at the surface of a liquid. With the capability to detect small AE signals or large AE signals such as are caused by brittle crack advance, AE technology warns of impending structural damage and can monitor costly and critical processes.

Sudden movements in solid materials generate AE waves that may be detected at distances of a few inches to distances of several hundred feet, depending on the properties of the material being tested. Sudden subcritical local failures in materials under stress are the classical sources of AE. AE technology provides an early warning system to prevent catastrophic failures, to assess structural integrity, and to enhance safety in a wide range of structures from fiberglass tanks to bucket trucks, from bridges and aircraft to high-pressure gas cylinders. AE is also used for a wide range of process applications such as leak detection, particle impacts, electrical discharges, and a variety of friction-related processes.

The recent high-tech explosion has opened up a whole new range of technical capabilities for the AE method. The frequency domain is readily accessible and full waveform capture capabilities enable advanced, real-time, and post-test analysis of waveforms that were previously reduced in real time to a small set of half-a-dozen features measured with bulky electronic circuits.

PAC's REACT department makes use of these new technological advances to develop customer-based AE applications.

2.22 APPLICATIONS

The following applications are broken down into four categories that demonstrate the wide-range use and popularity of the acoustic emission method.

2.22.1 Behavior of Materials—Metals, Ceramics, Composites, Rocks, Concrete

- Crack propagation
- Yielding
- Fatigue
- Corrosion, stress corrosion
- Fiber fracture, delamination

2.22.2 Nondestructive Testing During Manufacturing Processes

- Material processing
- Phase transformation in metals and alloys
- Detection of pores, quenching cracks inclusions, etc.
- Fabrication
- Deforming processes—rolling, forging, extruding
- Welding and brazing defects detection—inclusion, cracks, lack of penetration
- TIG, MIG, spot, electron beam, etc.
- Weld monitoring for process control

2.22.3 Monitoring Structures

- Continuous monitoring—metal structures, mines, etc.
- Periodic testing—pressure, pipelines, bridges, cables
- Loose part detection
- Leak detection

2.22.4 Special Applications

- Petrochemical and chemical—storage tanks, reactor vessels, offshore platforms, drill pipe, pipelines, valves, hydrotreaters
- Electric utilities—nuclear reactor vessels, piping, steam generators, ceramic insulators, transformers, aerial devices

- Aircraft and aerospace—fatigue cracks, corrosion, composite structures, etc.
- Electronics—loose particles in electronic components, bonding, substrate cracking

2.23 ADVANCED EQUIPMENT

2.23.1 PCI-2 AE Unit

A complete two-channel AE system on a card utilizing the PCI-2 is an 18-bit A/D converter with an analog input frequency response of 3 kHz to 3 MHz. The unit provides two channels of AE for simultaneous waveforms and feature processing. The PCI-2 provides a sampling rate 40 Msamples/sec with lower noise and higher speed. The unit provides a DSP system on a single full-size 32-bit PCI-card, which is ready for operation in a customer's PC or one of PAC's hardened PCs for multiple channel operation.

Superior low-noise and low-threshold performance is achieved by revolutionary AE system design, through the use of an innovative 18-bit A/D conversion scheme, with up to 40 Msample/sec acquisition and real-time sample averaging. This performance is obtained via the system's pipelined, real-time architecture without sacrificing AE throughput speed.

With these features and the PCI-2's low cost, it is ideal for laboratories, universities, and industrial turnkey systems, and any application where low noise, low channel count, and low cost are required, as well as where the use of an existing PC is desired.

Through the high-performance peripheral component interface (PCI) bus and direct memory access (DMA) architecture, significant AE data transfer speeds can be obtained, assuring a wide bandwidth bus for multichannel AE data acquisition and waveform transfer. In addition to the two AE channels, the system also has two parametric channels for other transducers, such as strain gage, pressure, temperature, load, etc.

Waveform data streaming is built with the board, allowing waveforms to be continuously transferred to the PC hard drive. The 32-bit PCI bus is standard in all PC computers currently being shipped. PCI-2 AE system cards can be implemented inside most standard PC computers or inside one of PAC's rugged, multichannel PAC system chassis, including the 8-channel benchtop chassis, the 12-channel portable AE system, or a 4-channel notebook-based chassis (μ-series).

2.23.2 Key Features

- Very-low-noise, low-cost two-channel AE system with waveform and hit processing, built on one full-sized 32-bit PCI card
- Internal 18-bit A/D conversion and process with resolution better than 1 dB at low-signal amplitude and low threshold settings

ADVANCED EQUIPMENT

- Forty MHz, 18-bit A/D conversion with real-time sample averaging for superior enhanced accuracy
- Built-in real-time AE feature extraction and DMA transfer on each channel for high-speed transient data analysis at high hit rates directly to the hard disk (HD)
- Built-in waveform procession with independent DMA transfer on each channel for high-speed waveform transfer and processing
- Designed with extremely high-density FPGAs and ASIC ICs for extreme high performance, minimal components and cost
- Four high-pass and 6 low-pass filter selections for each channel, totally under software control
- AE data streaming provides continuous recording of AE waveforms to the hard disk at up to 10 Msamples/sec rate on one channel or 5 Msamples/sec on two channels
- Up to two parametric inputs on each PCI-2 board with 16-bit A/D converter and update rates to 10,000 readings/sec. The first channel provides full instrumentation conditioning with gain control, offset control, and filtering for direct sensor input. The second channel provides a ±10 volt input for conditioned sensor outputs.

2.23.3 PCI-8, 16-Bit, 8-Channel AE Unit

This PCI-8 board provides eight channels of acoustic emission for simultaneous waveforms and feature processing. It is ideally suited for low-frequency portable applications. This cost-effective full-sized PCI card can achieve AE data transfer speeds up to 132 Mbytes/second. PCI-8 AE subsystem boards can be used with most PC computers in use today. The boards can also be used in the SAMOS™ 32, SAMOS™ 48, SAMOS™ and 64, μSAMOS™ chassis. The 16-bit PCI-8 board features eight complete high-speed AE channels featuring data acquisition, waveform processing and transfer, two analog parametric input channels, and eight digital input and output control signals. Figure 2.21 shows a schematic diagram of the PCI-8 AE board.

As shown in the schematic diagram, each AE channel on the PCI-8 AE board consists of a D/A converter (DAC) power supply, automatic sensor test (AST), and programmable gain amplifier. The DAC preamp power voltage controls a special preamplifier, which has a built-in voltage-controlled gain and test signal. By varying the DAC, the transducer preamp can be software controlled to change its gain and test signal. Four selectable high-pass filters, four selectable low-pass filters, a 16-bit A/D converter, and real-time feature extraction unit follow the programmable amplifier. In addition, the PCI-8 board provides up to two analog inputs, eight digital inputs, eight digital outputs, board, and PCI controllers.

The incoming signal is conditioned and passed to the selectable filter circuitry, where one of four high-pass filters and one of four low-pass filters is amplified in accordance with the user-programmed filter strategy. The filtered

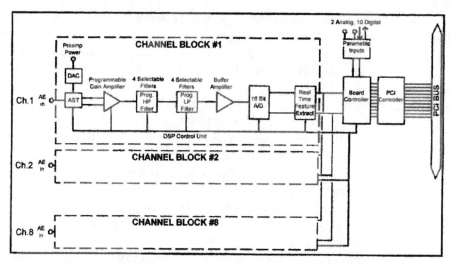

Figure 2.21 PCI-8 AE board schematic diagram. Courtesy of Physical Acoustics Corp.

signal is passed on to the 16-bit A/D converter module where the AE signal is digitized at rates up to 1.0 MSPS. The 16-bit digital AE output is then passed to the feature extraction circuitry, which converts the signal to AE hits and features instantaneously. The high-speed feature extracted data are transferred to the computer for further processing through the PCI bus. Waveforms are simultaneously processed and transferred to the computer using high-speed DMA access directly to the computer for collection, further processing, and storage.

2.23.4 MicroSAMOS™, Budget, Compact AE System

The μSAMOS™ system is a small, portable, battery-operated (optional) notebook operated acoustic emission system. It is similar in size to a notebook computer and contains up to three PCI-8 cards to provide a powerful 8-, 16-, or 24-channel AE system. Components used with the microSAMOS™ system are shown in Figure 2.22. Key advantages of the microSAMOS include small size, high channel density, portability, light weight (less than 10 pounds) and notebook computer connectivity. The microSAMOS provides very good acquisition performance while maintaining field ruggedness. An external connector enables connection of up to two parametrics and provides for control input and alarm output functions.

The notebook computer used with the microSAMOS can be used to analyze data and prepare reports at any location. There are two models of microSAMOS, one AC powered only and one designed for AC/DC operation. When used with internal battery, the microSAMOS can be powered for up to

ADVANCED EQUIPMENT

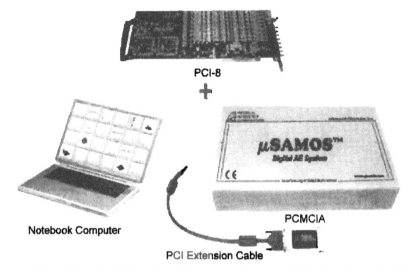

Figure 2.22 MicroSAMOS digital AE system. Courtesy of Physical Acoustics Corp.

2.5 hours of operation on a charge with one PCI-8 card. An external battery option can be used to extend operation to 8 hours.

The microSAMOS comes complete with a PC-card and cable to plug into most PCMCIA connector ports. The notebook computer must be compatible with Windows® 98 (Second Edition), ME, 2000, or XP to operate and run PAC's AEwin™ Windows software. Contact the manufacturer concerning compatibility questions.

2.23.5 DiSP Systems

The DiSP system is a technologically advanced AE system based on PAC's PCI/DSP-4 AE "4-channel system on a board." A single enclosure is used to house the entire 8-, 16-, 24-, or 56-channel AE system including PC computer and peripherals, such as CDROM, hard drive, and/or floppy disk. This consolidation of equipment makes it very efficient when compared with other AE systems, which require an external PC computer and multiple AE channel expansion enclosures to perform the same tests. Figure 2.23 shows a 24-channel portable DiSP system. The DiSP system results in fewer enclosures, substantial weight reduction, and lower costs. The use of a PCI bus and integrated PC computer also significantly enhances AE signal-processing performance.

2.23.6 PCI/DSP-4 Card

The PCI/DSP-4 card (Figure 2.24) is a four-channel, digital signal processing (DSP)-based, cost-effective AE data acquisition system on a single, full-size,

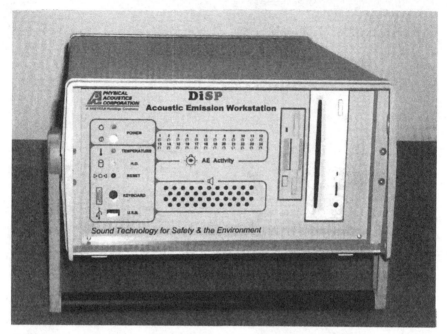

Figure 2.23 Portable AE channel field unit. Courtesy of Physical Acoustics Corp.

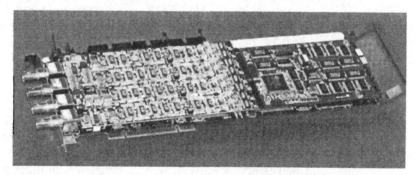

Figure 2.24 Four-channel PCI-DSP-4 board. Courtesy of Physical Acoustics Corp.

PCI card. Through the high-performance PCI (peripheral component interface) bus, AE transfer speeds up to 132 megabytes/second can be attained, assuring a wide bandwidth bus for multichannel AE data acquisition and optional waveform processing. The PCI bus has become the de-facto standard for PC computers currently being shipped.

PCI/DSP-4 boards can be implemented inside most standard PC computers or inside one of PAC's four rugged, multichannel PAC DiSP system chassis, housing 8, 16, 24, or 56 AE channels in a single enclosure with integral

PC computer. Recent advances in surface-mount technology and high-density programmable gate array devices enables PAC to provide a single board with four high-speed AE channels of data acquisition with optional waveform module and up to 8 parametric input channels (on the first board).

2.23.7 Features of PCI/DSP-4 System Board

- Four digital AE channels on one full-size PCI card complete with four-channel AE system.
- PCI bus provides AE data transfer rates up to 132 Mbytes/sec to PC computer.
- Sixteen-bit, 10 MHz A/D converter with >82 dB dynamic range w/o gain settings.
- Four high-pass, low-pass filter selections for each channel, totally under software control.
- High-performance 32-bit floating-point on-board DSP processor with up to 50 MFLOPS performance.
- Up to eight parametrics on the first PCI/DSP-4 master board with update rates of 10,000 readings/second when attached to hit data.
- Designed with multiple field programmable gate arrays (FPGAs) and application sensitive integrated circuits (ASICs) for high performance and low cost.
- Hit LED drivers are built into the board so hit LEDs can be directly attached.
- Optional waveform module for capturing and processing waveforms on all four AE channels with its own DSP and 40 MIPS DSP and waveform buffer memory.
- DSP circuitry virtually eliminates drift, thereby achieving high accuracy and reliability.

2.23.8 PCI/DSP-4 Board Operation and Functions

A simplified block diagram of the PCI/DSP-4 board is shown in Figure 2.25. Digital inputs from preamplifiers or integral preamplifier sensors are routed to the PCI/DSP-4 channel inputs via BNC connectors that are located on the rear PC metal hold-down bracket of the board. Phantom power for the preamplifiers and AST (automated sensor test) provides computer-controlled pulsing for each active AE channel to determine sensor coupling efficiency and provide system verification.

The incoming AE signal is conditioned and passed to the selectable filter circuitry where one of four high-pass filters and one of four low-pass filters is applied in accordance with user program strategy. The filtered signal is fed to the 16-bit A/D converter module where the AE signal is digitized at rates up

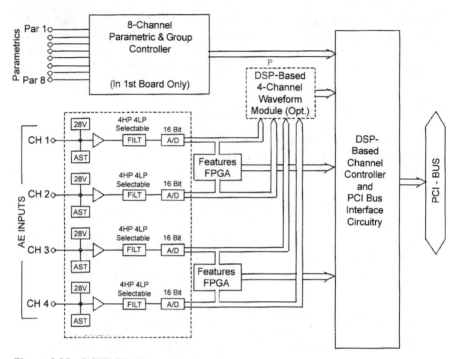

Figure 2.25 PCI/DSP-4 board block diagram. Courtesy of Physical Acoustics Corp.

to 10 MHz. The 16-bit, digital AE output is then passed to feature extraction circuitry, which converts the signal into AE hits and features instantaneously. It is also fed to the waveform option module for waveform recording and further digital signal processing. The high-speed feature extraction data are processed further by the on-board DSP signal processor. Then the signal is processed to form additional AE features, front-end filtering, time-sorting, and stored in the large on-board buffer and transmitted to the PCI bus where the system CPU reads and further processes the AE data.

The optional waveform module receives all incoming waveforms from the 16-bit A/D converter. The board allows the user to decimate the waveform for waveform compression purposes, select various pretriggering and waveform lengths, and process these waveforms in its own on-board DSP. The DSP provides extra capabilities and features as needed, such as FFT's (Fast Fourier Transform's) feature extraction, time of arrival extraction, partial powers, partial energy, etc. The module also has standard on-board memory storage for up to 256 waveforms. PAC's philosophy is to get the AE data to the PC's memory, display, and the hard disk drive, rather than hold waveforms in a buffer. Their goal is high-speed data transfer, over the PCI bus, to get the waveforms into the PC as soon as possible during the AE test.

2.23.9 DiSP System Block Diagram

The DiSP system, shown in Figure 2.26, is a 56-channel system. In this block diagram, the first PCI/DSP-4 card is shown in the top left of the figure. This group channel controller provides up to eight parametric inputs, one cycle counter, and control inputs and outputs. Therefore, there is a difference between the first board in a system and all other boards. The first board also provides master timing, synchronization, and communication signals to the remaining PCI/DSP-4's in the chassis. Synchronization to other AE system chassis is accomplished by adding an expansion interface board to the chassis as shown on the bottom-right side of the figure. A cable connected between the chassis transfers the synchronization and control to the other chassis.

A single-board CPU, shown at the top right, provides all the standard resources needed by the PC computer including RAM, hard and floppy disk and CD, video output, keyboard and mouse inputs, serial communications, parallel ports, and a 10/100 Ethernet network interface. The CPU board also provides a high-speed PCI bus and an ISA bus path.

DiSP system plug-in cards include PCI/DSP-4 cards, standard PAC audio/alarm board, and one or more spare slots for optional boards such as a modem card or network interface card.

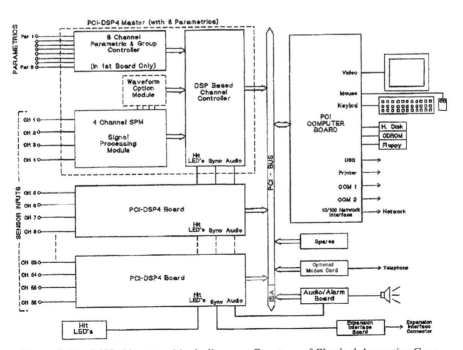

Figure 2.26 DiSP-56 system block diagram. Courtesy of Physical Acoustics Corp.

All DiSP chassis have a CPU board even if they will be part of a multiple-chassis system. It is desirable to utilize this "smart chassis" concept in the DiSP, with a standard high-speed network communications interface for the following reasons:

- Smart expansion chassis can be used either as multiple smaller systems or as part of a larger AE system.
- Smart expansion chassis allow the performance of the entire AE system to be scalable. Hit rate performance will increase with the number of chassis. In a single CPU system with multiple dumb chassis, the hit rate does not change with the number of chassis or expansions.
- Adding network interface is more reliable than extending the actual bus signals. A network costs less and is more proven as a high-speed transfer medium.

2.23.10 Other Company Products

Company products include:

- DiSP™ workstations with up to 56 channels
- AIMS—Acoustic Industrial Multichannel Systems
- Acoustic emission leak detector instruments and systems
- Pulsers, voltage time gates, AE calibrators, carrying cases, cables, and magnetic hold-downs
- AEwin™ real-time Windows-based software. (See 32-bit AE software display, Figure 2.27.)

2.24 CODES, STANDARDS, PRACTICES, GUIDELINES, AND SOCIETIES

The importance of AE methods for structural integrity assessment is expected to increase significantly in the near future. From an international viewpoint, there is still a considerable amount of work that needs to be done to formulate clear and concise codes, terminology, and guidelines that will lead to clear and concise international standards and procedures. This concern is discussed in much greater detail in an NDT.net article dated September 2002 and titled, "Acoustic Emission standards and guidelines 2002: A comparative assessment and perspectives." This work by A. J. Brunner, EMPA, Swiss Federal Laboratories for Materials Testing and Research, Dübendorf, Switzerland and J. Bohse, BAM, Federal Institute for Materials Research and Testing, Berlin, Germany, highlights some of the current problems.

Major organizations concerned with codes, standards, practices, and guidelines are:

CODES, STANDARDS, PRACTICES, GUIDELINES, AND SOCIETIES 51

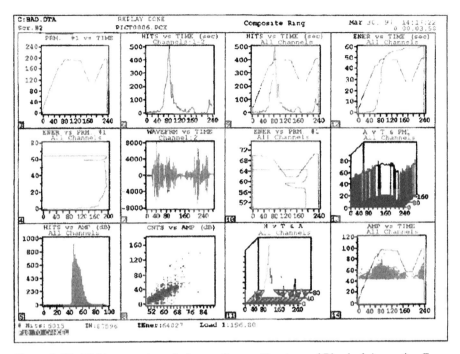

Figure 2.27 32-bit acoustic emissions software. Courtesy of Physical Acoustics Corp.

1. ISO—International Organization for Standardization
2. CEN—European Standardization Committee
3. ASTM—American Society for Testing and Materials
4. EWGAE—European Working Group on Acoustic Emission
5. AFNOR—French Standardization Society
6. DGZIP—German Society for Nondestructive Testing

Of the six organizations listed above, the ASTM is a leading organization concerned with codes, standards, practices, and guidelines related to AE. However, there are many other American and foreign organizations concerned with the development and adoption of standards. Their input needs to be solicited for AE codes, standards, practices, and guidelines related to their specific interests. Some additional American organizations concerned with standards are:

1. Association of American Railroads (AAR)
2. American Gear Manufacturer Association (AGMA)
3. American Petroleum Institute (API)
4. American Society of Nondestructive Testing (ASNT)
5. Compressed Gas Association (CGA)

6. Department of Transportation and Federal Aviation Administration (DOT & FAA)
7. Institute of Electrical and Electronics Engineers (IEEE)
8. Instrument Society of America (ISA)
9. Related Military Standards (MIL)
10. National Association of Corrosion Engineers (NACE)
11. Society of Automotive Engineers (SAE)
12. U.S. Nuclear Regulatory Commission (USNRC)

2.24.1 Sheer Numbers

The 18 above-named organizations have a proprietary interest in AE codes, standards, practices, and guidelines. Many also share an equal interest in other NDT methods.

Many manufacturing and testing organizations also have similar interests. This tends to lead to simultaneous proliferation and duplication of efforts, rather than a systematic development of well-thought-out codes, standards, and procedures for the international community.

The Brunner Bohse reference paper discusses the classification of AE standards and guidelines based on Class (1) terminology, Class (2) general principles, Class (3) measurement technique and calibration, and Class (4) applications and product-specific procedures as the most likely classifications. The total number of documents classified by the first six earlier-listed organizations listed totaled 55 with six documents in Class 1, eight documents in Class 2, sixteen documents in Class 3, and twenty three documents in Class 4. The ASTM was by far the most prolific document originator. ISO terminology appeared identical to ASTM documentation except for minor phrasing details.

2.24.2 Terminology

Considering only the four organizations providing English-language documents, there could have been a maximum of 44 common Class 1 terms, but in fact there were only 13 common terms. The 13 common terms found were *acoustic emission, signal duration, signal risetime, (sensor) array, acoustic emission count, continuous emission, acoustic emission event, arrival time (interval), source location, zone location, acoustic emission signal, (signal) peak amplitude, and (acoustic emission) waveguide.* There were 108 different terms listed in the four organizations providing English-language documents.

2.24.3 Common Term Definitions

Common term definitions vary from one document to another. For example, ISO and ASTM define *acoustic emission* as the class of phenomena that generates transient elastic waves by rapid release of energy from localized sources and as the waves themselves. The CEN document restricts the acoustic emission term to the transient elastic waves generated by the release of energy or by a process. The ASTM defines the *Felicity effect* as the presence of an acoustic emission (or detectable acoustic emission) at a fixed sensitivity level at stress levels below those applied previously. The CEN document requires the appearance of significant acoustic emission at loads below those applied previously. This terminology is confusing at best. Other important terms such as *acoustic emission intensity and severity* are not defined in any of the English documents. Therefore, there is an immediate need for a clarification of AE terms and definitions. This need may extend to other NDT methods as well.

2.24.4 General Principles

The problem appears to be that Class 2 documents tend to discuss the general principles of standard guides for agencies performing NDT without discussing the principles of the AE test method itself. CEN currently has a document in the draft stage that addresses this deficiency. Authors of this draft document recommend a second level of Class 2 documents for detailing test cases and giving detailed examples as guidelines for making appropriate selections. They feel this would help people with limited experience in the specific classes of material where standard guides are being developed.

2.24.5 Measurement Techniques and Calibration

Class 3 documents deal with primary and secondary sensor calibration in the lab and verification of sensor response in the field. Instrumentation can also be tested in the laboratory and field. Measurement and calibration standards are based on sensors and equipment state-of-the-art at any given time and are therefore constantly changing, requiring frequent updates. For example, high-speed, high-memory notebook computers can measure AE hits in real time and computer storage devices can store vast amounts of AE waveform information for later software analysis. The problem here is trying to keep up with the latest high-tech developments.

2.24.6 Areas of Opportunity

Areas of opportunity for greater AE development include:

1. Analyzing the structural integrity of space-age composite materials including the new superstrength carbon composites.
2. Determining the structural integrity of bridges, buildings, structural steel skeletons, heavy equipment, process pipes, and vessels such as reactors.
3. Determining the structural integrity of glass, ceramic, cement, and laminated structures.

4. Investigating in depth the effect of thin adhesive bonding layers on the overall integrity of AE waveforms.

2.25 APPLICATION AND PRODUCT-SPECIFIC PROCEDURES

AE testing is especially useful for testing large pressure vessels, storage tanks, and piping systems. In the past, this type of industrial equipment had to be shut down, emptied, and decontaminated. These procedures are expensive and can take several days to accomplish. After decontamination, it is usually considered safe for an inspector to enter the equipment to take X-rays, make a visual inspection, or ultrasonic measurements.

With AE technology, chemical and petrochemical reactors can be monitored as they are started up. Storage tanks and vessels can be monitored as they are filled. AE can be used to proof test and determine the structural integrity of fiberglass or metallic systems. Table 2.1 lists additional successful AE applications.

2.26 IMPACT-ECHO METHOD

2.26.1 Background

Traditionally, destructive methods such as coring, drilling, and removal of concrete sections for visual inspection have been used to evaluate failed or known defective concrete and masonry structures. In general, these methods are both time-consuming and very costly. The *impact-echo method* represents an improved, effective, less costly method for evaluating suspected faulty concrete and masonry structures.

TABLE 2.1. Acoustic Emission Applications

Bearing monitoring	Metal/GRP peiodic testing
Buried pipeline monitoring	Microelectronics manufacturing
Detection of leaking valves	Nuclear plant monitoring
Fatigue mechanics problems	Offshore platform testing
Fracture mechanics problems	Preventive maintenance
General weld monitoring	Punch press monitoring/control
Glass reinforced plastic (GRP) tank testing	Steel lamination detection
Human knee joint implants	Stress corrosion detection
In-flight AE monitoring	Tool wear monitoring
Laser material processing	

Source: Based on information supplied by Physical Acoustics Corporation.

IMPACT-ECHO METHOD

In the 1940s, the *ultrasonic pulse-echo method* was used extensively for measuring the thickness of and detecting flaws in metals, plastics, and other homogeneous materials. However, the pulse-echo method was found to have very limited use in concrete applications because of its relatively high frequency and poor transmission qualities. Ultrasonic AE transducer frequencies typically range from 100 to 900 kHz.

A lower-frequency impact-echo method for nondestructive evaluation (NDE) of concrete and masonry was developed by small groups of researchers from the U.S. National Bureau of Standards (NBS) in 1983–1986 and Cornell University from 1987–present. In the early 1980s research engineers from the NBS explored the use of short-duration mechanical impacts, produced by small ball bearings, as a source of stress waves for testing concrete structural elements, such as slabs. This work was reported by Sansalone (1986) and Sansalone and Carino (1986). Four key breakthroughs in the mid-1980s led to the successful development of the impact-echo method (Sansalone and Streett, 1995 and Sansalone, 1997). These key breakthroughs were:

1. Numerical simulation of the propagation of stress wave in solids using finite-element-based computer models
2. Use of ball bearings to produce impact-generated stress waves
3. Identification of a transducer that responds faithfully to small surface displacements
4. Use of frequency domain analysis for signal interpretation

Cornell researchers coined the term *impact-echo* to describe the method and set it apart from the *ultrasonic pulse-echo method.* In the mid-1990s the impact-echo method was extended to masonry structures (Williams and Sansalone, 1996 and Williams et al., 1997).

2.26.2 Finite Element Code

In the period from 1984 through 1987, a two-dimensional finite-element code was developed by Lawrence Livermore National Laboratories to carry out axisymmetric simulations of the propagation of transient stress waves in plates. This code made it possible to produce detailed, accurate computer simulations of stress waves in plates with and without flaws.

2.26.3 Ball Bearing–Generated Stress

The use of ball bearings to generate stress waves in concrete eliminated the need for a sending transducer and overcame all the problems that researchers had faced in trying to develop a transducer that could generate low-frequency, short-duration pulses with sufficient energy to penetrate concrete. Ball bearings ranging in size from 4 to 15 mm in diameter, with impact speeds of

2 to 10 m/s, produced impacts with contact times ranging from about 15 to 80 µs. These impacts provide sufficient energy for testing concrete structures up to about 1.5 m thick. Ball bearing size is selected based on customer application.

2.26.4 Impact-Echo Transducer Development

Impact-echo transducers are extremely sensitive, broadband receiving transducers that respond faithfully to small displacements normal to the surface. Transducers developed by Proctor in 1982 for the acoustic emission testing of metals were found to work well as impact-echo receiving transducers. This transducer is a small conical piezoelectric element attached to a brass backing. A thin lead disk placed between the transducer tip and test surface provides effective transducer coupling to rough surfaces such as concrete and masonry without the use of gels or other coupling agents. The Proctor transducer produces output signals of one volt or more for normal surface displacements of less than one micron.

2.26.5 Frequency Domain Analysis

The breakthrough in frequency domain analysis came about when it was realized that Fourier transforms could be used to overcome the inherent difficulties encountered when attempting to interpret time-domain displacement waveforms. The Fourier transform breaks the waveform down into a series of simple sine functions, thus providing a graph of amplitude vs. frequency or amplitude spectrum. The peaks in the spectrum represent transient resonances caused by multiple reflections of waves between surfaces and internal flaws. See Chapter 12 on vibration analysis for additional information.

2.26.6 Theory of Operations

The impact-echo method for NDE of concrete and masonry uses impact-generated sound stress waves that propagate through the structure and are reflected by internal flaws and external surfaces. This method can be used to make accurate nondestructive, ASTM-approved measurements (ASTM Standard C 1383–98a) of thickness in concrete slabs and plates. The method can be also used to determine the location and extent of flaws, such as cracks, delaminations, voids, honeycombing, and debonding in plain, reinforced and post-tensioned concrete structures. Figure 2.28 shows how the impact-echo method works. The method can also be used to determine the thickness or locate cracks, voids, and other defects in masonry structures where brick or block units are bonded together with mortar. Impact-echo testing is not adversely affected by the presence of steel reinforcing bars.

IMPACT-ECHO METHOD

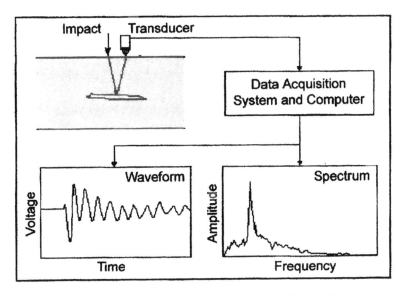

Figure 2.28 Impact-Echo theory of operation. Courtesy of Impact-Echo Instruments, LLC.

2.26.7 Propagation of Waves

Two types of elastic waves generated in solids are compression waves, primary waves, or P-waves and shear waves, secondary waves, or S-waves. A third type of elastic wave known as a surface wave, Raleigh wave, or R-wave can propagate along the surface of a solid. These waves are shown in Figure 2.29. P- and S-waves propagate into the solid along spherical wavefronts, and R-waves travel radially outward across the surface. Impact-echo testing relies mainly on the propagation and reflection of P-waves. For normal concrete, P-wave speed varies from about 3000 to 5500 m/s. The relationship among wave speed, frequency, and wavelength is given by Eq. (2.1).

$$C = f \times \lambda \qquad (2.1)$$

where C = speed of the stress wave
f = frequency in kHz
λ = wavelength in m

The interaction of stress waves with internal flaws is dependent on the relationship between wavelength and the dimension and depth of the flaws. For the detection of flaws, the wavelength of the P-wave should be equal to or smaller than the actual length of the flaw. Figure 2.30 shows the diffraction of P-waves at the concrete crack edges.

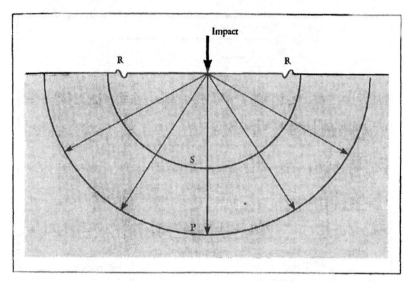

Figure 2.29 Impact-echo generated waves. Courtesy of Impact-Echo Instruments, LLC.

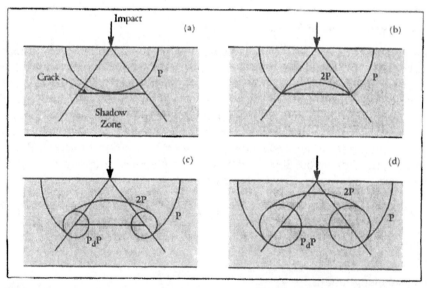

Figure 2.30 Diffraction of P-waves at edges of a crack. (a) Leading edge of P-wave reaches the crack. (b) Specular reflection of P-wave from surface of crack. (c) and (d) Diffracted waves (PdP) propagate outward from the crack tips. Courtesy of Impact-Echo Instruments, LLC.

2.26.8 Impact-Echo Instrumentation

Figure 2.31 shows two configurations for the instrument. The main components are hand-held transducer units (cylindrical devices on the left), a set of spherical impactors (hardened steel ball bearings on steel rods—barely visible in front of the computer), a high-speed analog-to-digital (A/D) data acquisition system (box at left of computer), and a notebook computer. The computer's software system guides and controls the tests and displays the results in graphical and numerical form. The two transducers on the right are separated for a fixed distance by their position in a spacer bar. These transducers are used for independent measurements of wave sound (photo at right). With a 7.5-lb notebook computer, the system on the left weighs about 13 lb (5.9 kg) and the system on the right weighs about 18 lb (8.2 kg).

Internal batteries provide power for the notebook computer and data acquisition system. A fully charged notebook battery provides 2 to 3 hours of operating time. The data acquisition system batteries provide 6 to 8 hours of operating time when fully charged. Power supplies for recharging the batteries from a 12 VDC or 1110/220 VAC source are included.

2.26.8.1 System Components

1. A/D data acquisition system—14-bit resolution, 2 MHz maximum sampling speed
2. Two transducer units—each with six spare snap-in protective disks
3. Impactors—set of 10 hardened steel spheres on spring rods
4. Spacer bar/connecting handle—for independent wave speed measurements using two transducers
5. Computer software CD—with Impact-E operating software and wave animation software (computer simulation of stress wave propagation and reflection)

Impact-Echo Testing

Wave Speed Measurement

Figure 2.31 Impact-Echo testing—wave speed measurement. Courtesy of Impact-Echo Instruments, LLC.

6. Two BNC cables—for connecting transducer units to data acquisition system
   7. Serial port cable—to connect data acquisition to RS-232 serial port of computer
   8. Output adapter—90–264 VAC input, 12 VDC output—to recharge batteries and provide power for data acquisition System
   9. DC to AC power inverter—10–15 VDC input, 110 VAC output, used with item #8 to charge batteries and power data acquisition system from 12 VDC source, such as a car or truck battery
   10. Printed materials:
      - Book, *Impact-Echo: Nondestructive Evaluation of Concrete and Masonry*, by M. J. Sansalone and W. B. Street, Bullbrier Press, Ithaca, NY, 1997, 339 pp.
      - *Impact-Echo Instrument Manual*—instructions for use of impact-echo test system.
      - *Impact-Echo User's Manual*—self-teaching course for use with Impact-E software.
   11. Roll-on carry-on case—wheeled computer case.
   12. Notebook computer—optional.

2.26.8.2 Heavy-Duty Carrying Case A heavy-duty carrying case, with padded pocket for the notebook computer, luggage-style collapsible handle, and in-line skate wheels are provided to house the impact-echo equipment. When fully packed, the overall size of the carrying case is about $18 \times 14 \times 10$ inches ($46 \times 36 \times 25$ cm) and its weight is about 30 lb (13.6 kg). The carrying case meets current airline size requirements for carry-on luggage.

2.26.8.3 Computer Recommendations A test system can be purchased from Impact-Echo Instruments, with or without a computer. Impact-E software requires an IBM-compatible computer with a Windows® 95 or later operating system and English-language option available for both "Regional Setting" and "Keyboard Language." The minimum recommended computer system capabilities are as listed:

- 100 MHz or faster processor
- 16 MB RAM
- 800 MB hard drive
- 10.3" TFT color screen
- 3.5" floppy disk drive and/or CDROM
- Serial or USB port

Virtually all IBM-compatible notebook computers meet or exceed the above requirements including Acer, Compaq, Gateway, Dell, HP, IBM, Micron,

Panasonic, Toshiba, Winbook, and others. The Panasonic Toughbook (a ruggedized notebook computer) has a moisture- and dust-proof magnesium alloy case, shock-mounted hard drive, spill-proof keyboard and other features that make it well suited for outdoor use.

2.27 TECHNICAL SPECIFICATIONS

2.27.1 Hand-Held Transducer Unit

Two types of hand-held transducer units are available: a cylindrical model (Figure 2.32) and a pistol grip model (Figure 2.33). The cylindrical model is especially useful in narrow or restricted spaces, while the pistol grip model is easier to use and more convenient for flat surfaces. These transducers include a piezoelectric crystal and a battery-powered on-board preamplifier. A replaceable snap-in assembly in the end cap contains a 0.004″ (0.1 mm)-thick nickel disk that protects the tip of the crystal where it contacts the test surface. These nickel disks provide good coupling to concrete surfaces, and last far longer than the lead disks used in early work on impact-echo instruments. Spare snap-in assemblies are provided.

Power requirements: 9 VDC battery inside stainless steel cylinder.
Controls: Automatically armed when handle is depressed—indicator light for low battery.
Cable connections: BNC connector for cable to data acquisition system.

Figure 2.32 Hand-held transducer. Courtesy of Impact-Echo Instruments, LLC.

Figure 2.33 Impact-Echo pistol grip transducer. Courtesy of Impact-Echo Instruments, LLC.

Figure 2.34 Impact-Echo data acquisition system. Courtesy of Impact-Echo Instruments, LLC.

Dimensions and weight: Overall length is 11.3″ (29 cm); main cylinder is 2″ in diameter and 6.8″ long (5 × 17 cm); handle is 1.5″ in diameter and 4.5″ long (3.8 × 11.4 cm); weight is 2.2 lb (1 kg).

2.27.2 A/D Data Acquisition System

The A/D acquisition system (Figure 2.34) is a 14-bit A/D converter. Maximum sampling speed for the unit is 2 MHz on each of two channels.

Power requirements: Internal, rechargeable battery (5 VDC) provides 5 to 6 hours of operating time on a full charge.

TECHNICAL SPECIFICATIONS

Software: Proprietary on-board software functions only with the Impact-E software accompanying the instrument.

Controls: On/off switch, indicator lights.

Cable connections: Two BNC cable connectors for transducer units; coaxial connector for power supply; one 9-pin serial port connector (for computers with USB ports a serial-USB port adapter is provided).

Dimensions and weight: 10.2 × 6.2 × 1.6 inches (26 × 16 × 4 cm) and 2.2 lb (1.0 kg).

2.27.3 Windows-Based Software

Impact-E is a Windows-based, interactive, user-friendly, software system for impact-echo testing and for examination and analysis of impact-echo test results. The software incorporates the required physics and mathematics, on which the impact-echo method is based. The software can be loaded on any computer and used to examine and analyze impact-echo test data, but it can be used only for testing with the instrument-supplied A/D data acquisition system. Figure 2.35 shows the "Main Menu" screen of the Impact-E software.

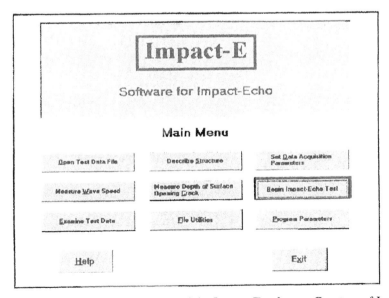

Figure 2.35 The "Main Menu" screen of the Impact-E software. Courtesy of Impact-Echo Instruments, LLC.

2.28 APPLICATIONS

Mary J. Sansalone and William B. Street discuss many of the following applications in *Impact-Echo: Nondestructive Evaluation of Concrete and Masonry*:

- Measurement of Plate Thickness
- Detection of Crack and Voids in Plates
- Detection of Shallow Delaminations
- Analysis of Honeycombed Concrete
- Evaluation of Surface-Opening Cracks
- Evaluation Plates in Contact with Soils
- Evaluation of Plates Consisting of Two Layers
- Determining Bond Quality at Internal Interfaces
- Evaluation of Plates with Asphalt Overlays
- Evaluation of Plates with Steel Reinforcing Bars
- Evaluation of Bonded Post-Tensioning Tendons
- Detection of Voids in the Grouting of Tendon Ducts
- Evaluation of Hollow Cylinders
- Evaluation of Mine Shaft and Tunnel Liners
- Evaluation of Circular and Square Cross-Sections
- Evaluation of Rectangular Cross-Sections
- Structural Diagnostics for Bridges, Parking Slabs, and Buildings
- Detection of Faulty Pile Heads
- Evaluation of Masonry

3

ELECTROMAGNETIC TESTING METHOD

This chapter describes eddy current, leakage flux, and magnetoelastic techniques. Eddy current testing involves the use of alternating magnetic fields and can be applied to any conductor. Leakage flux testing involves the use of a permanent magnet, DC or AC electromagnetic fields, and can be applied only to ferromagnetic materials.

In eddy current testing, the alternating magnetic field sets up circulating eddy currents in the test part. Any parameter that affects the electrical conductivity of the test area can be detected with the eddy currents. With the flux leakage technique, any discontinuity that produces lines of leakage flux in the test area can be detected. The magnetoelastic technique is used for characterizing and determining the amount of residual stress in magnetic materials by measuring magnetic or "Barkhausen noise."

These techniques can be combined with other methods, such as ultrasonic testing and laser dimensional measuring to achieve multifunction high-speed testing of oil field drilling pipes and other piping systems that are subject to stringent overall quality requirements. High-speed automatic testing is possible using multiple NDT methods because they can be electronically gated and discriminated to evaluate a large number of variables simultaneously with computers. A typical multichannel tubing inspection system might consist of 16 flux leakage probes, eight ultrasonic transducers, four eddy current coils, and two laser measuring devices. The flux leakage probes respond to surface flaws, the ultrasonic probes respond to internal flaws and wall thickness

---

Introduction to Nondestructive Testing: A Training Guide, Second Edition, by Paul E. Mix
Copyright © 2005 John Wiley & Sons, Inc.

changes, and the eddy current coils respond to surface defects, abrupt changes in wall thickness, and conductivity differences. Laser devices, separated by 90°, can measure tube concentricity. With this system, throughput speeds of 9 ft/s or 540 ft/min are feasible.

3.1 EDDY CURRENT THEORY

3.1.1 Surface Mounted Coils

When an alternating current is used to excite a coil, an alternating magnetic field is produced and magnetic lines of flux are concentrated at the center of the coil. Then, as the coil is brought near an electrically conductive material, the alternating magnetic field penetrates the material and generates continuous, circular eddy currents as shown in Figure 3.1. Larger eddy currents are produced near the test surface. As the penetration of the induced field increases, the eddy currents become weaker. The induced eddy currents produce an opposing (secondary) magnetic field. This opposing magnetic field, coming from the material, has a weakening effect on the primary magnetic field and the test coil can sense this change. In effect, the impedance of the test coil is reduced proportionally as eddy currents are increased in the test piece.

A crack in the test material obstructs the eddy current flow, lengthens the eddy current path, reduces the secondary magnetic field, and increases the coil impedance. If a test coil is moved over a crack or defect in the metal, at a con-

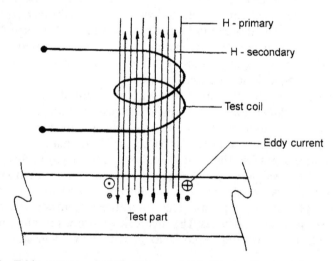

Figure 3.1 Eddy current principle. Primary field of test coil enters the test part, generates eddy currents that generate second field. Strength of the eddy currents decreases with depth of penetration.

stant clearance and constant rate of speed, a momentary change will occur in the coil reactance and coil current. This change can be detected, amplified, and displayed by an eddy current flaw detector. Changes in magnetic flux density may also be detected by *Hall effect* devices, amplified, and displayed on PCs and laptop computers.

A block diagram of a simple eddy current tester is shown in Figure 3.2. As shown in the figure, an AC generator is used to drive the test coil. As the test coil passes over various defects, the coil impedance and AC voltage changes. The AC voltage is converted to DC voltage by a diode rectifier and compared to a stable DC voltage of opposite polarity produced by a battery. With the meter properly zeroed at the start, changes in coil voltage can be measured. The block diagram represents the most rudimentary form of eddy current instrument. As such, it would not be capable of detecting minute discontinuities that can be reliably detected with today's more sophisticated instruments.

Figure 3.3 shows an eddy current test coil located at distance A above a conductive material. The coil is considered to be an "ideal" coil with no resistive losses. The impedance of the coil in the complex plane shown is a function of the conductivity of the material at distance "A". If the material in figure was an insulator, its conductivity (the reciprocal of resistivity) would be infinite. The coil's reactance would remain unchanged at point "$P1$". However, if the material is a conductor, eddy current losses will occur. The coil will signal this change by increases in resistive losses with a simultaneous decrease in reactance, and the operating point of the system will shift to "$P2$". When the conductivity of the material approaches infinity (a superconductor), the resistive losses will again approach zero. With very highly conductive materials, eddy current flow will be very high and the strong secondary field will reduce the reactance of the coil to point "$P3$". Since the complex plane approaches a semicircle as conductivity varies from zero to infinity, it can be concluded that

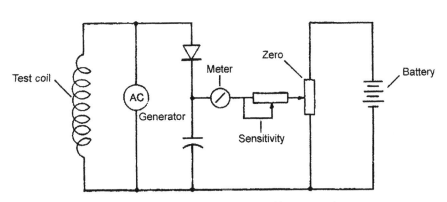

Figure 3.2 Schematic diagram of basic eddy current instrument.

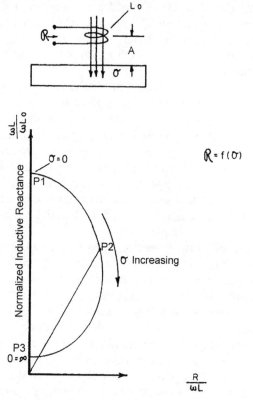

Figure 3.3 The effect of conductivity on coil impedance.

the conductivity of a material has the greatest effect on coil impedance. Coil impedance is dependent on the vector sum of the coil's inductive reactance and the test part's resistance to the eddy current field.

Another important influence on coil impedance is the clearance or lift-off between the coil and the conductive material surface. At great distances above the surface, the field of the coil does not reach the surface of the test piece or induce eddy currents in it. In this case, coil impedance remains unchanged regardless of any conductivity changes in the material. However, as the coil approaches the surface in the stepwise fashion illustrated in Figure 3.4, stronger eddy currents are induced in the material, producing the family of impedance plane curves shown. If A is held constant and conductivity varies, a circular curve is produced. As "A" approaches zero, the diameter of the circle increases. Due to the need for a wear surface, geometry, and finiteness of the coil, "A" cannot be actually zero.

If the conductivity of the material is held constant and "A" is changed, the straight line from point "$P1$" to "A_o" is generated. When attempting to mea-

EDDY CURRENT THEORY 69

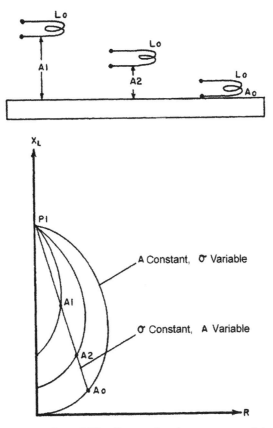

Figure 3.4 The effect of lift-off or probe clearance on coil impedance.

sure changes in conductivity, changes in spacing or lift-off are highly undesirable. In order to minimize variations in lift-off, eddy current coils may be recessed a short distance into the eddy current probe head, and the probe head may be spring loaded to maintain surface contact. However, since the lift-off effect is linear over a limited probe clearance range, eddy current probes can be designed to measure nonconductive coating thickness over uniformly conductive materials. Coil impedance can be calculated for any known combination of conductivity and probe clearance.

In many cases, we do not want to measure the effect of probe clearance or conductivity on coil impedance. Instead we want to locate and measure the effect of discontinuities on coil impedance and probe output. Figure 3.5 shows the effect that cracks and defects have on coil impedance. When the coil passes over a crack, the impedance of the coil varies by the value shown by the vector point "$P1$". A significant change in vector direction occurs and the vector points toward "P_o" when probe clearance changes. This change in vector direction is used to advantage by modern instruments as will be described later.

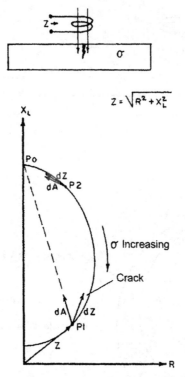

Figure 3.5 The effect of a crack on coil impedance.

The relationship shown at point "*P1*" applies to a specific value of conductivity. If the conductivity value decreases to point "*P2*", vector direction differences are less significant and it is harder to differentiate between the impedance change caused by the crack and the impedance change that is caused by probe clearance. The planar diagram shows that it is more difficult to distinguish between defect indications and lift-off indications with low-conductivity materials.

So far, we have described how eddy current resistance (heating) losses, conductivity, probe spacing, and defects affect coil impedance; no mention has been made of the effect of frequency on coil impedance. We know that conductive reactance and impedance of the coil are affected by test coil frequency in accordance with Eq. (3.1):

$$X_L + 2\pi fL \tag{3.1}$$

where X_L = the inductive reactance of the coil in ohms (Ω)
$\pi = 3.1416$

EDDY CURRENT THEORY

f = frequency in Hertz (Hz)
L = inductance in Henrys (H)

Equation (3.1) shows that both inductance and frequency directly affect coil impedance. Thus, conductivity and frequency have exactly the same effect on coil impedance. Figure 3.6 shows the effect of holding frequency constant and varying conductivity and vice versa. Assuming that material conductivity is reasonably constant, we can use the frequency relationship to our advantage. For a particular material conductivity, a test coil frequency may be selected that will create a favorable operating point for detecting flaws while differentiating against nonrelevant indications. The frequency "f_g" is the limiting frequency or the point where further increases in frequency will not increase the ohmic losses in the test material. When material conductivity is known, optimum test coil operating frequency can be calculated or determined experimentally.

3.1.2 Encircling Coils

Encircling coils are used more frequently than surface-mounted coils. With encircling coils, the degree of filling has a similar effect to clearance with surface-mounted coils. The degree of filling is the ratio of the test material cross-sectional area to the coil cross-sectional area. Figure 3.7 shows the effect of degree of filling on the impedance plane of the encircling coil. For tubes,

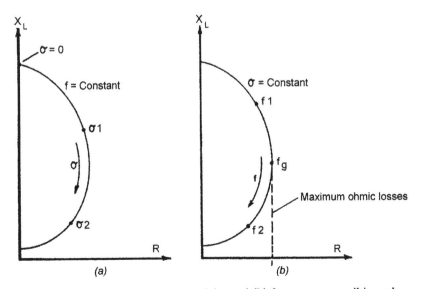

Figure 3.6 The effects of (a) conductivity and (b) frequency on coil impedance.

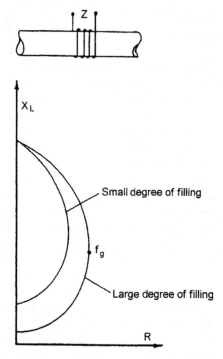

Figure 3.7 The effect of degree of filling on coil impedance.

the limiting frequency (point where ohmic losses of the material are the greatest) can be calculated precisely from Eq. (3.2):

$$f_g = 5056/\sigma d_i w \mu \qquad (3.2)$$

where f_g = limiting frequency
σ = conductivity
d_i = inner diameter
w = wall thickness
μ (rel) = relative permeability

For most applications, two coils are employed—the primary (field) coil generates the eddy currents and the secondary (pickup) coil detects the change in coil impedance caused by the changes in conductivity and permeability. Eddy currents are generated in the material in accordance with *Maxwell's law*, which states that every part of an electric circuit is acted on by a force tending to move it in such a direction as to enclose the maximum amount of magnetic flux. Furthermore, according to Lenz's law, these eddy currents must flow in the opposite direction to the current in the field coil. As previously discussed, the magnitude of the eddy current depends on frequency of the field current, conductivity and permeability of the test material, and geometry of the test part. Because of the skin effect (eddy current heating), the depth of penetration of eddy currents is relatively small and can be calculated from Eq. (3.3):

$$dp = 1/\sqrt{\pi f \sigma \mu} \qquad (3.3)$$

where dp = depth of penetration
$\pi = 3.1416$
f = frequency
σ = conductivity
μ = permeability

Eddy currents weaken the original magnetic field in the interior of the material while strengthening the magnetic field outside the material, which is in opposition to the test coil's magnetic field. If a defect is present in the sample, the magnetic field just outside the defect region is reduced and the magnetic flux through the test coil and the test coil voltage increases. Figure 3.8 shows a hypothetical defect dipole that can be used to illustrate the effect a defect has on the test coil. Most defects may be thought of as an infinite series of magnetic dipoles. The single dipole current path is represented by the infinitely small circular current whose direction is indicated by "x" going into the paper and "·" going out of the paper. Eddy currents generated by the test coil are diverted by the magnetic field of the dipole; the external magnetic field is weakened, the magnetic field of the coil is strengthened, and the coil voltage is increased.

3.2 MAGNETIC FLUX LEAKAGE THEORY

When ferromagnetic materials are magnetized, magnetic lines of force (or flux) flow through the material and complete a magnetic path between the pole pieces. These magnetic lines of flux increase from zero at the center of the test piece and increase in density and strength toward the outer surface. When the magnetic lines of flux are contained within the test piece, it is difficult if not impossible to detect them in the air space surrounding the object. However, if a crack or other defect disrupts the surface of the magnetized piece, the permeability is drastically changed and leakage flux will emanate from the discontinuity. By measuring the intensity of this leakage flux, we can determine to some extent the severity of the defect. Figure 3.9 shows magnetic flux patterns for a horseshoe magnet and flat bar magnet. Note the heavy buildup of magnetic particles is a three-dimensional pattern at the poles. All of the fine magnetic particles near the magnets are drawn to the pole pieces

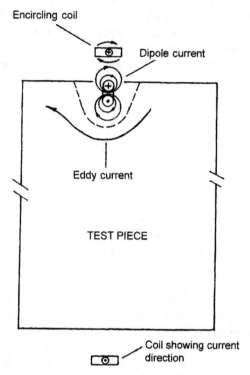

Figure 3.8 Simulation of a defect by a hypothetical defect dipole.

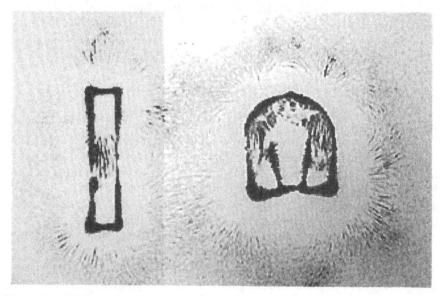

Figure 3.9 Magnetograph of two permanent magnets in close proximity. Magnetic lines of flux take the path of least resistance and bridge horseshoe magnet first.

MAGNETIC FLUX LEAKAGE THEORY 75

and sharp edges of the magnets where leakage flux is strongest. At a greater distance, the circular nature of the magnetic lines of force can be more easily seen. The pattern for the horseshoe magnet shows weaker poles near the back curved portion of the magnet. The weaker poles were probably created as a result of the magnetizing technique used to initially magnetize the ferromagnetic material. The ideal permanent magnet should be easy to magnetize and hard to demagnetize. The ideal ferromagnetic test piece, inspected with flux leakage equipment, should be easy to magnetize and demagnetize. In practice, these ideal relationships are hard to achieve.

Based on what we have learned about magnetic flux leakage, Figure 3.10 illustrates that a notch or defect distorts the magnetic lines of flux causing leakage flux to exit the surface of the ferromagnetic material. If the material is not too thick (<0.3 in.), some flux may also exit the far surface. Figure 3.11 illustrates that with DC magnetic flux leakage (DC-MFL) outer and inner cracks of equal magnitude produce similar, but opposite flux patterns and signals of differing width and amplitude when they are scanned from the outer surface of the test piece.

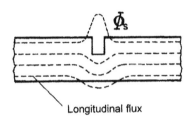

Figure 3.10 Effect of radial crack or notch on longitudinal flux pattern. Courtesy of Institut Dr. Foerster.

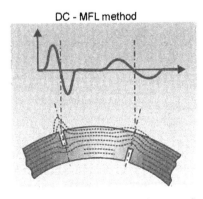

Figure 3.11 Effect on similar inner and outer defects on flux pattern and measurement. Courtesy of Institut Dr. Foerster.

Automatic flux leakage inspection systems use magnetic field sensors to detect and measure flux leakage signals. For longitudinal flaw detection on round bars and tubes, a rotational yoke DC-MFL system is used. The magnetic poles of the yoke are 180° apart with a series of magnetic sensors 90° from the poles as shown in Figure 3.12. The rotational yoke is fed with a direct current that produces a low-frequency AC field as the yoke rotates around the tube. By using a series of rotational heads, tubes with diameters of 0.4 to 25.0 in. can be tested.

Flux leakage sensors have small diameters, some as small as 0.02", in order to have adequate sensitivity for detecting short-length or small diameter defects. Because of their small size, the scanning head may have 16 or more sensors in order to achieve satisfactory throughput speeds. Probes are spring-loaded against the tube surface to provide fixed lift-off; they are lowered after the leading end of the tube is detected and raised just before the lagging tube end is reached. Signals from the probes on the inner and outer surfaces of the tube are transmitted through springs to the electronics unit where they are filtered and analyzed by a continuous spectrum analyzer. Inside and outside flaws are automatically marked by different-colored dyes that indicate the size and type of flaws detected.

Transverse flaws are detected by passing the tube through a ring yoke that produces longitudinal magnetization. In this case, the tube surface is surrounded by and scanned with a ring of small probes. Signal processing and flaw marking is the same as previously described.

When slower inspection speeds can be tolerated, a stationary yoke and spinning tube DC-MFL arrangement, shown in Figure 3.13, can be used. In this case, the inspection head is moved down the length of the tube to achieve a 100% surface inspection. It is relatively easy to combine other NDT techniques, such as ultrasonic testing, with this physical arrangement.

Since early 1978, the high-energy alternating field stray flux method has gained in popularity for testing round ferromagnetic bars from 1 to 4.5 in. in diameter. With the *Rotoflux AC magnetic flux leakage (AC-MFL) technique*, a rotating head (Figure 3.14) containing the magnetizing yoke and sensitive pickup coils rotates as the bar stock is inspected at traverse speeds of 180 to 360 feet per minute (fpm). Figure 3.15 shows the cross section of a ferromagnetic bar being exposed to an alternating field between the pole pieces. The frequency of the alternating field is about 1 to 30 kHz, so that penetration of the magnetic flux is only a few tenths of a millimeter or few hundredths of an inch. With very-high-intensity alternating fields, requiring exciting yokes using kilowatts of power, the area of the rod near the surface and near the sides of the crack is magnetically saturated. Increases in intensity increase the depth of saturation. The permeability of the saturated areas approaches the permeability of air (one) while the inner areas of the bar, identified by the "x's", have no magnetic flux and remain unchanged. From a magnetic point of view, both the crack width and depth has been increased by the amount of saturation. In effect, this magnifies the effect of the defect and results in a very high

MAGNETIC FLUX LEAKAGE THEORY

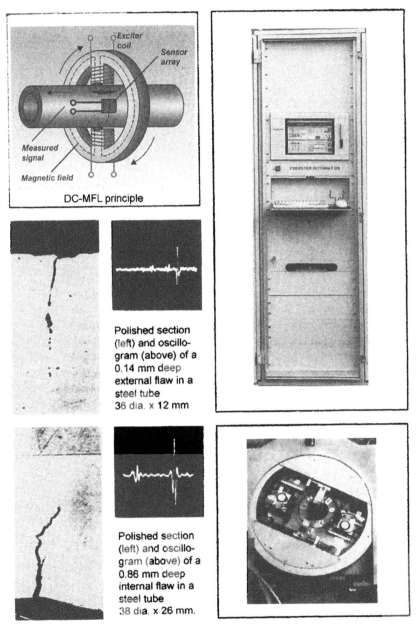

Figure 3.12 Rotating direct current magnetic yoke for establishing circular magnetic flux pattern to detect longitudinal defects—Rotomat method. Courtesy of Institut Dr. Foerster.

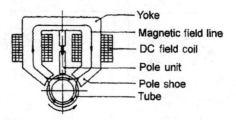

Figure 3.13 Direct current electromagnet scans length of rotating tube. Circular flux pattern detects longitudinal defects. Courtesy of Institut Dr. Foerster.

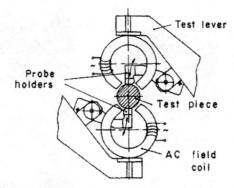

Figure 3.14 Rotating magnet arrangement for detection of AC magnetic flux leakage current. Courtesy of Institut Dr. Foerster.

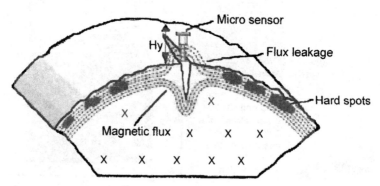

Figure 3.15 Measurement of AC magnetic flux leakage. Courtesy of Foerster Instruments, Inc.

EDDY CURRENT SENSING PROBES   79

signal-to-noise ratio that is easily detected by the pickup coil even on relatively rough bar surfaces. The probability of detecting a 0.01-in.-deep defect is 95% with both the Rotoflux and magnetic particle methods, but the magnetic particle test cannot be adapted for automatic, high-speed, in-line testing.

3.3 EDDY CURRENT SENSING PROBES

For reliable flaw detection with eddy currents, various forms of interference, such as coil clearance, must be reduced and suppressed. The signal-to-noise ratio of the eddy current system can be favorably enhanced through the use of:

- Probe design
- Vector analysis equipment
- Filtering techniques
- Elimination of permeability variations in ferromagnetic materials

In the case of the single surface-mounted (absolute) coil or single encircling (absolute) coil, changes in the coil-to-test piece clearance are a major interference. Even with a well-guided probe, the problem cannot be eliminated because of the surface variations of the material and vibrations normally encountered during testing. However, using a differentially connected probe arrangement as shown in Figure 3.16 can substantially reduce the interference problem. Note that with the differential coil (two-coil) system, the magnetic lines of flux in the coils (shown by directional arrows) oppose each other. With this configuration, changes in clearance affect both coils to the same extent and are therefore self-compensating. As the probe moves over the defect, it is

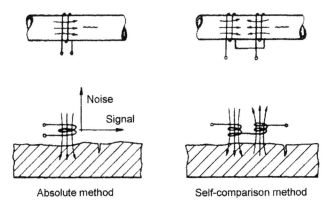

Figure 3.16 Absolute and differential (self-comparison) encircling and surface coils. Courtesy of Institut Dr. Foerster.

first sensed by one coil and then sensed by the other coil. If the defects shown are identical, each coil of the differential coil arrangement will sense the defect with the same sensitivity as the single coil.

Probe coils can be shielded with conducting or magnetic material to shape the field, increase sensitivity, or increase test resolution. Resolution determines the test probe's ability to distinguish between adjacent flaws or discontinuities. When RF noise or interference is a problem, copper can be used to shield the pickup coil and improve the signal-to-noise ratio. However, when magnetic shielding is used around the primary coil, the extension of the magnetic field and depth of eddy current penetration are reduced. Nonmagnetic materials such as plastic resins or epoxy compounds are typically used for permanently mounting probe coils in their holders.

With the differential coil arrangement, defect sensitivity is retained while clearance sensitivity is reduced by a factor of two or three. The differential method is also known as the self-comparison method because the adjacent sections of the material are compared to each other. Differential encircling coils are shown in Figure 3.16. As shown, one section is compared to an adjacent tube section, which provides automatic compensation for probe clearance and gradual changes in tube diameter.

Figure 3.17 shows a surface scanning probe with a differential detector coil located inside a larger excitation coil. Magnetic flux from the excitation coil passes through both coils of the differential detector into the test piece, and sets up circular eddy currents near the surface. The crack distorts the eddy currents, weakens the primary magnetic field, and increases the impedance and voltage output of the pickup coil. A number of these surface-mount coils can be used to scan irregularly shaped objects.

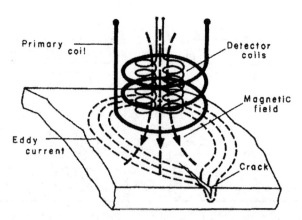

Figure 3.17 Differential surface coil showing that primary and detector coils share the same primary field. Note that crack distorts the eddy current path. Courtesy of Institut Dr. Foerster.

EDDY CURRENT SENSING PROBES

Figure 3.18 compares the conventional eddy current test method with the transmission method. Note that with the transmission method, the differential detector coil is positioned directly opposite the excitation coil. The wall thickness of the test material determines the applicability of this method.

Figure 3.19 shows an encircling coil arrangement with excitation coil and differential pickup coil. In this case, the circular or tubular test material is passed through the test coil arrangement. A large outer coil induces a magnetic field in the test material that in turn sets up circular eddy currents near the material's surface. A smaller diameter differential pickup coil is located inside the larger excitation coil. The differential coil detects the

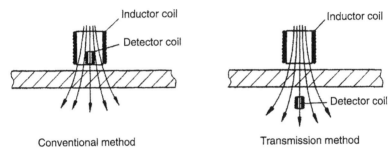

Figure 3.18 Conventional and transmission-type eddy current methods. Courtesy of Institut Dr. Foerster.

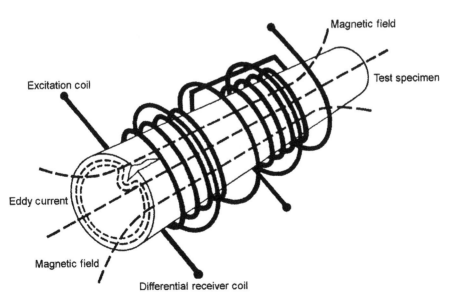

Figure 3.19 Encircling coil arrangement. Courtesy of Institut Dr. Foerster.

complex amplitude-phase changes caused by the crack while providing self-compensation for eccentricity and other nonrelevant indications.

The application of eddy current sensing elements is dependent on the geometry of the part to be inspected. In cases where ferromagnetic parts are to be inspected, large encircling coils or magnetic yokes may be incorporated to magnetically saturate the test surface. The type, depth, location, and orientation of the discontinuity are other important factors that affect probe selection. Large encircling coils would hardly be suitable for detecting small imperfections that can be readily detected with magnetic particle or small-diameter, surface-mount differential coils. An outer surface inspection of a tube might not even be employed if the majority of defects encountered were on the inside surface of the tube. In this case, inspection with an annular probe might be all that is required.

The speed of testing is also an important consideration affecting probe selection. Annular probes, with absolute and differential coils, have been designed to be pulled through the bores of heat exchanger tubing at a constant rate of speed (up to 3 feet per second—fps). Since heat exchangers may contain hundreds of small diameter tubes, some probes have been designed so that they can be blown through the heat exchanger tubes and withdrawn at a constant rate of speed or pushed through the tube at a constant rate of speed. When used with a strip chart recorder, indications of flaw location are easily obtained. For example, if recorder chart speed is 1 in./s, and a cable puller is used to withdraw the probe at the rate of 1 fps, a 30-foot-long heat exchanger tube would be represented by a 30-inch chart trace. In addition to accurately recording the position of baffles and tube sheets, a carefully calibrated system can actually detect defects near and under baffles and tube sheets in addition to defects between baffles and tube sheets. However, for this kind of sensitivity, considerable work must be done in advance of actual testing. For one thing, the bores of the heat exchanger tubes must be reasonably clean so that dirt and particulate matter do not produce nonrelevant indications. Heat exchanger testing is generally a two-person operation because one person must blow the probe through the heat exchange to another person on the other end who sets the starting position and logs the tube location on a map of the face of the heat exchanger. The first person also operates the cable puller and recording equipment, which can be partially automated. In some cases, probes may be pushed through a heat exchanges and retrieved from the same side. Periodically a reference tube standard will also be tested to assure that calibration of the test has not drifted. For high-speed operation in rolling mills and other manufacturing locations, encircling coils are frequently used.

Another important consideration for probe selection is the required percentage of surface area that must be inspected. Encircling and annular probes can provide 100% inspection of the outer and inner tube surfaces respectively. They can detect moderate-sized flaws, dents, gouges, and other major discontinuities. Wide coils are used to measure changes in conductivity and narrow coils are used primarily for flaw detection. For the detection of small defects, a small surface-mounted differential coil might be required. For a 100% surface inspection of a 30-foot-long tube with a $\frac{1}{4}$-inch diameter probe and helical scanning technique, a prohibitively long time would probably be required, depending on the rpm of the probe head or tube. However, it is easy to visualize how testing time can be reduced by a factor of eight by using eight $\frac{1}{4}$-inch diameter probes to simultaneously test 2 linear inches of tubing. The number of probe sensors and rotational and longitudinal test speeds must be adjusted to achieve the required percentage test surface scan.

3.4 FLUX LEAKAGE SENSING PROBES

Flux leakage sensors are designed to measure the leakage flux emanating from the surface discontinuities in magnetized ferromagnetic materials. The ferromagnetic material can be continuously magnetized or contain a residual magnetic field. The majority of these sensors are inductive coil sensor or solid-state Hall effect sensors. Magnetic powder, magnetic diodes, and transistors, whose output current or gain change with magnetic field intensity, and to a lesser extent, magnetic tape systems can also be used.

3.4.1 Induction Coils

For an induction coil to detect a magnetic field, the magnetic field must be alternating or pulsating, or the coil must be moved through the magnetic field at a reasonable rate of speed. Various absolute and differential surface and encircling coil arrangements are illustrated in Figures 3.16 through 3.19. Figure 3.20 illustrates the design and coil arrangement of a differential-type inside or annular coil.

Some coil parameters that would affect the coil's ability to pick up or detect small leakage flux fields are:

- Coil diameter
- Coil length

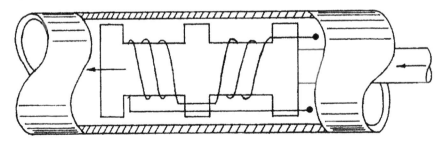

Figure 3.20 Bobbin-type differential coil for scanning inner surface.

- Number of turns of wire
- Permeability of core material
- Coil orientation

If the diameter of a surface coil is much larger than the defect size, the defect will not be reliably detected. Likewise, if the vertical coil length is longer than the vertical component of the magnetic field, the excess coil length is not utilized. If the vertical coil length is much shorter than the vertical component of leakage flux, a weaker signal will be induced into the coil and the signal to noise will be lower. A greater electromotive force (emf) will be induced into the coil if the number of turns of wire is increased; this increases the inductance, coil impedance, and signal-to-noise ratio of the system. There are, of course, practical limitations to the number of turns of wire for optimum coil design. Improved flaw detection can result from using a soft iron or powdered iron core in the direction of the coil. The iron core tends to concentrate the lines of flux more effectively, coupling them into the coil and favorably increasing both coil impedance and output voltage. Finally, coil orientation is as important as defect orientation with regard to detecting flux leakage from defects. Surface coils are particularly effective for detecting flux leakage because they are normal to the surface and all surface defects produce a strong vertical or normal flux leakage pattern (see Figures 3.10 and 3.11).

3.4.2 Hall Effect Sensors

Hall effect probes are popular magnetic flux measuring devices for the following reasons:

- They are long-life solid-state devices.
- The voltage output of the device can be alternating current or direct current depending on the current and magnetic field inputs.
- As high-speed switches, they provide bounce-free, contact-free logic level voltage transfer.
- They have a wide range of operating temperatures (−40 to +150°C) and are highly repeatable.

E. H. Hall accidentally discovered the Hall effect in 1879 while he was investigating the effect of magnetic forces on current-carrying conductors at John Hopkins University. At that time, he discovered that strong magnetic fields skewed the equal potential lines in a conductor, thus producing a miniature voltage perpendicular to the direction of current flow. With metal conductors, the voltage levels produced were so low that the phenomenon remained a laboratory curiosity until practical development of the semiconductor. In the early 1960s, F. W. Bell, Inc. introduced the first commercially available, low-cost, bulk indium arsenide Hall generators. In 1968, Micro Switch revolutionized the keyboard industry with their introduction of the first solid-state keyboard using Hall effect devices that incorporated the use of a Hall generator and its associated electronic circuit on a single integrated circuit chip. Today Hall effect devices are used in a host of products too numerous to mention.

Figure 3.21 illustrates the Hall effect principle. When a current-carrying conductor is placed in a magnetic field, a Hall voltage is produced, which is perpendicular to both the direction of current and the magnetic field. The Hall voltage produced is given in Eq. (3.4):

$$V_H = (R_H/t)(I_c)(B\sin\theta) \tag{3.4}$$

where I_c = input current
$B \sin\theta$ = perpendicular magnetic field in gauss (G)
R_H = Hall coefficient
t = thickness of the semiconductor

Equation (3.4) holds true only when the semiconductor has an infinite length-to-width ratio. For practical purposes, Eq. (3.4) reduces to Eq. (3.5):

$$V_H = k(I_c)(B\sin\theta) \tag{3.5}$$

where k is a constant that combines the Hall coefficient, temperature, and semiconductor geometry.

From this equation we can see that if the input current is held constant, the Hall voltage will be directly proportional to the normal component of the magnetic field. If either the magnetic field or input current is alternating, an alternating Hall voltage will be produced. When both the input current and magnetic field are direct current, a direct current Hall voltage will be pro-

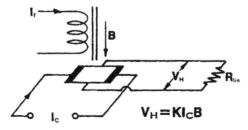

Figure 3.21 Schematic representation of Hall effect device.

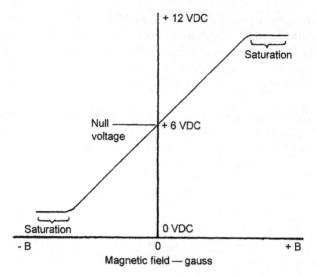

Figure 3.22 Effect of magnetic field on Hall output voltage. Courtesy of Micro Switch, a Honeywell division.

duced. Direct current magnetic fields are commonly produced by small, strong, permanent magnets.

Since a 1 gauss (G) field produces a Hall voltage of about 30 μV, signal conditioning is generally required. An external or internal regulator is needed to keep input current constant and a low-noise, high-impedance, moderate-gain amplifier is needed to amplify the Hall voltage to a practical level. Differential amplifiers are typically integrated with Hall generators using standard bipolar transistor technology. Figure 3.21 is a schematic representation of a linear output Hall effect transducer (LOHET). Figure 3.22 shows the output characteristic of the device. Both positive and negative magnetic fields can be sensed or measured by the Hall effect transducer. The null voltage point of the output curve (point where the magnetic field is zero) is positively biased so that only one power supply voltage is needed. The upper and lower saturation levels on the curve are caused by amplifier saturation as the Hall output voltage approaches the power supply voltage value. Increasing flux levels drive the amplifier to saturation.

With integrated Hall effect transducers, voltage or current regulation, temperature compensation, output voltage ratioing, current sinking, and digital (on/off) outputs from Schmitt triggers can be provided. In addition, these transducers can be used with biasing magnets and interfaced with other amplifiers, comparators, or computers in thousands of applications. Other methods of detecting magnetic flux leakage include the magnetic powder method, the oldest known method, and the magnetic rubber method, which is relatively new.

3.5 FACTORS AFFECTING FLUX LEAKAGE

The more highly magnetized the ferromagnetic object, the higher its leakage flux field intensity will be for a given defect. Both continuous and residual magnetic fields can be used. However, leakage flux testing should not be used with ferromagnetic objects that have weak residual fields and poor retentivity because test sensitivity may be too low to detect small discontinuities.

The amount of leakage flux produced depends on defect geometry. Broad, shallow defects will not produce a large vertical component of leakage flux; neither will a defect whose long axis is parallel to the lines of flux in the test object. For these reasons, longitudinal defects in tubes, rods, and bars will be more easily detected with circular magnetic fields.

Defect location is a very important consideration. Internal defects in thick parts may not be detected because the magnetic lines of flux will merely bypass the defect on both sides and produce little or no leakage flux. Tubular defects on the outer surface are easily detected by surface-mount probes scanning the outer surface and inside surface defects are more easily detected with surface-mount probes scanning the inner surface.

Defects oriented so that they are normal or perpendicular to the surface and normal or at right angles to the magnetic lines of flux will be more easily detected than defects lying at an angle with respect to the surface or magnetic lines of flux. Defects lying at shallow angles with regard to the surface and oriented so that they are parallel to the magnetic lines of flux produce the weakest lines of flux and will be difficult, if not impossible, to detect.

Probe velocity is dependent on but not limited by the frequency response of the probe. For typical Hall effect probes, operating speed is listed as 100 kHz maximum. Other flux-sensing probes have a typical response time of 10 kHz. In the worst case, the period of one full cycle would be 1/10,000 or 0.0001 second. With a $\frac{1}{4}$-inch diameter probe, we can move the probe or test object $\frac{1}{4}$ inch in 0.0001 second. Total test time for a 100% surface inspection 3-inch diameter by 30-ft-long tube, would be $(\pi \times 3 \times 4)(30 \times 12 \times 4 \times 0.0001) = 5.43$ seconds. The required rotational speed of the probe head or tube would be $(30 \times 12 \times 4)/5.43 = 265$ revolutions per second (rps) or 15,900 rpm, which is not practical. At a more practical rotational speed of 120 rpm, $(15,900/120) \times 5.43 = 720$ s or 12 min would be required to test the tube. By using a chain of 12 sensors, the total test time could be reduced to 1.0 minute. As shown, the mass and shape of the test object and physical design characteristics of the material-handling equipment limit the total test time per tube.

The distance between adjacent defects also affects test resolution. If the frequency response of the probe is 10 kHz, adjacent defects must be separated by a time interval of 0.0001 s to provide electrical separation of the signals. If the linear throughput of the automated line is 6 fps, defects that are $(6 \times 12)/10,000 = 0.0072$ inch apart can be resolved by a single probe element. However, it must be remembered that the typical probe has an effective diameter of

$\frac{1}{4}$ in. and is not a point source. Therefore, defects must be about $\frac{1}{4}$ in. apart for total signal separation.

3.6 SIGNAL-TO-NOISE RATIO

The signal-to-noise ratio can be defined as the ratio of signal amplitude from a relevant indication to the signal amplitude received from the background noise or nonrelevant indications. A major problem associated with eddy current testing is that there are a large number of known and unknown variables that can affect the output indication. One of the greatest sources of noise is caused by lift-off variations. Spring-loaded eddy current probes are used to minimize the effects of lift-off. Other sources of noise or extraneous indications are varying test speed, gradual dimensional changes, and unwanted high-frequency harmonics. Noise filtering or differentiation, phase discrimination, and signal integration can compensate for many of these sources of noise. For example, high-pass filters minimize the effects of gradual dimensional changes and low-pass filters are used to filter out unwanted high-frequency harmonics. Filters should not modulate the test frequency.

Coil design can be an important design consideration in maximizing the signal-to-noise ratio. A self-compensation differential coil compensates for minor variations in diameter, chemical composition, and hardness. Circuits that tend to minimize the effects of insignificant variables usually provide signal amplitude, phase, or frequency compensation. Phase (signal timing) analysis is frequently used to separate conductivity, permeability, and dimensional variables. Higher test frequencies tend to minimize the effect of varying test part thickness. Test frequency should be selected on the basis of part thickness, the desired depth of penetration, and the purpose of the inspection. Increasing circuit gain will decrease the signal-to-noise ratio if all other test parameters are held constant.

With flux leakage testing, signal-to-noise ratio is affected by surface noise (the sensor bouncing along the surface) and probe lift-off variations. Lift-off decreases the amplitude of the flux signal and changes its frequency. Spring-loaded probes can help minimize these effects. Too high a rotational test speed or too high a rotating probe head speed can also cause a loss of test indication by eddy current shielding. Eddy current shielding refers to the formation of direct current flow caused by the rapidly occurring flux changes in the part that is created by the rotation of the magnetic field. In flux leakage testing, the maximum permissible test probe or tube speed is about 300 fpm or 5 fps.

3.7 TEST FREQUENCY

In theory, the maximum eddy current testing speed is determined by test coil frequency. In turn, the test frequency selected determines the initial impedance of the eddy current test coil; as operating frequency increases, empty coil impedance increases. If test frequency increases and field strength is held constant, the surface eddy current density increases. Small discontinuities are classified as high-frequency variables because they are tested at high frequencies. The relationship between coil impedance and frequency is given by Eq. (3.6):

$$X_L = 2\pi f L \tag{3.6}$$

where X_L = inductive reactance of the coil in ohms (Ω)
π = 3.1416
f = test frequency in Hertz (Hz)
L = inductance in Henrys (H)

As shown in Eq. (3.6), test frequency affects the inductance of the coil. Lowering the test frequency increases the depth of eddy current penetration. Lower test frequencies are typically used with ferromagnetic materials because of their low permeability. Frequency, temperature, material hardness, and permeability affect the formation of the skin effect that limits the depth of eddy current penetration. At a fixed frequency, eddy current penetration will be the greatest in a metal with the lowest-percentage International Annealed Copper Standard (% IACS) conductivity.

For any given set of test conditions, there is a range of suitable frequencies centered on the optimum test frequency. In modulation analysis, conductivity, part dimensions, and defects modify frequency. Chemical composition, alloy, and heat treatment changes produce low-frequency modulation. The oscillator section of the eddy current instrument controls the test frequency. Proper selection of frequency, centering, and adjustment of phase obtain the optimum sensitivity to a known defect.

The ratio of test frequency (f) to limit frequency (f_g) provides a useful number for evaluating the effects of various variables based on their impedance diagram. The limit frequency, limit frequency equations, and impedance diagrams are different for solid rods and thin-walled tubing. A change in f/f_g ratio will cause a change in both the phase and magnitude of voltage developed across the test coil. The limit frequency is the frequency at which additional increases in frequency do not produce additional increases in eddy current losses. Limit frequency is defined when the mathematical function describing the electromagnetic field within a part is set equal to one. The limit frequency is also known as the "characteristic" frequency of the material. If the characteristic frequency is 100 Hz, the test frequency that is required for an f/f_g ratio of 10 is 1.0 kHz.

The characteristic frequency for a solid magnetic rod is calculated by Eq. (3.7):

$$f_g = 5060/\sigma\mu d^2 \tag{3.7}$$

where σ = conductivity
μ = permeability
d = diameter of the rod

The similarity law states that to test all parts of similar geometry, it is necessary only to choose a test frequency so that the f/f_g ratio lies at the same point on the impedance diagram for each specimen.

Multiple frequency tests are playing an increasingly important role in defect evaluation. For example, with eddy current testing of tubing, it is possible to set up one channel for conventional external flaw testing with phase-independent amplitude excitation. The second channel could be optimized for detecting internal flaws at high sensitivity. In this case, phase-selective evaluation in a small sector of the complex plane could be used. It also would be possible to set up absolute and differential test channels operating at different frequencies if it were advantageous to do so.

3.8 MAGNETIZATION FOR FLUX LEAKAGE TESTING

For flux leakage testing to be effective, ferromagnetic parts must be magnetized to saturation. To understand how leakage flux fields are established and measured, we must have an understanding of basic magnetic fields and magnetism.

As previously stated, magnetic fields flow through and surround permanent magnets. Magnetic lines of flux exit the north pole of a magnet and enter the south pole. Throughout this text, the units of flux density will be given in centimeter gram second (cgs) units or gauss (G). A conversion table for cgs and Système International (SI) are given in Table 3.1.

The characteristics of magnetic materials can be described by reference to the materials' magnetization or hysteresis curves. A typical magnetization curve is shown in Figure 3.23. Note that the magnetization curve is a plot of flux density (B) in guass on the vertical axis as a function of magnetizing force (H) in oersteds on the horizontal axis. Nonmagnetized ferromagnetic material is represented by the origin, point 0,0. As this material is gradually subjected to increasing magnetizing force, the magnetic flux density in the material increases from 0 to B_{max}. At this point, further increases in magnetic force will not increase the magnetic flux in the material and the material has reached magnetic saturation. The B_{max} condition represents one condition under which satisfactory leakage flux testing can be done.

If the magnetizing force is now gradually decreased to zero, the flux density decreases to point B_r, which represents the residual flux density remaining in the material after the magnetizing force has been removed. The ferromagnetic part now has a significant amount of residual induction or magnetism. This condition represents the second condition under which satisfactory leakage flux testing can be done.

TABLE 3.1. SI to cgs Conversion Factors

Quantity	Unit		Multiplier[a]
	cgs	SI	
Length, L	centimeter (cm)	meter (m)	100
Mass, M	gram (g)	kilogram (kg)	1000
Time, t	second (s)	second (s)	1
Magnetic flux, θ	maxwell (Mx)	weber (Wb)	10^8
Flux density, B	gauss (G)	tesla (T)	10^4
Field strength, H	oersted (Oe)	ampere/meter (A/m)	$4\pi/1000$
Magnetomotive force, F	gilbert (Gi)	ampere (A)	$4\pi/10$
Permeability (vacuum), μ_0	(unity)	henry/meter (H/m)	$10^7/4\pi$
Reluctance, R	gilbert per maxwell (Gi/Mx)	1/henry (1/H)	$4\pi/10^9$
Permeance, P	maxwell per gilbert (Mx/Gi)	henry (H)	$10^9/4\pi$

[a] SI units must be multiplied by multiplier factor to convert to cgs units.

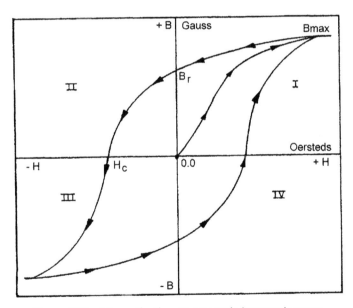

Figure 3.23 Magnetization or magnetic hysteresis curve.

If the magnetizing force is reversed in direction and gradually increased again, the residual magnetism can be reduced to zero point H_c. The force required to demagnetize the ferromagnetic part is known as the coercive force and the second quadrant of the magnetization curve is known as the coercive demagnetization curve. In general, ferromagnetic parts should be demagnetized after flux leakage testing to assure that they do not attract minute iron particles that might interfere with subsequent machining operations.

By taking the product of B and H for every point on the demagnetization curve and plotting it against B, an energy product curve similar to the one shown in Figure 3.24, can be obtained. The peak energy product in million gauss-oersteds provides one important way of comparing magnetic materials. The permeance or permeability (μ) of a magnetic material depends on many factors, but for any set of conditions is defined as the ratio of B/H.

In many cases, it is possible to control the quality of ferromagnetic components by automatic measurement of coercive field strength because the coercive force is affected by alloy composition, structure, particle size, nonferrous inclusions, manufacturing techniques, heat treatment, and internal and external magnetic stresses. As a general rule, hard materials have a high coercive force and are not easily demagnetized. For automatic measurement of coercive force, the ferromagnetic parts are first magnetized to saturation with direct current pulses and an impulse coil arrangement. Magnetizing current in the coil is gradually reduced to zero, reversed, and increased until the residual flux density passes through zero. The rate of change of the opposing field is exponential and slows down as the magnetic field of the test piece approaches zero; this increases the measurement accuracy of the Hall effect device or flux measuring probe. The coercive force required to reduce the

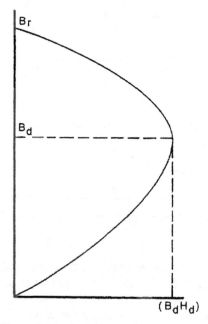

Figure 3.24 Peak energy product curve.

residual flux of the test piece is calculated, stored, and digitally displayed by the microprocessor-controlled electronics.

Sophisticated PC-controlled *BH* meters have been developed that automatically plot the *BH* curve of various ferromagnetic materials and provide a printable display of important magnetic characteristics. These instruments typically measure the *BH* characteristics of magnetic alloys, tapes, powders, toners, inks, floppy disks, rigid disks, and small samples.

Encircling coils for leakage flux testing typically magnetize tubes, rods, and bars. The number of turns of wire, current strength, and length of the coil determines the strength of the magnetic field inside the induction coil. For greatest effectiveness, the part to be magnetized should be near the inside wall of the coil. Relative motion between the test part and flux sensor is necessary to obtain a response from a flaw. One or more excitation coils can be used to establish the longitudinal flux pattern shown in Figure 3.25. In this case, the direct current leakage flux is detected by sensor (S); as shown, both outer and inner defects can be detected and easily distinguished from each other.

For circular magnetization, a direct current carrying conductor can be placed down the center of the tubular elements. The central current carrying conductor sets up a circular flux pattern in the cross section of a tubular element similar to the flux pattern shown in Figure 3.26. The recommended magnetizing current is 600 to 800 A/linear in. of section thickness for circular magnetization.

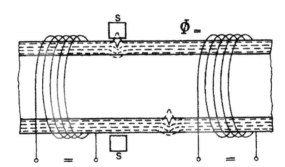

Figure 3.25 Encircling coils establish longitudinal field. Courtesy of Institut Dr. Foerster.

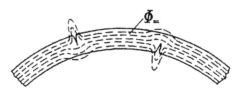

Figure 3.26 Effect of radial cracks on circular magnetic field. Courtesy of Institut Dr. Foerster.

Longitudinal and circular magnetization also can be induced in ferromagnetic parts with permanent magnets. Longitudinal fields can be established by orienting magnetic yokes parallel to the tube or bar's longitudinal axis and energizing the yoke with direct current while moving either it or the tube. The DC-MFL sensor is typically located between the pole pieces of the yoke. For longitudinal flaw detection, both rotating and fixed magnetic yoke heads can be used to establish circular magnetic fields as illustrated in Figure 3.27. The rotational yoke at the top is fed with direct current as it rotates around the tube, thus producing a low-frequency alternating current field capable of penetrating wall thickness of 0.6 in. or more. With this system, both inner and outer flaws are scanned from the outside surface. The scanning head glides on the surface of the tube to assure constant lift-off and sensor signals are transmitted over slip rings to the electronics unit for evaluation. As many as 16 sensors can be used with this configuration to achieve throughput speeds of 6 fps. With the fixed yoke at the bottom similar test results can be obtained by rotating

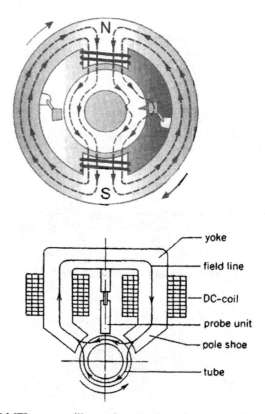

Figure 3.27 DC-MFL sensor illustrating the Rotoflux principle: top-rotating yoke, bottom-rotating tube. Courtesy of Institut Dr. Foerster.

the tube while the inspection head is moved down the tube to achieve a 100% inspection of the tube surface.

3.9 COUPLING

In eddy current testing, the test piece is coupled to the test coil by the coil's magnetic field, which is quite similar to transformer action. Coupling efficiency is intimately related to lift-off; coupling efficiency is 100% when lift-off is zero. Lift-off describes the change in electromagnetic coupling as a function of probe clearance. As lift-off or probe clearance increases from the test surface, coupling efficiency and eddy current probe output decreases. Lift-off changes both the amplitude and phase of the eddy current signal. Impedance changes produced by small lift-off variations are greatest when the coil is in contact with the test material. For this reason, spring-loaded probes and self-comparison coil or differential coil arrangements are frequently used. With eddy current testing, lift-off is a complex variable that can be detected and compensated for through frequency selection to achieve a desirable operating point on the complex impedance plane.

Fill factors apply only to encircling coils, and are likewise intimately related to coupling efficiency. The percentage fill factor is the percentage of secondary coil area occupied by the test part. The ideal fill factor for a feed-through encircling coil should approach 100%. As the fill factor decreases, the impedance variation of the pickup coil decreases for a given change in material conductivity. The fill factor for a $\frac{1}{2}$-in.-diameter bar in a 1-in.-diameter coil is 25%; fill factor decreases as the square of the diameter ratio, for example, $0.5/1.0 = 0.5$; $0.5 \times 0.5 = 0.25$ or 25%. With decreased coupling or lower fill factors, there is less eddy current flow, smaller voltages across the pickup coil, and inadequate electronic compensation.

With leakage flux testing, lift-off affects the flux leakage signal that decreases as the inverse square of distance from the test surface. For this reason, detector coils or Hall effect devices are designed to be spring-loaded and glide over the surface of the test piece. With encircling coils, the fill factor would have the same relationship in leakage flux testing as it has in eddy current testing.

3.10 EDDY CURRENT TECHNIQUES

Small-diameter surface coils or probe coils are primarily used to pinpoint and determine the magnitude of small discontinuities. Surface probes are not normally used to inspect small-diameter tubing because of relatively slow inspection speed and inherent mechanical problems.

Large encircling coils, which test the entire diameter of the test piece, are used to detect conductivity and dimensional changes. Round bars and tubing are generally inspected with encircling coils. The phase of the eddy currents varies throughout the conductor when a uniform conductor is tested with an encircling coil. With feed-through encircling coils, the material should be reasonably centered in the test coil to get uniform sensitivity. Centering of the material is verified by running a calibration standard, containing a known defect, through the encircling coils in various positions. When outside encircling coils are used for testing, the phase of the outer surface discontinuities will lead the phase of identical inner surface discontinuities. For best results with encircling coils, inspection coil length, the desired resolution, and test frequency are used to determine the maximum velocity of inspected tubing.

Direct current (DC) saturation, produced by DC saturation coils, is used when ferromagnetic materials are tested with encircling coils; this aligns the magnetic domains ahead of time and eliminates the heating effect caused by the work required to rotate the domains into the preferred magnetic direction. The magnetic domains in nonmagnetized ferromagnetic material are randomly oriented and tend to neutralize each other. When a coil's magnetizing force is applied to ferromagnetic materials, the flux density in the material is much greater than the flux density generated by the test coil. In effect, the eddy current test coil detects the leakage flux emanating from small surface and near-surface defects. Increases in alternating current field strength decrease the eddy current penetration in magnetic materials to some minimum value. Beyond this point, further increases in alternating current increase eddy current penetration. When magnetic penetration is maximized, the part is magnetically saturated.

Eddy current systems are generally calibrated with reference standards that contain natural, artificial, or, in some cases, no discontinuities. Holes, grooves, and notches are examples of artificial discontinuities that are frequently used to determine test sensitivity. The purpose of these calibration standards is to provide a check on amplitude and phase shift. The difference between the actual instrument output and a straight-line calibration curve defines the nonlinearity of the system. With automatic inspection systems, an actual part should be used for calibration and sensitivity adjustments.

3.11 INSTRUMENT DESIGN CONSIDERATIONS

Three important considerations in the design of eddy current instruments are amplification, phase detection and discrimination, and signal differentiation and filtering. With modern low-noise, medium-gain amplifiers, many manufacturers no longer list the sensitivity range or gain specifications of their instruments because they are more than adequate.

Modulation analysis (impedance/magnitude systems) shows only the magnitude variations associated with coil impedance changes. Coil current is held constant so that the output indications contain only flaw information and filters are used to eliminate the conductivity and permeability effects. Any test

INSTRUMENT DESIGN CONSIDERATIONS

coil can be used with this system; however, the test coils are somewhat sensitive to dimensional changes. Chart recorders are normally used to record flaw indications at probe or test part speeds of 40 to 300 fpm. The major disadvantage of modulation analysis is that the system requires a moving test.

Phase analysis is a technique that discriminates between variables based on phase angle changes of the test coil signal. Figure 3.28 shows some of the variables that can be detected and measured by phase analysis techniques. With phase analysis, the magnitude of the transmitted signal does not affect the phase of eddy currents in nonferromagnetic conductors. The vector point, ellipse, and linear time base methods are subdivisions of the phase analysis technique.

With the ellipse method, one variable is usually represented by the angle of the ellipse or line, and a second variable is represented by the size of the ellipse opening. With the ellipse display, the vertical and horizontal signal frequencies are the same. When voltage and current waveforms reach their max-

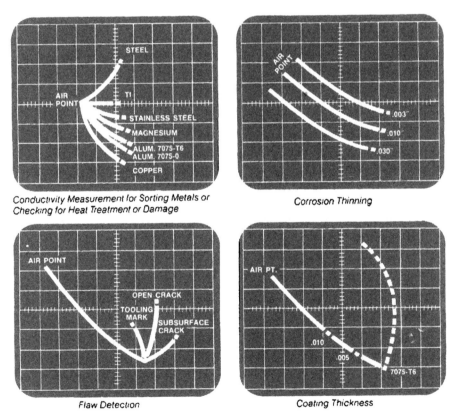

Figure 3.28 Cathode ray tube phase analysis presentations. Courtesy of Nortek/metrotek.

imums and minimums at the same time, the signals are in phase. A straight horizontal line is obtained on the screen when the normal indication for the test piece and reference standard are identical. The ellipse method can be used with feed-through coils for detecting surface and subsurface defects in rods, tubes, and wires; cracks produce significant phase shifts.

With the linear time base form of analysis, no voltage is applied to the vertical deflection plates of the CRT when conditions are balanced; a sinusoidal wave is applied to the vertical deflection plates when an unbalanced condition exists; sawtooth voltages are applied to the horizontal deflection plates. The phase control of the time-based system shifts the signal to the left or right on the CRT screen. With any of the phase analysis methods, storage scopes are useful for examining nonrepetitive flaw indications during high-speed testing.

Many state-of-the-art eddy current instruments now offer dual-frequency operation from 1 kHz to 10 MHz, storage scope display, absolute or differential detectors, and impedance balancing. Other features include high- or low-frequency filters for each axis of each impedance plane, 360° rotational control of each axis, instant replay of inspections, printing options, and computer interfacing.

Table 3.2 summarizes some of the currently available design features for eddy current instruments. The table and the photographic illustrations that follow clearly show there is a trend toward phase analysis, the incorporation of microprocessors into individual eddy current test instruments, and computer control of large automated lines.

Some general categories of equipment currently in use are:

1. Small, portable, battery-operated, multipurpose instruments primarily designed for maintenance applications.
2. High-tech, wide-range multipurpose instruments, primarily designed for laboratory investigations and use.
3. Compact units designed for running automatic inspection lines. These units have adjustable signal thresholds, marker and sorting devices, and some self-check diagnostics.
4. Multichannel instruments, with the same features as category 3, designed to run fully automatic inspection lines.
5. Complex, modular instrument systems designed to solve complex inspection problems.

3.12 UNIWEST US-454 EDDYVIEW™

UniWest makes several state-of-the-art ET instruments including the highly portable US-454 EddyView™, which combines eddy current, video, and strip chart recorder functions in a lightweight 5-pound (with batteries) instrument. (See Figure 3.29.) Main instrument features include:

TABLE 3.2. Comparison of Instrument Design Features

Purpose of Instrument	Operating Frequency	Electronic Gating	Phase Adjustment	Readout Device	Operating Mode
Crack detection and sorting	100 Hz–2.5 MHz	Dual box gates	0–360° rotation	CRT screen	Manual
Rod, tube and wire testing	1 kHz–2 MHz	Dual box gates	X and Y axis	CRT screen	5000 fpm
Crack detection and sorting	55–200 kHz	Level gate	Balance control	Analog meter	Manual or automatic
Hole scanner	55–200 kHz	Lift-off component	Filters	Meter and recorder	40–120 rpm
Metal comparator	37 kHz	Hi–Lo threshold	Auto tuning	Meter A or D	Sorting mode
Alloy sorting	Cold and hot junctions are formed with the test piece and the voltage produced by this temperature difference (Seebeck effect) is used for alloy identification.				
Portable multitester	37 kHz	Sensitivity adjustment for comparison with known reference standards. Digital readout.			
Universal computer controlled	Dual, 1 kHz–10 mHz	Dual, independent alarms	0–360°, filtering, 4 axis	CRT screen or printer	Outputs for automatic control
Universal computer compatible	100 Hz–6 MHz	Optional gates/alarms	0–399°, balancing	CRT screen, recorder, computer	Manual or automatic control

- Selectable eddy current, video, and strip chart recorder
- Programmable memory—stores up to 100 test setups
- Data storage—internal memory stores up to 10 data sets
- Hand-held scanner input
- Video input for camera or borescope
- Dual frequency—from 10 Hz to 10 MHz
- Sequential smart battery charger
- Bright 6.5″ flat-panel color LCD display
- Lightweight, portable, ergonomic, and user friendly

Control and Display. Instrument display settings are selectable with a continuously variable control knob. The scrolling menu permits quick instrument

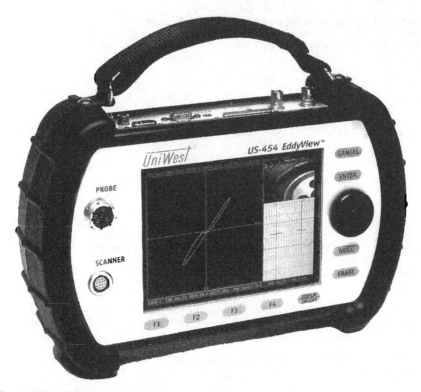

Figure 3.29 UniWest-454 EddyView multifunction instrument. Courtesy of UniWest.

setups with the programmable push-button function keys provided. Each function key, F1 through F4, is capable of being programmed for any of the following functions—Frequency, Gain, Rotation, Drive, Filters, Sensitivity, Alarm, Save Data, Recall Data, Data Burst, Print Screen, Save Setup, Recall Setup, Null Position, Chart Recorder, Video, EC Display, Null Value, Auto Clear, Persist, and Measure Video. By using the preset function keys and scrolling control knob, the operator can quickly adjust the operation of the US-454 to meet the requirements of the inspection.

The selectable mode display allows the operator to choose between eddy current impedance planes, video image, and X-Y strip chart recording, which can be displayed simultaneously or independently on the color LCD display. The video display has RGB output and NTSC video capture input.

Scanner operation uses a separate connector with power supply including a synchronized input to maintain constant rotation. Readout of null value can be displayed, indicating balance between probe coils.

Scaling sensitivity is 0.2, 0.5, 1.0, 2.0, or 5.0 volts per division and the O-scope sweep speed can be set from 10 ms/div to 1 ms/div. An auto clear and variable persistence of 0–10 seconds can be selected depending on application.

The null point can be adjusted for operator convenience. Phase rotation of 0–359° in 1° increments is also provided.

Frequency, Gain, Filters, and Probes. Frequencies of 10 Hz to 10 MHz with 7.0 V peak-to-peak output into 10 ohms impedance is provided for low-noise operation. Dual-frequency modes F1 and F2 may be used independently, added, or F2 can be subtracted from F1. Gain is adjustable from 0 to 99.9 dB in 0.1 dB steps. Low- and high-pass filters are selectable from 0 to 10 kHz in 1 Hz increments. Absolute, differential, reflection, and differential reflection probes may be used.

3.12.1 E-Lab Model US-450

The E-Lab Model US-450 (Figure 3.30) is a fully functional eddy current instrument that can be used with most customer-owned Windows-based PCs. The E-Lab is a versatile instrument suitable for laboratory, process control, or field use. Linked with a typical laptop, it is very portable and suitable for field use. Data can be taken in the field and analyzed in the lab or office. Linked with a conventional PC, field data can be integrated into powerful reports complete with strip chart recordings and graphs.

The instrument's frequency range of 100 Hz to 10 MHz (sine or square wave) permits both low-noise and high-frequency response optimization. Interchangeable adapters are provided for fast and easy connections to UniWest, Foerster, Stavely, Fischer, Zetec, and other common probes and cables.

The digital E-Lab provides the following benefits:

- Customer-owned Windows-based PCs can be used.
- Strip chart data can be saved and incorporated into reports.

Figure 3.30 E-Lab model US-450 full-function eddy current instrument. Courtesy of UniWest.

- Conductive and nonconductive coating thickness can be measured.
- Metal thickness and conductivity can be measured.
- Instrument can simulate dual-channel strip chart recorder.
- Provides direct readout of signal phase and amplitude.

3.12.2 ETC-2000 Scanner

Because of the need for a low-cost eddy current scanner to inspect engine hardware, the Federal Aviation Administration (FAA) funded the Engine Titanium Consortium (ETC), which included Allied Signal, General Electric, Iowa State University, and Pratt & Whitney, to provide the engineering and software development to accomplish this goal. This new ETC-2000 technology was transferred to and is currently available from UniWest.

A series of end effecters (robotic end-of-arm tools) and interface modules have been developed to complete the interface between the inspection site, inspection sensor, and the motions controlled by the scanner. This enables the ETC-2000 scanner (Figure 3.31) to adapt to a wide range of engine components including rotors, disks, cases, and seats. The ETC system, augmented by conventional and extended flexible sensors, provides the aircraft industry with a turnkey solution for complex needs associated with inspecting engine hardware.

ETC-2000 scanner features include:

- Continuous circumferential rotation
- Radial indexing for webs
- Coordinated motion for slots
- Axial indexing for bores
- Manual probe positioning

Field testing includes the testing of bores, broach slots, bolt holes, webs, scallops, and blades. Specifications include the following:

- Four-axis scanner
- Linear X-axis motion over 17″
- C-axis motion—continuous 360°
- Data acquisition—16 bit with 8 differential inputs and a maximum sampling rate of 50 K/sec
- Linear R-axis motion up to 5.8″
- C-axis resolution—0.01″
- X- and R-axis resolution—0.001″

Software. Control software provides simple programming for complex scan-plans with easy-to-use dialog boxes. Written scan-plans can be saved or edited for future use. Integrated display language software provides for both

Figure 3.31 ETC-2000 configured with large interface module, manual positioning, and scanning a jet engine part. Courtesy of UniWest.

prepackaged or user-developed signal processing and display routines. Scanner position and eddy current signal information are recorded and stored as data files and displayed as graphs.

3.13 INSTITUT DR. FOERSTER

Institut Dr. Foerster specializes in eddy current testing and magnetic flux leakage testing for longitudinal and transverse flaws in wires, bars, pipe, and tubing. The eddy current principles used with virtually all stages of wire and bar production are shown in Figure 3.32. Through-type encircling coils (upper left) are typically used to detect crack and hole-type flaws in the longitudinal direction. Longitudinal magnetic fields (upper right) enable differential through-type coils to be used for detecting holes and transverse cracks as well as estimating the depth of longitudinal flaws. Likewise, probes with small focal active areas (lower left) are frequently used to scan and determine surface finish on semifinished products. Finally, one or more rotating sensors (lower right) set up a circular flux pattern on the test surface and can be used to determine the severity of surface defects.

It is possible to provide 100% eddy current testing on hot wires and bars for quality assurance. High-temperature ET probes may be used to measure surface roughness on hot wire in rolling mills. These ET sensors must be capable of withstanding temperatures up to 1200°C, while operating at speeds up to 150 m/s. For hot wire applications, the T-60 sensor can be provided with water-cooled guides for hot wire ranging from 5 to 65 mm (0.20 to 2.56 in.). With rolling operations, large quantities of data can be processed in real-time so that rolled wire trends can be determined quickly and corrected if necessary during hot rolling operations.

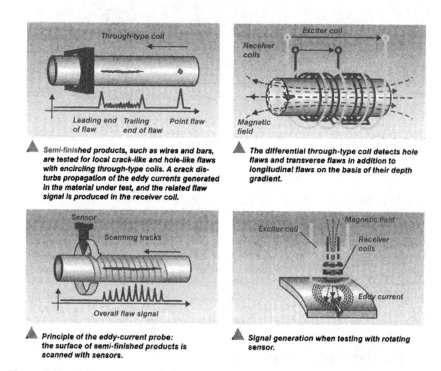

Figure 3.32 Eddy current techniques commonly used by Institut Dr. Foerster for the detection of flaws in wires, bars, pipes, and tubing. Courtesy of Institut Dr. Foerster.

For cast-rolling of copper wire, the Defectomat® with an additional Ferromat® channel is used to simultaneously detect flaws and ferrous inclusions with the same encircling through-type coil. With the two-channel system, simultaneous recordings are made showing surface damage and ferrous inclusions. For testing of hot material, Therm® coils are used for hot wire or cooling is provided so that standard encircling coils can be used. These two methods are shown in Figure 3.33. Defectomat systems are available as a single-channel model or dual-channel model operating on the absolute or differential method, alternately or simultaneously.

Through-type coils are being increasingly used in combination with rotating heads for stringent drawbench applications. Rotating heads with L (lever) sensor designs are held in their test position by centrifugal force. High-speed rotation guarantees 100% inspection of the wire/bar surface at speeds up to 8 m/s. The Circograph DS® electronic system (Figure 3.34) has also been designed for the most stringent applications, such as the testing of nonuniform valve spring wire. With this system, longitudinal flaw depths >30 μm for round wire and >70 μm for nonround, oval wire cross sections can be reliably detected. For maximum sensitivity, the Circograph provides constant test sensitivity regardless of the distance between the rotating scanning sensors and nonround wire circumference.

Automatic eddy current testing is typically used for all processing steps. For final inspection of polished stainless-steel bars made from ferritic, austenitic

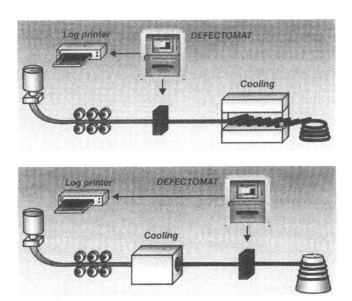

Figure 3.33 ET testing of hot copper wire using Therm coil (above) and standard coil. Defectomat with additional Ferromat channel used. Courtesy of Institut Dr. Foerster.

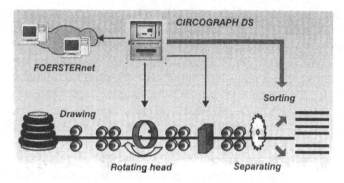

Figure 3.34 Optimum sensor location is between horizontal and vertical straightening systems. Courtesy of Institut Dr. Foerster.

steel or titanium, nickel, or cobalt alloys, stringent requirements are specified regarding surface flaws. To comply with these stringent requirements, combined eddy current testing with encircling through-type probes and rotating scanning probes are required. Precise guidance of the bar over its length is essential. For small-diameter bars, it is essential that final inspection be vibration-free.

3.14 MAGNETIC FLUX LEAKAGE TESTING

The magnetic flux leakage method is dry, fast, online, and recommended by the American Petroleum Institute for tubes with small to medium wall thickness. Magnetic flux leakage testing is of great importance for process reliability and quality control assurance in the production of oil field and boiler tubes. Magnetic flux leakage tests help assure the safety of nuclear and conventional power plants, offshore platforms, the oil and gas industries, and chemical and petrochemical plants.

DC field magnetization is used over the entire cross section of pipes transversely using Rotomat® and longitudinally using Transomat® (Figure 3.35), thereby providing simultaneous testing for internal and external flaws. With appropriate with state-of-the-art filtering and signal gating, separate indications are provided for internal and external flaws. When testing with external sensors, internal flaws have lower peak height and longer wavelength. ASTM E 570 describes magnetic flux leakage as the most efficient method for integration of an automated, nondestructive test for detecting rolling flaws during the production of seamless, hot-rolled tubes.

Referring back to Figure 3.12, the Rotomat principle is shown at top left, polished flaw sections and indications at bottom left, Rotomat DS electronics at upper right, and Rotomat sensor system at bottom right. The Transomat electronics unit is similar in appearance to the Rotomat unit. For detection of

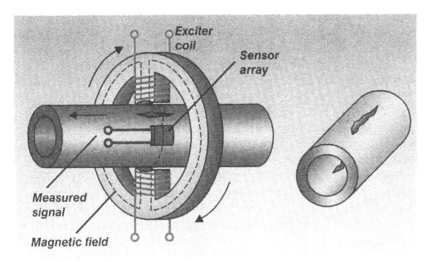

(a) Rotomat principle for the detection of internal and external longitudinal flaws.

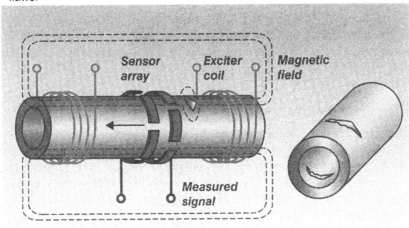

(b) Transomat principle for the detection of internal and external transverse flaws.

Figure 3.35 Rotomat/Transomat in-process flux leakage method. Courtesy of Institut Dr. Foerster.

longitudinal and transverse flaws, both Rotomat and Transomat electronic units and sensor systems would be combined and controlled via a Rotomat/Transomat power cabinet. Other testers also may be incorporated into automatic testing lines to measure tube diameter, wall thickness, etc. Contact manufacturer for specific applications.

Maximum testing speeds for Rotomat/Transomat systems depend on sensor design and tubing size. Rotomat Ro sensor systems have four rotating heads covering tubing diameters from 20 to 520 mm (0.79 to 20.5 in.). Transomat TR sensor systems have three sensor systems covering tubing diameters from 26 to 440 mm (1.02 to 17.3 in.). The Rotomat Ro 180 sensor system has a limited design range and can scan tubes with diameters ranging from 20 to 180 mm (0.79 to 7.09 in.) with throughput speeds of 2.6 m/s (8.5 f/s). The Transomat TR 180 sensor system can also scan tubes with diameters ranging from 26 to 180 mm (1.02 to 7.09 in.) with throughput speeds of 2.6 m/s (8.5 f/s). The maximum throughput speed for larger diameter tubing and larger Ro and TR sensing heads is about 2 m/s (6.6 f/s), depending on tube diameter. Large, fully automated testing lines, incorporating multiple test parameters, are extensively used in Europe for the inspection of critical piping systems. A schematic representation of a fully integrated material handling and transport system with automatic testing lines is shown in Figure 3.36.

3.15 APPLICATIONS

The following list of applications has been broken down into four sections; general and specific eddy current applications and general and specific flux leakage applications.

3.15.1 General Eddy Current Applications

General eddy current applications can be further subdivided into applications involving the measurement of physical property differences, such as flaws and thickness, applications involving the measurement of parameters relating more to conductivity, such as hardness, or applications involving the permeability changes in ferromagnetic parts. These are not clear-cut subdivisions because the physical parameters also affect coil impedance and material conductivity. Heat treatment also affects conductivity in nonferromagnetic parts and permeability in ferromagnetic parts.

Applications involving physical parameters first and permeability last are:

1. Detect and determine the severity of various surface cracks (stress, hardening, grinding, etc.), weld seams, laps, pits, scabs, porosity, voids, inclusions, and slivers.
2. Determine seam and seamless tubing integrity by measurement of % OD wall loss, intergranular corrosion, seam cracks, splits, and so on.
3. Measure flaws in graphite composites, aluminum, and titanium.
4. Detect and measure flaws in fastener holes.
5. Measure coating and plating thickness. Measure nonconductive coatings on conductive materials.

APPLICATIONS

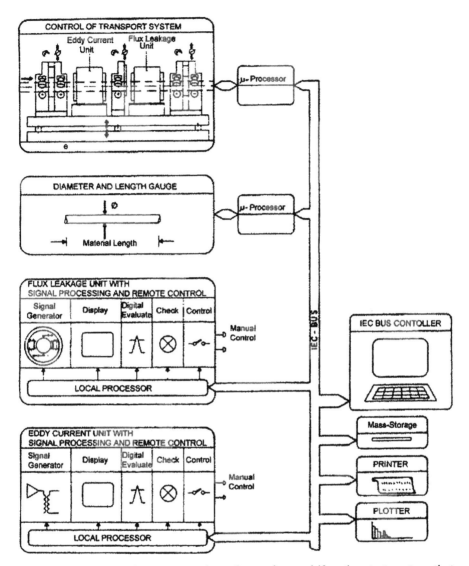

Figure 3.36 Schematic representation of complex multifunction test system that combines eddy current, flux leakage, and dimensional testing. Courtesy of Institut Dr. Foerster.

6. Measure nonmagnetic conductive sheet thickness. Measure dimensional differences in machined, formed, or stamped parts.
7. Determine the integrity of wire cable. Detect and locate broken strands.

8. Detect wanted or unwanted metals in nonmagnetic materials. There is a broad range of metal detectors or "treasure finders" that can be placed in this category.
9. Determine metal powder mixture ratios and the degree of sintering in metal powder parts.
10. Determine the hardness and depth of case hardening in bearing rings and other parts.
11. Determine the effects of corrosion thinning in pipes and vessels.
12. Provide alloy indication and sorting.
13. Sort metallic materials by microstructure or grain structure.
14. Measure electrical conductivity. Conductivity is related to tensile strength in magnesium and aluminum alloys.
15. Determine heat treatment condition, degree of annealing, and effects of aging.
16. Determine the carbon content of various steels.
17. Determine alloy composition of ferromagnetic materials based on permeability.
18. Measure nonmagnetic coatings over magnetic materials.
19. Measure magnetic permeability or the effect of heat treatment on magnetic permeability.

Inexpensive 60 Hz eddy current comparators are designed to measure eddy current variations caused by differences in part size, shape, grade, chemistry, or method of processing or manufacture. When eddy current testing is used to sort a known mixture of two alloys, instrument readings should be kept within the range of the readout device in order to detect the possible presence of a third alloy. Alloy variations in nonmagnetic materials primarily affect the conductivity of the part.

3.15.2 Specific Eddy Current Applications

Numerous NDT papers and articles have been written on specific eddy current applications. Many of these articles involve the detection and measurement of flaws in steering mechanisms, airplane landing gears, engine parts, reactor and steam generator tubes, aircraft wheels, aircraft wing structures, condenser pipes, and turbine blades. Other specific applications are listed below:

1. Determine the installed position of piston rings in motors using the differential probe technique.
2. Determine the extent of defects on the face of tungsten contact rivets using a hand-held, rotating probe instrument.
3. Detect missing needle bearing using a hand-held, rotating probe instrument.

APPLICATIONS 111

4. Measure the length of nonwelded seams in welded tubing using the absolute coil technique.
5. Test fine wire at maximum flaw sensitivity using absolute coils.
6. Test motor valve tappets using pointed eddy current test probe.
7. Test ball pins with a single pointed eddy current test probe.
8. Test motor valve seating rings using two moving probes while scanning a rotating test piece.

3.15.3 General Flux Leakage Applications

The main advantage of the flux leakage method over the eddy current method of flaw detection is that the flux leakage method has a much higher signal-to-noise ratio for even small flaws on rough surfaces. This greater inherent signal-to-noise ratio is due to the nature of signal generation that occurs when the flux sensors cut the magnetic lines of leakage flux. Signals from inner and outer surfaces differ in both phase and amplitude. Continuous spectrum analyzers can be used to separate signals and direct them to different channels where they are filtered and detected with threshold discriminators. Some general applications for the flux leakage method are:

1. Detect and determine the severity of internal and external flaws.
2. Measure ferromagnetic part thickness to 0.3 in.
3. Use differential coils for flaw detection and measurement in seamless, hot-rolled steel tubes.
4. Use absolute coils for flaw detection and measurement in cold-rolled or cold-drawn steel tubes.
5. Use combination absolute and differential coils for the detection and measurement of discontinuities, voids, slivers, and inclusions in seamless and welded tubes.

3.15.4 Specific Leakage Flux Applications

The measurement of stranded wire cable and rope integrity is an example of a specific leakage flux application that deviates somewhat from our preconceived idea that the leakage flux method is merely a reliable flaw-detection technique for ferrous materials.

Long lengths of wire rope can be easily inspected with the leakage flux method using encircling coils. Recorders are frequently used as the readout device for these high-speed tests. If the response time of the recorder is 60 Hz, test speed is limited to about 300 fpm. The response time of the test coil is dependent on the cross-sectional area of the broken wire, the location of the wire within the cable, and the gap or spacing between the ends of the broken wire. Permanent magnets are the preferred method for magnetizing wire rope because they have a constant magnetic field and do not require a flowing magnetic current. Frequently, the number of broken wires can be determined. However, in cases where there are overlaps or inside breaks the number of broken wires cannot be determined. The leakage flux method can also be used to determine the radial location of wire breaks by using two different diameter search coils, with different depths of penetration. Finally, the detection of consecutive wire breaks depends on the frequency response of the recorder and rope speed.

3.16 USE OF COMPUTERS

New multifrequency systems pose new calibration and data-handling problems. Each test frequency represents two key variables, phase and amplitude. In addition, each application is certain to introduce several unwanted parameters associated with inherent noise; these parameters must be suppressed to increase test reliability. The overall effect of computerization is to eliminate the number of subjective judgments an inspector has to make when dealing with a multifrequency, multiphase system. Ideally, we would input the test variables into a computer, push a start button, and then take a coffee break. Meanwhile, the computer would detect flaws and discontinuities, determine their number and severity, and mark them. Finally, a robot would separate the good and bad parts and stack them in neat piles. All of this would be accomplished at high speed with a 99.9% or better yield rate.

3.17 BARKHAUSEN NOISE/MICROMAGNETIC TESTING

The Barkhausen noise/micromagnetic testing method characterizes and determines the amount of residual stress in magnetic materials by measuring *Barkhausen noise* (*BN*). BN measuring instruments measure positive and negative stress units in N/mm^2, ksi, or relative *magnetoelastic parameter* (MP) units. MP units over the range of 0 to 140 correspond to a stress range of about -650 to $+450 N/mm^2$. Tensile stresses are positive and compressive stresses are negative.

3.17.1 Introduction

In 1911, Dr. Heinrich G. Barkhausen became the world's first professor in the Communications Branch of Electrical Engineering at Dresden's Technical Academy. He worked on the theories of spontaneous oscillation and nonlinear switching elements. He also formulated electron-tube coefficients and experimented with acoustics and magnetism, which led to his discovery of the Barkhausen effect in 1919. In 1920 Barkhausen and Karl Kutz developed the

EARLY APPLICATIONS

Barkhausen-Kutz ultrahigh-frequency oscillator, which eventually led to the development of the microwave tube.

Barkhausen learned that a slow, smooth increase in magnetic field applied to a ferromagnetic material caused it to become magnetized in a series of minute steps. These sudden discontinuous jumps in magnetism were detected by a coil of wire wound on a ferromagnetic core that amplified, producing a series of clicks in a loudspeaker. These sudden jumps in magnetism were interpreted as sudden changes in the size or rotation of magnetic domains.

There are a large number of potential applications for BN. Qualitative results can be obtained for determining steel composition along with hardening, grinding, turning, annealing, and tempering effects. However, because of the large number of variables that can be evaluated, simultaneous changes by more than one variable can be confusing.

3.18 EARLY APPLICATIONS

Stresses in rolling mill rolls are induced during manufacture and in-plant service. Casting, forging, heat treatment, and cooling all produce substantial compressive stresses in the case of the roll that are balanced by tensile stresses in the core. Some factors affecting mechanical stresses induced in rolling mill rolls are listed in Figure 3.37. Because of these mechanical factors and thermal stress factors that can be introduced, it is difficult if not impossible to predict stress on a theoretical basis.

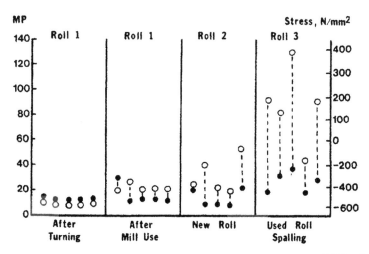

Figure 3.37 Effect of mill roll spalling on magnetoelastic parameter (MP) values. MP values and corresponding calibrated residual stresses on the surface of 43 forged steel rolling mill rolls in different conditions; ○, transverse; ●, longitudinal stress. Courtesy of American Stress Technologies, Inc.

Typical roll failures are caused by spalling, pinching, fracture, breakage, and roll explosion. Spalling is caused by insufficient fatigue strength of the roll case. Pinching occurs when the material being rolled wraps itself around the roll and is partially welded to the roll surface. Fracture, breakage, and explosions are the results of imbalance stress conditions caused by incorrect roll manufacturing or excessive applied loading during milling operations. Roll life can be improved substantially by proper fabrication and timely redressing.

The proper heat treatment followed by quenching can produce rolls with high case hardness, good compressive stress, and fatigue strength. However, even properly fabricated rolls can develop microcracks due to the high contact pressures during rolling. The highest contact pressures occur at the center of the rolls during cold rolling and at the barrel ends of the rolls during hot rolling. Redressing consists of removing the damaged material by turning and finish grinding. Finish grinding introduces a minimum amount of stress. Eddy current testing techniques are then used to verify that the microcracks have been removed. The redressed roll can then be heat treated and quenched to restore the proper case hardening.

The magnetoelastic technique involves the generation of Barkhausen noise (BN) that can be used to provide a fast, nondestructive evaluation of the new and used rolls in both static and dynamic tests. BN can be generated by a controlled magnetic field and detected by an electromagnetic probe. The average amplitude of the generated BN is proportional to the residual tensile stresses in the ferromagnetic materials. Increasing tensile stress increases BN; increasing compressive stress decreases BN. This technique has been found to be extremely useful in detecting grinding burns on injector lobes of diesel engine camshafts. Abusive grinding after final machining creates camshaft burns. Abusive grinding changes the induction-hardened compressive stresses of the injector lobes to tensile stresses.

3.19 PRINCIPLES OF MEASUREMENT

Magnetoelastic testing is based on the interaction between magnetostrictive and elastic lattice strains. If a piece of ferromagnetic steel is magnetized, it is also elongated slightly in the direction of the magnetic field. Likewise, if an applied load stretches the steel, it will be slightly magnetic in the direction it is stretched. With compressive loads, the same sort of domain alignment occurs except the resulting magnetism is 90° to the direction of the compressed load. When a magnetizing field is gradually applied to a magnetic material, any unaligned magnetic domains tend to jump into alignment with the induced magnetic field, generating BN, a measurement of compressive and tensile stress. The technique provides a method for rapid comparisons. A more accurate analysis may require calibrating procedures and known loads. Figure 3.38 shows how compressive and tensile stress levels affect BN signal levels.

EQUIPMENT

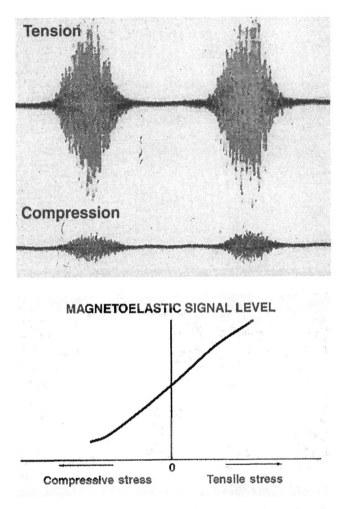

Figure 3.38 A comparison of Barkhausen noise levels (top) as a function of magnetoelastic signal level (bottom). Courtesy of American Stress Technologies, Inc.

3.20 EQUIPMENT

A number of commercial instruments have been developed for measuring internal stresses using BN. Figure 3.39 shows the Rollscan and Stresscan instruments developed by American Stress Technologies, Inc. The Rollscan instrument is a one- to four-channel instrument designed for semiautomatic testing when used with mechanical handling equipment that can rotate the part while simultaneously moving the measurement head so that the entire surface can be scanned. The Stresscan instrument is a lightweight, battery-

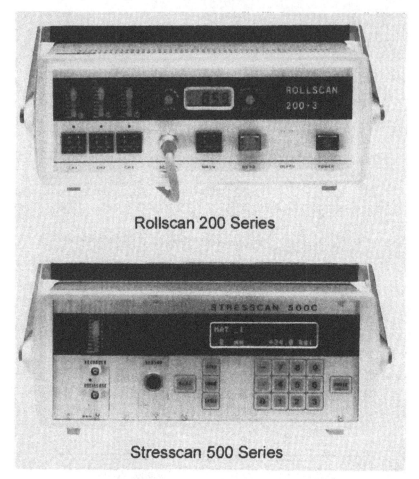

Figure 3.39 Three-channel Rollscan 200 instrument (top) and Stresscan 500C series instrument. Courtesy of American Stress Technologies, Inc.

operated, fully portable, programmable, single-channel unit with keyboard entry. A simplified diagram showing the basic BN operating principle is shown in Figure 3.40.

The appropriate depth of measurement with the Stresscan unit is 0.008 in. (0.2 mm), 0.003 in. (0.07 mm), and 0.0008 in. (0.02 mm), which corresponds to a BN frequency range of about 500 Hz to 10 kHz. Response time to full reading is about one second at a depth of 0.2 mm. A built-in alphanumeric display indicates stress in English (ksi) or metric (N/mm^2) units. Red and green LED display columns provide rapid visual indications of proportional stress values. RS 232 series digital communications port and 0 to 4 V analog recorder outputs are also provided.

CALIBRATION AND TESTING   117

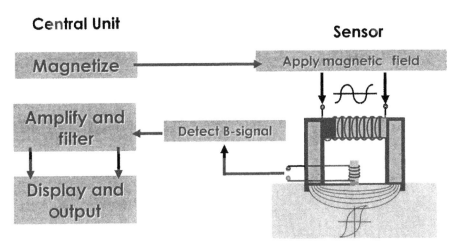

Figure 3.40 Basic functional schematic diagram for Barkhausen noise instrumentation. Central unit provides the drive for the electromagnetic sensor and provides the amplification, filter, display, and output functions. Courtesy of American Stress Technologies, Inc.

The Rollscan unit can automatically test parts with a cycle time of 5 to 60 seconds per part. The cycle time depends on test part size and the design of the material scanning system. Surface testing speeds of up to 3 ft/s can be used. Red LEDs provide quick visual indications of grinding burns. An oscilloscope and 0 to 2 V analog recorder output are also provided. The unit provides a rapid nondestructive alternative to chemical etching for the testing of grinding burns on camshafts, crankshafts, bearing races, drill bits, tool steel parts, rollers, piston pins, gears, and other items. Its response time is 30 ms at 0.02 mm depth, which is standard for the unit.

3.21 TECHNICAL SPECIFICATIONS

A partial summary of technical specifications is shown in Table 3.3.

3.22 CALIBRATION AND TESTING

Barkhausen noise measuring instruments are calibrated in terms of stress, N/mm^2 or ksi. Comparisons of stress levels in materials also can be made in relative MP units. Uniform samples, typically 0.5 in. wide × 4 in. long × 0.25 in. thick, can be used to obtain more accurate calibration curves. Sample

TABLE 3.3. Technical Specifications

		ROLLSCAN 200	STRESSCAN 500C
PROGRAMMABLE		No	Yes
DISPLAYS	Digital	3-1/2 digit LCD	2 × 20 character alphanum. FLC
	Light Column	9 × green LED 11 × red LED	10 × green LED 10 × red LED
NO. OF CHANNELS	(max)	4	1
ALARMS	Hi Limit, adjustable	Standard	Standard
	Lo Limit, adjustable	Optional	Standard
MEASUREMENT DEPTH in iron*		0.02 mm standard 0.2 mm optional	0.2 mm/0.7 mm/0.02 mm
RESPONSE TIME		30 ms at 0.02 mm 500 ms at 0.2 mm	60 ms–0.02/0.07 mm 500 ms at 0.02 mm
CONNECTIONS	Sensor	LEMO EGG 3B 314 CNL	LEMO EGG 3B 314 CNL
	RS 232	Optional	Standard
	Signal Monitor	±8 V BNC	±8 V BNC
	Recorder	0–2 V BNC	0–4 V BNC
BATTERIES	2 × 12 V/1.8 Ah Rechargeable	Optional	Standard
AC POWER	Various V and Hz	110–120 V, 50/60 Hz 220–240 V, 50/60 Hz	110–120 V, 50/60 Hz 220–240 V, 50/60 Hz
DIMENSIONS	W H D	365 mm 160 mm 272 mm	365 mm 160 mm 272 mm
WEIGHT		7.7 kg	8.2 kg

* Varies based on characteristics.
Courtesy of American Stress Technologies, Inc.

CALIBRATION AND TESTING

preparation consists of stress relieving the samples at 650°C and allowing them to slowly cool to room temperature. Then the samples are strained in tension and compression within their elastic limits using a tensile tester, while simultaneously monitoring the strain with strain gauges and a BN measuring instrument. Using this procedure, approximately 12 measurements were made with the samples in tension, and an additional 6 measurements were obtained with the samples in compression, to obtain the calibration curve shown in Figure 3.41. The three steels used to obtain this calibration curve had yield strengths ranging from 248 to 710 N/mm². As shown, the magnetic parameter is very sensitive to changes in internal stress. Magnetization settings are front panel settings.

Comparative studies by Kirsti Titto have shown excellent agreement in mill roll use and grinding burn studies. Figure 3.42 shows the effect of severe grinding burns on diesel engine camshafts. These studies seem to verify that BN

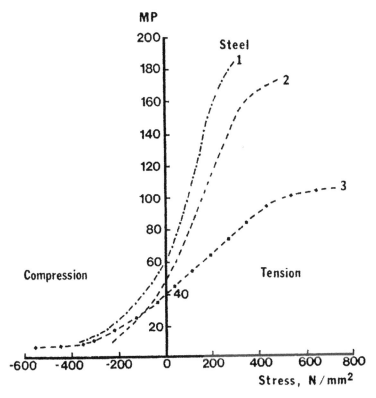

Figure 3.41 Magnetic parameter calibration curves for three different types of steel: 1, magnetizing 20, yield strength 248 N/mm² (36 ksi); 2, magnetizing 30, yield strength 399 N/mm² (58 ksi); 3 magnetizing 99, yield strength 710 N/mm² (103 ksi). Courtesy of American Stress Technologies, Inc.

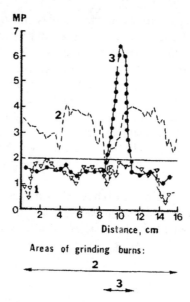

Figure 3.42 Evaluation of grinding burns on diesel engine camshafts; 1, no grinding burns; 2, grinding burns nearly all over the injector lobe; 3, severe local grinding burns. Courtesy of American Stress Technologies, Inc.

measurements and subsequent roll redressing can be used to extend the life of rolling mill rolls. In some cases it might be possible to detect and retire defective rolls prior to catastrophic failure. Testing for grinding burns permits the detection of parts with inferior fatigue strength. Once detected, these parts can be excluded from final assemble until they have been reheat-treated to restore the desired case hardness.

3.23 CURRENT APPLICATIONS

Barkhausen noise and magnetoelastic measuring techniques are typically used before and after heat treating, surface grinding, case hardening, carburization, welding, forming, irradiation, and cold deformation operations.

3.23.1 Applications in Aircraft/Automotive/Marine Industries

- Inspection of aircraft landing gear components
- Detection of grinding burns in aircraft/automotive/marine parts
- Inspection of high-strength aircraft/automotive/marine parts
- Providing grinding quality control for case-hardened camshafts
- Stress analysis to improve crankshaft grinding operations

- Optimization of grinding operations for carburized steel gears
- Determining heat treatment and deformation in gear steel properties
- Determining residual stress distribution in shipyard steel plates
- Determining the surface quality of marine parts/making improvements

3.24 GENERAL APPLICATIONS

- Determining the effects of prestraining noise vs. stress
- Determining residual stresses after heat treating/grinding
- Determining hardness and residual stress profiles
- Determining material characterization after forming operations
- Determining thermal damage in ground steel parts
- Determining causes/cures of part or component failure
- Determining causes/cures of ball bearing fatigue failure
- Determining the grindability of high-speed steels for drill manufacturing

3.24.1 Pipe/Tubing/Sheet/Plate Manufacturing

- Determining residual stress in magnetic sheet material
- Determining residual stress in welded steel tubing
- Evaluation of biaxial residual stresses in welded steel tubes
- Determining residual stress in carbon steel plates
- Determining residual full stress condition of welded plates

3.25 ELECTROMECHANICAL ACOUSTIC TRANSDUCERS (EMATS)

Weld Inspection Systems, Inc. (WIS, Inc.) applies EMAT technology to the nondestructive evaluation (NDE) needs of industry. EMATs were developed in the 1970s and have become the technique of choice for many real-time, high-speed, noncontacting, ultrasonic analyses in conductive metals. The EMATs system requires specialized high-current, high-voltage electronics.

Standard ultrasonic techniques have traditionally been used to detect cracks, thin spots, corrosion, disbonds, laminations, blowholes, and voids. The ultrasonic method has also been used to inspect welds and measure remaining wall thickness.

EMATs overcome many common ultrasonic problems associated with UT couplants by using the interaction between eddy currents and magnetic fields to generate UT waves directly in the test piece. When static or pulsed magnetic fields are applied to the test piece, the driven EMAT circuit induces eddy

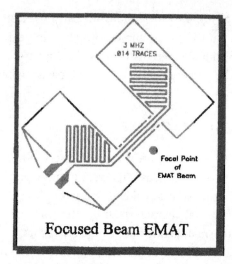

Figure 3.43 High-frequency, focused-beam EMAT. Courtesy of Weld Inspection Systems, Inc.

currents in the test piece. The interaction of the eddy currents with the magnetic field induces stresses in the test piece, which generates ultrasonic waves that are detected by the transducer. A high-frequency focused-beam EMAT for flaw detection on the ID of a cylindrical part is shown in Figure 3.43. In this case, the transducer generates angle beam shear waves.

3.25.1 EMATs Advantages Over Piezoelectric Transducers

- Less prone to operator error
- More repeatable test results
- Less affected by many types of coatings
- Useable at high temperatures
- Can generate horizontal and vertical shear, Lamb, longitudinal, and Rayleigh waves
- Automatically scans parts at high speeds
- Useful in most piezoelectric applications such as flaw detection and weld inspection
- A-, B-, and C-scans are possible

3.26 BASIC THEORY OF OPERATION

A magnetic field is applied to the test piece and the EMAT circuit is placed in the magnetic field. A high-power AC toneburst is driven through the EMAT.

This toneburst passes into the test piece and produces eddy currents. The eddy currents interact with the static magnetic field to produce stresses in the metal, generating ultrasonic waves. The frequency of the current, the shape of the EMAT transducer, its position in the magnetic field, and the thickness of the test piece combine to determine the mode of ultrasound generated. The wave mode and frequency determine the type of analysis to be performed. With Lorenz force generation, vertical and horizontal shear waves are typically generated. With magnetic materials, magnetostrictive forces, which have certain benefits compared to Lorenz forces, can also be generated and used to advantage.

3.27 RECENT APPLICATIONS AND DEVELOPMENTS

- *Invention of a revolutionary thickness gauge for detecting remaining wall thickness to less than 0.1" for rapid scanning of sheets or pipe.* The gauge performed well in an international blind testing at Southwest Research Institute.
- *Automatic Butt Weld Inspection System (ABWIS®).* As part of the production of steel, welders create flash butt welds in large coils of plate. These welds are then scanned with an EMAT system to analyze the validity of the welds.
- *A flaw detection/thickness gauge for oil field tubular goods.* This technology is gaining rapid acceptance as a vastly improved NDE method over existing technology. It has performed well in Arctic conditions at 20° below zero as well as in Saudi Arabian deserts.
- *A projectile analysis system for a large ordnance manufacturer.* This system uses three separate EMATs to scan for projectile cracks in three areas. The system has scanned over 5 million parts to date.
- *A girth weld inspection system for large pipes used to anchor off-shore oil-drilling platforms in the North Sea.* (See Figure 3.44, 36" girth weld inspection.)
- *Boiler tube pipe inspection.* This system uses two separate EMAT systems—one that scans for flaws in the tubes and another that is a thickness gauge capable of measuring wall thickness down to 70 mils.

Figure 3.44 had EDM notches cut into both the ID and OD sections of the crown. A single EMAT was used to analyze the ID and OD of the crown by varying the frequency. This caused the angle of the ultrasonic beam to change, thus directing the energy to either the ID or OD weld surface. Time to complete the scan was about 2.5 minutes.

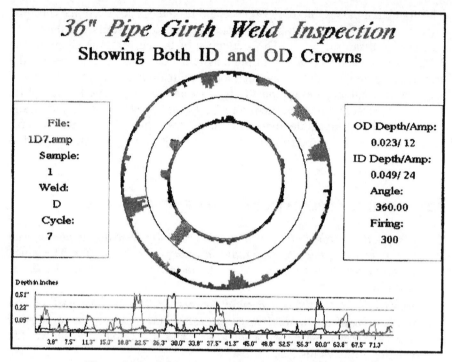

Figure 3.44 Courtesy of Weld Inspection Systems, Inc.

3.28 ALTERNATING CURRENT FIELD MEASUREMENT (ACFM) METHOD

This is a relatively new method originally developed for the undersea inspection of offshore platform piping and structures. The method has been found to be useful in providing an early warning of service life structural integrity problems so that corrective action can be taken prior to catastrophic failure.

The ACFM method can reliably locate surface cracks and determine the extent of cracking without removing paint and other coatings. This means that paint and other coatings do not have to be removed and replaced later. This also means that special scaffolding and other structures do not have to be assembled beforehand and disassembled after these operations. This usually results in substantial cost savings.

One advantage of the ACFM method is that it can be used underwater or in normal work environments by relatively unskilled operators. Windows-based software and laptop computers are used to process the data and determine the location and extent of the flaws. All data can be recorded for later display and analysis. For undersea operations, an unskilled operator can scan structural members while a more experienced topside operator views results, analyzes data, and relays information back to the undersea operator.

The method also lends itself to the inspection of heavy machinery and equipment, bridges, heavy rail equipment, and process pipes and vessels. In each case, the extent of the structural problem determines the corrective action that must be taken to extend useful service life. Applications for this method have been increasing exponentially since the 1990s.

3.29 ACFM PRINCIPLES OF OPERATION

The alternating current field measurement (ACFM) is an electromagnetic technique used for the detection and sizing of surface breaking cracks in metallic components and welds. ACFM combines the advantages of the alternating current potential drop (ACPD) technique, which can size defects without calibration, with the ability of eddy current techniques to work without electrical contact. This is accomplished by inducing a uniform field in the test object. The technique requires a moderate to high level of training in its exploitation. Figure 3.45 shows a half-cycle representation of the technique. Alternating current follows the "skin" or outer surface of a conductor. The expected skin depth of penetration is given by Eq. (3.8).

$$\delta = (\eta \cdot \sigma \cdot \mu_r \cdot \mu_o \cdot f)^{1/2} \qquad (3.8)$$

where σ = electrical conductivity of the material
μ_r = relative magnetic permeability
μ_o = the permeability of free space
f = the frequency of the applied alternation current

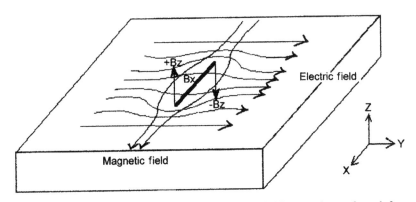

Figure 3.45 Current flow and induced magnetic field around a surface defect.

Materials with high permeability or conductivity have relatively shallow skin depths. At 5 kHz, the following materials exhibit the approximate skin depths shown.

- Mild steel ~0.1 mm
- Aluminum, tungsten, zinc ~1 to 2 mm
- Titanium, stainless steel, and Inconel ~5 to 8 mm

When there are no defects, uniform current flows in the Y-direction and the magnetic field is uniform in the X-direction perpendicular to current flow and other magnetic components are zero.

3.29.1 Bx and Bz Components

The presence of a defect diverts current away from the deepest part of the crack and concentrates it near the ends of the crack. This produces strong peaks and troughs in Bz above the ends of the crack, while Bx shows a broad dip along the entire defect length with its amplitude related to depth. Since current lines are close together at the ends of the crack, there are also positive Bx peaks at each end of the Bx trough. With a forward probe scan, Bz values decrease steadily from background level to the first trough half and subsequently increase to the maximum point near the end of the trough. Figure 3.46 shows typical plots of Bx and Bz. Bz starts at a trough and ends at a peak or the value of Bz starts at a peak and ends at a trough depending on the direction of current flow. The curve of Bz is insensitive to the speed and lift-off of the probe, and the length of the crack. Bz is used to determine if there are cracks in the surface of the part or weld, and calculate their length.

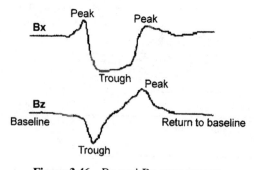

Figure 3.46 Bx and Bz components.

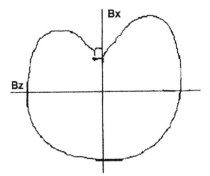

Figure 3.47 Typical butterfly plot.

3.29.2 Butterfly Plot

To aid interpretation, Bx and Bz components are plotted against each other—a closed loop indication confirms a crack is present. Figure 3.47 shows a typical butterfly plot with the time base removed. The butterfly plot may be flatter than shown, but with a similar butterfly shape. This means that whether a defect is scanned slowly, quickly, or even in a series of finite steps, the butterfly plot will still be formed. The butterfly plot is independent of probe movement speed. However, the butterfly plot always starts at the top of the illustration and returns to the same region of the screen when a defect is present.

Magnetic fields are strongest at the defect. Unexpected lift-off and any offset from the centerline contributes to inaccuracies in crack-depth estimation. These factors reduce the sensitivity of the sensing probe. Therefore, positive probe contact or carefully controlled probe lift-off would increase accuracy. A lift-off sensor may be incorporated into the probe design to provide some automatic lift-off compensation.

3.30 PROBE DESIGN

Standard ACFM probes contain both a field induction coil and magnetic field (voltage) sensor in a single probe head and multiple conductor cable. Probes are typically designed for specific applications, such as a "standard weld probe." Probe design is optimized to follow a toe weld and reduce probe noise caused by probe movement. With toe welds, a standard weld inspection would require two probe scans, one for each toe. Welds up to 20 mm wide can be scanned this way.

Because of the positional sensitivity of pencil-style probes, they should be used only in applications requiring pencil probes to gain access. For the various applications, the manufacturer's instruction manuals and recommended company test procedures should be consulted and followed.

Paintbrush array probes are popular because they cover a broad band of material in one placement. Consecutive placements are usually overlapped to guarantee complete coverage of the test area. When used in automatic scanning systems, they usually contain a position encoder to provide feedback on scanning speed, defect size, and location. Pick-and-place arrays cover rectangular areas and are often used with programmed robotic manipulators. Test results for probe arrays are often displayed as false-color contour maps of the inspection area. Windows-based software simplifies analysis and display functions.

The field induction coil typically operates at 5 kHz with standard probes and 50 kHz with special high-frequency probes. The 50 kHz high-frequency probes are typically used to obtain higher test sensitivity for nonmagnetic conductive materials. TSC Inspection Systems has a wide range of AMIGO probes designed for use with the AMIGO ACFM Crack Microgauge. Standard probes include weld, minipencil, and micropencil types with several variations and orientations. Figure 3.48 is a flowchart selection guide for common AMIGO probes.

Note: AMIGO probes cannot be used with other ACFM instruments because they contain a *probe identification chip* (*PIC*) that tells the instrument what probe is being used.

3.31 APPLICATIONS

One active encoder weld probe for ACFM inspection of welds, plates, and tubular materials features a position encoder to provide a continuous position reference, faster defect sizing, and lower probe wear. The probe nose contains a moving belt that is laid on the surface as the probe is scanned along the weld. The moving belt drives the position encoder. Crack end locations are determined by software, reducing inspection time. Probe wear is reduced by use of a replaceable belt and low-wear supporting wheels. This is just one example of a specialized probe.

Radiation-hardened array probes have been designed for robotic use in hazardous environments by the USNRC. This system is designed to locate and size fatigue cracks and corrosion pitting. Most robotic systems may also be used underwater.

In aerospace, ACFM applications include inspections of:

- Undercarriage components for civil and military aircraft
- Engine components such as crankshafts, cylinder heads, turbine discs, and blade roots
- Gearbox components and gear teeth

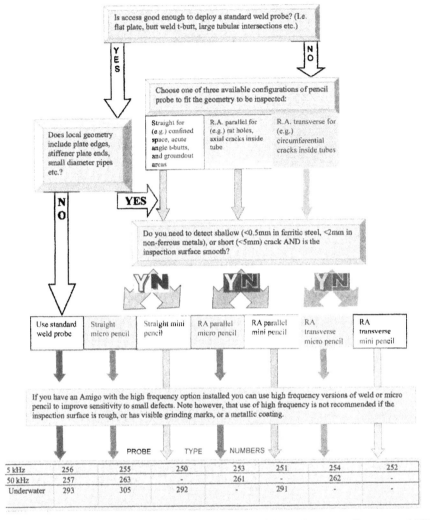

Figure 3.48 Flowchart selection of common AMIGO probe types. Courtesy of TSC Inspection Systems.

- Structural members, engine mounts, and metal skins
- Bolts, studs, and threaded bolt holes—wheels and hubs
- Gas, pneumatic, and hydraulic lines

Materials typically examined during these inspections include steel and stainless steels, duplex and super duplex steels, aluminum, and titanium. Again because of temperature and stress, weld areas are areas of prime concern.

Other applications include:

- Inspection of cracks and corrosion pitting in process pipes and vessels, structural steel supports, cable trays, etc.
- Civil structures such as bridges and buildings
- Virtually all automotive parts subject to stress, wear, corrosion, and fatigue

4

LASER TESTING METHODS

4.1 INTRODUCTION

Laser testing methods are among the fastest-growing nondestructive optical methods; they can be micro or macro in nature. Since they are also noncontacting, irregular objects and nonplanar surfaces can be inspected without the wear problems typically associated with contact probes, or the cleaning problems associated with coated surfaces. Full-field test areas are quite flexible, but they vary, depending on the application, sensitivity, and range of the test equipment.

Full-field areas typically range from several square centimeters in microapplications to several square meters in macroapplications. The full-field area approach facilitates high inspection rates and results in substantial timesaving, which hopefully more than offsets any increases in method costs. Since there are no consumables, such as plastic probe shoes, markers, dyes, and chemicals (in most cases), additional cost savings can be realized.

Easy-to-view, real-time imaging systems are possible because of charge-coupled device (CCD) optical detectors for TV cameras. Laser testing methods are popular in the aircraft and automotive industries.

Introduction to Nondestructive Testing: A Training Guide, Second Edition, by Paul E. Mix
Copyright © 2005 John Wiley & Sons, Inc.

4.2 DISADVANTAGES

Laser testing methods have some disadvantages or limitations. With shearography, areas of concern are:

- Special safety considerations must be taken into account when using lasers.
- Component loading is required to see results.
- Results depend on lateral shearing vector component.
- Interpretation of results is somewhat subjective.
- Slight corrections are necessary for precise location of defects.

One limitation of little concern today is that laser technology cannot be used to measure objects smaller than the wavelength of light or one micrometer (μm). In the nanobot world of the future other NDI techniques will have to be developed.

4.3 MAIN ADVANTAGES

The main advantages of optical NDT techniques such as speckle interferometry, shearography, and TV-holography are:

- *Noncontacting measurement.* In many cases, there are no surface preparation or marking requirements.
- *Full-field view.* Whether the measurement is macro- or microscopic, the full-field view of the area of interest can be examined and measured.
- *No consumables.* Unlike other NDT methods, no penetrants, marking materials, or coatings are required.
- *Real-time measurement.* With electro-optical sensors like CCD cameras, real-time measurements are possible.

4.4 LASER THEORY

The word *laser* is an acronym for *l*ight *a*mplification of *s*timulated *e*mission *r*adiation. A mathematical explanation of how laser light is produced and amplified is beyond the scope of this book; however a simplified written explanation follows.

As the acronym implies, laser light is amplified by stimulated emission radiation. This light is very intense, essentially monochromatic (single wavelength), plane polarized (unidirectional), and very stable. Today there are literally hundreds of lasers with emission frequencies covering the electromagnetic spectrum.

The classical helium-neon (He-Ne) laser consists of a crystal tube filled with He-Ne gas lasing medium. Two reflective mirrors are at each end of the tube; one mirror is flat, the other is semitransparent. Laser radiation is emitted through the semitransparent mirror.

When a high-voltage source is applied across the He-Ne tube, helium atoms are excited by the electron flow. These newly excited helium atoms collide with neon atoms that in turn decay to slightly lower energy levels and produce photons. Photon light is reflected back and forth by the mirrors, which stimulates the emission of even more photons. Some of the photons produced are absorbed by ground potential electrons, which in turn become excited and produce more photons. Some spontaneous photon emission also occurs. When the rate of photon emission exceeds the rate of photon absorption due to population inversion, stimulated emission occurs and a coherent beam of light forms with a fixed amount of light amplification.

The four processes involved in laser production are:

1. Stimulated emission
2. Pumping (excitation from an external source)
3. Spontaneous emission
4. Absorption of photons

The last two processes are considered competitive processes because they do not contribute to the overall gain of the system.

4.5 LASER SAFETY

The approach to laser safety is very similar to that for radiation safety or any other form of dangerous electromagnetic radiation. While in the eyes of some, the responsibility for laser safety may lie with the manufacturer, health physics, or medical department of an employer or institution, you, the laser operator, should be the most responsible for your own safety because you and your family have the most to lose. The laser operator should always be familiar with the laser system and qualified to use it. He or she should also have the required safety equipment on site and be wearing any required protective gear. The operator should not deviate from the manufacturer's recommendations or company-established procedures—these should be close at hand. The laser operator should never be in a hurry or take shortcuts to expedite anything. Most employers welcome suggestions for improvement of safety.

4.6 LASER CLASSIFICATION

Class 1 lasers. These lasers are considered incapable of producing harmful radiation levels and are therefore exempt from most control measures. Laser printers are an example.

Class 2 lasers. These lasers emit radiation in the visible portion of the electromagnetic spectrum, produce bright light, and are generally not directly viewed for extended periods of time.

Class 3a lasers. These lasers would not normally produce injury to the unaided human eye if viewed only momentarily. They may present a hazard if viewed using light-collecting optical devices. An example would be He-Ne lasers with up to 5 mw of radiant power.

Class 3b lasers. These lasers can cause "severe eye injuries" if viewed directly or if specular reflections are looked at. A Class 3 laser is not normally a fire hazard. An example is He-Ne lasers above 5 mw, not exceeding 500 mw of radiant power.

Class 4 lasers. These lasers are hazardous to the eye from both direct and reflected beams. In addition, they can damage the skin and start fires.

Note: Irreparable retinal damage can occur instantly with Class 3b and Class 4 lasers. Seek medical assistance immediately if eye damage is suspected.

4.7 TRAINING

On-the-job training programs should include basic information on the following pertinent subjects:

- Biological damage from exposure to laser radiation
- The physical principles of laser operation
- A review of laser classification
- Rules for control of laser-use areas (posting and warning systems)
- Medical precautions and emergency instructions
- A review of basic safety rules (public access availability)
- A review of protective equipment familiarization
- Control of laser-related hazards—fire, electrical, chemical, hazardous gases, cryogenic materials, laser dyes, X-rays, and UV lasers
- Acquisition, inventory, transfer, and disposal of hazardous materials
- A review of emergency response procedures
- A review of medical exam surveillance and procedures

4.8 PROFILOMETRY METHODS

When surface shape, roughness, or finish is of prime concern, some form of surface profilometry is usually the inspection method of choice. A manufacturer may be concerned with the smoothness of roller bearings, machined and plated parts, optical parts, or the uniformity of solder coatings or insulation layers during the electronic chip manufacturing process, or the final profile geometry of the finished chip, machined, and plated parts. The vertical and spatial (lateral) sensitivity and range of the profilometer are of prime importance when considering the profile method to be used.

4.8.1 Stylus Profilometry

For many years, mechanical stylus profilometry was the industrial standard and it is still a popular, relatively inexpensive, high-accuracy method for measuring the surface roughness of many mechanical parts.

The physical limitations of a mechanical stylus can best be visualized by trying to imagine what happens when you use an old 78 rpm record stylus to play a 45 rpm record or vice versa. The larger-diameter 78 stylus will not properly track the narrower, slower-speed 45 rpm record grooves, resulting in groove peak record damage, and the smaller-diameter, faster-speed 45 rpm stylus quickly damages the 78 rpm record groove sides. In both cases, the stylus diameter and operating speed do not match the vertical or spatial sensitivity requirements of the record.

Assuming the mechanical stylus and surface roughness are appropriately matched, it may still take a long time to make a 100% inspection of the test part surface, depending on surface area. Again the main disadvantage of the mechanical stylus is that it requires surface contact, which in time can damage the stylus tip, contact surface, or both.

4.8.2 Optical Profilometry

Optical profilometry overcomes most of the disadvantages of stylus profilometry and has many advantages. It provides exceptional lateral resolution, full-field analysis, high-speed measurement, and high-accuracy inspection of surface shapes for the test object. Currently, the two most popular methods used are *electronic speckle pattern interferometry (ESPI)* contouring for rough surfaces, using coherent light, and *white light interferometry*, using white or low coherent light for smooth discontinuous or stepped surfaces. The ESPI method is sensitive to out-of-phase displacements when the object surface is illuminated by two wavelengths of coherent light that differ by about 10 nm. In turn, this small phase shift limits the method's vertical sensitivity. However, the advantages of both techniques can be combined to produce a surface profilometer with much greater vertical range.

4.8.3 White Light Interferometry

Since ESPI contouring and phase-shifting interferometry cannot measure surface height variations greater than the wavelength of the light source, *white light interferometry (WLI)* was developed to permit surface profiling over a wide range without phase errors. Using this technique the entire surface is scanned at one time, producing a large amount of information to be analyzed. To minimize data acquisition and analysis times, coherent radar techniques have been developed. This technique uses two algorithms and simple computer programs to reduce these times. One algorithm is used for profiling optically smooth surfaces, while the other is used for profiling optically rough surfaces. A resolution of 60 nm is typical for smooth surfaces and a resolution of a few mm is typical for rough surfaces.

4.9 BASIC TV HOLOGRAPHY/ESPI INTERFEROMETRY

4.9.1 Single Laser Operation

Holographic interferometry became a reality after the 1960 invention of the laser, a monochromatic, coherent light source. It is noncontacting, nonintrusive, and therefore nondestructive in nature. It has the advantages that it is not dependent on surface preparation, object size, or positioning. With special recording equipment, a moving, full-view image of an object can be presented.

Until the 1960s, holographic interferometry was one of the most underutilized inspection tools because procedures were complex and equipment was costly, complicated, and unreliable. However, a free exchange of information by researchers and manufacturers has turned this problem around. Today holographic interferometry is relatively easy to understand, apply, and interpret with regard to inspection results.

4.9.2 Camera Operation

In the mid-1980s, holographic camera recording systems, similar to the one shown in Figure 4.1, were considered to be reliable, easy to use, relatively inexpensive, and readily available. This system operates as follows: A variable attenuator beam splitter is installed in the He-Ne laser beam that divides the beam into an "object beam" and "reference beam." The object beam illuminates the object and the reference beam is projected directly on the film. With the He-Ne laser, a vibration isolation table and darkened room are required. The object and reference beams are adjusted so that they are in-phase and the same length from the beamsplitter to film. The beams can be independently directed, expanded, or collimated by mirrors and lenses in each beam. The light intensity for each beam (beam ratio) is determined and exposure time is calculated based on the beam settings and film speed rating. After exposure the film is processed, using standard photographic techniques, in about 20s.

The reference beam reilluminates the processed film and the light is diffracted into the exact waveforms that had been originally reflected by the object. The holographic image of the projected object contains all the visual information of the original object. When this image is superimposed on the

BASIC TV HOLOGRAPHY/ESPI INTERFEROMETRY 137

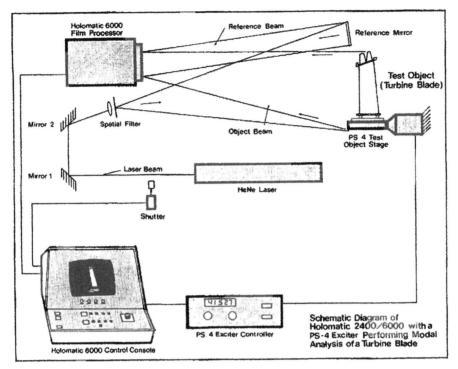

Figure 4.1 Early schematic diagram of automatic hologram recording system. Courtesy of Laser Technology, Inc.

object, real-time holography is produced. Any differences between the image of original object and present object can be immediately seen.

Another holographic camera, developed by Honeywell, used reusable thermoplastic plates to record images. With this technique, holograms were developed in about 10 s without wet chemical processing. Figure 4.2 illustrates how the thermoplastic recording/erasing process worked. With this camera, objects as long as 4 in. on a side were placed inside the camera and the holographic images were projected anywhere. More powerful, continuous or pulsed, external lasers were used for projecting holographic images of moving objects or objects of 6 ft or more in diameter.

4.9.3 Applications

Early applications with equipment of this type included:

- Determining the vibrational modes and characteristics of rotating objects such as wheels, gears, and early disk drives.
- Determining mechanical deformation in structural materials.

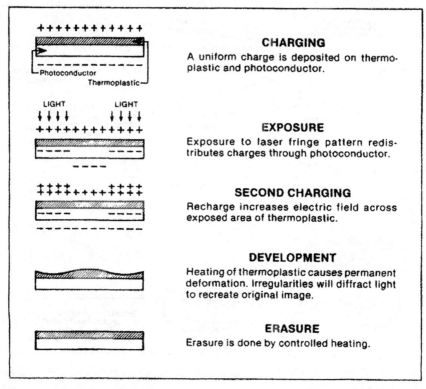

Figure 4.2 Early Honeywell/Newport thermoplastic optical recording/erasing system. Courtesy of Newport Corp.

- Determining the integrity of bonds and laminations in composite materials.
- Determining mechanical stress-strain and thermal stress-strain relationships in a multitude of materials.
- The detection of cracks in steel channel boltholes, ceramics at high temperature, concrete, glass tubes and plates, railroad wheels, pressure vessels, welded steel plates, and honeycomb panels. Improved laser systems are used for many of these same applications today.

The holographic interference patterns caused by stress relationships resemble the contour lines of a topographical map. Bright lines on a hologram represent nodal areas of zero motion, while dark fringe areas represent areas of maximum peak-to-peak displacement. These lines can be evaluated both qualitatively and quantitatively. Known stress, both thermal and mechanical, can be induced into an object during holographic inspection, providing a basis for these evaluations.

4.9.4 Thermal Stresses

Induced temperature changes of 1 or 2°C cause changes in the interference patterns of composite pipe sections. Figure 4.3 shows the interference for a uniformly heated composite pipe and similar pipe with an internal defect. Similarly, minor differential pressure changes can produce interference

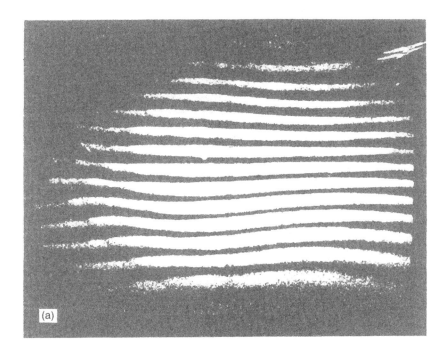

Figure 4.3 Early interference patterns in composite pipe sections. (a) Composite pipe section without major flaw indications. (b) Shows an internal defect. Courtesy of Laser Technology, Inc.

pattern changes in the hologram of a pressure vessel. In the mid-1980s, interference holography had three basic forms—real-time holography, double exposure holography, and time average holography.

4.9.5 Quantitative Aspects of Mechanical Stress

Simple quantitative analysis can be illustrated by a bending beam test, where one end of the beam is fixed and the other end of the beam is free to move. When the beam is bent, stress or fringe lines are closer together at the free end as movement increases because the light waves are in and out of phase. Each successive contour line represents one-half wavelength of light or about 12.4 µin. of deflection when using a He-Ne laser. Therefore, the strain can be calculated at any point on the beam using Young's modulus, applied force, and distance factors.

In cases where strain gages had to be used, holography was used first to determine where the strain gages should be placed. Initially, a real-time holographic view of the object was used to determine the best grid pattern spacing for strain gage placement. This saved computer time—no longer a significant consideration. Holography was also used during the tests to verify strain gage results and determine when shifts in strain gage calibration occurred.

Real-time holography involved recording the hologram of an unexcited object and superimposing the image on the object itself. Any movement of the part produced destructive interference in the viewing plane. When the part was excited through a range of frequencies, its movement at critical frequencies was readily observed. The amplitude of motion with sinusoidal excitation was determined from Eq. (4.1).

$$Z = Sn \times L/4\pi \qquad (4.1)$$

where Z = peak to peak vibration amplitude in microinches (µin.)
Sn = zero-order Bessel function associated with a given fringe count
L = laser light wavelength (24.91 µin. for a He-Ne laser)

Equation (4.1) can be reduced to:

$$Z = 1.9825 Sn \qquad (4.2)$$

By knowing the zero-order Bessel function associated with a given fringe line, the amplitude of motion was calculated from Eq. (4.2) as shown in Table 4.1.

By recording the vibrational mode patterns throughout a time exposure, a permanent record of a time average hologram was made. Double exposure holography was similar to real-time holography, except in this case, the image of an unstressed object was recorded and then a second image of the object was made on the same film, but in a different stress state.

TABLE 4.1. Amplitude of Motion as a Function of Bessel Zero

Fringe Number	Bessel Zero[a]	Amplitude[b]
1	2.4048	4.77
2	5.5200	10.94
3	8.6537	17.16
4	11.7915	23.38
5	14.9309	29.60

[a] Zeros of the Bessel function, J_0.
[b] Amplitude of motion in microinches (μin.).

The development of high-speed CCD cameras and computers has made it easy to observe, measure, and record the vibrational modes of an object in real time. These data can be saved and recalled later for detailed analysis.

4.9.6 Qualitative Aspects

Interference holography shows real-time stress-strain patterns in objects. If harmful stress patterns are noted, the design of the object can be changed. New stress patterns can be instantly recorded, and if the design is satisfactory, the object can be manufactured in the factory, minimizing test and evaluation times. This provides a qualitative test to determine how well the new design relieved the undesirable internal stresses of the object. Time-averaging holography involved stressing the object throughout the entire film exposure, thus producing a record of accumulated stress change. Time-average holography was commonly used to study bonding and lamination integrity in honeycomb panels and aircraft turbine seals.

Holographic images can be permanent, full-view images of an object. Any surface imperfection or defect causing deformation or movement can be seen. The images can be viewed with video cameras and displayed on monitors or videotaped and photographed. Any two-dimensional photographic image can be converted to a three-dimensional holographic image with current technology. Using a computer, visual images can be enhanced, digitized, stored, or recalled for comparison, subtraction, and other data processing.

4.10 NANOMETRIC PROFILING MEASUREMENTS

4.10.1 Introduction

Noncontacting 3D nanometric surface measurements can provide real-time quality control for chip manufacturing, and other industries concerned with making or inspecting miniature parts during various stages of manufacture. Customers benefit from faster development cycles and higher product yields.

Nanometric laser profilometry typically uses point sensors and high-precision stages to create surface profiles and 3D topographies. Samples are placed under sensors and precisely moved by stages as the sensor transmits height data to the measurement control unit. To improve lateral or spatial accuracy, the sensors are synchronized to the stage movement. This sychronization eliminates errors caused by stage acceleration and deceleration.

The three main types of Solarius™ Development sensors are designated as the autofocus sensor AF2000, holographic sensor CP1000, and confocal point sensor LT8010. Laser profilometers are flexible and easy to automate. They are noncontacting and therefore nondestructive in nature. They can operate at high speeds, making them ideal for production environments. The accuracy of Solarius™ profilometers is verified by in-house calibration. Certified standards are used in association with developed test procedures.

4.10.2 Autofocus Principle

Condensed light is focused from a laser diode onto the specimen surface. Reflected light is directed to a focus detector, which measures deviations from the ideal focus to within a few nanometers. The deviation in focus generates an error signal that is used to refocus the movable objective lens of the optical stage. The position of the movable objective lens at the measurement point represents the absolute measurement of specimen surface height. The autofocus sensor has high vertical resolution. With a measurement range of 1.5 mm, it resolves down to 25 nm. The system is capable of simultaneous fast form and roughness measurements and performs best on surfaces with varying reflection coefficients.

4.10.3 Specifications

4.10.3.1 Sensor
Spot size: 1.5 um
Vertical resolution: 0.025 um
Measurement frequency: 10,000 Hz
Stand off: 2 or 5 mm
Linearity: <0.08%

4.10.3.2 Camera
Type: integrated in-axis camera
Field of view: 0.6 × 0.8 mm
Laser diode: Class I
Wavelength: 630 nm

4.11 CONOSCOPIC HOLOGRAPHY

4.11.1 Theory

In conoscopic holography the separate coherent beams of conventional holography are replaced by ordinary and extraordinary components of a single beam traversing a uniaxial crystal. This produces holograms even with incoherent light, with fringe periods that can be measured precisely to determine the exact distance to the point measured. (See Figure 4.4.)

The unique design of the CP1000 sensor is ideal for the measurement of narrow, deep holes and sharp edges. Because it is collinear, it can easily measure blind corners and sharp angles up to 85 degrees. This characteristic of the sensor allows the sensor to measure screw threads and solder bump volumes.

4.11.2 Specifications

Focal length: 25, 50, or 75 mm
Absolute accuracy: <3, <6, or <10 um
Repeatability: <0.4, <1.0, or <2.0 um
Minimum working range: 1.8, 8.0, or 18.0 mm
Stand off: 15, 42, or 65 mm

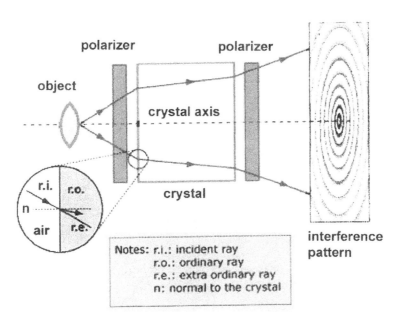

Figure 4.4 Measurement principle of the CP1000 sensor. Courtesy of Solarius Development.

Linearity: <0.3% for all ranges
Spot size: 22, 45, or 65 um
Lateral resolution: 12, 15, or 25 um
Laser type: Class II

4.12 CONFOCAL MEASUREMENT

The confocal point sensor uses a point light source and detector pinhole to discriminate depth. The laser beam from the point light source is focused on a specimen through an objective lens that rapidly moves up and down. The maximum light intensity occurs when the specimen lies within the focal plane of the objective lens. As the objective moves closer to or further from the specimen, the reflected light reaching the pinhole is defocused and does not pass through it. As a result the amount of light received by the detector quickly decreases. A detection signal is only generated when the maximum light intensity goes through the pinhole. A precise height measurement is achieved by continuously scanning along the z-axis.

The confocal measurement assures high reliability on steep edges and resistance to optical artifacts (false indications). The LT8010 sensor can measure complex surfaces as well as structures with high aspect ratios, which occur in *micro-electromechanical systems (MEMS)*. The sensor can be used for measuring tasks from single profiles to fast and precise measurement of large surfaces.

4.12.1 Specifications

4.12.1.1 Sensor
Spot size: 2 um
Vertical resolution: 0.1 um
Measurement frequency: 1400 Hz
Stand off: 5 mm
Linearity: ±0.5 mm

4.12.1.2 Camera
Type: integration off-axis camera
Magnification: 200×
Laser: Class II
Wavelength: 670 nm

4.13 NANOSURF CONFOCAL MICROSCOPY

4.13.1 Introduction

This microscope is especially designed for R&D and industrial quality assurance as an independent 3D optical measurement system. It is based on white light confocal microscopy.

This microscope can measure optically complex surface structures while maintaining high vertical and lateral resolution. The confocal physical filtering leads to a significant reduction of optical artifacts and a strong and predictable height signal in most cases, where other optical methods fail.

Measuring results using surface roughness standards show a good correlation to the most accurate tactile instruments. The Nanosurf confocal microscope is the only nondestructive optical measurement system that provides a high level of correlation for noncontact characterization of complex surface structures.

The instrument, shown in Figure 4.5, consists of a compact confocal module that is mounted to a solid stand. The module is moved in the vertical direction. The sample is placed on an x-y precision slide. Because of the precise motion control, extended measurement areas can be measured by combining basic fields via stitching. The microscope is controlled by software using the

Figure 4.5 Nanosurf confocal microscope. Courtesy of Solarius Development, Inc.

MS Windows® NT operating system. With software, surface topography data can be presented and analyzed in various ways.

4.13.2 Standard Components

Confocal image sensor: operates on the principle of white light confocal microscopy

Vertical measurement range: limited by working distance of the objective lenses

Laser resolution: 512 × 512 pixel in the basic field

Operating system: MS Windows (NT)

4.13.3 Options

- Stitching (automated measurements of larger areas)
- Long working distance
- Piezo for higher vertical resolution

4.14 3D CONFOCAL MICROSCOPY

4.14.1 Principle of Operation

In white light confocal microscopy, light emitted from a point light source is imaged into the object focal plane of the microscope objective. The reflected light is detected by a diode behind a pinhole. This arrangement leads to a partial suppression of light from the defocused planes. However, the area scanning technology used by the Nanosurf confocal microscope replaces the detector pinhole with a rotating disc in the intermediate image plane. This Nipkow disc consists of a spiral-shaped multiple-pinhole mask as shown in Figure 4.6.

When combined with CCD image processing, the rotating Nipkow disc affects the x-y scan of the object field in real-time video. An additional z-scan is used for the evaluation of highly precise height data.

4.14.2 Advantages

An increase of about 20% lateral diffraction limited resolution is achieved as compared to conventional microscopy because of the physical pinhole filtering. The system has the ability to measure steep edges, complex structures, or transparent coatings.

4.14.3 Specifications

Magnification: 10×, 20×, 50×, or 100×

Resolution (z direction): 50, 30/20*, 20/10*, or 20/5* nm

SCANNING LASER PROFILOMETRY

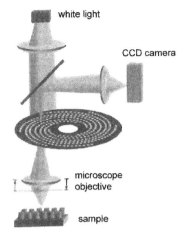

Figure 4.6 Schematic diagram of the NanoSurf confocal microscope. Courtesy of Solarius Development.

Field size: 1600 × 1510, 800 × 755, 320 × 302, or 160 × 151 um
Maximum stitching size: 121, 36, 5.7, or 1.4 mm × mm
Working distance: 10.1, 3.1/12**, 0.66/10.6**, or 0.31/3.4** mm
Numerical aperture: 0.30, 0.46/0.40**, 0.80/0.50**, or 0.95/0.80**

Note: *piezo, **long distance.

4.15 NANOMETRIC PROFILING APPLICATIONS

With high vertical and lateral resolution, laser profiling is well suited for the following measurement applications:

- Thickfilm—height, width, and area analysis
- IC-packaging—height, width, and volume calculation as well as warpage and coplanarity
- Tribology—roughness and 3D parameter of mechanical engineering components
- MEMS—form, roughness, and defect analysis
- Engineering surfaces—roughness, quality control, form, and coplanarity

4.16 SCANNING LASER PROFILOMETRY

The Laser Techniques Co. (LTC) uses laser-based profilometry (LP) sensors and probes operating on the principles of optical triangulation. The profiling probe takes a succession of single point distance measurements, acting as a "laser caliper," as the probe scans the target surface. The result is a 3D data cloud that can be displayed using Windows™-based computer graphics. Features such as denting, ovalization, corrosion pitting, etc. can be accurately mapped and analyzed using laser-based profilometry sensors.

4.16.1 Optical Principle

Optical triangulation relies on the use of a light source, imaging optics, and a photodetector as shown in shown in Figure 4.7.

The light source and focusing optics generate a focused beam of light that is projected on a target surface. The imaging lens captures the scattered light and focuses it on a single element photodetector. The photodetector generates a signal that is proportional to the position of the spot in its image plane. As the target surface distance changes, the imaged spot shifts due to parallax. To generate a 3D image of the test part surface, the sensor is scanned in two dimensions, thus generating a set of distance data that represents the surface topography of the part.

Because of their high brightness, monochromatic wavelength, and phase coherence, diode lasers, which operate in the visible range (650 nm), are typi-

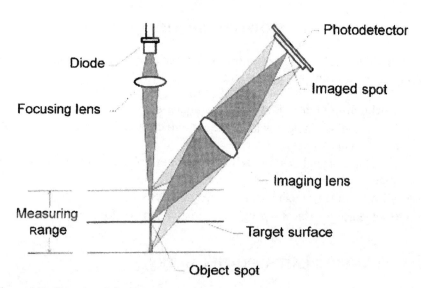

Figure 4.7 Functional diagram of scanning laser profilometer. Courtesy of Laser Techniques Company.

SCANNING LASER PROFILOMETRY

cally used in LTC applications. In most cases, customers are provided with Class 1 (eye safe) laser products. Both transmitting and receiving optics are typically custom designed multi-element lens systems. For both standard and custom probes, the optical system is customized to achieve the desired measuring range and resolution for the specified application. For some applications, standard optical components are used. For more demanding applications, proprietary optical designs and materials are used. Because of their small size and minimal need for local circuitry, the photodetector is usually a single-element *position-sensing device (PSD)*. When required, a *charged-coupled device (CCD)* or other detectors can also be furnished.

4.16.2 Probes

LTC uses two main types of LP probes, namely rotary and fixed, which are shown in Figure 4.8. Rotary probes are used for the inspection of components such as boiler and heat exchanger tubes. Standard probes can be used in tubes as small as 7 mm (0.275 in.) inner diameter. Fixed probes are intended for use with larger components such as pipes, flat surfaces, and complex structures. They are normally attached to robotic arms or other scanning mechanisms, such as the LTC 3D profiler.

4.16.3 3D Profiler

The 3D profiler provides high-resolution, noncontact maps of small parts. It can also be used for scanning replica molds to provide quantitative 3D images and documentation on difficult-to-measure features. The 3D profiler, shown in Figure 4.9, can be used with a variety of laser-profile (LP) sensors for appli-

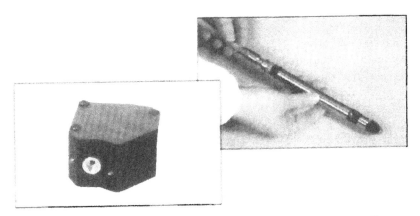

Figure 4.8 Fixed (left) and rotating (right) laser-based profiling sensors. Courtesy of Laser Techniques Company.

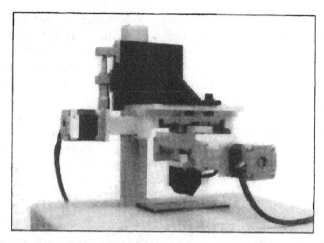

Figure 4.9 3D profiler for small parts. Courtesy of Laser Techniques Company.

cation customization. LP sensors are interchangeable and easy to install. Standard LP sensors are available with resolutions to 2.5 μm (0.0001 in.) and measurement accuracy better than ± 10 μm (0.0004 in.). Laser spot size can be less than 0.025 mm (0.001 in.), providing very high spatial scan resolution. Information can be exported to a number of other software packages compatible with this laser technique.

4.16.4 LP-2000™ Control Unit

The LP-2000™ inspection system rapidly, accurately, and quantitatively maps the surfaces of tubes, pipes, flat plates, and complex surfaces. The system uses Windows™-based control and analysis software to provide a variety of options for motion control, data acquisition, analysis, and display.

The system, shown in Figure 4.10, is modular, flexible, expandable, and can be used with any commercial or custom probe delivery system. This permits a variety of probe pushers, pipe crawlers, and custom manipulators to be used, thereby minimizing costs. Probes can also be used with complimentary eddy current and ultrasonic technologies.

The system is driven by proprietary software under license from UTEX Scientific Instruments. This software provides a wide array of options for data acquisition, display, and analysis. Inspection results can be displayed in contour, axial (side view), and polar (cross-section) plots. Operators can display two or more data formats at the same time or superimpose two cross-sectional images and quantitatively compare them. Figure 4.11 shows analysis software displaying laser-profiling data, representing a tube with two dents.

SCANNING LASER PROFILOMETRY

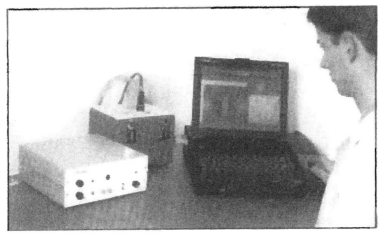

Figure 4.10 The LP-2000 laser profiling system is compact and portable. Courtesy of Laser Techniques Company.

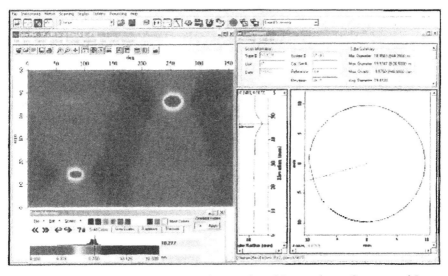

Figure 4.11 Analysis software displaying a tube with two dents. Courtesy of Laser Techniques Company.

4.17 LASER-SCANNED PENETRANT INSPECTION (LSPI™—PATENT PENDING)

LTC has developed an innovative way to obtain high-resolution images of features such as surface breaking cracks on critical part surfaces using a laser-based *fluorescent penetrant inspection (FPI)* method. With the LTC control unit, it is possible to get digital images of features on critical parts, including tubing, flat plates, or complex surfaces remotely and automatically. Cracks less than 1 mm (0.039 in.) can be located and mapped in near real-time.

With standard FPI techniques, liquid penetrant is applied to a clean part surface, which is allowed to penetrate and absorb into surface defects. Then the excess penetrant is rinsed from the surface and exposed to UV light, causing the trapped penetrant to fluoresce. Typically, a hand-held UV lamp is used to excite the penetrant. Observation of the fluorescent indications is then conducted either visually or with a CCD camera. LTC's laser-scanning FPI probe uses a focused ultraviolet laser beam and a single element photodetector in a confocal arrangement. This concentrates a large amount of laser energy on small features such as intergranular stress corrosion cracking. A complex crack found by LSPI™ is shown in Figure 4.12.

A conceptual diagram showing the LSPI™ is shown in Figure 4.13. The small ultraviolet laser diode operates at 375–405 nm to generate the excitation light beam. Violet and ultraviolet diodes represent a significant breakthrough in diode laser technology and have only recently become commercially available. Their high power and compact size result in very small package LSPI™ scanner designs.

The focused laser spot may be translated over a treated surface, either in a helical path (tube scan) or with a raster motion in the case of a flat plate scan. The high-resolution image is a function of the size of the focused laser spot and the data sampling density. With this configuration, focal spot sizes as small as 0.025 mm (0.001 in.) can be obtained, enabling operators to achieve high-resolution imaging of small cracks in part surfaces that might otherwise go undetected.

An LSPI™ sensor for scanning the bores of small-diameter tubing to detect cracks is shown in Figure 4.14. When the laser beam strikes a penetrant-holding defect on the part, it fluoresces at about 560 nm. This light radiates as a point source and is captured and collimated by the confocal lens. A high-pass filter removes any 405 nm reflected light from the collimated beam, and then a final lens focuses the 560 nm light on the surface of the photodetector. This optical configuration allows for the detection of very small amounts of fluorescence, resulting in signal-to-noise ratios that normally exceed 10:1 on a well-cleaned surface.

Once scan data have been acquired, the user can quickly manipulate and analyze the data using dynamic customizable color palettes and an extensive array of analytical tools. Operators can accurately measure feature geometry,

LASER-SCANNED PENETRANT INSPECTION (LSPI™—PATENT PENDING)   153

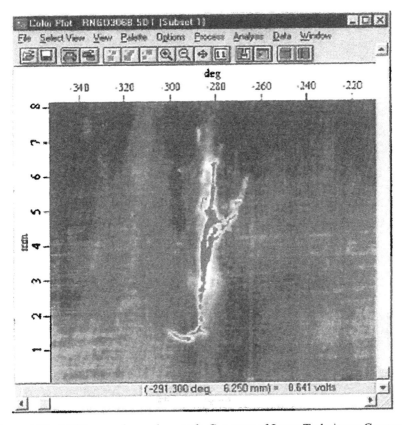

Figure 4.12 LSPI scan of complex crack. Courtesy of Laser Techniques Company.

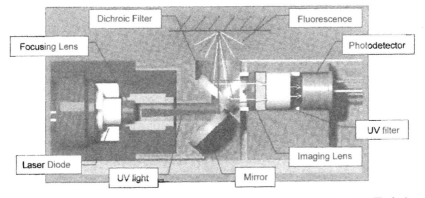

Figure 4.13 Conceptual layout of LSPI optics unit. Courtesy of Laser Techniques Company.

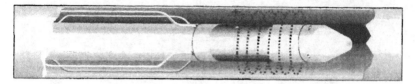

Figure 4.14 LSPI sensor scan of bore surface. Courtesy of Laser Techniques Company.

print out hardcopy images, and archive data for future analysis and comparison.

4.17.1 Applications

Applications include, but are not limited to:

- Safety critical tubing
- Aircraft engine parts
- Automotive parts
- Internal welds
- Internal threaded parts
- Internally finned or rifled tubes

4.18 ADVANCED TECHNIQUES

Advanced 3D-ESPI systems use several laser illumination directions or cameras to produce 3D information about displacement and deformation as well as contour information. From these data, strains, stresses, vibration modes, and many other values can be derived. A typical 3D-ESPI camera system is shown in Figure 4.15. In this case, the laser beam is split into illumination beams 1 and 2. The CCD camera views in-plane and out-of-plane reflections from the object surface and compares them to the reference beam, producing a 3D indication of deformation on the test object surface.

4.19 NATURAL AND EXTERNAL EXCITATION

Objects that are already pressurized, evacuated, heated, or subjected to internal or external vibrations are in an excited state. For objects at rest, environmental chambers, some as large as walk-in rooms, may also be used to subject objects to pressure, vacuum, moisture, and temperature variations within the chamber's design range.

Specific machines have also been designed to apply tension, compression, torsion, and bending forces to mechanical parts, thereby inducing stress and

STRAIN/STRESS MEASUREMENT 155

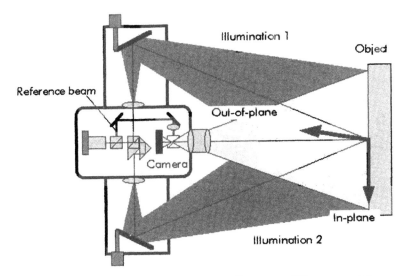

Figure 4.15 3D-ESPI camera system. Courtesy of Ettemeyer AG.

strain. Automated shaker tables and impact hammers also can be used to apply shock and vibration excitation to test objects. Lasers can be used to induce surface heat. See also Chapter 12 on vibration analysis.

4.20 STRAIN/STRESS MEASUREMENT

While working for Ettemeyer AG, team leader Dr. Lianxiang Yang, Ph.D. (Dr.-Ing.) and his team developed and improved a novel 3D *digital speckle pattern interferometry (DSPI)* sensor, initially known as MicroStar and later designated as the Q-100 Strain Measuring System by Ettemeyer AG. This sensor uses two laser diodes to successfully integrate both contour and deformation measurements in one compact instrument. The instrument has received wide acceptance in the automobile and aerospace industries.

4.20.1 Theory of Operation

The Ettemeyer Q-100, shown in Figure 4.16, is a compact video laser unit using 3D (ESPI), which provides a 1.4 by 1.4 in. field of view. The instrument allows in-plane strain gradients to be viewed and out-of-plane flexion to be measured. The instrument is placed on the part to be tested and the system gathers data like a strain gauge with the video detector providing about 0.5 million pixels, acting as individual strain gauges to image the strain. Little or no surface preparation is required. Since complete 3D information on the surface is recorded, tensile and bending components of strain can be separately calcu-

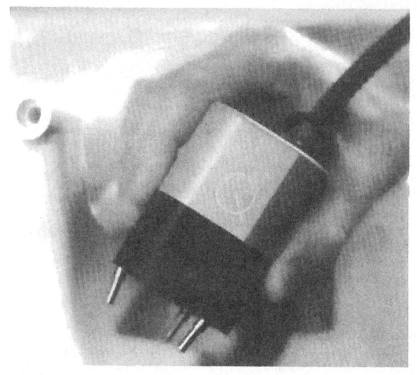

Figure 4.16 Ettemeyer Q-100 strain measuring system. Courtesy of Ettemeyer AG.

lated and displayed. Data collection time with this system is typically reduced by 80% when compared to convention methods of strain measurement.

4.20.2 Technical Data

Measuring sensitivity: deformation 0.03 to 0.1 μm; strain 5 to 20×10^{-6}
Measuring area: 35×25 mm ($1.4 \times 1.0''$)
Working distance: 27 to 40 mm (1.1 to 1.6'')
Spatial resolution: 0.5 mm
Sensor head dimension: $54 \times 54 \times 59$ mm ($2.1 \times 2.1 \times 2.3''$)
Sensor weight: 370 g
Length of sensor cable: 4.5 m (9 ft.)
Data output: various strain/stress options*
Data interface: TIFF, ASCII, Windows metafile, VRML

---

* To calculate stress information, material properties have to be input to the system.

Processor: Pentium II PC
Operating system: Windows 98/2000/NT

4.21 LONGER RANGE 3D SPECKLE INTERFEROMETRY SYSTEM

The Ettemeyer Q-300, shown in Figure 4.17, is a longer range, noncontacting 3D-ESPI system incorporating many of the features discussed above. The compact sensor head is portable and easy to set up. The system is best suited for testing complex-shaped components in the electronics, automotive, machining, and materials research industries. Researchers can also use the system to verify analytical and numerical calculation techniques.

The longer range 3D-ESPI system is symmetrically designed with interchangeable laser illumination arms to obtain different illumination base lengths and sensitivities. It can be operated in the 1D, 2D, or 3D mode without loss of accuracy. Zoom and macro lenses permit both reasonably large and very small areas to be analyzed. Windows-compatible software packages can be provided for automatic or manual measurement as well as full-field quantitative data analysis of displacement and strain fields. The ISTRA for Windows® software takes into account the varying sensitivity at different object coordinates for maximum measuring accuracy for every object point.

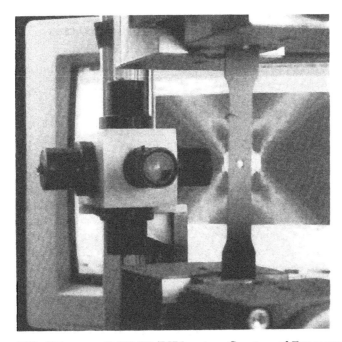

Figure 4.17 Ettemeyer Q-300 3D-ESPI system. Courtesy of Ettemeyer AG.

4.21.1 Technical Data

Measuring sensitivity: 0.03–1.0 µm, adjustable

Measuring range: 1–20 µm per measuring step with serial measurement

Measuring area: up to 200 × 300 mm (8 × 12"), larger areas with external laser

Working distance: variable, 0.1 to 1.5 m (4 to 60")

Operation modes: automatic, manual

Data interface: TIFF, ASCII, Windows metafile

Data acquisition speed: 2.5 seconds for 3D analysis

Data analysis: automatic serial analysis or manual at any loading step

Base length of illumination: minimum 120 mm (4.7")

Weight: 2.5 kg (5.5 lb.) for total sensor

Built-in laser: diode, 2 × 50 mW

Processor: Pentium PC

Operating system: Windows 98/2000/NT

4.21.2 Hardware and Software Options

- Extension of illumination arms for increased sensitivity
- Automatic geometry measurement for strain analysis on complex structure
- Interface to loading devices such as testing machines, etc.
- 3D-vibration analysis with stroboscopic illumination
- Glass fiber coupling of external NdYAG laser for illumination of large areas up to 1 m^2 (10 square feet)
- Interface to *finite element method (FEM)* software

4.21.3 Applications for 3D-ESPI Systems

Applications for these systems appear to be limited only by one's imagination and perhaps the cost/benefit ratio. An ESPI system provides information about deformation, displacement, strains, and stresses. With the appropriate software, the materials industry can use these systems to measure Young's modulus, Poisson ratio, crack growth, true strain/true stress functions and many other parameters used to describe new materials. Figure 4.18 shows the full-field strain distribution for a specimen in a tensile test.

In the automotive industry, ESPI can be used to analyze fatigue behavior of the auto chassis, power train, engines, gearboxes, wheels, and other components that are highly stressed items that are essential for automobile safety. High-speed measuring systems can provide dynamic material values obtained from crash tests and crash simulations.

Full-field strain distribution on a specimen in a tensile test

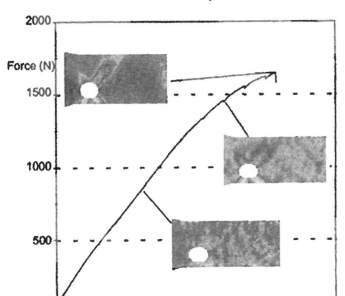

Figure 4.18 Force (N) vs. elongation (mm). Courtesy of Ettemeyer AG.

Noise vibration harshness (NVH) problems are solved by pulse ESPI techniques where a pulsed laser fires two laser pulses with variable time delays and one to three high-speed ESPI cameras record images. Measuring results show operation deflection that can be used to eliminate sound sources, optimize damping systems, eliminate break squeal, etc.

ESPI techniques are equally important in all transportation industries such as railroad, marine, and aerospace because of the advantages of full-field view, 3D presentation, and noncontacting measurement capabilities.

Other applications include:

- Investigations on aluminum alloys, 3D-ESPI
- Deformation and strain field analysis on spot-welded joints, 3D-ESPI
- Deformation and strain at teeth in biomedicine, 3D-ESPI
- Characterization of metallic alloys during tensile tests, 3D-ESPI
- Strain analysis on a T-branch pipe connection, 3D-ESPI and FEA
- Potential for displacement analysis on small areas, 3D-microscope-ESPI
- Strain distribution on knitted fabrics

- Stress distribution at laser weld seams
- Stress concentration analysis at fractures

Note: A low-cost, offline speckle-tracking-system (Ettemeyer Q-390) with almost unlimited microscopic and macroscopic measuring area and wide measuring range (nanometers to meters) for elastic, plastic, and rigid body shifts is also available. This standalone system uses a high-resolution CCD camera with zoom lens for data acquisition. Displacement vectors are shown on a user-selected grid of measuring points in the x- and y-direction. Data can be exported to standard calculation programs for postprocessing. Animated video displays of displacement are possible. Figure 4.19 shows the Q-390 system being used for high-speed crash test measurements. Calculated stress displacement vectors (arrows) are generally shown pointing down in this illustration.

4.21.4 Technical Data

Measuring area: microscopic and macroscopic, practically without limitations

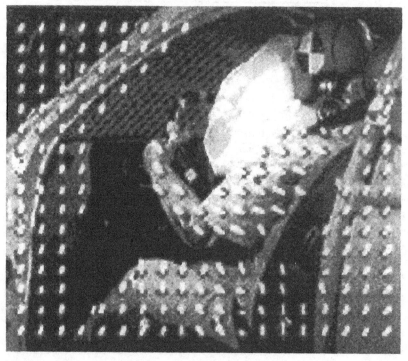

Figure 4.19 Q-390 system being used in high-speed crash tests. Courtesy of Ettemeyer AG.

Measuring range: unlimited (elastic, plastic, rigid body shifts) typically 1/10 of pixel distance, in special cases up to 1/100 pixel distance

Data recording speed: standard: video frequency, higher image acquisition rates with special equipment

Camera: standard, 768 × 576 pixel or high resolution (e.g., 1300 × 1030 pixel)

Optics: zoom lens (e.g., f = 12.5 to 75 mm, c-mount)

Features: without marking, full field, noncontact, high-resolution, multidimensional

Processor: Pentium PC

Operating system: Windows 95/98/2000, Windows NT

Options include:

- Mounting to tensile testing machines (U.S. patent no. 8/644.752)
- 3D measurement
- High-speed measurement

4.22 NONDESTRUCTIVE TESTING (NDT)

NDT and nondestructive inspection (NDI) is where the laser shearography method is most often applied. During the production of composite materials, many different components are bonded together during the manufacturing process. NDT and NDI techniques are used to verify product reliability and assure quality control following all manual assembly operations.

4.23 SHEAROGRAPHY

4.23.1 Principle of Laser Shearography

Figure 4.20 shows a schematic illustration of laser shearography operation. In this drawing a coherent laser beam illuminates the test object and a CCD shearing camera with appropriate optics registers the reflected light. The optics in the shearing camera doubles and shears the image of the object, producing an image known as a *speckle interferogram*, which is a unique space map of the surface. The interferogram can be made only when coherent divergent light is reflected from a diffuse surface.

By comparing interferograms from the surface at rest to an interferogram of a loaded or excited surface, a fringe pattern will be produced. Pressure, vacuum, heat, or vibration can be used to load the surface of the test object. Software typically provides image enhancement, analysis, and report functions. The deformation of each point on the full-view surface can be measured and recorded.

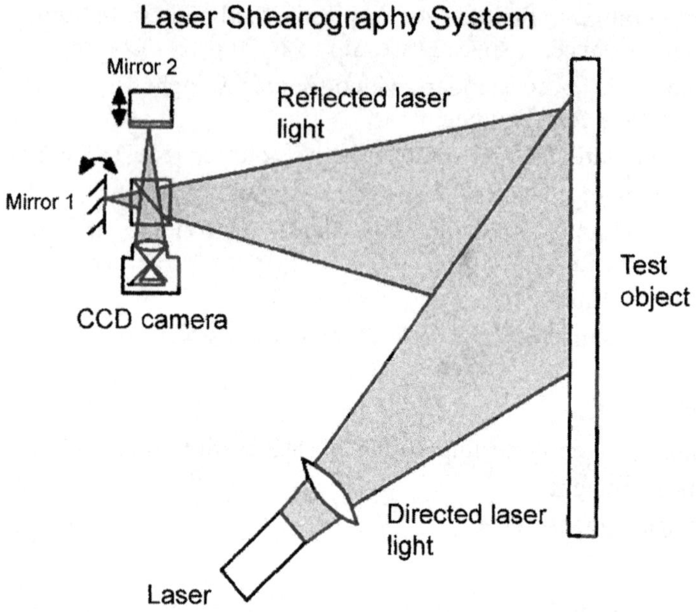

Laser Shearography System

Figure 4.20 Functional diagram showing operation of laser shearography system. Courtesy of Ettemeyer AG.

Advanced phase-stepping optical systems, used for aircraft inspection, increase the sensitivity and signal-to-noise ratio of the interferometry method. A phase-stepping optical configuration incorporates real-time phase stepping and phase calculation and display in a relatively compact portable vacuum head.

Phase-stepping shearography produces a warped phase map of measured surface deformation that needs to be processed further to extract continuous data pertinent to the surface flaws. The processing consists of nonlinear filtering and two-dimensional phase unwrapping. Additional filtering smoothes and corrects for local and global tilting and bending effects.

4.23.2 Compact Shearography System

The Ettemeyer Q-800 laser-shearography system is a fully portable compact measuring system for full-field, noncontacting nondestructive inspection of any material. It is especially useful for inspection of complex material and material combinations such as composites, fiberglass reinforced plastics, and bonded composite materials. The inspection readily shows delaminations, cracks, voids, impact damage, failed repaired areas, and other type of defects or stressed components. Various loading systems such as vacuum or pressure, heating, and vibration excitation can be used to stress or excite the components being inspected.

SHEAROGRAPHY

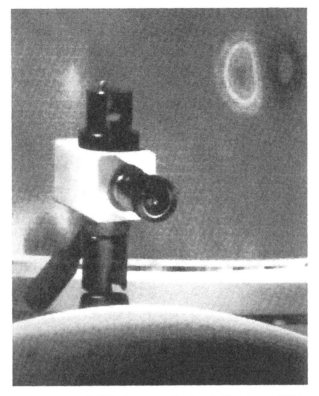

Figure 4.21 Ettemeyer Q-800 shearography head. Courtesy of Ettemeyer AG.

This laser-shearography system consists of a miniaturized shearing sensor with integrated high-resolution CCD camera and variable shearing system, an integrated powerful diode laser, and computer control system with analysis software ISTRA for Windows™.

The shearography sensor head, shown in Figure 4.21, either can be mounted on a tripod or, due to its light weight and size, the sensor head can be easily integrated in a fully automatic robotic production inspection system. The system can be operated in normal daylight conditions. For illumination of laser areas a NdYAG laser can be coupled to the sensor head with fiber optics cables. The system can inspect areas up to $1\,m^2$ (10 sq. ft) and deliver quantitative data on deformation and/or stress distribution on the surface of the component.

4.23.3 Technical Data

CCD-resolution: 768 × 582 pixels
Standard C-mount lens: zoom f = 1.4/6 ... 12 mm

Shear angle: progressive adjustable 0.3°
Shear direction: progressive adjustable 360°
Measuring area: up to 1.2 × 0.8 m (with external laser)
Measuring sensitivity: 0.03 m/shear distance
Sensor head dimension: W × H × D = 60 × 60 × 120 mm
Sensor head weight: 850 g, including zoom lens
Laser: diode 50 mW

Contents in Standard System

- Portable shearography sensor head
- Illumination with diode laser (50 mW)
- Control electronic (PC)
- Software package for ISTRA for Windows™

Options Include

- Motor for pan/tilt of tripod
- NdYAG laser up to 5 W
- Loading systems (vacuum chamber/hood)
- Robotic inspection system
- Solutions for special inspection purposes

4.24 PORTABLE SHEAROGRAPHY SYSTEM

The Ettemeyer Q-810 portable shearography system is a compact inspection system for fast inspection of components made from composite structures like aircraft fuselages and components, rocket shell parts, marine ship hulls, pipes, etc. Figure 4.22 shows a crack in a CFRP aircraft panel (upper left), impact damage to a rudder (upper right), the complete Q-810 system including transportation box (bottom left) and a technician inspecting an aircraft fuselage (bottom right).

The Q-810 inspection system consists of a vacuum hood with flexible sealings, which can be attached to the component to be inspected. Inside the hood a vacuum up to 150 mbar can be pulled in order to stress the material underneath the hood. The integrated shearography camera will detect any debonding or structural defect. The camera image is analyzed by ISTRA for Windows™ and displayed on a computer monitor and heads-up display.

4.24.1 Technical Data

Inspection area: ca. 70 sq. ft
Typical pressure difference: up to 100 mbar

PORTABLE SHEAROGRAPHY SYSTEM

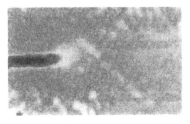

Crack in CFRP-panel

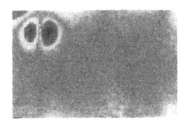

Impact damage in rudder

Complete system including transportation box

Portable Shearography Inspection System Q-810 in use at fuselage testing

Figure 4.22 Defects found and equipment used for inspecting aircraft. Courtesy of Ettemeyer AG.

CCD resolution: 768 × 582 pixels
Measuring sensitivity: 30 nm/shear distance
Portable hood dimensions: W × D × H = ca. 400 × 200 × 315 mm
Portable hood weight: 7 kg
Control rack dimension: W × D × H = ca. 540 × 405 × 835 mm
Control rack weight: 70 kg
Option: heads-up display

4.24.2 Other Applications

The aerospace industries use shearography systems to test composite materials such as fiberglass-reinforced plastics (FRP), carbon fiber–reinforced plastics (CFRP), GLARE, foam and aluminum composites, etc. GLARE is a new aircraft plymetal material consisting of one or more layers of strong fiberglass sandwiched between layers of aluminum and bonded together. Each layer in the sandwich is about one millimeter thick. GLARE is safer than aluminum alone in the advent of a fire, more resistive to damage, and less sensitive to metal fatigue.

Fully automatic inspection systems have been developed for Ariane 5 equipment as well as helicopter rotor blade inspection. Ariane 5 is a satellite launch vehicle for commercial satellites. For maintenance inspections, portable shearography inspection systems have been developed using vacuum loading or heat loading to reveal defects. Recently, shearography has been validated for maintenance repair of Concorde parts. Pratt & Whitney jet engine abradable seals are also inspected with laser shearography systems using vibration excitation.

In the automotive industries, tire testing and dashboard inspections are major laser testing and inspection applications.

4.25 FELTMETAL INSPECTION SYSTEM

The Ettemeyer Q-830 feltmetal inspection system is designed for rapid inspection of abradable seals in jet engine components. The system is used for production and maintenance inspection of Pratt & Whitney type PW-4000 engines. The system inspects the soldered bond between the felt metal insert and the base structure of the engine segment. Each segment is inspected in an automatic process within a few seconds. A clear picture of the test results is displayed on a large screen and printed out in the test report. Figure 4.23 shows the feltmetal inspection system and typical debonding defects.

4.25.1 Setup and Technique

The Q-830 uses the technique of laser speckle interferometry to detect debondings. A high-frequency vibration excitation shows debonded areas in a second. The feltmetal segment is placed on a positioning ring in the working platform of the system. For different stages of the engine, different positioning rings are used. In each measuring position the felt metal segment is excited by a piezo shaker system in a predefined frequency range and the vibration amplitudes are measured with the shearography system. In case of missing adhesion between the feltmetal and base material, the unbonded area will start to vibrate and show typical amplitude on the video image of the shearography system.

4.25.2 Technical Data

Test procedure: according to Pratt & Whitney, "Nondestructive Test Method Specifications BTM-15"
Test object: felt metal rings, diameter 700 to 1000 mm
Diameter of test object: 700 to 1000 mm
Height of test object: 100 to 200 mm
Weight of test object: maximum 30 kg

FELTMETAL INSPECTION SYSTEM

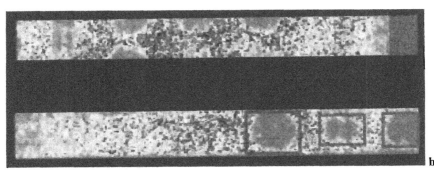

Figure 4.23 (a) Feltmetal inspection system for rapid inspection of abradable jet engine seals. (b) Inspection result with edge bondings (top) and center debondings (bottom): inspection time 1 second. Courtesy of Ettemeyer AG.

Inspection area: 30° segment, 2 seals simultaneously

Inspection speed: less than 1 second (online)

Total inspection time: 3 minutes (12 segments including handling and report) with shearography image

Dimension: L × W × H = 2000 × 1500 × 1600 mm

Weight: 700 kg

Power supply: 200 V/50 Hz, others on request

Air supply: 5 bar

Laser (built-in): diode Class I laser system

Processor: Windows 98/2000/NT: Pentium PC

Operating system: Windows 98/2000, Windows NT

4.26 OPTIONAL APPLICATIONS

- Similar designs are available for inspection of all types of bonding.
- See our robotic inspection systems for inspection of components with complex geometries.

4.27 OPTICAL INSPECTION SYSTEMS

4.27.1 ARAMIS

ARAMIS technology provides a unique device for noncontact measurement of displacements in material in full 3D, using high-resolution dual-video 3D image correlation, measuring deformation and strain from 100 microstrains through plastic deformation. This technology is nearly as sensitive as ESPI, but is substantially more robust, operates from static conditions to high speed, and has a much greater dynamic range. The ARAMIS HR system and a typical strain pattern are shown in Figure 4.24.

ARAMIS is Trilion Quality Systems' powerful new noncontacting, material-independent tool for the determination of deformation and strain. 3D video correction methods and high-resolution digital CCD cameras are used for:

- Testing materials and material characterization of metals, composites, ceramics, biomechanics, fabrics, films, etc.
- Finite element method (FEM) confirmation and boundary condition checking
- Fracture mechanics
- Estimating stability
- Dimensioning components using 3D coordinates

OPTICAL INSPECTION SYSTEMS 169

ARAMIS (3D) Image Correlation

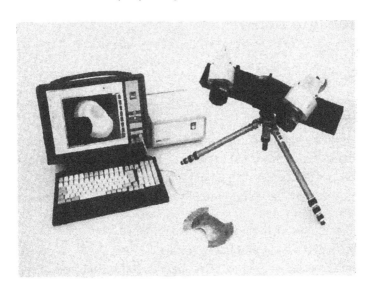

3-D Deformation Strain and Stress Patterns

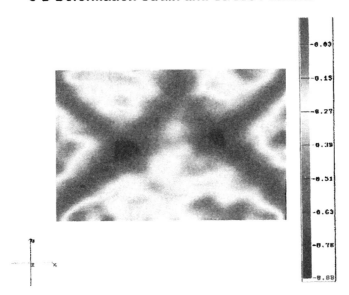

Figure 4.24 3D deformations—strain and stress measurement (by GOM). Courtesy of Trilion Quality Systems, LLC.

- Examining nonlinear behavior
- Characterizing creep and aging processes
- Determining residual sound stress
- High-speed deformation and strain measurements

4.27.2 Industry-Specific Applications

- Aerospace—composite and advanced material studies, on-aircraft structural load studies
- Automotive—manufacturing quality analysis and FEM confirmation
- Microelectronics—thermal expansion, PCB NDT and alignments
- Photonics—optics train alignment analysis
- Composites manufacturing—quality control, failure analysis, material and structural studies
- Metal forming—forming limit measurements
- Biomechanics—bone and tissue strains
- Civil engineering—road material analysis, soil loading, etc.

4.27.3 Measuring Procedure

One or more CCD cameras view the object under load. A random or regular structure marking is applied to the object's surface that deforms along with the object. The CCD cameras record the deformation of the structure under different loading conditions. The initial image processing defines the macro-image facets. These facets are tracked in each successive image with subpixel accuracy to 0.01 pixel. Using photogrammetric principles, the 3D coordinates of the surface of the specimen, which are related to the facets at each stage of the load, can be precisely calculated.

4.27.4 Measurement Results

On the basis of the 3D coordinates, the 3D displacements, the strains, and shape of the specimen can be calculated with a high degree of accuracy and resolution. The results are graphically displayed and the visualization is outputted in TIFF or JPG graphic formats. Further the measured results can be exported in a user-defined ASCII format.

4.27.5 Measurement Advantages

- Simple preparation of the specimen—ARAMIS works on both random and regular patterns, thus simplifying the preparation of the specimen.
- Large measuring area—the same sensor measures both small and large object (1 mm to 1000 mm) and strains in the range of 0.05% to several 100%.

OPTICAL INSPECTION SYSTEMS

- Full-field graphic results—the results comprise a large number of data points. The high data point density and graphical display of the results provide a better understanding of the component behavior.
- Mobility—due to the simple tripod camera system, the compact measurement setup, and highly developed analysis software, the system is highly efficient and flexible.

4.27.6 Comparison of ESPI and 3D Image Correlation

ESPI systems illuminate the test object with one or more coherent beams to generate a granular pattern that is dependent on surface microstructure. The microstructure changes as the surface deforms and the relative deformation can be determined by comparing images captured before and after the deformation.

The advantage of the ESPI system is that it is sensitive to displacements of about 0.1 microns. However, the ESPI system is also sensitive to equally small ambient vibrations within the measuring environment such as thermal currents. The dynamic range of ESPI is severely limited because each speckle on the measuring surface must be directly compared. Combined movements of more than one speckle diameter of the layer, object, and camera are not tolerable. Measurements at multiple load steps can be combined but the upper limit of deformation per step is a few microns. In addition, the FOV is limited by laser diode power. The typical maximum FOV is about 6×8 inches.

Pulsed ESPI systems overcome environmental problems but cost more and are relatively large. Pulsed 3D-ESPI systems calculate the in-plane deformations based on comparisons of out-of-plane images from three sensors. An imperfection, such as a glare spot or unresolvable fringe, can cause significant local in-plane deformation errors.

3D image correlation systems based on photogrammetry use ordinary light instead of coherent laser light. The system tracks a stochastic pattern applied to the measurement surface with subpixel accuracy. This means that as long as the test object remains in the field of view of the cameras, the deformations can be tracked. Therefore large deformations, up to 10 millimeters, can be tracked and analyzed by a single measurement. Massive rigid body motions have no effect but they can be calculated based on the original pixel measurement. Measurements can be completed after the part has been removed, processed, and replaced within the camera's viewing zone. Similarly, large strains exceeding 100% can be easily tracked.

The minimum sensitivity of 3D image correlation with 1.3 megapixel cameras is 1/30,000 the field of view. With a 3 cm field of view, sensitivity is 1 micron, and with a 30 cm field of view, sensitivity is 10 microns. Fields of views up to several meters are possible as long as deformations of several 10s of microns are present. The system intrinsically measures 3D shape and therefore 3D deformations are measured simultaneously, rather than sequentially. Pulsed high-intensity illumination provides dynamic deformation analysis with results superior to pulsed ESPI. The complete deformation measurement occurs at up to 20 Hz with standard cameras or 20,000 fps with standard high-speed cameras or down to 20 ns frame duration with pulse illumination or framing cameras.

Note: For typical tensile tests a two-layer stochastic pattern is easily applied. White dye penetrant is used for the first layer. Black spraypaint with 50% coverage is used for the second layer.

4.27.7 ARAMIS HR Specifications

Measuring area: mm^2 up to >m^2

Camera resolution: 1280 × 1024 pixels, 12-bit digital or 2000 × 2000 pixels

Exposure time: 0.1 ms up to 1 s, computer operated and asynchronous trigger (pulsed illumination to 6 ns)

Dimensions: 500 × 190 × 125 mm

Weight: 3.5 kg

Strain measuring range: 0.02% up to >100%

Results: 2D/3D displacement fields, strains, and contour

4.28 ARGUS

ARGUS is an advanced form of optical photogrammetry that measures and visualizes the quality and degree of the stamping process on sheet metal with high precision. For measurement, the sheet is marked with a rectangular grid of dots, using chemical etching, laser marking, or other marking technique. The sheet is then stamped or formed by hydroforming, magnetic forming, etc. For the measurement, one or two scale bars and some coded markers are placed directly on the part or close around it. A digital CCD camera is then used with real-time viewing of the actual view on the PC to take shots from different viewing directions. The ARGUS system and a formed panel with defect are shown in Figure 4.25.

After image acquisition, ARGUS defines in each image the exact center point of all marked dots to within a hundredth of a pixel. Then, using photogrammetry techniques, the images are virtually assembled to represent the object as a 3D model. From this virtual assembly, the center of each marked dot on the object is defined in true 3D coordinates. These calculated 3D points define the exact form of the stamped sheet part.

From the local distortion of the regular grid pattern that was applied on the flat sheet metal, the local strain introduced by the stamping process is calcu-

Forming Analysis

ARGUS System

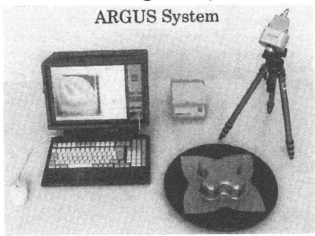

Major Strain Display

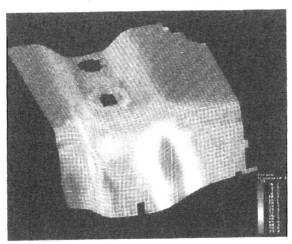

Figure 4.25 Forming analysis with ARGUS System. Courtesy of Trilion Quality Systems, LLC.

lated in many thousand local data points. These strain values define the degree of forming, which are typically viewed as major strain (principle strain 1) and minor strain (principle strain 2), and especially thickness reduction. These values are graphically displayed and validated in relation to the *forming limit curve (FLC)* of the selected material or in relation to the maximal allowed reduction of the sheet thickness by the stamping process. The accuracy of the strain values is better than ±0.5%, thus also hardening investigations on parts with a low forming degree (car hoods and roofs) can be done.

For ARGUS measurement of a simple stamped sheet metal part, only a few images are taken and a typical measurement to validate the stamping process is finished in a few minutes. To measure the stamping quality of complex and large parts, additional views are needed to cover the complete surface and all of the detail areas. For these applications, ARGUS accepts and integrates many images into a virtual assembly (3D model) and is therefore able to measure the complete range of formed shapes used in sheet metal stamping. Even cylindrical form hydroforming or other processes are ideal applications for the technology. With an additional software module, the ARGUS data can be directly compared to the 3D CAD drawing.

Trilion/GOM measurement systems are used daily in many steel and aluminum producing companies and by automotive enterprises and their suppliers. The systems deliver exact and relevant data of the quality and degree of the stamping process.

4.29 TRITOP

4.29.1 Photogrammetric Offline System

Two different concepts are currently employed:

1. Analog to the tactile measured point layout, where retroreflective targets are stuck onto the object features so that the feature positions and orientations can be calculated from their coordinates.
2. Corner, edge, and bore-hole adapters are provided with measurable retro targets that are placed on the object features. The coordinates of the retro targets automatically lead to the coordinates of the object features.

4.29.2 Measurable Object Size

The measurable object size is typically in the range of 0.5 to 10 meters. Depending on object size, measurement accuracy is in the range of 0.02 to 0.1 mm. The system can handle thousands of object points per measurement without any problem.

4.29.3 Digital Photogrammetry Functionality

Photogrammetry has been successful for many years in the fields of land survey, architecture, ship-building, and aircraft construction. Modern computers and digital cameras that now form the basis of complete systems increase its significance and acceptance. The offline system TRITOP operates by performing the following steps:

1. Object features are marked as described above.
2. The object is recorded with a digital reflex camera from different views.
3. The images are transferred from the digital camera to the notebook computer via the PCMIA hard disk.
4. The image coordinates of the adhesive retro targets are automatically measured with high precision.
5. From the image coordinates the object coordinates are determined by calculating the intersection of the rays of different camera positions. Because these intersection points result from the intersection of several rays, the standard deviation of the measured value can also be given (beside the object coordinates).
6. During the calculation of the object coordinates, the positions and characteristics of representation of the cameras are determined simultaneously.
7. Finally, the measured data are presented in standard data format.

4.30 SYSTEM ADVANTAGES

With TRITOP it is possible to record the actual 3D state of the object. Then the recorded state can be compared to original object geometry from a sketch or CAD data. The system provides a simple, easy-to-use, economical way to capture the deformation of an object under load by recording two epochs. In many cases, the system can replace traditional measuring equipment such as mechanical 3D-coordinate measuring systems, theodolite systems, or dial gauges.

In modern surveying, a theodolite is a telescope pivoted around horizontal and vertical axes so that it can accurately measure both horizontal and vertical angles. The angles are read from circular plates calibrated in degrees and smaller intervals of 10 or 20 minutes. A vernier or micrometer is used to provide more accurate readings of the larger graduations.

4.31 PORTABLE MEASURING SYSTEM TECHNIQUE

The portable measuring system consists of the following:

- Digital reflex camera with PCMCIA hard disk
- Notebook
- Scales, adhesive retro targets, and adapters

TRITOP Camera

Adhesive Retro Targets

Figure 4.26 Tritop camera and adhesive retro targets. Courtesy of Trilion Quality Systems, LLC.

The TRITOP camera and adhesive retro targets are shown in Figure 4.26.

The 3D coordinates of the adhesive retro targets, stuck on the object or specific adapters, are determined with high accuracy. The coordinates of these points and adapters finally lead to the coordinates of the selected object features to be studied.

The object marked in this way is recorded by a reflex camera from different predefined views freehanded and sequentially. The images are then stored on the internal PCMCIA hard disk. After all the images are taken, the PCMCIA hard disk is inserted into the corresponding slot of the notebook. The TRITOP software calculates the 3D coordinates automatically from the image coordinates of the retro targets.

The coordinate system is fixed according to the 3-2-1 rule (analog to tactile coordinate measuring technique). Using a highly precise scale bar, placed beside the object, the measurement is properly scaled. For postprocessing purposes, the obtained data can be exported to a number of standard data formats.

The TRITOP 3D measuring technique provides a fast, accurate, and economical method for determining the coordinates of object features like corners, edges, bore-holes, etc.

4.32 DYNAMIC TRITOP

A two-camera system tracks deformation and 3D coordinates in real time for systems analysis, engine dynomometer studies, and wind tunnel measurements. High-speed cameras facilitate the tracking of high-speed events from car crashes to explosive impacts.

4.33 OTHER LASER METHODS

4.33.1 Measurement of Hot Spots in Metal/Semiconductor Field-Effect Transistors

NASA's Jet Propulsion Laboratory (JPL) in Pasadena, CA has developed a technique for measuring hot spots in high-power gallium arsenide metal/semiconductor field-effect transistors (MESFETs). Hot spots in these transistors are undesirable because they adversely affect transistor operation and life. The NASA system consists of a helium-neon (He-Ne) laser, beam expander, mirror, chopper, filter, focusing lens, and IR spectrometer, shown in Figure 4.27. This technique has a spatial resolution of about 0.5 μm compared to former systems having a 15 μm spatial resolution.

The technique uses a modified apparatus known as the *Microelectronic advanced laser scanner (MEALS)*. The He-Ne laser beam is aimed at the tran-

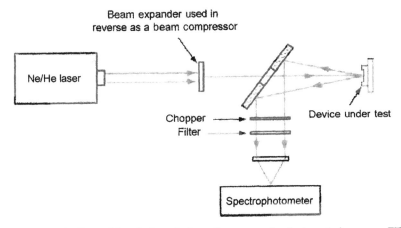

Figure 4.27 Jet Propulsion Lab technique for measuring hot spots in power FETs.

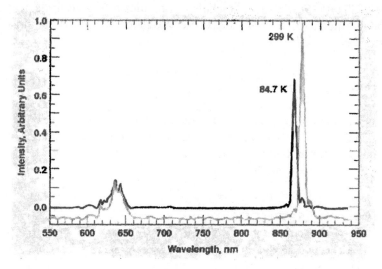

Figure 4.28 JPL study of FET junction temperature on wavelength.

sistor's gate using a micromanipulator stage having a resolution of 0.1 μm. Light emitted from the illuminated spot on the MESFET gate is reflected back to the mirror. Light reflected by the mirror is then chopped, filtered by a low-pass filter, and focused on the cooled photomultiplier tube detector of the IR spectrometer. The low-pass filter blocks the laser light emitted at its normal wavelength. The output signal from the PM tube is locked and synchronized by an internal chopper.

This technique relies on the temperature dependence of the wavelength of the peak of the simulated IR emission spectrum. Peak wavelength varies almost linearly with increasing temperature as the energy band gap of the transistor decreases with temperature. Therefore, the gate temperature of the MESFET can be determined at different operating power levels. The curves shown in Figure 4.28 were taken with no power applied to the MESFET at temperatures of 84.7 and 299°K. As shown, the peak wavelength at the higher temperature exceeded that of the lower temperature by about 10 nm.

5
LEAK TESTING METHODS

5.1 INTRODUCTION

Leak tests are required to assure that hermetically sealed electronic parts, valves, high-pressure tubing, and piping systems and welds do not leak. An item cannot be guaranteed to be "leak-proof" if it has not been tested. The four general classes of systems commonly leak tested are listed here:

1. Hermetic enclosures and components are tested to prevent the entrance of contamination or preserve internally contained fluids. Examples are electronic devices, integrated circuits, sealed relays and motors, pulltab can ends, and connector multipin feedthroughs.
2. Hermetic systems are leak tested to prevent the loss of contained fluids or gases. Examples include hydraulic and refrigeration systems. The chemical and petrochemical industries also do extensive leak testing to assure "no leak" integrity of their plant valves, piping, and vessel systems.
3. Evacuated enclosures and components are leak tested to assure that there is not too rapid a deterioration of the vacuum system with time. Examples would include electronic tubes, including TV picture tubes, sensing bellows, and vacuum-packaged items.
4. Vacuum systems are tested to assure that leakage has been minimized so that optimum gas removal can be achieved at any given vacuum (absolute pressure) rating.

Introduction to Nondestructive Testing: A Training Guide, Second Edition, by Paul E. Mix
Copyright © 2005 John Wiley & Sons, Inc.

5.2 FUNDAMENTALS

Leak rates are measured in atm cc/s units. One atm cc/s is one cubic centimeter of gas per second at a pressure differential of one standard atmosphere (760 Torr at 0°C). One Torr equals 1 mmHg; units of Torr are commonly used in vacuum work.

Over the years, many methods have been developed for detecting and measuring leaks. We can detect leaks in the order of 10^{-1} atm cc/s or larger with our ears. We can detect bubble-forming leaks at 10^{-4} atm cc/s with our eyes. Smaller leaks require more sophisticated methods of measurement and detection.

In order to put the subject of leak testing in proper perspective, it should be pointed out that most leaks at joints are about 5×10^{-7} atm cc/s or about 1 std cm^3/month, a very small leak rate. This average leak rate applies to leaks at ceramic-to-metal seals, plastic-to-metal seals, soldered, brazed, and welded joints. The moisture received from human breath easily plugs leaks of this size for days and atmospheric particles can permanently plug them. These facts also help point out the care that must go into the manufacture of reproducible permeation-type standard leaks that are used for calibration purposes.

5.3 ULTRASONIC LEAK TESTING

Ultrasonic leak testing has been specifically developed for detecting gas leaks in high-pressure lines. High pressure is defined differently in different industries; high pressure can be as low as 50 psig or as high as 50,000 psig. Depending on the nature of the leak, escaping gas produces ultrasonic sound that can be detected with an approximate sensitivity of 10^{-3} atm cc/s.

5.3.1 Ultrasonic Leak Detectors

Ultrasonic leak detectors have been accepted by a number of industries for detecting gas leaks in internal combustion engine valving and piston blow-by, gaseous piping, and ducting; air brake systems; bearings; seals in refrigerated van bodies, clean rooms, and air ducts; various hydraulic components, and other items. Leaks can also be detected in overhead piping systems and at other locations from a considerable distance. The ultrasonic leak detector has the advantage that it is not sensitive to audible background noise.

The hand-held probe and meter or headphone can be used for locating leaks in most air seal applications. After inflation, tires can be swept with the hand-held probe to find leaks before they cause highway downtime later. It is relatively easy to detect air escaping from a pressurized system.

However, for the detection of leaking intake valves, headphones are typically used with the engine running. All valves should emit a similar sound pattern when the contact probe is placed on the intake manifold opposite the intake port. The valve or valves out of pattern can be checked when the engine is not running, by bringing that cylinder to full compression and placing the probe on the valve stem. A hissing sound across the valve seat indicates that the valve is leaking. When ball or roller wheel bearings are in good condition and adequately lubricated, they normally produce a soft whirling sound when using headphones. Lack of lubrication produces scraping sounds and flat spots or nicks on bearings and races produce grating, grinding, or clicking noises. These are just a few truck and automotive applications. When audible background noise is a distraction to the operator, the contact probe and headphones should be used.

Ansonics Son-Tector 110M uses an ultrasonic microphone and associated electronics, which is sensitive in the frequency range of 35,000 to 45,000 Hz. These signals are amplified by self-contained circuitry and converted either to sounds that can be heard or a reading on a meter.

Figure 5.1 shows the Son-Tector 110M equipment. Equipment includes:

- The small leather-cased *basic unit* with built-in meter, probe, and headphone connectors.
- The *hand probe* that hears airborne leak or malfunction sounds and transmits them to the basic unit.

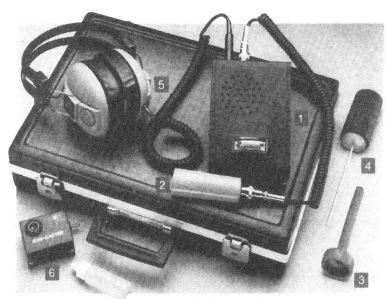

SON-TECTOR 123 package
1. main unit 110M in case
2. hand probe
3. sound concentrator
4. contact probe
5. headphones
6. ultrasonic noise generator

Figure 5.1 Modern ultrasonic leak detector. Courtesy of Ansonics, Inc.

- The *sound concentrator* consisting of a small rubber funnel that can be placed on the hand probe to reduce the sensitive area to about ¼-in. in diameter.
- The *contact probe* to pick up sound conducted inside a pipe, gear, or bearing housing.
- *Headphones*—these are essential in noisy industrial or commercial environments.
- The *Son-Caster II-S ultrasonic noise generator*, which is used to assist the operator in finding nonpressure leaks. The unit can be placed inside a tank, vehicle body, or watertight compartment to fill the area with ultrasonic sound that can be detected with the basic unit.

Figure 5.2 shows the sensitivity of the Ansonic ultrasonic detector to leaks ranging in size from 0.001 to 0.010 inches in diameter as a function of

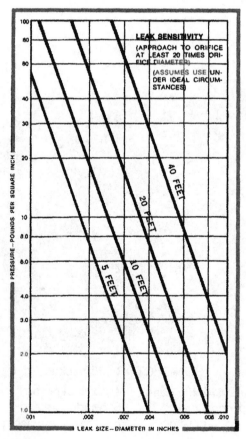

Figure 5.2 Son-Tector leak sensitivity. Courtesy of Ansonics, Inc.

pressure (psi) on a vertical log scale at various distances of 5 to 40 feet from the leak under ideal conditions. The ultrasonic leak detector can be an economical alternative to provide an early warning for many leak and bearing problems.

5.4 BUBBLE LEAK TESTING

One of the simplest and most economical tests is the water immersion or air bubble observation test. Bubble leak tests are frequently performed in pressurized instrument and process tubing systems to detect costly air, nitrogen, and oxygen losses to the atmosphere. In pressurized process lines and vessels, leak detector solution can detect hazardous (flammable, explosive, or toxic) leaks to the environment. Generally, these pressurized systems must be shut down, depressurized, and drained before the leaks can be corrected. After depressurization, leaking valves, flanges, gaskets, or tubing fittings can be replaced. Attempts to tighten leaking fittings under pressure can result in a blowout in the face of the operator. Bubble leak testing is more qualitative than quantitative. Small leaks produce numerous small bubbles, but it is difficult to determine the actual leak rate. The sensitivity of bubble leak tests has been estimated at 10^{-3} to 10^{-4} atm cc/s.

5.5 DYE PENETRANT LEAK TESTING

Dye penetrant also provides an economical leak testing method. Special dyes can be sprayed on the high-pressure side of suspected leak areas. If a leak is present, the differential pressure of the system will cause the dye to seep through the leak and appear on the low-pressure side of the object. This method can take an hour or more for a leak test sensitivity of 10^{-4} atm cc/s. Because of the long time involved, this test method is infrequently used.

5.6 PRESSURE CHANGE LEAK TESTING

Pressure change leak tests are performed to determine if acceptable leak rates exist or if hazardous conditions exist, and to detect faulty components and equipment. By knowing the volume and pressure of the pressurized system, and being able to time pressure changes due to leaks, a relatively accurate indication of leak rate can be obtained.

Some advantages of the pressure change measurement test are:

- Total leakage rate can be measured on evacuated or pressurized systems.
- Total leakage rate can be measured on any size system.
- No special tracer gas is required.

5.7 HELIUM MASS SPECTROMETER LEAK TESTING

Many people consider the helium mass spectrometer leak detector (MSLD) the most versatile of the industrial and laboratory leak detector methods. Developed as a method to meet the strictest requirements in the development of nuclear devices in the early 1940s, these methods have proved useful in many other applications.

Helium is an inert, nontoxic, noncondensable gas that is plentiful and relatively inexpensive. It is also a very small molecule and light, and therefore easily slips through very small leaks. Fortunately, there is only a low concentration of helium naturally present in the atmosphere (about 5 parts per million), so normally occurring background levels are manageable.

Nothing is considered absolutely leak-tight and light gas molecules can permeate many so-called "solid materials." In fact, in ultrahigh vacuum a major source of gas is hydrogen permeating the main body of a steel or aluminum vacuum vessel.

Helium MSLD leak rates are expressed in terms of flow rates such as atm cc/second, mbar liters/s and Pascal liters/s. Levels of sensitivity range from 10^{-1} down through 10^{-11} atm cc/s. In terms that may be more intuitive, a 10^{-1} atm cc/s leak rate equals the loss of 1 cubic centimeter of gas at atmospheric pressure every 10 seconds, while a 1×10^{-9} leak indicates a loss of one cubic centimeter of gas at atmospheric pressure over 30 years.

Due to the extremely sensitive nature of helium mass spectrometry, care must be taken to distinguish between real leaks that transgress wall or barrier, and permeation of helium through many elastomers (o-rings, gaskets, seals) in a given part or system. Generally speaking, real leaks will appear and decay fairly quickly, in a matter of seconds, whereas permeation takes many minutes to occur and remains somewhat constant before slowly degrading.

5.8 MASS SPECTROMETER LEAK DETECTOR

The maximum sensitivity of a mass spectrometer leak detector (MSLD) is determined by the minimum mass flow of helium tracer gas that produces a measurable signal. This can be limited by the level of ambient helium (background), electronic noise, and/or the minimum capabilities of the detector to read current imparted on it by the ion beam. The sensitivity of the detector when using a "sniffer" probe in pressurized volumes is orders of magnitude less sensitive than when the leak detector is used to evacuate parts due to helium background entering the sniffer probe.

5.9 MSLD SUBSYSTEMS

Helium mass spectrometer leak detectors are available in many different brand configurations. All of them, however, consist of three major subsystems, namely the spectrometer tube, the vacuum system, and the electronics required for operating the system.

5.9.1 Spectrometer Tube (Figure 5.3)

The spectrometer is a device that acts as a partial pressure gauge for helium. It consists of an ion source, a magnetic sector, and a detection device. There may also be some type of gauge to measure the total pressure or vacuum level in the spectrometer and used to protect against overpressure. Though all of these components can differ from one manufacturer to another, they operate in a similar manner. The ion optics is designed in such a way as to separate helium (atomic mass 4) from the other adjacent masses. Some mass spectrometers can be tuned to detect hydrogen (mass 2) or hydrogen (mass 3) also known tritium.

The ion source creates ions by bombarding the incoming gas with electrons generated by a hot filament. Once ionized, these charged particles can be manipulated. They are then collimated and accelerated into the magnetic sector portion of the spectrometer.

In the magnetic sector, the ionized particles are separated as a result of their atomic mass and the force imparted on them by the magnetic field. Lighter atomic mass ions are deflected more than the heavier ones. In general, the helium atoms are bent at a predetermined angle, typically 90°, 135°, or 180°, to strike a detector. The greater the angle of curvature, the greater the mass separation, but a loss in sensitivity may also occur.

Helium ions are then collected and "counted" by some type of detector, such as a faraday detector. Whatever the technology, the ions that are detected

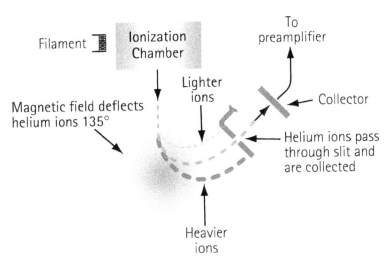

Figure 5.3 Basic spectrometer tube operation. Courtesy of Varian, Inc.

on the far side of the magnetic sector are proportional to the helium at the inlet of the leak detector.

5.9.2 Vacuum Systems

The vacuum system of a helium MSLD consists of one or two (when a gross leak option is required) primary vacuum pumps. These pumps can be rotary vane pumps, diaphragm pumps, scroll pumps, or other types of pumps. In a single primary pump system, their job is to reduce the pressure at the test port or inlet of the leak detector and act as a backing pump for a high-vacuum pump. In a dual primary pump system, one pump evacuates the test port and the other is dedicated to backing the high-vacuum pump. This arrangement allows for extended roughing times without having to periodically switch the pump to a backup mode.

The high-vacuum pump creates a sufficiently low pressure in the spectrometer tube, usually at or below the 10^{-4} Torr range, to support proper operation. Additionally, there is the accompanying tubing and valving required, creating a system that allows helium to flow to the spectrometer tube in one or more paths as described under Section 5.10, Vacuum System Configurations.

5.10 VACUUM SYSTEM CONFIGURATIONS

The vacuum systems of the mass spectrometer leak detectors can be designed so that the helium flows to the spectrometer tube in different paths. The oldest and inherently most sensitive is the conventional or direct flow leak detector. This system often required a liquid nitrogen trap as they were originally designed with vapor jet vacuum pumps. The next innovation was the contraflow design that allows helium to pass through the high-vacuum pump in a "reverse flow" direction, moving from the foreline or outlet of the high-vacuum pump to the inlet. This design uses the high-vacuum pump as a filter of sorts, pumping most gases away while being somewhat transparent to helium. The most recent development is a hybrid of the above-mentioned systems. It uses a "midstage" port in the high-vacuum pump (typically a turbo or turbo drag pump) to introduce helium to the spectrometer tube. Various combinations of two or more of the above can be employed in one design to widen the usable ranges and performance characteristics.

5.10.1 Conventional (Direct) Flow (Figure 5.4)

A Conventionally pumped leak detector system has the spectrometer tube located on the top of the high-vacuum pump. The early systems always had liquid nitrogen traps to prevent the migration of diffusion pump fluid to the test piece. Turbo-pumped systems do not require these traps. The test piece is

VACUUM SYSTEM CONFIGURATIONS

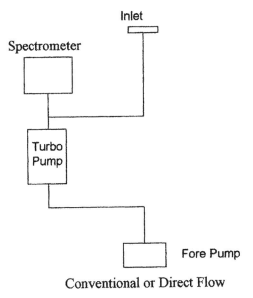

Conventional or Direct Flow

Figure 5.4 Direct flow schematic. Courtesy of Varian, Inc.

evacuated by a primary pump down to a pressure sufficiently low that the test piece can be transferred into the high-vacuum side of the vacuum system. If the part is not pumped low enough when it is valved in, the pressure in the spectrometer tube could rise to a level that would degrade its performance. This transfer pressure is typically below 10 milliTorr. The liquid nitrogen trap adds pumping speed for condensables to help keep the spectrometer tube pressure down. In older systems that are pumped by vapor jet pumps, the liquid nitrogen trap protects the spec tube and test piece from contamination resulting from backstreaming vapor jet pump fluid.

5.10.2 Contraflow (Reverse) Flow (Figure 5.5)

A contraflow or reverse-flow vacuum system takes advantage of the differences in the compression ratios of the high-vacuum pump for different gases. Helium can travel in a "contraflow" or reverse direction through the high-vacuum pump and arrive at the spectrometer tube on the high-vacuum side of the high-vacuum pump. In this scenario, a primary pump evacuates the test piece until it is below the tolerable foreline pressure of the high-vacuum pump. The foreline tolerance for a diffusion pump system may be in the range of 100 milliTorr, whereas a turbo drag high-vacuum pump may be able to tolerate pressures as high as 10 Torr or higher at the foreline. This is the maximum pressure that the foreline, or outlet of the high-vacuum pump, can withstand without negatively impacting the pressure on the inlet side of the high-vacuum pump. Above this limit, which is a function of the compression ratio of the

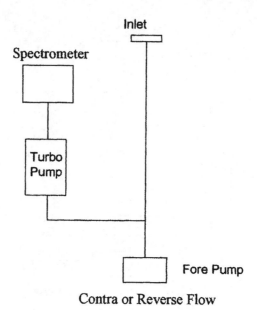

Figure 5.5 Reverse flow schematic. Courtesy of Varian, Inc.

high-vacuum pump, the pressure in the spectrometer tube rises above the level critical for acceptable ion beam optics. Contraflow systems tend to have inferior helium sensitivity compared to conventional systems, but can transfer into test at a much higher pressure at the inlet. They are also somewhat less sensitive to contamination.

5.10.3 Midstage Flow (Figure 5.6)

The evolution of system designs lead to the introduction of systems that sense helium via a "midstage" port. This port is located somewhere between the inlet of the high-vacuum pump (as with a conventional flow design) and the foreline of the high-vacuum pump (as with a contraflow design). As would be expected, this configuration would have less sensitivity but a higher inlet pressure tolerance than a conventional flow design, but lower than a contraflow design. This design also helps protect the spectrometer from direct exposure to possible contaminants from the test port.

5.10.4 Multiple Flow (Figure 5.7)

Leak detectors are available that utilize some or all of the above-mentioned vacuum system designs. In a unit that incorporates all of the above, typically the unit would transfer into test at the high-vacuum pump's highest tolerable inlet pressure in the contraflow mode, then switch into a midstage mode, and

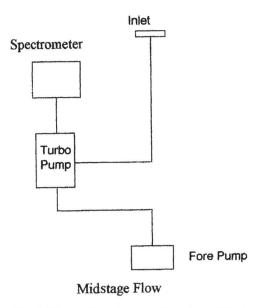

Figure 5.6 Midstage flow diagram. Courtesy of Varian, Inc.

finally into the conventional flow mode. This system design would realize the benefits of high-pressure testing in contraflow at one end of the test specimen, and high-sensitivity testing at the low test port pressures on the other end.

5.11 ELECTRONICS

Leak detection instruments have power supplies that control and regulate the ion source, amplifier, display, CPU, and all other electronics within the unit. The spectrometer typically has a high-vacuum gauge operated by a gauge controller. The high-vacuum gauge indicates when the source filament needs to be shut off due to excessively high spectrometer pressure. One or more rough vacuum gauges are also needed to monitor manifold pressures and drive valve sequences.

Modern leak detectors are microprocessor controlled and can be used to perform multistep processes with complex calculations and algorithms. This allows the unit to be set up for optimization of the leak detection cycle for specific test sequences that can run automatically. Automatic sequencing can efficiently reduce test cycle times and can incorporate go/no-go limits associated with them.

The DC amplifier provides a signal proportional to the concentration of helium in the spectrometer and an A/D converter supplies digital signals for PC displays and alarms.

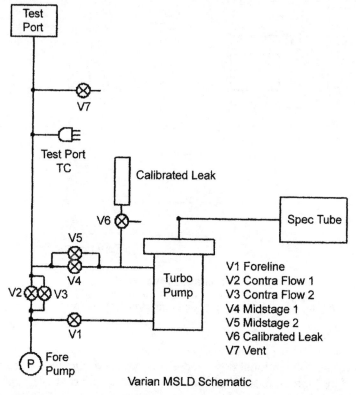

Figure 5.7 Varian MSLD flow diagram. Courtesy of Varian, Inc.

The power control circuit provides test status information and opens and closes the valves in the vacuum system. System valves provide for the evacuation and testing of the test piece, control maintenance sequences, and protect the spectrometer and high-vacuum pump in the event of sudden overpressure or power failure.

5.11.1 I/O Functions

Signal I/O (input/output) is used to determine the status and control operations of a leak detector, drive an auxiliary recorder, or provide a signal to a host computer. The I/O interface contacts are used to establish set-point alarms and indicate filament status. Leak rate outputs can also be recorded.

5.12 METHODS OF LEAK DETECTION

There are many different ways to leak test parts using helium as a tracer gas. In general, the leak detection method is selected based on the actual working conditions of the part being tested. It is recommended that during leak testing, the same pressure differential be maintained in the same "direction" as exists during the actual use of the part. For example, a vacuum system is tested with a vacuum inside the chamber, while a compressed air cylinder should be tested with a high pressure inside the cylinder.

There are also two general concerns when leak testing. One is the location of the leaks and the other is the measurement of the total leakage rate of the part, as some leakage may be acceptable. In many cases, parts may be first tested to determine if they pass an acceptable level, and if not, the part may be taken offline and subjected to a second test with the intent of locating the leak. Additionally, many parts may be tested in batches. If a batch fails, the individual parts in that batch may then be tested separately to identify the leaking part(s).

5.13 VACUUM TESTING METHOD (OUTSIDE-IN)

The part to be tested is evacuated with a separate pumping system for large volume, or with only the leak detector itself for small volumes. When the appropriate transfer pressure is reached, the leak detector transfers into the test mode and the part is tested using one of the following methods.

5.13.1 Locating Leaks (Figure 5.8)

To determine the total quantity of leakage (but not measure the total leakage rate), helium is administered to the suspected leak sites of the part using a spray probe with an adjustable flow. In some cases suspected leak sites may

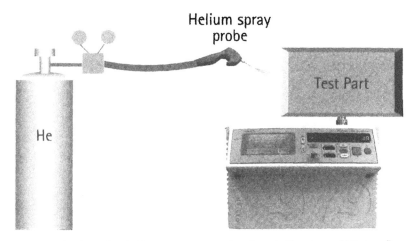

Figure 5.8 Spraying helium on evacuated test part. Courtesy of Varian, Inc.

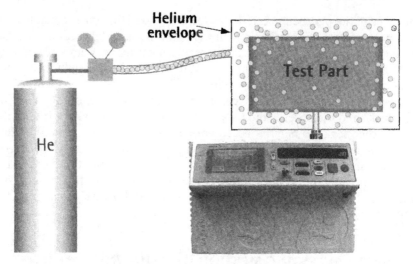

Figure 5.9 Testing part in helium envelope. Courtesy of Varian, Inc.

be "bagged" to prevent helium from drifting to another potential leak site nearby.

5.13.2 Measuring Leaks (Figure 5.9)

To determine the total quantity of leakage (but not the number or location of leaks), the part is connected to the leak detector and shrouded by a helium environment. This helium environment can be contained by various methods ranging from a simple plastic bag to more complex bell jar arrangements. To guarantee accuracy, steps can be taken to remove all gas and then backfill with 100% helium. Alternatively, a known mixture of helium and nitrogen, for example, can be used. In this case, the leak rate value will be proportionally lower.

5.14 PRESSURE TESTING METHOD (INSIDE-OUT)

In this technique, the part is pressurized with helium or a mixture of helium and air. To best simulate the physical stresses of real life, it is suggested that the parts are pressurized to the same pressure as they will see when in use. The parts are then tested by one of the following methods.

5.14.1 Locating Leaks (Figure 5.10)

To pinpoint the location of leak(s), but not measure the total leakage, the likely potential leak sites of the part are scanned using a sniffer probe connected to

PRESSURE TESTING METHOD (INSIDE-OUT)

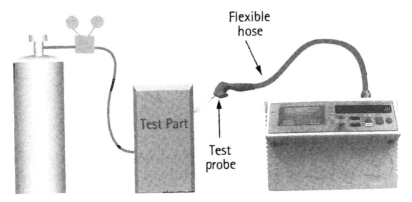

Figure 5.10 Test probe sniffing pressurized test part. Courtesy of Varian, Inc.

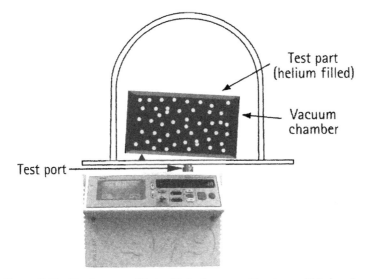

Figure 5.11 Testing part filled with helium gas. Courtesy of Varian, Inc.

the inlet of the leak detector. Attention should be given to the speed and proximity at which the sniffer probe is moved along the potential leak sites. If the probe is moved too quickly or too far away from the leak sites, less helium will enter the probe and leaks could be missed. Additionally, a longer hose between the probe and the leak detector will result in a longer time for the machine to respond to the presence of helium at the probe.

5.14.2 Measuring Leaks (Figure 5.11)

To determine the total quantity of leakage (but not the number or location of the leaks), the part is pressurized with helium (or a mixture of helium and air or nitrogen). This can be done by bombing or backfilling small hermetically sealed parts. Larger parts can be actively pressurized using a hose or tubing to deliver the helium. The part is placed in a volume that is then evacuated by the leak detector. All the helium escaping from the part is captured in the volume surrounding the part, pumped by the leak detector, and quantified.

5.15 ACCUMULATION TESTING METHOD (Figure 5.12)

This method can both locate and quantify leaks. Some type of shroud or hood is placed in such a manner as to envelop a potential leak site. A certain amount of time is given to allow leaking helium to accumulate in the shrouded area, increasing the helium concentration. The leak detector is then valved into the shrouded volume. If many potential leak sites exist in a manifold or if many parts are to be tested at the same time, they can be sequentially valved in to determine which site is leaking.

5.16 VACUUM SYSTEMS (Figure 5.13)

In general, vacuum systems are tested with a portable leak detector. Typically the leak detector is connected by means of a "tee" connected in between the foreline of the high-vacuum pump and the inlet of its backing pump. A system should be capable of maintaining a foreline pressure low enough to operate

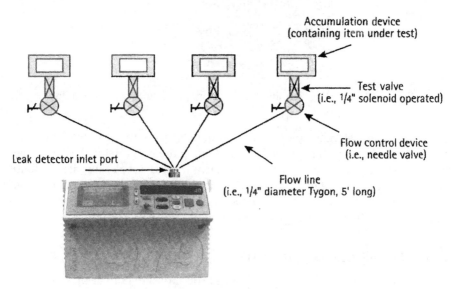

Figure 5.12 Testing multiple devices. Courtesy of Varian, Inc.

PRESSURIZED SYSTEMS

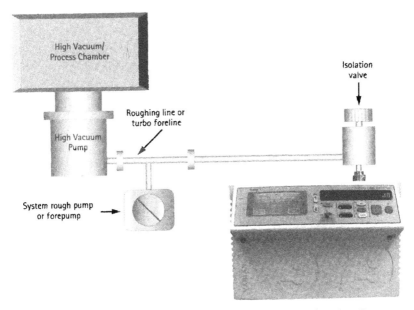

Figure 5.13 Testing high vacuum system. Courtesy of Varian, Inc.

the leak detector at this location. Helium is supplied to a potential leak site using a spray probe or by bagging suspected areas. If a leak exists, helium will enter the system and rapidly pass through it. The leak detector should respond within several seconds or less. Note that the leak detector sensitivity will be diminished in systems with large backing pumps. If a system is using a cryopump as a high-vacuum pump, it must be valved off before helium is introduced as cryopumps have limited pumping capacity.

5.17 PRESSURIZED SYSTEMS

Many different types of pressurized systems also need to be leak-free. These systems can be charged with helium or some mixture of helium and another gas such as nitrogen. If a diluted helium mixture is used, the helium signal will be diminished proportionally. For example, if a mixture of 10% helium and 90% nitrogen is used, the signal will read 10% of the actual value of the leak, or a decade lower. This may be acceptable in many cases as system leak checking is usually to locate rather than quantify leaks. Once the system has been charged with an appropriate amount of helium, leak checking can be performed by means of a sniffer probe, or by bagging suspected leak sites so that leaking helium will accumulate to a detectable level.

5.18 MSLD CONFIGURATIONS

There are many different ways that leak detectors are packaged to cater to the many types of applications. They range from cabinet or workstation models, to portable or mobile units, to component or integratable units. The subsystems are fundamentally the same, but these packaging options make their use a great deal more convenient. Any of these systems may be configured with "wet" or "dry" pumping technologies.

5.18.1 "Wet" or "Dry" Pumps

Most modern-day leak detectors utilize a turbo pump, a drag pump, or a combination turbo-drag pump as a high-vacuum pump. For this reason, the high-vacuum pump is not a major source of potential hydrocarbon contamination. The most frequently used primary pump, however, has traditionally been the oil-sealed rotary vane pump. There is a potential for hydrocarbon contamination migrating from the primary pump to the inlet of the leak detector and possibly the test piece. This is especially true with contraflow configurations.

Most suppliers offer wet and dry versions of their leak detectors. Wet versions usually employ oil-sealed rotary vane pumps as previously noted. This is a tried-and-true technology that offers good helium-pumping characteristics. There are a great variety of options with respect to choices of dry primary pumps. In general, the cost per unit of pumping speed goes up significantly with dry pumping technologies. These technologies range from diaphragm pumps, to dry scroll pumps, to multistage blower-type pumps and molecular drag pumps. In some cases the choice of dry primary pump can negatively impact the ultimate sensitivity of the leak detector. This may be due to the compression ratio of the dry pumps and their propensity to allow helium to flow backward through them, effectively appearing as a leak. Different technologies can be used in conjunction with one another to improve pumping characteristics.

5.18.2 Cabinet or Workstation Models

These units are typically used for applications that require a great deal of repetitive testing and/or the utility of a large work surface. They are typically larger than other units and parts are usually brought to the leak detector. As testing often requires short cycles for high quantities of parts, the vacuum system is designed accordingly. This may include large, high-speed pumps, and a vacuum system that is optimized to deliver high speed to the test port. Additionally, some consideration should be given to long-life valves and components due to the high cycling rate.

Cabinet or workstation-type units may also include autosequencer functions that allow for programming of test times and acceptable reject levels.

This is useful when similar parts are being repetitively tested and therefore have consistent and predictable cycles. If a part does not meet the reject level allotted in the time allotted, the test can be aborted. Some units open a test fixture automatically when a part passes, but require operator intervention for failed parts, while capturing the reject leak rate.

5.18.3 Portable Units

Portable units traditionally were intended to be brought to the part to be tested, such as a large piece of manufacturing equipment or a semiconductor process tool. They are smaller and more mobile than cabinet or workstation models. Some have the primary pump within the unit, while others mount the pump on the lower shelf of a cart. Generally speaking, the models that have a primary pump mounted with the leak detector need to use a relatively small primary pump for it to fit. A small primary pump can significantly reduce sensitivity as less helium is "pulled" into the test port.

5.18.4 Component or Integratable Units

In addition to the above-mentioned configurations, a somewhat newer innovation is the component or integratable leak detector. This design allows the unit to be designed into a larger, multistep manufacturing line. High-production environments, such as in the automotive components manufacturing industry, often use this configuration because leak detection may be just one step in a high-speed production line. This allows the vacuum system/spectrometer to be located close to the test parts in a complex production system. Typically, the electronics are located remotely from the vacuum system/spectrometer as is convenient for the customer or system integrator. The electronics are designed to communicate information to a control room.

5.19 CALIBRATION

Various electronic tuning parameters of the spectrometer tube can be optimized to produce the greatest helium signal. These voltages basically effect the focusing of the helium ion beam and are adjusted in an effort to maximize the signal response to a calibrated helium leak source. Once this tuning process has been performed, a calibration step can be initiated to bring the actual leak rate indication to the exact value of the calibrated helium source. This process is done by manually adjusting potentiometers, or by a microprocessor-driven, automated tuning/calibration process. The latter of the two is more convenient these days, as it requires less training for the operator. It is also more expensive.

5.19.1 Calibrated Leaks

Standard leaks permit a known amount of tracer gas to enter the MSLD for calibrating purposes. These calibrated helium sources can be connected to the inlet of the leak detector as needed, or can be built into the unit in a permanent manner. Most leak detectors that have automatic calibration features have several standard leaks built in. They may also have an option to tune an internal or external standard leak. Because all parts leak at some level, many parts have an acceptable leak rate threshold. For these parts calibration of the leak detector may be critical. In applications in which the objective is only to locate a leak, the calibration accuracy may be much less critical.

MSLDs primarily use two types of standard leak technologies, glass permeation types or capillary types. The glass permeation technology consists of a glass membrane that is surrounded by a reservoir of helium. The glass membrane separates the helium reservoir from the port that is connected to the leak detector. The helium permeates the glass at a constant rate. The leakage rate is affected by the temperature of the device and the pressure of the helium in the reservoir. The temperature coefficient is approximately 3% per degree Celsius. The approximate loss associated with the changing pressure in the helium reservoir is about 2 or 3% per year for standard leaks in the 10^{-7} and 10^{-8} atm cc/s range. One disadvantage of permeation leaks is the fragile nature of the glass membrane, which can break if dropped.

Capillary standard leaks also have a pressurized helium reservoir. The helium leaks through a thin capillary tube to the connecting flange of the standard leak. These leaks are less sensitive to temperature fluctuations, with an approximate temperature coefficient of 0.2% per degree C. Capillary leaks can be more rugged than permeation leaks, but they may be prone to problems associated with clogging from solids or condensables.

5.20 RADIOISOTOPE TRACER LEAK TESTING

Short-life radioisotopes also can be used to leak test hermetically sealed cavities and closed piping systems. In gas systems, a radioactive tracer gas can be added to the pressurized gas system and tracked through the system. The process path and flow rate can be determined by placing detectors downstream of the injection point and measuring the time from injection to detection. The unexpected loss of flow or the detection of the tracer gas at an unexpected location could indicate a leak. In liquid systems, radioactive barium tracers can be added to some process liquids and similarly tracked. These tests have about the same sensitivity as the helium mass spectrometer leak detector, but they are more expensive. The supplier or customer must also provide a radiation safety officer and radiation-monitoring devices for personnel working in the area.

5.21 BUBBLE LEAK TESTING

Bubble leak testing is frequently used to detect small leaks in pressurized tubing, pipes, and vessels that can be isolated under pressure. Leaks in larger depressurized pipes and vessels can also be detected with the aid of a bubble solution, a vacuum box, and an air injector or other rapid evacuation system. In some cases, small parts may be tested and simultaneously cleaned by immersing the part in a solvent or cleaning solution; this is known as immersion bubble leak testing.

Before bubble tests are applied, all test surfaces must be cleaned and inspected. When welds are to be bubble leak tested, weld slag should be removed and the weld joint should be cleaned with a wire brush. If necessary, dirt and grease should be removed with a solvent-soaked rag. Before pressure bubble or vacuum box leak tests are performed, the calibration of the pressure, vacuum, or combination pressure/vacuum gage should be checked. The range of the pressure gage should be roughly double the desired system pressurization. The gage should carry the date of calibration, calibrator's initials, and due date of next calibration. A gage calibrated in psia units is calibrated in pounds per square inch absolute; a gage calibrated in psig units is calibrated in pounds per square inch gage. The relationship between these units is shown in Eq. (5.1).

$$\text{psig} = \text{psia} + 14.7 \tag{5.1}$$

5.21.1 Leak Detector Solution

Leak detector solution, consisting of water with a wetting agent, is applied to a test surface by flowing, brushing, squirting, or spraying the bubble solution. When this is done small leaks under pressure or vacuum will cause the solution to form a multitude of small bubbles. The size and number of bubbles generated will be proportional to the pressurized gas leak rate. Leak detector solution should be tested periodically against a known leak to assure that the solution is functioning properly. If the bubble solution dries, the test surface must be recleaned and the bubble solution reapplied. The ideal leak detector solution would have low surface tension for detecting very small leaks and high surface tension for detecting large leaks. Because of this impossibility, some compromise in surface tension must be reached. The leak detector solution should be chemically inert and have a neutral pH of 7.

5.21.2 Vacuum Box Bubble Leak Testing

A vacuum box is a box equipped with a transparent viewing window, gasketed sealing surface, rapid evacuation equipment, and a combination pressure/vacuum gage. Specifically designed vacuum boxes contoured to the geometry of the test part are used for testing corner welds.

With vacuum box testing, the box should be applied within a minute, or while the bubble solution is still wet. The force required to initially seat the box and seal its gasket should be minimal. The sensitivity of the test depends on the amount of differential pressure created with the vacuum box, and is defined as the smallest amount of leakage that can be detected. The sensitivity of the test increases as differential pressure increases. Sensitivity is dependent on operator technique, alertness, and skill. For optimum results, adequate lighting of the test area is essential and the operators should be able to place their eyes within 24 in. of the test surface. As the vacuum box is positioned and initially evacuated, the operator should observe the action of the leak detector solution to make sure it is not instantly blown away by large leaks. The test area is considered leak free when the operator cannot detect any bubble formation in the specified period of time for the test. Under ideal conditions (1 atm of differential pressure) leakage as small as 10^{-4} atm cc/s can be detected. Under normal field conditions, leakage in the order of 10^{-2} to 10^{-3} atm cc/s can be detected with 1 atm of differential pressure. A pressure differential of at least 2 psi should be maintained during vacuum box bubble leak testing examination.

To assure 100% inspection of the test surface, the vacuum box inspection area should be overlapped by at least 2 in. If a large area is to be tested with a vacuum box, the work should be divided among the various work crews or inspectors to minimize worker fatigue. Limited skill and training is required for successful bubble leak testing.

If the vacuum box is improperly designed, it may implode. In order to prevent an implosion, vacuum boxes should be designed to withstand a full atmospheric pressure differential. Atmospheric pressure is typically given as 14.7 psi, 30 in. of mercury, or 760 mmHg. When properly designed for full atmospheric differential pressure, relief valves are not necessary. The vacuum box is considered to be in good condition when it can maintain slightly more than the required test vacuum.

5.21.3 Pressure Bubble Leak Testing

With pressure bubble leak testing, "soak time" is the elapsed time from completion of pressurization until the application of the leak detector solution. The purpose of the soak time is to provide sufficient time for the pressurized gas to escape through the long, irregular leak path. Care must be taken to assure that the test specimen, piping, or tubing is not overpressurized. If the test specimen is overpressurized, it can rupture or explode. Because of its explosive properties, hydrogen gas must never be used for pressurization. It is a good practice to use redundant pressure gages when pressurizing equipment for a bubble leak test. When pipes and vessels are tested, one indicating and one recording pressure gage should be used. Pipes are frequently plugged or blanked prior to pressurization by using expandable stoppers or bladders.

When these devices are used, a stop bar should be installed downstream of the plug to prevent a possible blowout.

With pressure bubble leak testing, very small leaks and very large leaks can go undetected. Very small leaks go undetected because the operator does not observe the test area for a long enough period of time. Very small leaks are best detected by applying a light coating of solution that is relatively free of bubbles. The required observation time is specified by the respective standard or test procedure. Large leaks can go undetected because the bubble solution is blown away before the bubbles have a chance to form. When large leaks are suspected, a mixture of thick suds or foam can be used. Sometimes large leaks produce audible noise or air jets, which can be felt.

When pressure bubble leak testing a vessel with reinforcing pad plates, the end of the pipe nipple of the pressurizing assembly is often notched to assure pressurization of the pad plate in the event that the end of the nipple seals itself against the shell of the vessel.

With immersion bubble leak testing, sealed test specimens are immersed in a preheated liquid having low surface tension; this is a simple, rapid, and economical test. When methyl alcohol is used as the fluid for immersion bubble leak testing, the alcohol acts as a cleaning agent, providing a secondary benefit. Ordinary tap or process water is a poor solution for immersion bubble leak testing because of its high surface tension.

5.21.4 Indications

A continuous flow of bubbles will be produced by a hole or crack in the test area. With a vacuum box system, very large leaks may briefly produce very large bubbles or the bubble may be blown away instantly. Small leaks produce a series of small bubbles at regular intervals or a slowly growing bubble. When enclosed heated areas are bubble leak tested, the reading of the pressure or vacuum gage over the specified period of time provides the most reliable indication of a small leak.

Air, grease, or dirt trapped in surface defects can produce false indications of leakage. If static bubbles appear in the area of interest during a vacuum box bubble test, the bubbles should be removed and the test area should be recleaned to determine if the bubble was a false indication. A false or virtual leak is formed when a few bubbles form rapidly and then quickly disappear. Bubble leak tests on large outdoor equipment should not be attempted in freezing weather. Calm, cloudy, moderately warm weather conditions are ideal for testing large outdoor equipment. For vacuum box bubble leak testing, surface temperature should be in the range of 40 to 100°F. In hot weather, evaporation of the leak test solution may be a problem.

Topics covered by bubble test specifications include:

Scope of work
Applicable documents

Design criteria
Description of work
Personnel qualifications
Procedures to be used
Leak testing details
NDT operator qualifications
Engineering verification
Submitted records and documents
Final checks

5.22 PRESSURE CHANGE LEAK TESTING

5.22.1 Principles

Pressure change measurement tests, conducted at pressures above atmospheric pressure, are also called pressure hold tests, pressure decay tests, or pressure loss tests. Pressures above atmospheric are frequently monitored with pressure gages that read out in gage pressure (psig).

Hydrostatic tests should not be conducted prior to pressure change measurements because hydrostatic tests can introduce water in the system that will destroy the sensitivity of the pressure change tests.

The effects of pressure change measurements can be understood by considering a pressurized, rigid, closed system that does not leak. If external (barometric) pressure decreases, internal gage pressure will decrease, and if internal gas temperature decreases, internal gage pressure will decrease.

5.22.2 Terminology

Absolute pressure can be defined as gage pressure plus barometric pressure. As previously stated, standard atmospheric pressure is equal to 14.7 psia, 29.96 in. of Hg, or 760 mmHg. As elevation above sea level increases, the barometric pressure decreases. This is one reason that aircraft cabins must be pressurized above 10,000 ft. Pressure change measurement on large vessels, such as nuclear containment vessels, is usually described in terms of leakage rate tests.

Temperature compensation for pressure change is calculated in terms of absolute temperature, degrees Rankin (°R) or degrees Kelvin (°K), depending on whether temperature is measured in terms of degrees Fahrenheit (°F) or degrees Celsius (°C). These relationships are shown in Es. (5.2) and (5.3).

$$\text{degrees R (°R)} = \text{degrees F (°F)} + 460 \tag{5.2}$$

$$\text{degrees K (°K)} + \text{degrees C (°C)} + 273 \tag{5.3}$$

PRESSURE CHANGE LEAK TESTING

If it is desirable to convert degrees F to degrees C or degrees C to degrees F, Eqs. (5.4) and (5.5) can be used.

$$\text{degrees C (°C)} = 5/9(\text{°F} - 32) \qquad (5.4)$$

$$\text{degrees F (°F)} = 9/5(\text{°C} + 32) \qquad (5.5)$$

If internal system volume is so small that the internal temperature sensors cannot be used to measure system temperature, a stable external temperature can be maintained around the system or an insulated surface thermometer can be used. The surface thermometer must be insulated from the external temperature so that it is representative of internal system temperature. Surface thermometers can be attached to the test surface with tape, adhesive, magnets, or clamps. If external temperature cannot be maintained and surface thermometers cannot be used, the measurement of mass flow leakage rate should be considered.

5.22.3 Equipment

Absolute pressure gages measure any pressure above the zero value that corresponds to a perfect vacuum. In large volume systems, they are used to measure test pressure independently of barometric pressure.

The range of Bourdon tube pressure gages, used to monitor pressure change tests, should be neither <$1\frac{1}{4}$ nor >4 times the test pressure because gage accuracy is very poor in the lower 10–20% of their range. Pressure gages with mirrored scales help improve the reading accuracy because they help correct for parallax. Quartz Bourdon tube pressure gages are sensitive, accurate absolute gages used with systems pressurized above atmospheric pressure. Readability, or resolution, and reproducibility are two of the most important factors to consider when selecting test gages. Resolution is the least discernible unit of measurement that can be read. A typical accuracy range of ±0.25 to ±0.33% is adequate.

When U-tube manometers are used to measure gage pressure during pressure change leakage tests, both legs of the manometer must be carefully read to correct for any changes in zero reading of the gage. Errors in zero reading can be caused by failure to calibrate the manometer or evaporation of liquid in the manometer.

When mercury manometers are used for pressure gage measurements, the top of the meniscus should be used as the reading point of the pressure because of the convex surface of the liquid. With a water manometer, the bottom of the meniscus should be the reading point of the pressure because of the concave surface of the liquid.

Ionization gages are used to measure pressure change in evacuated systems with an absolute pressure range of 10^{-4} to 10^{-6} Torr.

Bimetallic thermocouples are used to measure dry bulb temperature in pressure change measurement tests. Two metals, such as iron-constantan or copper-constantan, generate an electromotive force (emf) or voltage that varies with temperature. The calibration curves for thermocouples are nearly linear and thermocouple calibration tables can be supplied from thermocouple manufacturers. For greatest accuracy, thermocouples are individually calibrated.

Resistance temperature detectors (RTDs) are also used to measure dry bulb temperature in pressure change test systems. Resistance temperature detectors, such as the $100\,\Omega$ platinum resistance bulb thermometers, are preferred in high-accuracy applications because of their inherent stability, reproducibility, accuracy, and high-melting point. The $100\,\Omega$ platinum resistance bulb has a resistance of $100\,\Omega$ at $0°C$ and $139.2\,\Omega$ at $100°C$. The number of dry bulb temperature sensors used in a test depends on the contained free air volume, configuration, and required redundancy of the system.

5.22.4 Pressurizing Gases

Because of its availability, nonflammability, and nontoxicity, air is one of the most practical gases for use in pressure change measurement tests. If air cannot be used as the pressurizing medium, nitrogen is a good second choice for a pressurizing gas. One precaution should be observed when using nitrogen or other inert gases in a confined area. It should be remembered that these gases could replace air in the lungs and cause suffocation. Breathing nitrogen, for example, is painless, and the victim of suffocation may pass out without realizing the danger.

Any gas used for leak testing should follow the laws for ideal gases. The ideal gas law shows the relationship of pressure, temperature, and volume of the ideal gas in accordance with Eq. (5.6):

$$PV = nRT \quad \text{or} \quad P_1 T_2 = P_2 T_1 \tag{5.6}$$

where P = pressure
V = specific volume
n = number of moles
R = gas constant
T = absolute temperature
P_1, V_1, and T_1 = pressure, volume, and temperature at condition 1
P_2, V_2, and T_2 = pressure, volume, and temperature at condition 2

The ideal gas law states that the pressure changes directly as a function of temperature when volume is held constant. Volumetric changes due to thermal expansion or contraction of the fixed volume are considered insignificant. If the pressurizing gas is different from the in-service gas, differences in viscosity need to be taken into consideration when determining the in-service gas leakage rate.

5.23 PRESSURE CHANGE MEASUREMENT TESTING

In pressure change measurement testing, lines or vessels are pressurized. An isolation valve is used to trap the pressure in the system and the pressurizing line is then disconnected. The pressurizing line is disconnected so that the isolation valve can be leak tested to assure that there is no in-leakage or out-leakage through the isolation valve's seat. The purpose of pressure change leak testing is to verify that the component of interest meets its minimum intended service requirements. Systems often have a specified allowable pressure loss per unit time at design pressure. It is important to know the contained volume at design pressure in order to calculate allowable leakage rates, if any, and determine the most appropriate leak testing method. However, a high-precision calculation of the enclosed constant volume system is not necessary if the leakage rate is calculated as a percentage of the total enclosed mass change per unit time.

Pressure, temperature, and time are all parameters that change during a pressure change measurement test. With long-duration pressure change tests, barometric pressure must be measured or compensated for because it tends to vary. Extending the duration of the test increases test reliability. For short-duration tests of >1 h, it is usually not necessary to consider barometric changes. For a short-duration pressure change test, where temperature and barometric pressure remain reasonably constant, the rate of pressure change can be calculated in accordance with Eq. (5.7):

$$dp/dt = (p_1 - p_2)/dt \qquad (5.7)$$

where dp = change in pressure
dt = change in time
p_1 = starting pressure
p_2 = pressure at end of test

Pressure change measurement tests are more sensitive for small, contained volumes at test pressure. In other words, a small leak in a small, pressurized container will cause a more rapid drop in pressure than the same size leak in a similarly pressurized large container. In large-volume systems, water vapor partial pressure readings must be subtracted from the absolute pressure readings to improve accuracy. Measuring the internal gas dew point temperature and using steam tables to determine the water vapor partial pressure does this. Figure 5.14 shows moisture content as a function of dew point temperature. Water vapor pressure can be determined at atmospheric conditions by dividing the volumetric moisture content percentage by 100.

Moisture Content of Saturated Air or Other Gas at Various Temperatures (Dew-Points) and at 1 Atmosphere Absolute Pressure (14.7 PSIA)

Dew-Point Temperature °F.	°C.	Moisture Content Per Cent by Volume*	Dew-Point Temperature °F.	°C.	Moisture Content Per Cent by Volume*
110	43.3	8.70	16	—8.9	0.308
108	42.2	8.20	14	—10.0	.282
106	41.1	7.75	12	—11.1	.258
104	40.0	7.30	10	—12.2	.236
102	38.9	6.90	8	—13.3	.216
100	37.8	6.50	6	—14.4	.198
98	36.7	6.10	4	—15.6	.180
96	35.6	5.75	2	—16.7	.165
94	34.4	5.40	0	—17.8	.150
92	33.3	5.05	—2	—18.9	.136
90	32.2	4.75	—4	—20.0	.124
88	31.1	4.46	—6	—21.1	.113
86	30.0	4.18	—8	—22.2	.102
84	28.9	3.92	—10	—23.3	.093
82	27.8	3.68	—12	—24.4	.084
80	26.7	3.44	—14	—25.6	.076
78	25.6	3.22	—16	—26.7	.0685
76	24.4	3.02	—18	—27.8	.0619
74	23.3	2.84	—20	—28.9	.0558
72	22.2	2.65	—22	—30.0	.0503
70	21.1	2.47	—24	—31.1	.0452
68	20.0	2.31	—26	—32.2	.0407
66	18.9	2.16	—28	—33.3	.0364
64	17.8	2.02	—30	—34.4	.0328
62	16.7	1.88	—32	—35.6	.0294
60	15.6	1.75	—34	—36.7	.0264
58	14.4	1.63	—36	—37.8	.0235
56	13.3	1.51	—38	—38.9	.0210
54	12.2	1.40	—40	—40.0	.0188
52	11.1	1.30	—42	—41.1	.0167
50	10.0	1.21	—44	—42.2	.0149
48	8.9	1.12	—46	—43.3	.0132
46	7.8	1.04	—48	—44.4	.0117
44	6.7	0.966	—50	—45.6	.0104
42	5.6	.894	—52	—46.7	.0092
40	4.4	.827	—54	—47.8	.0082
38	3.3	.765	—56	—48.9	.0072
36	2.2	.707	—58	—50.0	.0063
34	1.1	.653	—60	—51.1	.0056
32	0.0	.602	—65	—53.9	.0041
30	—1.1	.553	—70	—56.7	.0029
28	—2.2	.511	—75	—59.4	.0021
26	—3.3	.472	—80	—62.2	.0015
24	—4.4	.434	—85	—65.0	.0010
22	—5.6	.398	—90	—67.8	.0007
20	—6.7	.367	—95	—70.6	.0005
18	—7.8	.337	—100	—73.3	.0003

*Vapor pressure in atmospheres at various dew-point temperatures can be obtained by dividing the values for percent by volume, given in this table, by 100.

Figure 5.14 Table of moisture content vs. dew point temperature.

The dew point temperature is the temperature at which the internal gas is saturated with water and dew begins to form. When dew or condensation forms system volume decreases and pressure increases independently of temperature. When leak testing water-cooled reactors, a sudden change is dew point temperature could indicate water leakage into the system.

Pressure change measurement tests performed at pressures less than atmospheric are called pressure rise tests, vacuum retention tests, or pressure gain tests. They measure the ability of the system to hold a vacuum. With vacuum testing, a rapid initial rise in pressure will be observed due to outgassing of absorbed gases on the test area surface. After this initial outgassing, a much slower pressure rise, representative of the leak characteristics, will be observed. The amount of test area surface, material of construction, and cleanliness of the internal surface influence outgassing rates in a vacuum system. For evacuated systems, measurement of pressure is in units of Torr that are equal to millimeters of mercury (mmHg). For leak rate calculations, absolute pressure, time, temperature, and volume must be known.

In large systems, the effects of outgassing make it difficult to determine the actual leak rate. Data comparison must be used for large evacuated systems that are exposed to the weather and wide temperature variations. The test data can be compared to test data taken during stable temperature periods or during similar cyclical temperature periods.

5.23.1 Reference System Technique

One variation of the pressure change measurement technique incorporates the use of a reference chamber having a known pressure and atmosphere. The reliability of this technique depends on the accuracy and resolution of the differential pressure gage that monitors the differential pressure between the reference chamber and the test system. In large-volume systems, the reference chamber technique is less dependent on internal dry bulb temperature and dew point temperature weighting factors than the absolute pressure technique because these temperatures are based on the internal volume of each sensor location relative to the total system volume.

When pressure change leakage tests are combined, plots of absolute temperature and absolute pressure versus time are used to provide early determinations of test validity and acceptability of test results. The plots also make it easier to detect errors in the recorded data.

5.24 LEAKAGE RATE AND FLOW MEASUREMENT TESTING

For a pressure change leak test, the maximum allowable leakage rate depends on test procedure, volume, and time. The accuracy of any pressure change leak test depends on the accuracy of system volume calculations, the accuracy of the internal temperature measurement, and the accuracy and resolution of the system pressure measurement. In a rigid-volume system, the mass flow leakage rate is calculated from Eq. (5.8):

$$Q = V\, dP/dt \qquad (5.8)$$

where Q = leakage rate
V = system volume at test pressure
dP = pressure change during the test
dt = test time

With the flow measurement technique, volume is fixed by the rigid volume system. Leak rate is determined by measuring the quantity of gas added or removed from the system in order to hold absolute pressure constant. When the flow measurement technique is applied to a variable volume system, such as a tank with flexible diaphragm, system pressure can be held constant and gas added to maintain pressure is measured. Using this technique, the flow measurement is independent of system volume and there is no need to measure or calculate system volume. The sensitivity of the flow measurement technique depends on the accuracy of the flow-measuring instrument. However, if pressure is not held constant and system volumes can be calculated, Eq. (5.9) can be applied:

$$Q = (P_1 V_1 - P_2 V_2)/(t_1 - t_2) \qquad (5.9)$$

where Q = leakage rate
P_1, V_1, and t_1 = pressure, volume, and temperature at condition 1
P_2, V_2, and t_2 = pressure, volume, and temperature at condition 2

5.25 NUCLEAR REACTOR SYSTEMS

Both absolute and reference methods are acceptable test methods for performing a Type A integrated leakage rate test (ILRT) of a primary nuclear reactor containment system. Preoperational Type A leakage tests must be conducted after preoperational structural integrity tests and Type B and C tests. Types B and C tests involve the testing of containment system components, such as gaskets, valves, and seal interspaces. These components are readily tested using short-duration vacuum retention and water collection techniques. When a leakage rate test follows a structural integrity test, the test pressure for the structural integrity test shall be reduced to 85% of the peak acceptable design pressure (Pac) for a minimum of 24 h, prior to pressurizing to Pac for ILRT.

After pressurizing a primary nuclear reactor containment system for a Type A ILRT, the system must stabilize for a period of at least 4 h. The temperature of the containment system is considered stabilized when the weighted average temperature for the last hour does not deviate by more than 0.5°F/h, compared to the rate of temperature change for the last 4 h. At least three dew point sensors are needed to calculate the weighted average temperature. After stabilization, the duration of the Type A leakage test shall be a minimum of 8 h, and have a set of at least 20 data points taken at roughly equal time intervals. During this period, the test pressure shall not be permitted to fall below the peak containment internal design pressure (Pac) by more than 1 psi.

The measured leakage rate is then determined from a linear regression analysis using least square fit. The upper confidence limit (UCL) for the measured leakage rate of a primary nuclear containment system is calculated at a probability of 95%. The leak rate at the upper confidence limit, including required local leakage rate additions, shall be <75% of the daily allowable leakage rate (La). For verification of Type A leakage rate test accuracy, a mass step change verification test is used. This test calls for accurate metering of air injected into or removed from the containment vessel. Verification tests usually take at least 4 h and require a minimum of 10 data points. The metered mass change of air in 1 h must be within 25% La.

The selection of instruments used for Type A ILRT is based on an instrument selection guide (ISG). With this guide, combined instrumentation errors can be calculated. The ISG also provides guidelines covering loss of sensors during an ILRT and subsequent leak rate calculations following the ILRT. For additional information on this subject, see R. C. McMaster (1982).

Topics usually covered in pressure change leak test specifications are:

Scope and description
Applicable publications
Gasket materials
General requirements for design and fabrication
Primary containment design considerations
Pertinent drawings, procedures, and instructions
Inspection and testing containment vessel steel liner work
Pneumatic proof testing of pressurized components
Documentation

5.26 HALOGEN DIODE LEAK TESTING

5.26.1 Principles

Halogen-rich refrigerant gases are detectable by a halogen diode leak detector as they pass through a leak. The halogen concentration, type of halogen tracer gas, and differential test pressure are factors affecting the sensitivity of this method. The rate at which positive ions are formed is proportional to the halogen concentration of the gas passing through the detector. Standard leaks of known size are used to calibrate halogen diode leak detectors.

The thermal conductivity method of halogen leak detection measures the difference in the heat transfer of the two gases. For example, air can be used as a reference gas and a comparison can be made between the thermal conductivity of a tracer gas and the reference gas (air).

5.26.2 Terminology

By definition, a halogen is any of five very active, nonmetallic elements, namely, astatine, bromine, chlorine, fluorine, and iodine. Fluorine is the most widely used element in refrigerant gases used for halogen diode leak testing. Materials that contain halogens are called *halides*. The halogen-rich refrigerant gas used with a halogen diode leak detector is called *tracer gas*.

The size of the smallest detectable leak determines test *sensitivity*. A dynamic leak test is performed by evacuating the interior of the test object, while applying a tracer gas to the outside surface of the object, and monitoring the evacuated gas with a leak detector probe. Pressurizing the interior of the test object with the tracer gas, then sniffing or scanning the outside surface with a leak detector probe performs a static test.

5.26.3 Gases and Equipment

Various refrigerants are commonly used as tracer gases. When a procedure specifies the use of refrigerant R-12, Freon 12, Genetron 12, or Ucon 12 may be used. These are tradenames used by the various manufacturers of refrigerant R-12.

Sensitive, good-quality halogen diode leak detectors are capable of detecting leaks as small as 10^{-8} to 10^{-9} atm cc/s. The halogen diode leak detector sensor is potentially more sensitive than the halide torch (a thermal conductivity sensor), halide sensitive tape, or halide sensitive paint sensors. The positive ion current between the sensor's emitter and collector increases proportionally to halogen concentration. This ion current is amplified and displayed on a front panel meter as illustrated in Figure 5.15. Sophisticated halogen diode detectors incorporate heater control adjustments to regulate heater voltage, and built-in automatic zeroing circuits (not shown) to compensate for local, relatively steady, atmospheric halogen background signals.

5.26.4 Calibration

Standard leaks with refrigerant reservoirs are used to calibrate halogen diode leak detectors. Table 5.1 shows the standard leak setting required to obtain various test sensitivities with known volumes of halogens.

When a reservoir-type standard leak is being recharged with halogen-rich refrigerant, personnel should wear safety glasses with side shields to protect

HALOGEN DIODE LEAK TESTING

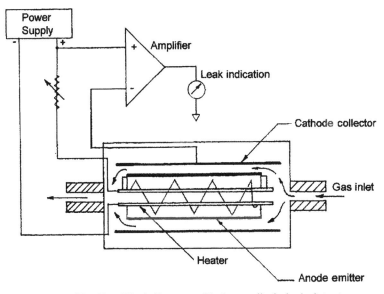

Figure 5.15 Simplified diagram of halogen diode leak detector.

TABLE 5.1. Halogen Standard Leak Setting (std cm³/s)

Percentage Halogen	Required Test Sensitivity (std cm³/s)				
	1×10^{-2}	1×10^{-3}	1×10^{-4}	1×10^{-5}	1×10^{-6}
1	1×10^{-4}	1×10^{-5}	1×10^{-6}	1×10^{-7}	
2	2×10^{-4}	2×10^{-5}	2×10^{-6}	2×10^{-7}	
5	5×10^{-4}	5×10^{-5}	5×10^{-6}	5×10^{-7}	
10	1×10^{-3}	1×10^{-4}	1×10^{-5}	1×10^{-6}	1×10^{-7}

their eyes. As the surrounding temperature increases, the maximum pressure attainable from a halogen-rich refrigerant bottle also increases. If the indicated leak rate continues to go up after the *increase* valve is closed, the increase valve is defective or not firmly closed. If the *decrease* valve is opened when the increase valve is not fully closed, refrigerant will vent from the standard reservoir, requiring refilling or recharging of the reservoir.

When a periodic instrument calibration indicates a decrease in test sensitivity, the instrument should be checked and repaired if necessary, and all tests performed since the last known good calibration should be repeated to assure reliable test results. After a satisfactory retest, other untested areas can be inspected.

5.27 "SNIFFER" TECHNIQUES

When drafts are noted during testing, the probe should be initially operated on the upstream side of the draft. When sniffer tests are being conducted on large objects, sniffing or scanning should proceed from the highest point to the lowest point in the system because refrigerant gases are heavier than air. Test procedures and/or various codes often specify the maximum allowable distance between the sniffer probe tip and test surface. In order to satisfy these requirements, notched plastic spacers are often placed over the probe tip to control the probe tip to test surface spacing.

Leak tests often call for a mixture of halogen with air or inert gas. In these cases, care must be taken to assure that bottled oxidants, corrodants, fuels, or toxic gases are not used. Carbon dioxide is one gas that can be safely used for pressurization if compressed air is not available. Tracer gas mixtures should be flushed out of the equipment and test areas after leak tests are completed.

Many times, sniffer tests must be performed on totally enclosed or "dead-ended" systems. In these cases, the enclosed space must be partially or totally evacuated before introducing the refrigerant, in order to provide a uniform mixture of tracer gas throughout the system.

5.27.1 Equipment Operation and Servicing

Halogen diode leak detectors should never be used in explosive atmospheres because high-voltage circuits and higher heater temperatures could ignite explosive mixtures. The occupational safety and health association (OSHA) standards and various plant safety rules and regulations generally prohibit the use of halogen diode leak detectors in hazardous environments.

Even though refrigerant gases are generally considered harmless and inert, refrigerant R-12 in the presence of high temperatures, such as hot welds, can break down into hydrogen chloride, hydrogen fluoride, chlorine, and phosgene (mustard gas), which are all toxic at relatively low levels. Refrigerant R-12 can also become corrosive in the presence of moisture. Finally, it should be remembered that all nontoxic gases could be asphyxiants by displacing air in the lungs. Therefore, halogen leak testing should be done only in well-ventilated areas having an ample air supply for operating technicians.

Test connections, used for pressurizing the halogen sniffer test system, should be halogen leak tested first, to detect and eliminate any leakage that could contribute to background interference. The sniffer probe or gun should never be placed directly in a stream of pure refrigerant because it will shorten the life of the sensing element.

5.27.2 Normal Operation

Good practice dictates that halogen diode leak detectors be allowed to warm up in accordance with the manufacturer's recommendation. When pipe, objects, vessels, and so on are being prepared for halogen diode leak testing, they must be free of moisture, grease, paint, solvents, and dirt that could temporarily plug leaks. When test objects are pressurized with a combination of air and refrigerant, the refrigerant is added first so that the air will tend to mix and disperse the refrigerant throughout the system. Specification or code defines the time allowed for dispersion of tracer gas. The sensitivity of the halogen leak detector test should be checked before and after testing and at intervals of every 2 h during routine tests.

Operators who smoke while performing halogen leak detector tests may note erratic signals caused by a high background of halogen-rich cigarette smoke. Another cause of erratic signals could be high background levels of halogens from a large leak. Other possible causes of erratic operation include excessive heater voltage, shorted element, or too high a sensitivity range setting.

Care should be taken when venting systems down after leak testing in order to prevent halogens from reentering the test area. Operating procedures must be reviewed, and possibly revised, when there are changes in the concentration of the test gas, or change of tracer gas or type of equipment used. Care should also be exercised when using the halogen leak detector methods around plastics and rubber because these materials readily absorb halogens. When absorbed, the materials may emit other gases, causing false indications.

Topics typically covered in halogen diode leak test specifications are:

Scope
Applicable documents
Design criteria
Description of work to be done
Personnel qualifications
Types of leakage tests to be performed
NDT operator requirements
Engineering verifications
Required record systems
Document submittal and approval
Responsibility for final inspection

5.28 VIC MSLD LEAK DETECTORS

The VIC MD-490S is a portable state-of-the-art helium leak detector (Figure 5.16) by Vacuum Instrument Corporation. It has a response time <0.5 s and a sensitivity of 5×10^{-8} atm cc/s as an atmospheric sniffer with the ability to zero out background. The instrument is considered the most aggressive portable leak detector designed specifically for industrial use. Its companion Vibra-

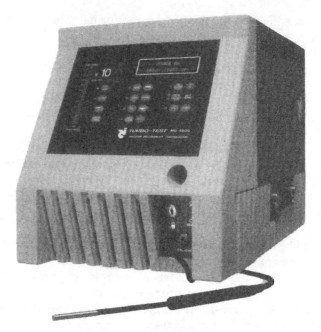

Figure 5.16 Model MD-490S The Real Sniffer™ with standard probe. Courtesy of Vacuum Instrument Corporation.

Leak™ probe is equipped with an auto-zero button and tactile feedback when the leak rate exceeds the reject set point. Therefore, the probe is ideally suited for noisy environments where audio alarms cannot be heard or where leak detectors are in close proximity.

Figure 5.17 shows the construction and helium ion flow within a calibrated MD-490S mass spectrometer tube. Helium ions are focused on the collector while heavier and lighter ions are deflected as shown. When a test object is pressurized with helium and sniffed, any leakage increases both the helium flow and the collector current. The increasing collector current is amplified and the actual leak rate is read on the left side of the front panel display.

Additional instrument features include:

- Total startup time of less than 3 minutes
- MD-490S pushbutton suppression of helium background
- Most sensitive sniffer available with 10^{-8} sensor capability
- Fully interactive system diagnostics of critical parameters
- Graphical and numerical displays of leak rate values
- Sealed membrane keyboard for long life in harsh environments

VIC MSLD LEAK DETECTORS 215

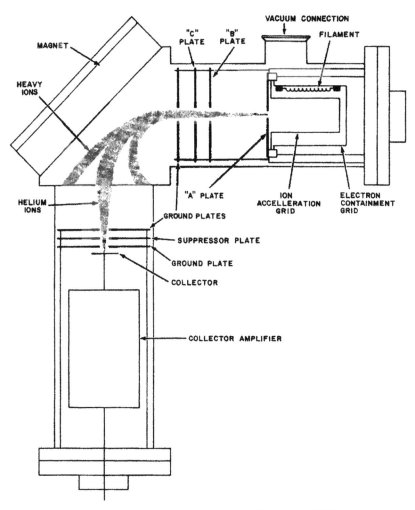

Figure 5.17 MD-490S mass spectrometer tube. Courtesy of Vacuum Instrument Corporation.

- Fully automatic calibration sequence with built-in 10^{-5} atm cc/s leak that prompts operator to insert probe into built-in leak and adjusts the machine's leak rate reading to match the leak value

Other features include dry (oil-free) vacuum pumping, probe blockage alarm, front panel key lockout, multilingual alphanumeric prompting display, and built-in calibrated gas leak accessible through a side panel door.

5.29 MSLD SUBSYSTEMS

Figure 5.18 shows a state-of-the-art Model MS-40 leak detector system by Vacuum Instrument Corporation. This instrument is fully automatic, easy to use, and very accurate. Ease of startup, calibration, and maintenance make it suitable for a broad range of applications.

Both the MS-40 and MS 40-Dry dry leak detectors feature:

- Simple one-button startup in less than 3 minutes
- High inlet pressure (7.5 Torr) for fastest startup to test times
- Built-in diagnostics for trouble-free maintenance
- Fully automatic tuning and calibration

A helium mass spectrometer leak detector (MSLD) consists of three major subsystems and an operator interface. The first major subsystem consists of the spectrometer tube consisting of an ion source, fixed or adjustable source magnets with permanent magnetic field, preamplifier assembly, built-in high-vacuum ion gage, and heater. The second major subsystem is the vacuum system with the high-vacuum turbo pump, one or two mechanical pumps,

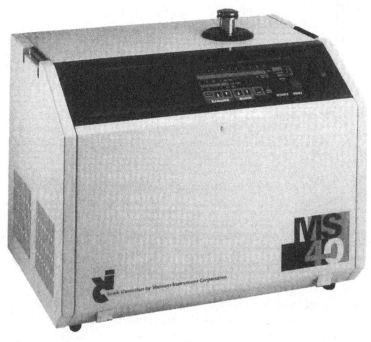

Figure 5.18 MS-40 mass spectrometer. Courtesy of Vacuum Instrument Corporation.

vacuum valves, and Pirani gauges. Pirani gauges measure vacuum system pressure in the range of 0.5×10^{-3} Torr.

The third major subsystem is the microprocessor-controlled electronics unit that automatically displays error codes and messages on an alphanumeric display. This unit also has a real-time clock with automatic backup. Other features of the electronic unit include an RS-232 communications port, printer port, and service/maintenance panel. Finally, the operator interface consists of a control panel that is a user-friendly and highly reliable unit with a sealed-membrane switch panel. Other features include single-button tuning and single-button automatic gain control. An optional hand-held remote control operates start/vent switches and ranging controls and provides leak rate readout up to 100 feet from the leak detector.

5.29.1 Spectrometer Tube

The dual-sector spectrometer tube is shown in Figure 5.19. The ion source sector consists of a pair of filaments (one a spare) that provide the source of electrons that are beamed into the ion chamber where they strike gas molecules and create positive ions. Adjustable dual source magnets direct the helium ion beam and adjust the electron beam for maximum ionization and sensitivity. Repeller electrodes repel the positive ions, forcing them out of the helium beam path. The focusing plate directs the ion beam through slits in the ground potential plates.

The magnetic fields in the spectrometer tube separate helium atoms from the other atoms. Lighter atoms are directed downward and heavier atoms are

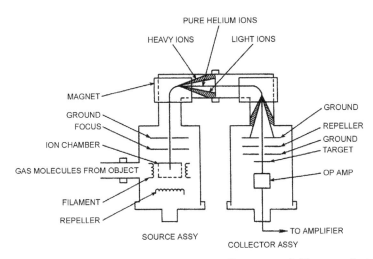

Figure 5.19 Dual-sector mass spectrometer. Courtesy of Vacuum Instrument Corporation.

directed upward by the first magnet. Only the helium atoms, guided by the ground potential electrodes, strike the target or ion collector plate. The ion current is then amplified by the solid-state operational amplifier.

The high-vacuum ion gage monitors the absolute pressure (vacuum) and triggers a protective system if the pressure exceeds 2×10^{-4} Torr. During normal operation, the ion source filaments keep the spectrometer tube hot, regardless of tube pressure.

5.29.2 Vacuum System

There are four ranges of vacuum, which are *rough* or *low vacuum* = 1.0 to 10^{-3} Torr, *medium vacuum* = 10^{-3} to 10^{-5} Torr, *high vacuum (HV)* = 10^{-6} to 10^{-8} Torr, and *ultrahigh vacuum (UHV)* = $>10^{-9}$ Torr, where one Torr = 1 mmHg.

The vacuum system, shown in Figure 5.20, consists of a turbo pump, one or two mechanical pumps, valves, and interconnecting piping. The high-capacity turbo pump removes 52-l/s of gas from the spectrometer tube to reduce pressure to $<2 \times 10^{-4}$ Torr. The two mechanical pumps serve as a roughing pump and forepump. An optional external roughing pump operates at 7 or 16 cfm capacity to evacuate the test port and test piece to the transfer pressure of the unit. The rough/forepump is a 1.5 cfm unit that maintains the proper pressure required for the discharge of the turbo pump.

There are three major electrically controlled valves in the system. They are opened electrically and closed by spring action. The *gross* and *fore/rev* valves

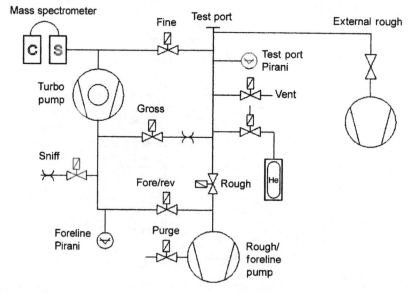

Figure 5.20 MS-40 vacuum diagram. Courtesy of Vacuum Instrument Corporation.

isolate the inlet of the turbo pump from the rest of the system. The rough valve isolates the test port from the *rough/foreline* pump. Other solenoid-operated valves are used for the vent purge, sniff, and helium.

Other features include:

- Three modes of operation—direct mode for higher sensitivity, reverse mode for faster cycle time, and combination mode for automatic switching between reverse and direct flow.
- Minisniffer port—allows simple sniffing operation through a dedicated sniffer port.
- Turbo pump with liquid-lubricated ceramic bearing—greatly increases turbo life; provides less heat and better lubrication than grease-lubricated bearing.
- Solenoid valve designed specifically for vacuum use—valve life up 20 million cycles.
- Dual magnetic sector mass spectrometer with resolving power of 14—lower background levels during testing; leak detector does not mistake adjacent peaks for helium.
- De-scan switch—allows for positive identification of helium at any time during testing.
- RS-232 port—for the integration of automated control systems and personal computers.

5.30 OPERATING SEQUENCE (MS-40 AND MS-40 DRY)

1. Connect the test object to the test port located at the top of the MS-40.
2. Press the green START pushbutton.
3. The reverse/foreline valve closes, the rough valve opens, and the test object is evacuated.
4. When the test port reaches *crossover 1* (7.5 Torr maximum), the reverse/foreline valve opens.
5. If a vacuum testing procedure is underway, spraying of the test object with helium should now commence.
6. Leak testing begins in the 10^{-4} atm cc/s range.
7. The unit will continue to search for a leak, ranging downward until it reaches 10^{-9} atm cc/s range.
8. If a leak is still not found, the unit will automatically begin direct mode testing when the test port pressure goes below *crossover 2*. The rough valve will close, the fine valve will open, and the leak detector will range downward to the 10^{-10} atm cc/s range.

9. If a leak is not found, the unit will continue to test at this level until the VENT pushbutton is pressed. If a leak is found, the unit will range upward to the proper range. The unit stays in direct mode when ranging upward until it reaches the 10^{-4} range, where it crosses back into reverse mode testing.
10. When leak testing has been completed, pressing and holding the VENT pushbutton closes all vacuum system valves except the foreline valve, opens the vent valve, and vents the test port and test object.

5.31 CALIBRATION SEQUENCE (MS-40 AND MS-40 DRY)

1. Calibrator valve opens to the mass spectrometer.
2. Auto ranging of anode, filament DC, focus plates, grid, and repeller voltages.
3. Calibrator valve closes; amplifier offset and system background is monitored.
4. Calibrator valve opens; variables are subtracted and the amplifier gain adjusted to the temperature-compensated calibrator value.

The following topics should be thoroughly reviewed by MSLD technicians:

Background information
    Scope and nature of work
    Applicable publications
    Engineering responsibility and expected performance
Test part requirements
    Surface finish
    Acceptable cleaning techniques
Testing requirements
    Leak rate test procedure
    Test result documentation
    Personnel qualifications
    Testing of subassemblies
    Final acceptance tests
Vacuum pump specifications

6

LIQUID PENETRANT TESTS

6.1 INTRODUCTION

The "oil and whiting" method, formerly used by the railroad industries in the 1920s, was the forerunner of today's liquid penetrant testing (PT). Although widely used by the industry on steel parts, the technique lacked sensitivity and was not applicable to a wide variety of surface defects. With the introduction of magnetic particle testing in the 1930s, the oil and whiting method faded into obsolescence. However, the need continued to exist for a simple, nondestructive test that would detect surface discontinuities in nonmetals and nonferromagnetic materials.

The liquid penetrant method was revived in 1941 when Robert and Joseph Switzer discovered that the addition of visible and fluorescent dyes to the penetrant greatly improved the technique. Continued development work in this field has led to today's improved dyes, penetrants, emulsifiers, and developers. Today liquid penetrant testing is more reliable than radiographic testing for locating minute surface discontinuities. Penetrants are classified or subdivided by the method used to remove the excess penetrant, such as "water washable" and "solvent clean." A good penetrant must penetrate very fine openings, remain in relatively coarse openings, have good wettability, and be easily removed from the surface after testing. It should also be inert with respect to the materials being tested, be nontoxic, have a high flashpoint, and be relatively inexpensive.

Introduction to Nondestructive Testing: A Training Guide, Second Edition, by Paul E. Mix
Copyright © 2005 John Wiley & Sons, Inc.

Since dye penetrant is carried into the surface defects by capillary action, it follows that the technique is limited to the detection of surface defects or subsurface defects with surface openings. Liquid penetrant testing is not an acceptable method with highly porous materials such as unglazed ceramic. In the case of subsurface defects with surface openings, the technique is primarily sensitive to the surface opening and relatively insensitive to the extent of the subsurface defect. Subsurface defects may act as reservoirs for the penetrant and some indication of this may be obtained if excessive "bleeding" is noted after developer is applied. Bleeding refers to the absorption of the dye penetrant by the developer.

Liquid penetrant testing is a simple method to apply. First, a liquid penetrant containing a dye is applied to the surface of a part and allowed to stand for a period of time. During this penetration or dwell time, the dye penetrant is absorbed into the surface discontinuities by capillary action. After the predetermined dwell time has elapsed, excess penetrant is removed from the surface of the part. Finally, a light-colored developer is applied to the surface that draws some of the dye penetrant out of the discontinuity. As the developer absorbs the penetrant, it spreads out, creating indications that are considerably wider and more easily seen than the actual surface defects.

Liquid penetrants are selected on the basis of their penetrating ability and dyes are selected for their brilliance. Colored dyes must be easily seen in visible light and fluorescent dyes must be easily seen in ultraviolet or "black light." In reality, only a small amount of dye is drawn out of the discontinuity during development. Therefore, the dye must be highly visible to the eye and provide a high degree of contrast with the background developer color.

6.2 PROCESSING

Generally, some preparation of the parts is required prior to liquid penetrant testing. Initially, the parts should be clean and free of foreign objects and coatings that could conceal surface defects or cause false indications. Chemical treatment or other suitable means should remove solid contaminants, such as paint, varnish, core and mold materials, and carbon coatings. Rust, scale, lint, and dirt must also be removed from steel plates because these materials tend to trap the penetrant, producing false indications. A gentle brushing with soft wire brushes will usually clean the surface of ordinary steel parts. Sandblasting, sanding, scraping, and grinding are not recommended as cleaning methods because these techniques may peen over small defects and conceal them. When necessary, etching can be used to remove smeared metal from parts that have been previously subjected to one of these cold-working techniques.

Vapor degreasing is the preferred method for removing petroleum products such as oils and greases; a hot-water rinse is not recommended. Acids and chromates should also be avoided when possible because they weaken most penetrants and reduce the ability of fluorescent penetrants to fluoresce. Therefore, steel, aluminum, and magnesium should be tested prior to passivating, anodizing, or chromate treating. Oils and greases also need to be completely removed from test part surfaces because these materials are also good penetrants and many of them tend to fluoresce under black light. Oils and greases tend to fill discontinuity spaces, preventing the entrance of the desired dye penetrant. They can also fluoresce on good surface areas during fluorescent penetrant inspection, thus causing false indications.

It may be surprising to some that all traces of water must be removed prior to liquid penetrant testing. Water by itself is a good penetrant that enters defect cavities and prevents entrance of the desired dye penetrant. Water also dilutes water-soluble dyes. The circulating hot air dryer used with most fluorescent penetrant equipment provides an excellent way of drying parts prior to liquid penetrant inspection.

Liquid penetrant may be applied by dipping, pouring, spraying, or brushing. In large operations, test parts are frequently dipped in a tank containing the desired liquid penetrant. Heating the penetrant prior to dipping is not recommended. Spraying with spray nozzles in a recirculating penetrant system is also used with fixed station equipment. For portability, kits containing aerosol cans are popular. With large parts or equipment, pouring or brushing the penetrant on a specific test surface may be adequate. In most cases, satisfactory test results will be obtained as long as a continuous film of penetrant is applied to the test surface. After dipping or otherwise coating the surface, parts should be set aside and drained during the dwell time period. The dwell time, or time required for the penetrant to seep into the discontinuity, depends on the desired test sensitivity, the type of penetrant, and the characteristics of the defects and test material. Dwell time may vary from a few minutes to as much as one hour.

Removal of the surface penetrant is accomplished in one of two ways. With water-soluble penetrants, a low-pressure, coarse water spray is used to remove excess surface penetrant. With oil-based penetrants, a soft cloth soaked in solvent may be used to hand clean the surface. In both cases, only excess penetrant is removed from the test surface; penetrant is not removed from surface cracks, pores, or other discontinuities.

The third step in the liquid penetrant process is the application of developer to the test surface. The developer acts like a blotter, drawing some of the dye penetrant out of the discontinuity and spreading it above and around the defect surface opening. Developer time, or the time required for a reliable image to appear, is comparable to the penetration time and can therefore range from a few minutes to as much as an hour. Improperly used developers may obscure defect indications, be difficult to remove, and become contaminated.

After a suitable developer time has elapsed, the parts can be inspected. When visible dyes such as red dyes are used, parts are inspected under white light and the operator looks for a red dye contrasting against a white developer background. When fluorescent dyes are used, parts are inspected under black light at a wavelength of 3650 Å or 365 nm, and the operator looks for a bright yellow-green color against a deep blue-violet background. In both cases, the operators are looking for small amounts of penetrant that indicate actual discontinuities. Penetrant on the test table or hands of the operator, or a combination of developer with penetrant can cause false indications.

Post-cleaning should be done as soon as possible after inspection to simplify removal of penetrant materials. Post-cleaning operations are important because it is necessary to remove entrapped penetrant residues that can attract moisture and cause corrosion. In other cases, the removal of residual penetrant and developer is required to prevent interference with subsequent processes or part service. Aluminum specimens must be thoroughly cleaned to prevent wet alkaline developers and emulsifiers from causing pitting. On parts where residual sulfur or chlorine are harmful, low-sulfur and -chlorine materials should be used and parts should be post-cleaned in an automatic detergent wash.

Skilled operators are the key to reliable liquid penetrant testing. Parts must be carefully prepared, the proper dwell time must be determined, care must be exercised when removing excess penetrant, and thoughtful evaluation must be applied to determine the significance of defect indications. The rejection or approval of parts should be based on the design of the part and its intended application. In many cases, thorough post-cleaning must be done to prevent corrosion.

Comparing two sections of artificially cracked specimens can check the overall performance of a liquid penetrant test system. When properly done, liquid penetrant inspection is a valuable NDT tool for isolating parts with defects that could lead to premature equipment failure.

6.3 TEST METHODS

6.3.1 Water Washable Fluorescent Penetrant Process

Water washable fluorescent penetrants have emulsifiers added to an oil base so that they can be easily removed from test parts with a water rinse. They have very good penetrating ability and produce a bright yellow-green fluorescence when exposed to black light. The sensitivity of the fluorescent dyes varies from medium to high. Superbright fluorescent penetrants may be more sensitive in dim light than color-contrast penetrants. Ultraviolet (UV) sensitive spectrometers are used to evaluate the quantitative values of light emitted by fluorescent materials. The relative amount of light emitted by fluorescent materials is compared to the amount of light emitted by other penetrant materials.

Fluorescent penetrants can be applied in a number of ways. Small parts can be placed in a basket and dipped in penetrant tanks. The dipping and draining process can be manual or automatic. The only requirement is that a coating of penetrant must cover the entire surface of the test part to locate possible discontinuities. Generally, drained penetrant is recovered or allowed to drain back into the penetrant tank and recovery systems do not need agitation because penetrants are homogeneous. However, care must be taken because it is relatively easy to contaminate the penetrant with water by subsequent processing.

Dwell time for the penetrant may vary from 5 to 60 min depending on the nature of the defects. Severe discontinuities require less dwell time; minute discontinuities require the longest dwell time. When dwell time exceeds 30 min, some of the penetrant may evaporate, making it difficult or impossible to remove the excess penetrant with a simple water spray. To prevent the penetrant from drying out, parts can be redipped and drained for an additional 5–10 min to restore the washability of the penetrant.

After the proper dwell time has elapsed, the remaining surface film of penetrant is removed by rinsing. A coarse water spray should be used in preference to a solid water stream. Care should be taken to assure that excess penetrant is removed from holes, threads, internal grooves, and sharp corners of machined parts. It is recommended that black light is used during the rinsing operation to assure that the part surfaces and cavities are completely cleaned. Insufficient rinsing of fluorescent penetrants will result in excessive background fluorescence. Fixed spray nozzles can be used for washing parts on large-scale, automatic production lines. For best results, wash water temperature should be 90 to 110°F. Higher temperatures tend to remove penetrant from shallow defects and lower temperatures tend to increase the required washing time. Shallow and broad discontinuities most likely will be missed if parts are overrinsed.

Wet (aqueous), dry, and nonaqueous wet developers can be used to locate defects. Developers blot penetrant, improve the image of the discontinuity, and help control bleedout. Wet developer is applied immediately after rinsing. For the greatest sensitivity in detecting fine cracks, a smooth, thin layer of wet developer is preferred. A newly mixed batch of wet developer should be allowed to stand for 4 to 5 h prior to use. Parts are dipped into a wet developer tank, withdrawn immediately, drained for a few seconds, and placed in a dryer to remove water. A circulating hot air dryer is used to heat parts to 170 to 225°F. Parts should be left in the dryer just long enough to remove water from their surface because excessive heating can also evaporate the penetrant and reduce the resolution and sensitivity of the process.

For the same reason, dryer temperature should be limited to 250°F. The drying operation heats the part and penetrant, helps produce a uniform developer coating, and reduces the viscosity and surface tension of the penetrant. This action enables the developer to more easily draw the penetrant out of the discontinuities, thereby increasing the sensitivity of the process. A good dryer should heat the part to the optimum temperature in a minimum time so that the part is both dry and warm as quickly as possible.

When dry developer is used, parts are dried thoroughly immediately after washing. Dry developer then can be applied to individual parts or several small parts can be placed in a basket and dipped into a tank containing dry powder. When a large number of parts are to be inspected by dipping them in dry developer, and exhaust system should be used. Developer powder may also be applied using a hand-operated powder bulb or air-operated powder gun. Developer should be evenly applied, preferably as a light dusting, and left on the part for a period of time equal to about half of the dwell time; this usually provides ample time for the developer to blot the penetrant in the discontinuities. Blotting is the term used to describe the action of a particular developer in soaking up the penetrant to get maximum bleedout for increased contrast and sensitivity. The time from developer application to inspection is the development time.

Parts are then inspected in a darkened room under black light. Indications will glow brightly, drawing the inspector's eyes to the defect areas. For best results, the room should be as dark as possible and the inspector's eyes should have ample time (not <30s) for dark adaptation. In the past, it was felt that 5 to 10 min was required for dark adaptation; recent military studies indicate 1 min is probably adequate. Reflected black light should be minimized because it impairs the inspector's effectiveness. If it is necessary to inspect parts outdoors, work should be done after dark. Post-cleaning of water washable penetrant can be achieved by using a fine, forceful water spray.

6.3.2 Post-Emulsification Fluorescent Process

This process is more sensitive than the water washable process previously described because it can more easily detect hairline, shallow, or contaminated discontinuities. It is also the most effective process to use if there is a likelihood that some parts may have to be retested. Because of its excellent sensitivity and water washability, the post-emulsification fluorescent method is recommended for testing investment castings.

Post-emulsifier penetrant has an oil base and contains a brilliant fluorescent dye additive. It cannot be removed from the part surfaces by water washing because emulsifiers have not been added. The penetrating quality of this penetrant is excellent. Dwell time with this penetrant typically varies from 2 to 30 min depending on the characteristics of the parts and their discontinuities. Fifteen minutes is a good dwell time to initially use with this penetrant when attempting to determine optimum dwell time.

After a satisfactory dwell time has been determined, emulsifier is applied by dipping or flowing it over the part. Emulsifier should not be applied by brushing it on a part because the brushing action mixes the emulsifier and penetrant irregularly and prematurely, making it difficult to control the emulsification time. The emulsifier combines with surface penetrant and makes the mixture water washable. Penetrant absorbed by surface discontinuities is not emulsified. Ideally, after emulsification, surface penetration can be washed off the surface while retaining all the penetrant in the discontinuity. Emulsification time is dependent on surface condition; smooth surfaces require shorter emulsification times than rough surfaces. Rough surfaces tend to prevent the emulsifier from properly mixing with the penetrant.

Emulsifier time is critical with respect to detecting shallow, hairline discontinuities such as are found with forging laps that may be partially fused. The detection of shallow discontinuities can be lost as a result of excessively long emulsification times. Generally, emulsification time should be kept as short as possible to provide the greatest sensitivity to shallow defects; it must be long enough to provide a good water wash of the surface. Average emulsification time is 3 min, but it can vary from 10 s to 5 min depending on surface conditions. In most cases, the best method for establishing emulsification time is by experimentation.

After emulsification, excess penetrant is removed by water washing with a strong forceful spray. Again, black light should be used to assure that all excess penetrant is removed. When difficulties are encountered during washing, the part can be completely reprocessed using a longer emulsification time. Subsequent development, drying, and post-cleaning is the same technique as previously described.

6.3.3 Reverse Fluorescent Dye Penetrant Process

With this method, the penetrant contains a dark cobalt blue or purple dye. Excess surface penetrant is removed first by hand wiping with a clean dry rag, and then wiping with a solvent-dampened rag. The parts are thoroughly dried with soft, clean, dry rags, and finally, developer, containing a low-intensity fluorescing agent, is sprayed on the part, producing a uniform, even coating. After a suitable developing time, the part is examined under black light. Defect indications appear as dark spots or lines against a lightly glowing background.

This technique is suitable for field testing, providing there is a source of electrical power or battery-operated black light. It has the advantage that a water supply is not required and it produces an indication similar to a weakly glowing radiograph. It is not as sensitive as the post-emulsification penetrant process.

6.3.4 Visible Dye Penetrant Process

Color contrast or visible dye penetrants are available in solvent clean, water washable, and post-emulsification types. Visible dye penetrant methods do not require a source of electricity because the part can be inspected under natural lighting conditions. However, when artificial lighting is used to aid inspection, high-powered spotlights should never be used because they produce too great a glare. Parts processed by any of the visible dye penetrant methods should not be reprocessed by a fluorescent penetrant method because visible dyes tend to kill the fluorescent characteristic of these dyes. As with fluorescent dyes, the visible dye penetrant process is identified by the method of penetrant removal.

6.3.5 Water Emulsifiable Visible Dye Penetrant Process

Water emulsifiable penetrant becomes emulsified when a remover-emulsifier spray comes in contact with it. After a suitable dwell time, the excess surface penetrant is emulsified and rinsed prior to development. Since the rinse does not come in contact with the penetrant in the discontinuity, it is not removed. After removal of the excess penetrant, parts can be dried by heating them in a drying oven at temperatures of 120 to 140°F. Clean, hot air may also be used to dry parts.

When parts are dry, aerosol cans or spray guns having a nozzle pressure <30 psi can apply a thin, even coat of developer. The developer provides a white background coating that contrasts with the colored dye. A thin, even coating of developer is required for highest sensitivity and greatest resolution.

Large flaws will become visible almost immediately, but sufficient development time should be allowed so that small surface discontinuities will also be revealed. Developer time should be approximately equal to dwell time with this process. Red defect indications will appear on a smooth, even white background coating as the developer dries. Some indication of defect depth will be apparent from the intensity of the color and the observed speed of bleedout.

Inspection can be carried out under natural or artificial light and defects will remain visible until the developer is removed. In some cases, penetrant indications are left on the part to show the location and extent of the defect; this provides valuable information to persons attempting repairs.

Dry developer film can be removed by manual wiping or other postinspection cleaning operations. Steel parts may be coated with a good rust preventative because the penetrant process leaves them free of oil and very dry, thus susceptible to corrosion. The total test time is longer with the water emulsifiable visible dye penetrant process.

6.3.6 Water Washable Visible Dye Penetrant Process

Water washable dye penetrant does not require a rinsing aid and therefore it can be easily removed with a simple water spray. This technique provides a simple, fast production technique suitable for processing large numbers of parts. However, it is the least sensitive of the liquid penetrant test methods. Contamination of water washable penetrant will reduce the penetrating quality of the penetrant and adversely affect the washability of the penetrant.

Penetrant is applied by any of the methods previously described. After a suitable dwell time, excess penetrant is easily removed using a wide, fan-shaped water spray at a temperature of 60 to 90°F. Water spray rinse should be continued until the colored penetrant on the surface of the part is removed.

TEST METHODS 229

Excess rinsing and high-pressure rinsing should be avoided to prevent washing penetrant out of the surface discontinuities.

After excess penetrant has been removed and parts have been dried, development and inspection can be accomplished as previously described.

6.3.7 Post-Emulsifiable Visible Dye Penetrant Process

This process is identical to the fluorescent post-emulsifiable penetrant process with the exception that the visible dyes are added to the penetrant.

When wet developers are used with this process, they are usually applied by spraying. The loss of water in a wet developer mix or excessive over concentration of development powder can cause cracking of the developer coating during drying.

Indications from shallow discontinuities are easily lost when emulsification time is excessive. The method has higher sensitivity to tight cracks than water washable methods.

6.3.8 Solvent Clean Visible Dye Penetrant Process

This highly portable test method is the preferred method for field inspections where there is no water supply for the water washable or post-emulsification penetrants. It is also the preferred method where there is no electrical supply for black light operation, which is required for any of the fluorescent penetrant methods.

Parts must be clean and surface defects must be open to be able to accept the penetrant. Prepared solvent cleaners are frequently used to remove oil, grease, and other impurities that could cause false indications. The parts must also be free of paint, rust, scale, carbon, and other solid coatings. After the parts have been cleaned, they are wiped with clean, dry paper towels or rags. If additional dirt is picked up on the towels or rags, the parts should be recleaned.

Penetrant is applied with aerosol cans, coating the entire surface to be inspected. Penetrant should not be allowed to stand on the surface for too long or it will be difficult to remove. When this does occur, the surface should be resprayed with penetrant and cleaned shortly thereafter. When retesting a specimen, the dried penetrant may not completely dissolve and test results may be misleading.

Dwell time depends on the sensitivity of the penetrant, surface condition of the part, type of material, nature of defects, and temperature; the nature of the defect is the most important variable. Heat cracks and fatigue cracks tend to require the shortest dwell time. Forging laps and seams tend to require the longest dwell times. When part temperature is low, the penetrant may become viscous and not amply penetrate the defects. When part temperature is high, the penetrant may flash or evaporate and produce weak indications. Liquid penetrants become sluggish and lose sensitivity at temperatures below 50°F. Viscosity is the most important factor that determines the speed with which the penetrant enters a surface flaw. With this method, recommended dwell time ranges from about 1 to 20 min.

Excess penetrant is removed by hand wiping the surface of the part with a clean dry rag first, and then, a clean rag that has been dampened with the solvent cleaner. Only enough cleaner should be used to obtain a reasonably clean surface; parts should not be flushed with liquid cleaners.

Spray cans of developer should be thoroughly shaken to assure that the developer is uniformly suspended. A light, thin film of developer should be applied to wet the part surface because thick coatings of developer tend to conceal defect indications. Spraying is the most effective way of applying nonaqueous developers; nonaqueous wet developers are preferred when it is important to obtain as smooth and as even a coating as possible.

Almost immediately, large cracks will begin to appear as red lines against a white background. The contrast ratio of red dye penetrant is about 6:1. The contrast ratio is the ratio of light reflected by the background compared to the light reflected by the dye penetrant. It may take several seconds for smaller defects to begin to appear. Developing time depends on the type of developer and the nature of the discontinuities.

6.4 ADVANTAGES AND DISADVANTAGES OF VARIOUS METHODS

Some of the advantages and disadvantages of the various liquid penetrant methods are summarized in Tables 6.1 and 6.2.

TABLE 6.1. Advantages of Various PT Methods

Variable	Water Washable Fluorescent	Post-emulsification Fluorescent	Water Emulsification Visible	Solvent Clean Visible
High sensitivity		X	X	
High visibility	X	X		
High speed	X	X		
Good retest		X		X
Portability			X	X
Intricate parts	X			
Shallow defects		X		
Large parts			X	X
Contaminated parts			X	

TEST EQUIPMENT

TABLE 6.2. Disadvantages of Various PT Methods

Variable	Water Washable Fluorescent	Post-emulsification Fluorescent	Water Emulsification Visible	Solvent Clean Visible
Black light required	×	×		
Poor retest	×			
High staffpower			×	×
Rinsing aid required		×	×	
Low speed			×	×
Poor on shallow defects	×		×	×
Poor on rough surfaces		×	×	×

Figure 6.1 Visible dye penetrant sprays—Cleaner Remover, Developer, and Penetrant. Courtesy of Sherwin, Inc.

6.5 TEST EQUIPMENT

Liquid penetrant test equipment varies from highly portable kits containing aerosol cans to large, high-speed stationary production units, to highly specialized units for testing of custom parts.

Portable inspection materials consist of visible or fluorescent dye penetrants, a penetrant remover, and a developer as shown in Figure 6.1. Complete portable inspection kits also provide a storage case, application materials, cleaning rags, and a suitable set of instructions. When fluorescent dye penetrant is supplied, a small, portable UV lamp may also be included.

Small general-purpose, stationary test units are completely self-contained. Equipment consists of a basket for dipping a number of small parts; a drain area is located next to the penetrant tank so that the parts can sit until the proper dwell time has elapsed; then, a rinse tank is provided with a black light for monitoring the rinsing operation; a drying cabinet with circulating hot air is mounted near the rinse tank. Finally, a developer tank and inspection table with hood and second black light complete the design. Figures 6.2 and 6.3 show fluorescent and visible penetrant process flow diagrams recommended by the ASM committee on liquid penetrant inspection.

In the past, it was common for the metal processing industries to use rows of semiautomatic bath equipment in well-lit rooms that were strictly dedicated to liquid penetrant testing of various parts. Some of this equipment is still in use, but current trends seem to incorporate the use of higher technology featuring fully automated, real-time testing techniques employing ultrasonics, eddy currents, radiography, or a combination of these methods. These techniques are cleaner, do not involve the use of consumable chemicals, and are less personnel intensive.

For color contrast visible dye penetrant inspection, white light illumination levels of 100 foot-candles (fc) at the surface are recommended for detecting

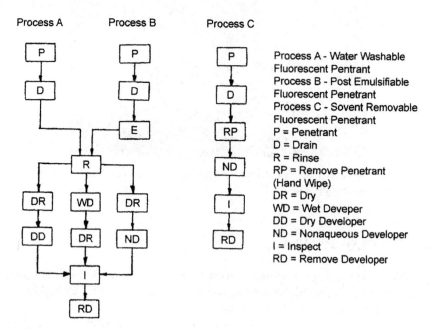

Figure 6.2 ASM flow diagrams for fluorescent penetrant processes, Type I.

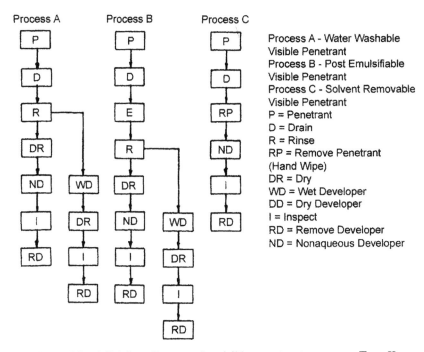

Figure 6.3 ASM flow diagrams for visible penetrant processes, Type II.

small discontinuities. For any of the fluorescent penetrant inspection methods, the inspection area should be as dark as possible to detect small flaws. Extreme care should be taken to avoid contaminating the inspection area with penetrant in order to prevent false indications. For optimum results, inspectors should allow sufficient time (not <30s) for dark adaptation of their eyes. During inspections, inspectors should avoid black light reflection when possible. Reflected black light can cause the inspector's eyes to fluoresce internally, clouding his or her vision. The effect is only temporary, but it adversely affects the inspector's vision and ability to detect defects. Safety goggles with UV absorbers are recommended to reduce eye fatigue.

Since the late 1960s, considerable advances have been made in the development of black light standards, black light lamps, and black light radiometers. Most of this development was stimulated by publication of a memorandum by McKelvey. A summary of this memorandum is shown in Figure 6.4. As a result of this work, today's most common source of black light is the high-intensity discharge (HID) mercury arc lamp with visible light absorption filter.

The UV region of the electromagnetic spectrum is divided into three sections. The UV-C section is the shortwave or far-UV range of 180 to 280 nm, the UV-B section is the medium wave or middle-UV range of 280 to 320 nm, and the UV-A section is the longwave or near-UV range of 320 to 380 nm. All

Research and Technology Division
Air Force Systems Command
United States Air Force
Wright-Patterson Air Force Base, Ohio

TECHNICAL MEMORANDUM Materials Application Division
MAA 67-1 Air Force Materials Laboratory
JANUARY 1967 Research and Technology Division

ULTRAVIOLET LIGHT INTENSITY MEASUREMENT

I. **PURPOSE:**
To present a correlation between footcandles and microwatts per square centimeter ($\mu W/cm^2$) for the measurement of ultraviolet light intensity. The data presented will be used as a basis for revising nondestructive testing methods and equipment specifications.

II. **FACTUAL DATA:**
The data is presented as APPENDIX I.

III. **CONCLUSIONS**
1. The microwatt per square centimeter is the correct unit of measure for ultraviolet light intensity
2. No known government facility is capable of calibrating ultraviolet light intensity meters.
3. A considerable amount of variability can be expected in the intensity of ultraviolet lights.
4. Ultraviolet light intensity meters should be filtered to exclude visible light from the photo cell.
5. The meter filter should provide peak transmission at 3650Å ± 50Å to match filters in Air Force standard ultraviolet lights.

IV. **RECOMMENDATIONS:**
1. Ultraviolet light requirements in Specifications MIL-I-6866, MIL-I-6868 and MIL-I-25135, and Technical Orders 3331-1 and 42cl-10 should be revised to require a minimum intensity of $865 \mu W/cm^2$ at 15" from the filter.
2. Ultraviolet light requirements in Specifications MIL-I-9445, MIL-I-6867 and MIL-I-9909 should be revised to require a minimum intensity of $1020 \mu W/cm^2$ at 15" from the filter.
3. An ultraviolet light intensity meter having a filter to exclude visible light and calibrated in $\mu W/cm^2$ should be listed in Table of allowances 455 after a means of calibration for these meters have been provided by Newark AF Station.

MAA TM 67-1

Project Engineer, Mr. Edward W. McKelvey, Air Force Materials Laboratory Project #7381, Task #738107 "Corrosion Control and Failure Analysis"

Figure 6.4 Air Force memorandum on UV light intensity measurement.

NDT fluorescent magnetic particle inspection and fluorescent liquid penetrant inspection is done in the UV-A section of the electromagnetic spectrum.

Military and other studies continue to confirm that the higher the black light intensity and the lower the ambient light intensity, the smaller the detectable indication. Military specification MIL-L-9909 specifies that a minimum of $800 \mu w/cm^2$ of black light should be present at the surface of the part with a maximum ambient white light of <1 fc. These black light intensity levels are easily achieved with 100w medium-pressure mercury arc lamps, fitted with a visible light-absorbing filter, at a distance of 15in. from the filter to the test surface.

Black light filters are designed to remove white light, protect the human eye, and filter out undesirable UV wavelengths. It should be kept in mind that dirty filters could allow harmful UV radiation to reach the eyes. Dirty filters are the most common cause of decreases in lamp output during the useful life of the lamp.

Figure 6.5 shows a 100w black light that produces a high-intensity spot and a tubular fluorescent black light (15, 30, or 40w units available) that produces large-area floodlight-type lighting. It takes about 5 min. for the black light bulb to warm up and it may go out if lamp voltage drops below 90 V. Both lamps are relatively portable and can be carried to various field inspection areas. The output of present-day mercury arc lamps remains reasonably constant over their useful life span. However, fluctuating line voltages can significantly decrease the useful life of the black light bulb.

Figure 6.5 Spot and fluorescent types of portable black lights. Courtesy of Spectronics Corporation and ESC/Econospect Corporation.

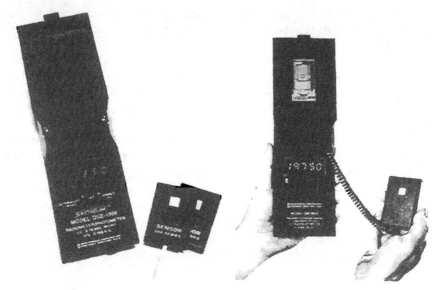

Figure 6.6 Radiometer and radiometer/photometer for measuring black light. Courtesy of Spectronics Corporation.

Figure 6.6 shows a Model DSE-100X radiometer/photometer capable of accurately measuring UV intensity in microwatts per square centimeter and visible light in foot-candles. UV intensity is measured by the Model DM-365X. Meters such as these are invaluable for measuring the amount of light produced by black lights and reflected by fluorescent penetrants.

6.6 PENETRANT MATERIALS

Liquid dye penetrants should have good penetrating characteristics or be able to enter fine cracks. They should also be easy to remove from part surfaces after a suitable dwell time has elapsed. In combination with their dyes, penetrants should have high visibility. The liquid carrier of the penetrant should have high visibility. The liquid carrier of the penetrant should hold the dye in suspension, distribute evenly over the surface of the part, and carry it deeply into the surface defect. Penetrants should have a moderate body, some volatility, and moderate viscosity. They also need to have good wetting action so that they will spread evenly and completely over part surfaces. Ideally, penetrants should be inert and noncorrosive with respect to test part materials.

From a safety standpoint, penetrants should have a high flashpoint and not readily form a flammable or explosive mixture. However, most aerosols found in portable dye penetrant kits are flammable and contain warning labels. When these flammable materials are used, they should not be used around open flames or other possible sources of ignition such as welding machines and sparking devices.

Penetrant materials should be nontoxic. However, most penetrants contain detergents (wetting agents) that can remove oils from the skin causing drying, irritation, and perhaps even cracking with prolonged contact. Some persons may also be allergic to solvents in penetrants. For these reasons, hands and arms should be washed with warm soapy water when exposed to penetrant. After cleaning, a good skin cream or oil should be applied to minimize irritation and replace body oils.

The most important characteristics of penetrants are brightness and contrast. With color contrast penetrants, large quantities of dye with dense colors and high contrast are sought. The white background color of the developer is used to enhance the color contrast ratio of the dye. Although color contrast ratios of 20:1 are theoretically possible, a good red dye produces a color contrast ratio of about 6:1 with the white developer background.

With fluorescent penetrants, the bright color of dye contrasts with darkness and color contrast ratios of 100:1 are possible. Some manufacturers offer electronic instruments specifically designed to measure the fluorescent of penetrants in accordance with military standard MIL-I-25135. The instrument, shown in Figure 6.7, features 10-turn zero and span potentiometers, negligible drift, built-in 4w 366nm UV lamp, primary and secondary filters, and sample holder with window as specified in the reference standard. This photometer measures color contrast ratios as high as 99.9:1.

The presence of moisture in water washable penetrants is the most likely cause of failure. As water washable penetrants become contaminated with

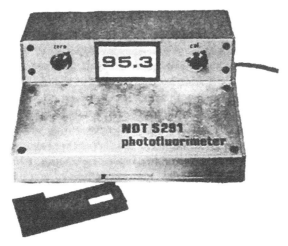

Figure 6.7 Photometer for measuring the fluorescence of fluorescent dye penetrants. Courtesy of NDT Italiana, S.A.S.

water, their cloudy appearance and viscosity increase with increasing water content. As viscosity increases, penetrants become more difficult to remove by water washing. The water content of water washable fluorescent dye penetrant can be determined by the ASTM D95 water test. Other materials that can contaminate penetrants are cleaning solvents, oils, acids, chromates, dirt, and salt.

When visible dye penetrants are used, care must be taken to prevent contaminating solvent cleaners with dye materials. Solvent cleaners are highly volatile and should be kept in closed containers to minimize evaporation.

Observing the ease with which penetrant can be water washed provides one indication of emulsifier contamination. If the rinsing of excess penetrants from the test surface becomes difficult, the emulsifier should be checked for water contamination. ASTM D95 specifies one water test, which uses water-free xylene. Penetrant materials can also contaminate emulsifiers, reducing their effectiveness.

Wet and dry developers are normally used with water washable fluorescent penetrants; solvent-based developers are normally used with visible color contrast penetrants. Dry developers are relatively trouble free as long as parts are thoroughly dried prior to dipping. Moisture will cause lumps to form in the dry powder, decreasing its effectiveness. Mild contamination of dry developers can also occur if excess developer soaked with penetrant drops back into the developer tank after part coating.

Wet developers are made by mixing dry powders with water. When too much water has been lost by evaporation or there is an excessive concentration of developer, the developer coating can crack. Freshly made developer should be allowed to stand 4 to 5 h then tested with a hydrometer. Developer concentration can be adjusted upward by adding more powder or downward by adding more water. The uniformity of wet developer should be checked periodically. Wet developers contain wetting agents to promote surface wetting. If the developer fails to evenly coat the part surface, the developer should be replaced and the developer tank should be thoroughly cleaned. Developers that have been allowed to stand on part surfaces for long periods may be difficult to remove. When this occurs, detergent should be added to the wash water to facilitate removal.

Solvent developers are widely used with visible penetrants and portable field kits. Since they are used only once, the only precaution (other than flammability) that needs to be taken is to avoid contaminating the developer with colored penetrant. Contamination of developer, like contamination of solvent cleaner, can create false indications and/or mask true dye penetrant indications.

6.7 SYSTEM COMPARISONS

Thermal cracks can be induced in aluminum blocks to produce standards for comparison of liquid penetrant system sensitivity. These cracked blocks are produced by heating the center of a 3 × 4 × 5/16-in. type 2024-T3 aluminum block from 950 to 980°F and then quenching it in cold water. The block is then turned over and heating and quenching procedure is repeated. The 4 in. dimension of the block should be in the direction of rolling. Temperature is checked with a 950 to 980°F Tempilstik or Tempilac, or equivalent. After both sides of the block are cracked, a 1/16-in.-wide × 1/16-in.-deep groove is cut along the 3-in. direction across the center of the heat-affected zone on both sides of the block. This produces two similar test areas on each side of the block. Half of the block on one side can be treated with a familiar liquid penetrant system and the other half of the block on the same side can be treated with a new penetrant system for comparison. Cracked aluminum blocks are equally effective for evaluating color contrast or fluorescent penetrant systems along with their required emulsifiers and developers.

Cracked plated strips are sometimes used to compare the sensitivity of liquid penetrant systems. The plates have brittle iron plating that cracks when the plates are bent on cantilever and radial bending dies. When cantilever bending dies are used, the stress cracks are closer together and tighter near the clamped end.

Both techniques provide relative comparison data. Cracked aluminum blocks will not have identical cracking on both halves and cracked plated strips will not have identical cracking patterns. However, the cracks produced should be similar enough in appearance for comparison purposes. If the differences in penetrant systems are not obvious, their sensitivities may be considered comparable for cracks of this nature.

6.8 APPLICATIONS

Some proven applications for liquid penetrant testing are:

- The inspection of tools and dies
- The inspection of tanks, vessels, reactors, piping, dryers, and pumps in the chemical, petrochemical, food, paper, and processing industries
- The inspection of diesel locomotive, truck, and bus parts, particularly axles, wheels, gears, crankshafts, cylinder blocks, connecting rods, cylinders, transmissions, and frames
- The inspection of field drilling rigs, drill pipe, casings, and drilling equipment
- The inspection of aircraft engine parts, propellers, wing fittings, castings, and so on

Some of the defects that can be detected with liquid penetrant testing are shown in Table 6.3.

Procedures for liquid penetrant testing typically cover the topics given in the following outline.

TABLE 6.3. Detectable Discontinuities and Indications

Type of Defect	Description
Casting porosity	Spherical surface indications
Porosity (glass)	Same as above
Casting cold shut	Dotted or smooth continuous lines
Cracks	Straight or jagged continuous surface lines
Hot tears	Ragged line of variable width, numerous branches
Sand casting shrinkage	Crack indications where part thickness changes
Forging lap (Al)	Sharp cresent-shaped indication on aluminum
Forging lap (partial)	An intermittent line indication
Rolling lap	Continuous line on rolled bar stock
Crater crack (Al)	Dish-shaped indication with spoke propagation
Crater crack (deep)	Rounded indication
Laminations (rolling)	Seams on rolled plate
Inclusion (rolling)	Broad elongated indications in rolled plate
Heat-treat cracks	Multiple irregular lines in finished goods
Thermal cracks (glass)[a]	Jagged interconnecting lines fired ceramics
Lack of fusion (welds)	Broken line of varying width near centerline
Fatigue cracks	Continuous lines in parts that have been in service

Nonrelevant Indications

Indications caused by part geometry or design
Those caused by contamination of inspection table or inspector's hands
Faint indications that do not appear when reprocessed

---

[a] For fine cracks in glass, the filtered particle method is best. With this method, penetrant enters the line cracks, leaving larger dye particles behind, thus concentrating the dye and making the indication more visible.

Nondestructive Examination Specifications and Procedures

Title

Scope

Parts to be inspected
Materials to be used
Definition of area to be examined

Personnel Qualification

Personnel certification requirements

Penetrant Materials Approval

Materials certification
Applicable ASTM codes
Temperature considerations

Nondestructive Examination Specifications and Procedures (Cont.)

Title
Surface Preparation

Visual examination
Precleaning
Drying
Specialized cleaning methods

Penetrant Application

Manufacturer's recommended procedure
Temperature requirements

Excess Penetrant Removal

Wiping
Flushing
Checking

Drying

Normal evaporation
Hot air application

Developing

Manufacturer's recommended procedure
Temperature requirements
Agitation before application
Application techniques

Evaluation of Indications

Definitions of relevant and nonrelevant indications

Acceptance Standards

Definitions of acceptance

Repair and Reexamination

Compliance with applicable codes

Postcleaning

Requirements

Nondestructive Examination Specifications and Procedures (Cont.)

Title
Procedure Qualifications
Deviations from established procedures
New procedures
Penetrant Examination Record
Acceptable forms
Information to be included
Safety
Flammability and toxicity considerations

6.9 MEASUREMENT OF UV AND VISIBLE LIGHT

The measurement of UV and visible light is very important for proper defect detection and resolution. When using fluorescent penetrants, high-intensity UV lamps are frequently used for inspection and visible light is significantly restricted. For daylight inspection with high-contrast penetrants, bright daylight or high-intensity white light is preferred. Therefore, modern light meters, capable of measuring the intensity of both visible light and UV light, have been optimized for NDT and other applications.

Gigahertz-Optik specializes in the measurement of light by measurement with light. The functional mode of the X9 8 Optometer (Figure 6.8) is typically limited by Gigahertz-Optic for NDT applications. Front-panel labeling is available for most standard applications and private labeling is available for OEM applications. As a result, the X9 8 has only three front-panel buttons for setup of specific applications. This simpler setup and operation enables inexperienced operators to become familiar with the optometer in a minimal time.

Main features of series X9 Optometer include:

- Meter setup for specific applications
- Multiple functions
- Easy-to-use ergonomic design
- Large, easy-to-read display
- High reliability
- Economical price
- OEM labeling
- Battery operation

MEASUREMENT OF UV AND VISIBLE LIGHT 243

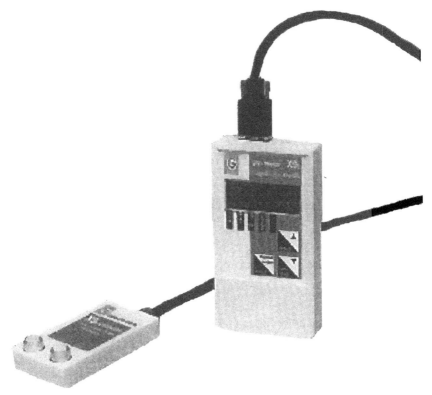

Figure 6.8 Gigahertz-Optik X9 8 Optometer for NDT applications. Courtesy of Gigahertz-Optik Company.

Model numbers for some standardized applications are listed below:

X9 1—photometer for illuminance and luminance measurements
X9 2—UV curing irradiance meter
X9 3—laser power meter
X9 4—radiant and luminous flux meter
X9 6—UV-A, UV-B and irradiance meter
X9 7—general irradiance meter
X9 8—illuminance and UV-A irradiance meter for nondestructive testing application
X9 9—illuminance and UV-A irradiance meter for photostability application
X9 10—optical power meter for telecommunications testing

Note: Please contact the manufacturer for OEM applications.

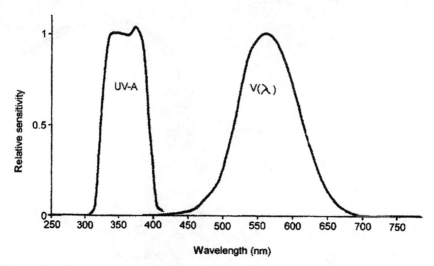

Figure 6.9 Spectral responses of both X9 8 detectors. Courtesy of Gigahertz-Optik Company.

The portable, hand-held, lightweight X9 8 Optometer is ideally suited for nondestructive testing and visible inspection programs. UV-A light irradiance over the range of about 310 to 400 nm and visible light illumination over the range of about 460 to 680 nm is measured as shown in Figure 6.9.

Additional features for the X9 8 hand-held meters include:

- Dual detector housing
- Flat UV-A & DIN Class B lux detectors
- RS232 remote control of data collection
- Traceable calibration
- Compact design for field use
- Wide measurement range

Functional operation modes for the X9 8 Optometer are:

- UV-A irradiance measurement
- Illuminance $V(\lambda)$ measurement
- Two detectors in multiplex operation
- Menu detector cal—factor selection
- CW snapshot hold function
- RS232 remote control

The X9 8 is especially useful for liquid penetrant testing (PT), magnetic particle testing (MT), and visual testing or inspection (VT) or (VI). UV-A illu-

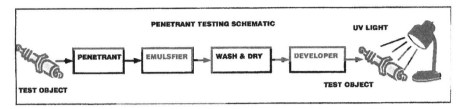

Figure 6.10 Penetrant testing schematic. Courtesy of Gigahertz-Optik Company.

mination (Figure 6.10) and inspection is becoming increasingly favored in liquid penetrant testing. ASTM, MIL, and DIN standard practices help ensure uniformity of these examinations.

6.10 AUTOMATIC AND SEMIAUTOMATIC PENETRANT TESTING METHODS

With new machine vision and laser techniques, rapid processing and display of fluorescent penetrant indications are currently being developed. As mentioned in Chapter 4, new procedures have been developed to obtain high-resolution images of surface breaking cracks for critical part surfaces using a laser-based fluorescent penetrant inspection method. The technique is applicable to tubing, flat plates, and complex surfaces, which can be automatically and remotely inspected. The test can reliably detect crack widths of less than 1 mm (0.039 in.) and map them in near real-time.

A focused laser spot can be translated over a treated surface, either in a helical path (tube scan) or with a raster motion, in the case of a flat plate scan. The high-resolution image is a function of the size of the focused laser spot and the data sampling density. With the configurations described, focal spot sizes as small as 0.025 mm (0.001 in.) can be obtained, enabling operators to achieve high-resolution imaging of small cracks in part surfaces that might otherwise go undetected. Once scan data have been acquired, operators can measure feature geometry, print out hardcopy images, and archive data for future analysis and comparison.

7

MAGNETIC PARTICLE TESTING

7.1 MAGNETIC PRINCIPLES

Materials containing iron, nickel, and cobalt are strongly attracted to themselves and to each other when magnetized; they are called ferromagnetic materials. Other materials, such as oxygen, which are weakly attracted by magnetic fields, are called paramagnetic materials. Diamagnetic materials, such as evacuated hollow glass spheres, are slightly repelled by a magnetic field.

As schoolchildren we first learned about the earth's magnetic field, its poles, and their effect on a compass needle. It is believed that the Chinese first discovered the magnetic effects of lodestone (iron oxide that had been magnetized by lightning strikes) around 2700 B.C. They are also believed to have constructed crude compass needles, which were used in early navigation. According to the Book of Mormon, the lost tribe of Israel found a magnetic object, built a ship, and sailed to the new continent.

What are some of the basic magnetic principles we learned in school? We learned that a compass needle points to "magnetic north," which is not the same as the "true north" that the earth rotates about. If we were lucky enough to have a teacher who had access to a handful of bar magnets, we quickly learned that like poles repelled each other and opposite poles attracted. By restraining their horizontal motion, we could float one magnet above the other if their like poles were properly aligned. Without touching the second magnet, we could push and repel it until it spun around and violently crashed into the

---

Introduction to Nondestructive Testing: A Training Guide, Second Edition, by Paul E. Mix
Copyright © 2005 John Wiley & Sons, Inc.

magnet we were holding. We learned how strong the relatively small magnets were as we attempted to pull opposite poles apart or force like poles together. Perhaps we even visualized magnetic cars floating over magnetic highways at superspeeds.

Later, we were allowed to experiment with iron fillings or magnetic powders, and we learned even more about these invisible lines of force that were called *flux*. We learned that the lines of flux ran through the magnets from south to north, exiting the north pole and reentering the south pole. By placing a bar magnet under a piece of white paper and sprinkling iron powder above it, we could observe the patterns formed by the magnetic lines of flux. Figure 7.1 shows the magnetic flux pattern for a rectangular bar magnet. We also learned that one line of flux per square centimeter was called "one gauss" and that the lines of flux formed closed loops that never crossed. The flux density could not be measured or observed as long as it was contained in the magnetized object. Only when the lines of flux exited the magnetized material could the effects of magnetism be observed. The magnetic lines of flux are most dense at the poles of a magnet and they seek the path of least magnetic resistance.

We probably made at least one more observation about magnetism at a relatively early age. That is, if a magnet is cut or broken into two parts, we end up with two magnets, neither of which is as strong as the original magnet. As a matter of fact, we can continue to break the magnet into smaller and smaller parts and the parts will still retain their magnetism. From this observation, we can conclude that a large magnet is constructed of many smaller magnetic

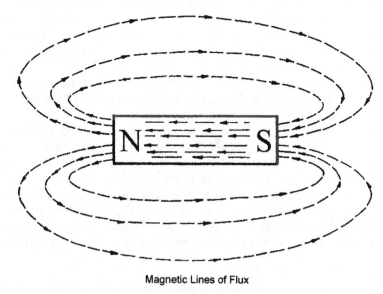

Magnetic Lines of Flux

Figure 7.1 Magnetic lines of flux from a bar magnet.

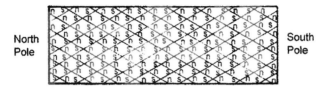

Figure 7.2 Illustration of the theory of magnetic domains.

domains, as illustrated in Figure 7.2, and can define a magnetic domain as the smallest particle capable of retaining a magnetic field and preferred magnetic orientation. It also follows that the strength of the magnetic domains varies as the magnetic properties of the material vary.

Magnetic particle testing (MT) makes use of the magnetic properties of ferromagnetic materials to locate surface and subsurface discontinuities in manufactured materials and parts. The permeability of the part determines how easily it can be magnetized; parts with low permeability are hard to magnetize. Surface and subsurface discontinuities, such as cracks and seams, cause a break in the magnetic uniformity or sudden change in the permeability of a magnetized part. These breaks in magnetic uniformity set up minute magnetic poles and leakage flux paths attract fine magnetic powder particles to the discontinuity. The fine magnetic powder accumulates at the discontinuity, lowering the magnetic reluctance of the leakage flux and forming a reliable visual indication of the discontinuity.

7.2 MAGNETS AND MAGNETIC FIELDS

There are three types of permanent magnet materials in common use today. The Alnico series of metallic and the Indox family of ceramic magnets are brittle because of their composition and fabrication by the sintering process. The typical composition of an Alnico V magnet is 8% aluminum, 14% nickel, 24% cobalt, 3% copper, 1.25% titanium, and the balance iron. Because their Curie temperature is 1625°F, they are extremely stable at relatively high temperatures. The Curie temperature is the transition point where ferromagnetic properties disappear and the magnetic domains become randomly oriented; it is a temperature below the melting point. Nonconductive ceramic magnets are produced by calcination of barium, lead, or strontium ferrite, compaction, and final sintering. They are lighter in weight and more resistant to demagnetization by external or coercive forces. The Cunife family of magnets consists of a ductile copper base alloy with nickel and iron. Their preferred direction of magnetization is in the direction of rolling. Cunife magnets are the easiest to machine and the least expensive.

There are also magnetic rubbers and plastics, usually containing barium ferrite, and so-called "rare earth" magnets such as platinum-cobalt and samarium-cobalt. Rare earth magnets have extremely high-energy products, about 20 million gauss-oersteds (G-Oe). Some of these are discussed briefly later in this book. Present-day permanent magnets are unaffected by shock, vibration, and temperatures below their Curie temperature. The brittleness of Alnico and ceramic magnets can be compensated for to some degree by various coating and encapsulation techniques.

Magnetic particle testing incorporates the use of both longitudinal and circular magnetic fields. Bar magnets are magnetized along their length and are longitudinally magnetized; circular magnets contain a circular magnetic field and are circularly magnetized.

Figure 7.3 shows how cracks form localized magnetic poles and affect the longitudinal field of a bar magnet. It should be noted that magnetic fields are three-dimensional in nature and hard to illustrate in two-dimensional sketches. To magnetize a bar magnet, it is first placed in a coil (Figure 7.4) and then sufficient current is applied to saturate it. A material is considered

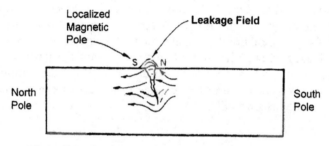

Figure 7.3 Leakage flux from a cracked bar magnet.

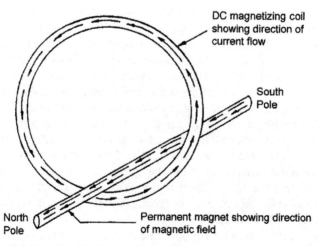

Figure 7.4 Longitudinal magnetism induced by a current-carrying coil.

MAGNETS AND MAGNETIC FIELDS

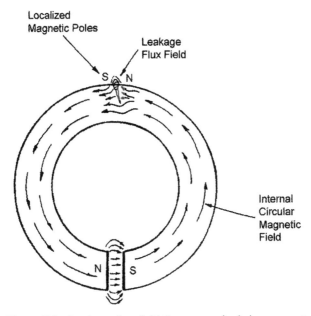

Figure 7.5 Leakage flux field from a cracked ring magnet.

completely saturated when further increases in current have no further effect on increasing magnetism. Using a modification of the right-hand rule, if you curve the fingers of your right hand in the direction of coil current, the thumb of your right hand will point in the direction of the north pole of the magnet. Coil or longitudinal magnetization is especially effective in locating transverse cracks and discontinuities. Longitudinal cracks cannot be easily detected.

Figure 7.5 shows the effect of a crack on a circular magnet. Again, localized magnetic poles are formed and lines of flux follow the path of least resistance. Placing a current-carrying conductor through the longitudinal axis of ring magnets or tubular elements as shown in Figure 7.6 can circularly magnetize these materials. Sufficient current is passed to saturate the parts. Applying the right-hand rule, if you point the thumb of your right hand in the direction of the magnetizing current, the fingers of your right hand will curl in the direction of the circular magnetic field. Circular magnetization is especially effective in locating longitudinal cracks and discontinuities. Transverse cracks cannot be easily detected.

Since air has a permeability of one and a relatively high magnetic reluctance, larger air gaps between magnetic pole pieces result in weaker magnetic fields. Horseshoe and circular magnets with small air gaps have very strong magnetic fields. Most meter movements incorporate this feature in their design. Soft iron "keepers" (Figure 7.7) are frequently placed across the pole pieces of strong magnets prior to shipment. This minimizes the leakage flux of

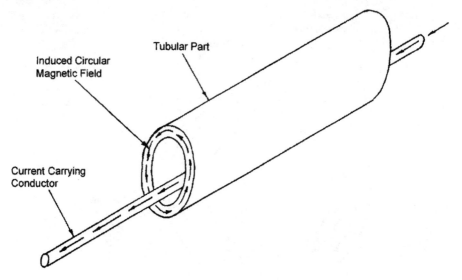

Figure 7.6 Induced circular field in a hollow ferromagnetic part.

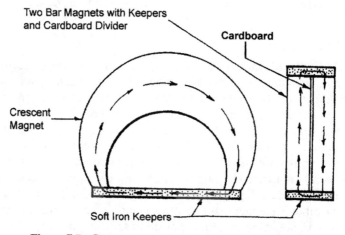

Figure 7.7 Permanent magnets with soft iron keepers.

the individual magnets and reduces the external attraction of the magnets being shipped.

7.3 DISCONTINUITIES AND DEFECTS

Magnetic fields are induced into ferromagnetic parts to detect discontinuities and defects. A defect may be thought of as a discontinuity severe enough to

DISCONTINUITIES AND DEFECTS

warrant rejection of the part. For example, let's say that a part may be rejected when crack depth is three or more times the crack width dimension in a thick part or more than one-third the thickness of a thin part. The manufacturer needs to clearly define what constitutes a rejectable defect in the manufacturing process. One important question that should be answered is, "What is the maximum size discontinuity that can be tolerated and still guarantee that premature failure of the part will not occur?"

When the magnetic field is at a right angle, 90° to the discontinuities, it will cause an abrupt change in permeability of the part. Leakage flux will emanate from the discontinuity and magnetic poles will be established at opposite edges of the discontinuity. These leakage fields attract finely divided magnetic powders used in wet and dry magnetic particle testing, causing visual indications to form. When magnetic flux lines are parallel to a discontinuity, little or no indication is formed. Magnetic particle testing is most effective on surface discontinuities, somewhat less effective on discontinuities lying just beneath the surface, and ineffective on deeper subsurface defects. Defects open to the surface produce sharp, clear indications, whereas subsurface discontinuities produce wide, fuzzy magnetic particle indications. Defects open to the surface are the most detrimental to service life of the part because they are localized weak spots that promote fatigue cracking or trap corrodants.

Magnetic particle testing should not be so sensitive that superficial surface scratches confuse the technician's interpretation of particle indications. However, if scratches are of concern, it should be remembered that they would be extremely hard to detect if there are other defects in the part oriented in the same direction. This aspect of magnetic testing will be covered in more detail in text sections dealing with wet and dry magnetic particle inspection techniques.

Earlier we learned that longitudinal defects in magnets are easily detected by circular magnetic fields and transverse defects in magnets are easily detected by longitudinal magnetic fields. With manufactured parts, we have several added complications. How can we set up an appropriate magnetic field in an odd-shaped object? How do we detect radial cracks, cracks penetrating in a radial direction toward the centerline of the part? The direction of the magnetic field and relative location of the discontinuity are of prime importance for reliable magnetic particle testing. In some cases, complicated-shaped parts are first tested with a longitudinal magnetic field and then reinspected with a circular magnetic field. In other cases, parts may be inspected using simultaneously induced circular and longitudinal magnetic fields.

Generally, we do not want to make permanent magnets out of manufactured parts when we induce magnetic fields in them. Magnetic fields are induced for the sole purpose of detecting discontinuities or defects that are of sufficient magnitude to justify further inspection or part rejection. In most cases, the parts will have to be demagnetized after inspection so that they will not have an adverse effect on the performance of the part or final assembly.

There are many ways to induce magnetic fields in ferromagnetic parts. Each technique has specific advantages and disadvantages.

7.4 INDUCED MAGNETIC FIELDS

With the exception of superconducting electromagnets, the same techniques used with charging or magnetizing magnets can be used to magnetize ferromagnetic parts for subsequent magnetic particle testing. These devices fall into four general categories, namely, permanent magnet magnetizers, direct current (DC) electromagnets, half-cycle impulse chargers, and capacitor discharge magnetizers. In addition to these DC or rectified alternating current (AC) methods, continuous AC magnetization can be used with magnetic particle inspection because most detectable discontinuities are near the surface and permanent magnetization is not desired.

Circular or U-shaped Alnico V permanent magnets with air gaps of $<1\frac{1}{4}$ in. can be used to produce longitudinal magnetization in short bar-shaped parts. This technique is seldom used in magnetic particle testing because of the size limitations, the difficulty in removing parts from the continuous field, and the likelihood of contaminating the magnetic pole pieces with magnetic powders.

Direct current magnetizers or electromagnets have one or two coils wound on a laminated soft iron C-shaped frame. They produce a longitudinal field of about 300 oersteds (Oe) across the air gap when direct current is applied to the coils. Pole diameter ranges from 1 to 3 in., with air gaps ranging from $1\frac{1}{2}$ to $3\frac{1}{2}$ in. Coils are usually rated at 10,000 to 50,000 ampere-turns (AT), but they find limited use because of the rapid heating caused by passing continuous direct current through the coils. The electromagnet is also restricted to the inspection of flat or slightly curved surfaces. Direct current and rectified AC magnetizers provide the deepest penetration of magnetic fields.

Half-cycle or half-wave magnetic chargers are also known as rectified AC magnetizers or half-wave direct current (HWDC) chargers. Power requirements for this type of magnetizer are generally quite high, dictating the need for a fixed unit location near a 220 or 600 V power line and in close proximity to a power transformer or electrical substation. Half-cycle chargers are generally used with current transformers having a single-turn secondary loop. Water-cooled, line-operated ignitrons also can be used with pulse durations of up to 8 ms. Rapid operation of the half-cycle charger is possible because there are no capacitor banks to charge and recharge. Magnetizing forces of 15,000 to 20,000 Oe are possible using this equipment. Half-wave magnetizers are most frequently used with dry magnetic particle testing. Three-phase rectified AC magnetizers are frequently used in horizontal units with wet magnetic particle testing.

Capacitor discharge impulse magnetizers (Figure 7.8) operate on stored energy. For several seconds, energy is stored in a capacitor bank. After charging to a predetermined voltage, the energy is discharged through a

INDUCED MAGNETIC FIELDS

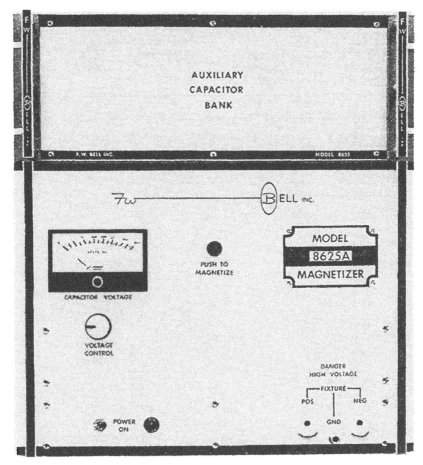

Figure 7.8 Capacitor discharge impulse magnetizer. Courtesy of F. W. Bell, Inc.

unidirectional switch, ignitron, or silicon-controlled rectifier (SCR). The unit is powered by conventional 115-VAC power and can be easily transported from one location to another. A simplified diagram for basic capacitor discharge impulse circuitry is shown in Figure 7.9. The shape and duration of the discharge pulse (Figure 7.10) is dependent on the capacitance of the storage bank, resistance and inductance of the current transformer or magnetizing fixture. Pulse duration of one to several tenths of a millisecond is typical. The high-current, short-duration pulses produce very little heating and power line requirements are quite reasonable, varying from 1 to 10 A at 115 VAC. With 0.5 ms pulse every 5 s, the effective operating time of the unit is 0.01%.

The superconducting electromagnet has a coil that is maintained at liquid helium temperature and a room-access Dewar container. Magnetic material is placed in the center of the magnetizing coil without being exposed to

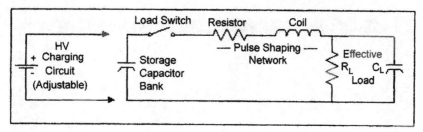

Figure 7.9 Basic capacitor discharge impulse generator.

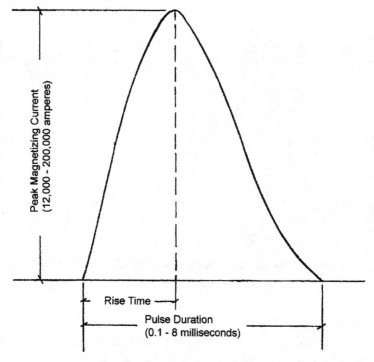

Figure 7.10 Current pulse from a capacitor discharge impulse generator.

the extremely cold temperature. Superstrength rare earth magnets can be made using this equipment because continuous magnetic fields of 60,000 to 90,000 Oe can be achieved. However, the superconducting electromagnet has no practical application in magnetic particle testing.

Alternating current magnetization at line frequency is the most effective magnetic particle testing method for detecting surface discontinuities such as fatigue cracks. It has the added advantage that parts are easily demagnetized

7.5 CIRCULAR AND LONGITUDINAL FIELDS

A singular conductor with an electric current passing through it is surrounded by a circular magnetic field as shown in Figure 7.11. According to the right-hand rule, if we were to grab the conductor with our right hand, with our right thumb pointing in the direction of the current flow, our fingers will be pointed in the direction of the circular magnetic lines of force or magnetic field. The flux within and surrounding a current-carrying conductor is its magnetic field. If the same current is passed through 1- and 2-in.-diameter bars, the circular magnetic field will be twice as strong at the surface of the 1-in.-diameter bar. The magnetic field at the center of both bars will be zero.

If alternating current is used, both the direction of current flow and the direction of the circular magnetic field will reverse or change direction every 180° or every half cycle. With ordinary 60 Hertz (Hz) line frequency, the current flow and circular magnetic field reverses its direction 120 times a second. Alternating current is less effective in locating subsurface discontinuities because of the so-called "skin effect" that limits the depth-of-field penetration. The magnetic field within the part is strongest when current is flowing. With DC magnetization, relatively strong residual magnetic fields may exist after the current stops flowing.

Circular magnetic fields may be induced in ferromagnetic parts by three techniques, namely, direct induction heads (head shot), direct induction prods, or a central conductor. The central conductor is the most effective technique for magnetizing hollow cylinders or tubular-shaped parts.

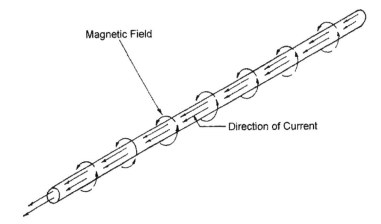

Figure 7.11 Circular magnetic field produced by a current-carrying conductor.

Long, solid, cylindrical objects can be easily circularly magnetized by passing current through them. The ferromagnetic part is usually placed between headstocks having large, soft contact areas and a high current is passed through the part. With this type of direct induction, surface cracks or discontinuities parallel to the length of the part are readily detected as illustrated in Figure 7.12.

Irregularly shaped parts can be inspected with circular magnetic fields in a number of ways. First, current can be passed through the part in more than one direction if possible. Soft lead or copper braid contact plates are used with irregularly shaped parts to minimize the possibility of burning the part. Each time the direction of current flow is changed, a circular magnetic field with a new direction is established. With small irregular objects, contact clamps, similar to battery clamps with a soft braid lining, can be used to pass the current through the part. In other cases, such as when the object is large as well as irregular, the part can be tested with direct induction prods as shown in Figure 7.13. The amperage used in prod testing depends on the distance

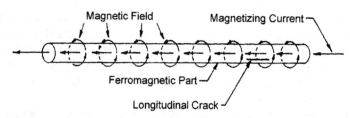

Figure 7.12 Detection of longitudinal crack with circular magnetization.

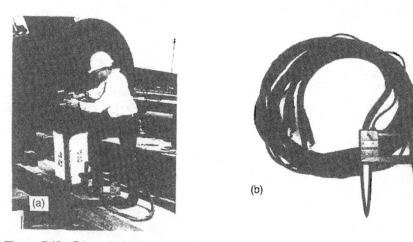

Figure 7.13 Direct induction prods for testing large objects. (a) Magnetic particle testing for large objects. (b) Closeup of production prods. Courtesy of ESC Econospect Corp.

between the prods. Current requirements in circular magnetization typically range from 400 to 800 A per linear inch of material. The minimum current, adequate to detect the expected discontinuities, should be used. With direct induction prods, a prod spacing of 6 in. is typical; this spacing will produce a distorted circular magnetic field.

When a central current-carrying conductor is placed inside a hollow cylinder, a circular magnetic field is produced on both the inside and outside surfaces as well as within the part. With this type of magnetization, the magnetic field strength will be more intense on the inside of the cylinder. The circular magnetic field surrounding is the same if the current is the same, whether the conductor is magnetic or not. If a copper conductor is placed inside a ferromagnetic cylinder, the magnetic flux density will be greater in the cylinder. Circular magnetization will detect lengthwise discontinuities on both inner and outer surfaces. Another advantage of circular magnetization is the ability to simultaneously magnetize and test a group of ring-shaped parts, such as gears, by passing a single conductor through them. Testing with a central current-carrying conductor (Figure 7.14) eliminates the possibility of burning parts because the parts do not come in contact with the conductor.

The most common method of inducing longitudinal magnetization in a part is by coil magnetization (Figure 7.15), which is achieved by placing the part inside a fixed or portable magnetizing coil. Magnetic parts concentrate the lines of magnetic flux along their length or in a longitudinal direction. The induced magnetic field in a coil extends 6–9 in. beyond either end of the coil. Therefore, exceptionally long parts require several magnetizing "shots" to effectively test the entire part. The strength of the magnetic field produced depends on the ampere-turns of the coil and is unaffected by part length. However, the magnetic field induced in the part can be reduced by reducing

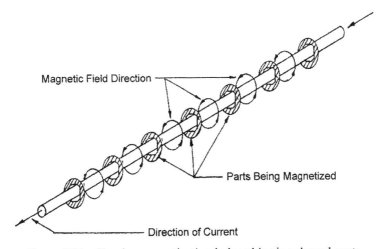

Figure 7.14 Circular magnetization induced in ring-shaped parts.

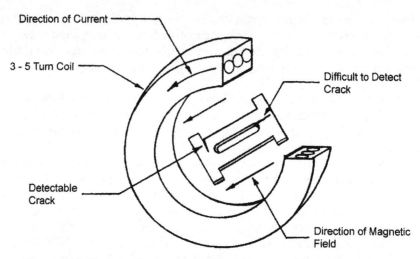

Figure 7.15 Longitudinal magnetization induced by coil magnetization.

the current, moving the part farther away from the coil, or moving the coil farther away from the part. Circular and irregularly shaped objects can be rotated 90° and retested to insure a thorough inspection capable of detecting radial cracks in multiple directions.

Longitudinal magnetization can also be induced in ferromagnetic parts by yoke magnetization. The yoke is generally a C-shaped laminated soft iron bar with a coil wound around it. When the ends of the yoke are placed on the part to be tested, a longitudinal field is set up in the part as shown in Figure 7.16. This field is identical to one that would be produced by a C-shaped permanent magnet.

For optimum coil magnetization, the length-to-diameter (l/d) ratio of the part being tested should be 2 or 3 to 1. The length-to-diameter ratio of the part should never exceed 15. For best results with a single magnetizing shot, the length of the part should not exceed 15 in. and its cross-sectional area should be one-tenth or less than the cross-sectional area of the coil. The part will receive the strongest induced magnetic field when its long axis is parallel to the magnetic field and it is positioned near the inside edge of the coil.

When test variables have been optimized, the optimum magnetizing force and current for DC coil magnetization can be calculated from Eq. (7.1):

$$\text{Ampere-turns} = 45,000/l/d \quad \text{or} \quad 45,000 \times d/l \quad (7.1)$$

In practice, the number 45,000 is divided by the length-to-diameter ratio of the part to determine the magnetizing force in ampere-turns. The ampere-

CIRCULAR AND LONGITUDINAL FIELDS

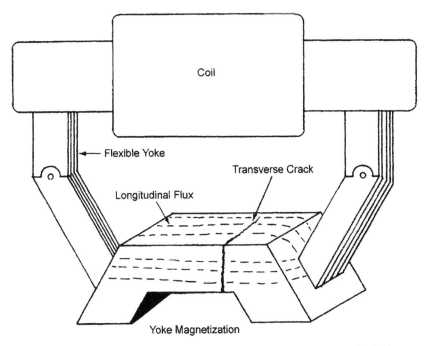

Figure 7.16 Fixture for yoke induction of longitudinal magnetic field.

turns number is then divided by the number of coil turns to get the correct amperage.

Portable equipment often incorporates the use of both prod or clamp contacts and flexible cable. On large parts, a prod spacing of 6 to 8 in. and current of 600 to 800 A is adequate for circular magnetization. A heavy accumulation of magnetic particles may cause a "banding" appearance, indicating that the field strength is too great, and the magnetizing current should be reduced or the prod spacing should be increased. For longitudinal magnetization, a coil of three to five turns can be formed by wrapping the flexible cable around large parts. Heavy accumulations of magnetic particles at the ends of a longitudinally magnetized part, indicates that the field strength is too great and that the magnetizing current should be decreased. If the correct magnetizing current is unknown for a specific method, it should be determined by experimentation and then recorded on a technique chart.

Hundreds of fixtures have been designed by suppliers and end users of magnetic particle inspection equipment to induce circular and longitudinal magnetic fields in various-shaped ferromagnetic parts. Figure 7.17 shows a fixture designed to magnetize pairs of horseshoe magnets.

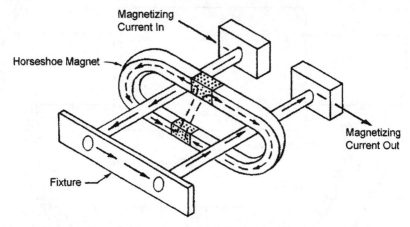

Figure 7.17 Fixture for simultaneous magnetization of two horseshoe magnets.

7.6 SELECTION OF MAGNETIZING METHOD

Many factors affect the selection of the magnetization method preferred for testing various ferromagnetic parts. The metallurgical composition or alloy of the part, its shape, and its physical condition are three important factors. The degree of heat treatment or cold working are additional factors affecting the permeability of the part. Cold working refers to bending or shaping a part without the addition of heat for annealing.

The end use of a part determines why a specific alloy has been chosen. However, the alloy chosen may be easy or difficult to magnetize and easy or difficult to demagnetize. A part may be easy to magnetize and easy to demagnetize, easy to magnetize and difficult to demagnetize, difficult to magnetize and difficult to demagnetize, or difficult to magnetize and easy to demagnetize. The inherent characteristics of the material, such as hardness or ductility, and its fabrication history, such as forging or heat treating, are determining factors.

From a practical standpoint, the size and shape of the part and the availability of equipment determines if a satisfactory magnetic particle test can be set up in a specific facility. Alternatives to in-house testing include sending the parts out for inspection or hiring a subcontractor to come on site to do the testing. If a large number of parts are involved, it may be economical to purchase new equipment.

The physical condition of the part can aid or hinder interpretation of magnetic particle inspection. Although it may be difficult to get ideal results, it is possible to inspect ferromagnetic parts that have been plated, painted, or even corroded. This is a major advantage of magnetic particle testing compared to liquid penetrant testing. On plated parts, the plating thickness should not exceed 0.004 in. Surface scratches and blemishes can cause nonrelevant indications that mask or conceal relevant defects.

The type of magnetizing current and the method of inducing magnetic fields should be carefully considered because the magnetizing method selected must be capable of detecting the discontinuities or defects that could be produced by the manufacturing process being used. As mentioned earlier, direct current or rectified current techniques provide the deepest penetration. Direct current is used to detect subsurface cracks because the skin effect (heating losses) reduces the effective depth of alternating current. If a defect is detected with direct current, it can be retested with alternating current to determine if the defect is a surface or subsurface defect.

Three-phase rectified alternating current is used with wet magnetic particle testing in large horizontal units. Half-wave rectified alternating current (HWAC) is typically used in dry powder units because the rectified current causes the magnetic particles to pulsate, making them highly visible. Continuous alternating current is the most sensitive and provides the highest resolution to surface defects such as fatigue cracks.

By far the most important consideration for obtaining reliable indications of discontinuities and defects is the direction of the magnetic field. For greatest sensitivity, a leakage flux field must be created at right angles to the discontinuities or flaws. If the magnetic lines of flux are parallel to the defect, there will be little or no indication of the defect. For this reason, parts are often rotated and retested or multidirectional magnetization is used. Multidirectional magnetization is the most effective method for testing large castings. A sufficient number of lines of leakage flux must be produced for reliable defect indication, but the direction of the magnetic field is much more important than the magnitude of the current or the strength of the induced magnetism.

7.7 COMMERCIAL EQUIPMENT

Examples of commercial equipment used for magnetic particle inspection (MPI) include the following:

- Battery-operated portable and AC-powered UV lamps
- Test meters—UV and white light photometers
- Hand-held permanent magnets and electromagnets
- Wheel and bar test pieces—natural and artificial calibration standards
- Magnetic flux field strength and direction meters (magnetometers/gaussmeters)
- Magnetic ink settling tube and stand
- Various water and liquid spray guns including electrostatic type
- Pressurized sprayer and manual rubber powder applicators

- Pie gauge, Ketos steel ring, magnetic stripe cards, and magnetic flux strips
- Automatic and semiautomatic inspection units.

7.8 WET AND DRY PARTICLE INSPECTION

Magnetic particle testing uses both wet (magnetic particle baths and magnetic inks) and dry magnetic particles to inspect for surface and subsurface discontinuities. Each technique has advantages and disadvantages. Nondestructive testing technicians should have a good understanding of these basic principles in order to optimize test results and develop satisfactory applications. With the wet method, magnetic particles are suspended in a liquid such as water or oil; with the dry method, magnetic powder is dispersed in air. In order to optimize test results, the size, shape, permeability, retentivity, and color of the particles are carefully controlled. Particles are colored so that they will provide a good color contrast with the part being inspected. Wet or dry magnetic particles and their coating agents should also be nontoxic.

Wet particles are primarily used for the detection of hairline surface cracks. They are typically used in stationary machines where they can be continuously agitated, pumped, and recycled through a spray nozzle until the liquid becomes contaminated with dirt, oil, or foreign magnetic particles. Magnetizing current is left on after the flow of material is stopped so that the flow of material does not wash away defect indications. Wet particles can be used in smaller, portable equipment if a constant source of agitation is provided. Dry magnetic particles are designed for use with portable equipment. They are especially effective on large parts with rough surfaces. Dry magnetic particles are used on a once-through basis and are seldom recovered for reuse.

Wet particles are ground in oil and purchased as a thick paste. They are available in red, black, and fluorescent colors that are equally sensitive to magnetic flux leakage. Color is used to improve the visibility of the fine magnetic particles on the part being tested. This facilitates faster inspection speeds. Red, black, and fluorescent pastes are also made for use with water solutions containing a mild wetting agent. The wetting agent promotes surface wetting of the part being inspected. The use of water eliminates potential fire hazards and minimizes potential dermatitis problems caused by skin contact with oil in some individuals. The disadvantages of water are that it is hazardous around electrical equipment and it may corrode the test equipment or parts.

In order to achieve the longest practical oil bath life, test equipment cleanliness is a must. After collecting the old oil bath suspension and cleaning the test equipment, the new oil bath suspension is made by adding oil to the paste in a separate container until $1\frac{1}{2}$ oz of paste is thoroughly mixed with every gallon of oil. Bath oil should be light in color and viscosity; it should be highly refined oil with low sulfur content and a high flashpoint. Paste should never be added directly to the test equipment because it could clog the circulating system. A thorough mixing and uniform suspension is required for reliable test indications. If the suspension is not uniform in wet testing, the strength of an indication may vary, causing erroneous indications. Particle concentration and suspension can be accurately controlled with microprocessor-controlled units.

The correct magnetic concentration is important in optimizing test sensitivity. If the concentration of particles is too low, reliable indications will not form; if particle concentration is too high, defect masking will occur. The correct concentration is determined by allowing the magnetic particles to settle for 30 min in a graduated centrifuge tube. After settling, from 1.5 to 2.0 cm^3 of red or black particles or 0.2 to 0.4 cm^3 of fluorescent particles should be observed in a 100 cm^3 sample. Bath concentration should be checked daily and corrected as necessary to maintain an acceptable range. With high-volume heavy usage, the bath should be changed weekly. If inspection volume is low, a monthly change will probably be sufficient.

Black and red wet magnetic particle indications are viewed in good outdoor light or under ordinary lighting conditions. Fluorescent particle indications are viewed in a darkened room with a 100 watt (W) black light similar to the one shown in Figure 7.18. Fluorescent particles are 100 times more visible than regular black and red particle indications. Because of their lower concentration and greater visibility, fluorescent particles are ideal for finding hairline

Figure 7.18 Typical 100-W ultraviolet black light for fluorescent magnetic particle inspection. Courtesy of UVP, Inc.

Figure 7.19 Magnetic powders, manual shaker and blower. Courtesy of ESC Econospect Corporation.

cracks in complex parts. With simple parts, fluorescent particle inspection can be performed at much higher inspection speeds.

Dry magnetic particle powders (Figure 7.19) are available in red, gray, black, yellow, and fluorescent colors. The fluorescent dry powder is light green under white light and fluorescent green under black light. They are equally sensitive to defect indications and are usually dispensed or gently floated into place with mechanical powder blowers, rubber spray bulbs, or manual shakers. An electric powder blower, with metered powder delivery, is shown in Figure 7.20. The goal is to apply a uniform cloud of particles with a gentle air stream. A mixture of globular and long, slender-shaped magnetic particles is best for dry powder inspection. For greatest penetration, continuous half-wave rectified alternating current is used with dry magnetic particle inspection. The dry

Figure 7.20 Electric magnetic powder blower. Courtesy of Parker Research Corp.

method is generally preferred for field inspection with portable equipment. Photographs, transparent tape, and transparent lacquer are used to preserve dry magnetic powder patterns. When indications are not formed, the part could be free of defects, the current could be off or too low, or the part could be nonmagnetic.

7.9 MT IMPROVEMENTS

Circle Systems, Inc. is a major manufacturer of particles for magnetic particle testing (MT). In 2002 the company conducted an extensive study of magnetic gradients and their effects. As a result of this study, magnetic stripe cards were developed, which can be used as a tool to grade MT materials.

A magnetic gradient is defined as the change of magnetic field over a distance (dH/dx). It is significant because it represents the force that attracts a particle to a discontinuity. The magnitude of the gradient is dependent on the applied magnetic field, the metal being tested, and the nature of the discontinuity.

Figure 7.21 shows the Type 2000 magnetic stripe card used for magnetic particle inspections. The card utilizes a unique and sophisticated magnetic encoding process that is highly controlled and extremely consistent from card to card and production run to run. Magnetic stripe cards are recognized as a tool for evaluation of magnetic particle inspection materials in ASTM E-709-01.

To use the card, it is first gently wiped with a clean soft cloth and then exposed to a magnetic particle suspension or puff of magnetic powder. More sensitive magnetic powders display a greater number of observable gradients than less sensitive materials. Both tracks on the card should display about the

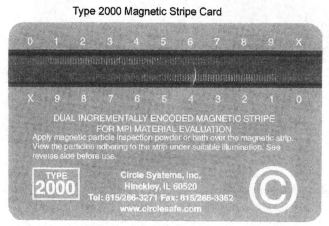

Figure 7.21 Front of Type 2000 magnetic stripe card and example of magnetic material status report. Courtesy of Circle Systems, Inc.

same number of indications if the magnetic inspection material is evenly applied.

In the case of baths, excess fluid is allowed to evaporate, or in the case of powders, excess powder is removed with a gentle puff of air. The card is then read and a permanent record can be made. The permanent record is made by placing transparent tape over the magnetic particles and transferring the magnetic particle pattern to an inspection card, form 93-MSC-1, as shown in the figure.

Recent work using magnetic finite element modeling (FEM) combined with the knowledge of the "threshold" gradient for attracting a particle, about $5\,MA/m^2$, can be used to aid industry in developing MT procedures, comparing AC to DC units, and for other related magnetic purposes.

7.9.1 Remote Magnetic Particle Inspection

For remote MPI, hand-held electromagnetic probes are generally used. Typical equipment and facilities being evaluated include tanks, bridges, power plants, refineries, pipelines, and petrochemical plants. Weld inspection is of prime importance. The main advantage of MPI is that it provides a reliable surface inspection, relatively quickly and at a low cost.

Disadvantages in setting up a remote inspection site include that it can be somewhat cumbersome since heavy AC and DC yokes are typically used and 120 VAC power may be required. If a fluorescent particle inspection is to be made, an AC-operated or battery-operated UV lamp must also be provided. When AC power is required, a motor-generator set with adequate load rating and extension cables may have to be provided. If conventional yellow-green fluorescent powders are used, visible light must also be limited to less than 22 lux or two foot-candles.

Circle Systems (Chedister, 1997) instituted two major advances in remote MPI. These are:

1. A different type of UV illumination
2. The use of dual light particles

7.9.2 Probe Power

AC yokes are proficient in locating surface discontinuities, but DC yokes are capable of finding both surface and subsurface discontinuities. MPI is considered a surface technique, while subsurface inspections are performed mainly by radiographic, ultrasonic, or other techniques. MPI is therefore normally focused on the test part's surface.

Power requirements and magnetic strength for AC and DC probes are very different and are measured on the basis of simple lifting power at specified leg spacing. According to standard guide ASTM E-709, the AC probe must lift a

4.5 kg (10 lb) certified test weight and the DC probe must lift a 13.6 to 22.6 kg (30 to 50 lb) certified test weight.

7.9.3 Lightweight UV Lamps

Conventional UV lamps are 150 W mercury vapor lamps, favored for their high output power and efficiency. However, 50 W incandescent units, which use the same battery as a portable DC probe, can also be used. The lightweight, portable lamp provides greater accessibility to the inspection area while still meeting required output requirements. The lamp features "instant on" operation, virtually eliminating "warmup" time, and less heat buildup during continuous operation.

Two rechargeable 12 VDC lead acid batteries are available for use, with 7 or 14 ampere-hour rating. The batteries weigh 30 or 60 kg (6.5 or 13 lb), respectively. The smallest battery and portable DC probes weigh about as much as a conventional heavy-duty probe. The portable DC probe has a built-in battery indicator (LED) that lights up when the probe is turned on to indicate adequate battery power. Failure to light indicates the battery must be recharged. Charge life (1 to 20 h) varies with load current and ambient temperature.

7.9.4 Dual Light (UV/Visible and Visible) Particle Indications

Dual light powders are detectable in visible light as well as fluorescent light under longwave (365 nm) UV radiation. Particles in the red color range are especially effective. It is well known that conventional yellow-green fluorescent materials lose their visible characteristics in the presence of visible light.

Red UV powders provide more striking indications in visible light than conventional red dry method powders. The indications can be highlighted with the use of a UV lamp, even when working in a nondarkened atmosphere.

7.10 APPLICATIONS

Fossil-fueled power plants use giant fans and blowers to control both air and gas flows. MT can help ensure fan reliability leading to a desired service life of 20 to 25 years. However, to achieve these goals, fans are typically inspected visually and with MT after their initial installation, during periodic inspections or scheduled shutdowns, and in accordance with any established preventive maintenance programs.

Figure 7.22 shows typical defects that can be detected by MT and a toe weld crack that was found along the edge of a weld (Chedister and Long, 1995). Large fans and blowers are subject to many daily and seasonal operating stresses. Operating temperatures for fans can reach 850°F and rotational

APPLICATIONS

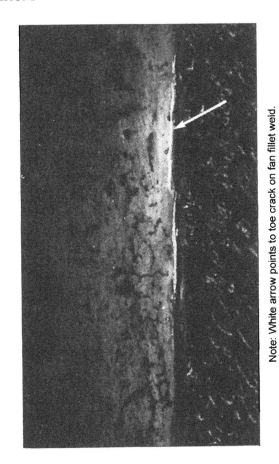

Note: White arrow points to toe crack on fan fillet weld.

Figure 7.22 At left, possible defects detected by MT of fan fillet welds. At right, high-contrast MT indication of toe weld crack. Fan welds are typically inspected after initial installation, during scheduled outages, and periodic preventive maintenance. Courtesy of Circle Systems, Inc.

speeds can range from 1800 to 3600 rpm. Catastrophic failure at these conditions can endanger both operating personnel and plant equipment.

Fan inspectors are typically certified for Level II work in both VT and MT methods. VT inspects for improper weld profiles and other discrepancies after initial installation. *E-709 Standard Guide for Magnetic Particle Examination* from ASTM focuses on MT as it pertains to field inspection of installed fans. Local plant and company safety rules, regulations, and procedures apply as well.

Previous advances (Erlanson, 1987) introduced the Mi-Glow 800 Series "Extended Life" materials that assure complete, durable pigment bonding to magnetic particles and continued process utility. This proprietary technique involves combining a new organic resin with a new high-UV-excitation level pigment and results in higher test reliability. Figure 7.23 shows a connecting rod with fluorescent flaw indication, without confusing background indications.

Connecting rods are inspected following an automated, water-based procedure using a computer vision test. Connecting rods are conveyed through a magnetic particle bath, magnetized in the bath, and then carefully spray rinsed to remove excess particles. The parts are viewed by four cameras using strobe lamps. The digital image is analyzed by a computer that provides a go/no-go status. Acceptance or rejection is based on geometry and intensity of the flaw indication.

Figure 7.23 Water-based medium is used for connecting rod inspection. Defects stand out with no confusing background. Courtesy of Circle Systems, Inc.

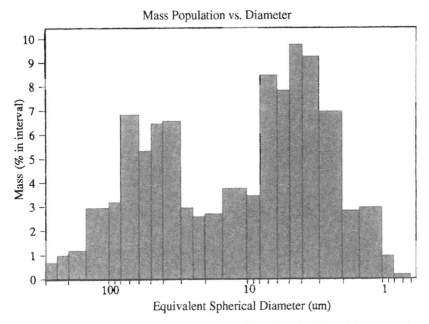

Figure 7.24 Typical distribution of UW #1 particles showing bimodal nature of particle distribution. Courtesy of Circle Systems, Inc.

Mi-Glow UW #1 underwater magnetic particles were designed for use in a variety of underwater inspections, including offshore structural welds, pipeline inspection, and ship husbandry. It has a broad particle range and is a combination of water conditioners and fluorescent magnetic particles. It is visible in both UV and visible light. The *bimodal nature* particle distribution is shown in Figure 7.24. Particle sizes less than 10 microns reveal finest defects and particle sizes over 10 microns provide proper buildup of particles in larger cracks.

Figure 7.25 shows a diver using UW #1 particles with an underwater electromagnetic yoke. Underwater MT particles can be premixed at higher-than-normal particle concentration and applied with squeeze bottles. As saltwater replaces the mixture used, it is diluted toward normal concentration. Magnetic gradient cards or magnetic flux strips can be used to determine when the mixture is no longer effective. Underwater particles are visible under all lighting conditions and flaw indications can be identified and photographed with underwater cameras.

7.11 RESIDUAL FIELDS AND DEMAGNETIZATION

After ferromagnetic parts have been magnetized, they will retain a certain amount of magnetism; this is known as "residual magnetism." The magnitude

Figure 7.25 Underwater weld inspection using UW #1 particles and an underwater electromagnetic yoke. Courtesy of Circle Systems, Inc.

of this residual field depends on the magnetic characteristics of the material, such as alloy composition and retentivity. Other important factors are strength, direction, and type of magnetizing force used, and the shape or geometry of the part. With the *residual method* of magnetic particle inspection, the part is magnetized, the current is removed, and then the magnetic medium is gently applied. Residual magnetism can be beneficial in the interpretation and evaluation of defect indications. For example, if the defect indication also exists with residual magnetism, the defect indication is probably relevant. To prevent magnetic writing, a nonrelevant indication, parts that have been magnetized should not be allowed to touch each other. Residual magnetic particle inspection is used when parts are highly retentive.

Magnetic chucks, welding machines, high-current circuits, and accidental contact with permanent magnets can induce unwanted residual magnetic fields in ferromagnetic parts. If the residual field is of little concern, there is no need to demagnetize the part. However, the residual magnetic field is considered harmful when it interferes with subsequent machining operations, or when the part itself will not operate satisfactorily because the magnetic particles attracted to it increase friction and wear. Parts should be thoroughly cleaned after demagnetization to assure that magnetic particles are removed. Demagnetization is also required when leakage flux emanating from a part interferes with nearby instrumentation.

Leakage fields from residual magnetization can be quite strong when parts have been longitudinally magnetized. However, the leakage flux field is easily detected and measured with a Hall-effect probe and DC gaussmeter or other flux-measuring device. Parts that have been longitudinally magnetized are easier to demagnetize than parts containing residual circular fields.

7.12 MAGNETIC FLUX STRIPS

NDE Information Consultants is the exclusive distributor of the NDE Consultants *magnetic flux strips* in the United States. Magnetic flux strips are especially useful for determining whether similar-sized surface imperfections in parts will give satisfactory magnetic particle indications.

Two types of strip are available with similar thicknesses but differing flaw widths. Each has three milled slots in high permeability magnetic material as shown in Figure 7.26. For both types, the top and bottom brass plates are 0.002 in. thick. The bottom brass layer positions the active high-permeability, low-retentivity steel element a small fixed distance above the test part. The brass is nonferromagnetic, merely protecting the steel slots and providing a controlled lift-off of 0.002 in.

The Type G (general use) strip has slot widths of 0.0075″, 0.009″, and 0.010″. The Type A (aerospace) strip has slot widths of 0.003″, 0.004″, and 0.005″. Since the G strip slots are about twice the width of the A slots, the A strip is more sensitive, but the magnetic flux level from the G strip slots will be about 4× greater than that of the A strip. Therefore, the G strip provides greater holding power for magnetic particles. Residual fields of three gauss can be detected when the strips are properly applied. The magnetic strips conform to the requirements of BS 6072. Type A strips designed for aerospace and other critical inspections conform to RPS 700 (Rolls Royce).

In an active magnetic field, wider slots of the same depth have larger amounts of magnetic flux leakage than narrower slots, and therefore provide more pronounced magnetic particle indications. Use of these flux indicators enables magnetizing field levels to be set for different magnetizing conditions such as internal (central) conductors, coils, yokes, cable wraps, etc. The type G strips are especially useful for the inspection of tubular oilfield pipes. They are

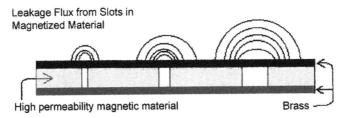

Figure 7.26 NDE Consultants magnetic flux strip. Courtesy of NDE Information Consultants.

flexible and easily conform to the curvature of the pipe. For maximum sensitivity, the strip slots must be positioned at a right angle to the induced magnetic field. The strips can typically detect a leakage flux of about three gauss.

When field strips are used in conjunction with a Hall-effect gaussmeter, measuring the tangential surface field strength, they provide assurance that the magnetizing current and field levels measured by the gaussmeter are sufficient for MPI without overmagnetization of the part. For additional information contact NDEIC.

When parts are circularly magnetized, they can be very difficult to demagnetize because the residual field is highly localized and very strong. There is little leakage flux because the field is almost completely contained within the part. If the magnetized area is to be subsequently machined, the residual field might be harmful. To demagnetize the part, it may be necessary to remagnetize it with a strong longitudinal field prior to demagnetization. Parts may also be demagnetized by heat treatment above the Curie temperature, by the use of diminishing AC fields, or by reversing and diminishing DC fields.

7.13 HALL EFFECT GAUSSMETER

The Trifield DC gaussmeter is designed for strong magnetic fields. It has full-scale ranges of 200, 2000, or 20,000 gauss. The meter, shown in Figure 7.27, is provided with a universal axial/transverse magnetic sensor and battery. A positive readout indicates a north pole; a negative readout indicates a south pole. An offset control permits the operator to adjust the "zero" field level by ±10

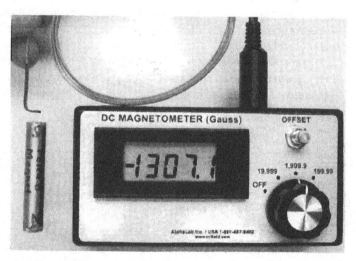

Figure 7.27 Trifield DC gaussmeter with full-scale ranges of 200 to 20,000 gauss. Courtesy of AlphaLab, Inc.

gauss. The gaussmeter is monotonic to the last digit (±0.005%), linear to ±1% at 20,000 gauss, and has an overall accuracy of ±2% from 30 to 110°F.

The DC magnetometer has a minimum resolution of 0.01 gauss on the minimum range of 200 gauss. By comparison, the strongest permanent magnets have an internal flux density or gap-magnetic field of about 14,000 gauss or about 7000 gauss on the face of their poles. The standard gaussmeter can be used to monitor the strength of most permanent magnets and detect when they have lost their strength. It can be also used to check magnetic field patterns.

The universal Hall sensor is 1.1 mm thick by 4.3 mm wide, on a one-meter-long plug-in cable. The standard sensor tip can be used straight to measure axial fields or repeatedly bent at a right angle (as shown) to measure transverse magnetic fields. An optional rigid-axial-only probe is available at the same cost. The meter is updated 3 times per second, but on the most sensitive range, a 1 second time-constant filter slows the response to minimize the noise level to several hundredths of a gauss. For faster and more stable readings at low signal strengths, a high-stability version of the meter is available.

Typical applications for the standard unit include measuring the magnetic fields of:

- DC motor magnets
- Electromagnets
- Permanent magnets

as well as measurement of high currents.

The higher sensitivity, high-stability version of the DC magnetometer is preferred for measuring residual magnetism in steel parts and for magnetic field mapping around large magnets. Undesirable residual magnetism in steel parts can range to 50 gauss. The high-sensitivity DC gaussmeter has a single measuring range of 0.01 to 200 gauss, but it features improved stability and faster update-speed.

The basic field strength meter, shown in Figure 7.28, also can be used to measure the amount of residual magnetism remaining in the part or to verify that demagnetization is complete. The meter is a small mechanical device with soft iron vane that is deflected by the magnetic field. When the vane is deflected, the meter needle rotates, indicating residual flux. The meter shown is calibrated for a range of −20 to 0 to +20 gauss.

7.14 THE HYSTERESIS CURVE

When ferromagnetic parts are subjected to a magnetizing current, magnetic domains tend to align themselves in the direction of the magnetic field. After the magnetizing force is removed, many of the domains remain aligned in the

Figure 7.28 Magnetic field indicator. Courtesy of Econospect Corporation.

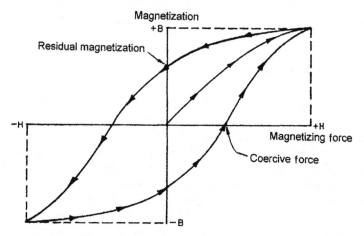

Figure 7.29 Magnetic hysteresis curve.

magnetized direction until a magnetic force in the opposite direction causes them to return to their original random state. The force required to return the magnetic domains to their random orientation is known as the *coercive force*. The coercive force represents the reverse magnetizing force necessary to remove residual magnetism. The *retentivity* of a material represents its ability to retain a portion of the magnetic field set up in it after the magnetizing force has been removed.

Figure 7.29 shows a typical magnetic hysteresis curve. Note that as magnetizing force (H) increases, magnetization (B) increases until the saturation point is reached. As magnetizing force is reduced back to zero, some residual

THE HYSTERESIS CURVE

magnetism (B_r) remains. Therefore, the magnetizing force must be reversed to reduce B_r to zero. If the reversed force is increased to the saturation point and then reduced to zero, a residual magnetism of equal magnitude and opposite direction will exist. Reversing the magnetic force again and increasing its strength can return the material to its original saturation point.

Parts can be satisfactorily demagnetized by subjecting them to a diminishing AC field greater than the coercive force but less than the magnetizing force, as shown in Figure 7.30. This systematically reduces the residual magnetism and coercive force required for demagnetization and the part is gradually returned to a nearly demagnetized state. Small parts requiring demagnetization should be processed through the demagnetization cycle one at a time. For demagnetization, the maximum required time is about 30s.

Materials that have higher residual magnetism have higher retentivity. Generally, hard materials have high retentivity and are hard to demagnetize; softer materials have lower retentivity and are easier to demagnetize. This direct relationship does not always exist between retentivity and coercive force. For example, magnetic powder materials are selected to have high retentivity and low coercive force. It is important that they are easy to demagnetize in order to simplify part cleaning.

The most popular way to demagnetize parts is to pass them through a 115 VAC line coil. Parts are demagnetized by setting the demagnetizing current higher that the original magnetizing current and slowly passing them through the bottom-inside edge of the coil for maximum effectiveness. The part being demagnetized is slowly conveyed away from the magnetic field in a straight line for a distance of about 4ft. Alternative methods of demagnetization include leaving the part in the alternating field while gradually reducing the

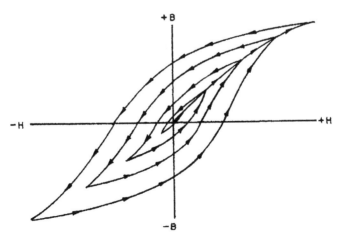

Figure 7.30 Demagnetization by diminishing AC field.

current with an autotransformer or similar device. In both cases, the longest dimension of the part and the axis of the AC coil should be positioned in an east–west direction. One disadvantage of AC magnetization is the relatively shallow depth of penetration.

When parts have been circularly magnetized with an alternating current, alternating current may be applied directly to them. As the current is gradually decreased to zero, the parts are demagnetized. Demagnetization often immediately follows magnetic particle inspection so that the magnetization–demagnetization cycle is one continuous smooth operation.

When parts have been magnetized by AC/pulsed and DC magnetic yoke, they may be demagnetized by using the AC mode and passing small parts between the poles in an east–west direction for a distance of about 2 ft before turning off the power. Large parts may be demagnetized by placing the probe in contact with the part in the test position, turning it on, and lifting it from the work for a distance of about 2 ft before turning it off.

For parts that are difficult to demagnetize, reversing DC demagnetization may be used. The part to be demagnetized is left in the head stocks, with circular magnetization, or DC coil, with longitudinal magnetization, and a direct current is applied and gradually diminished with each reversal of direction. This type of demagnetization applies deep penetration for hard-to-demagnetize parts. Direct current yokes provide the deepest magnetic penetration and parts that have been magnetized with a DC yoke may also be demagnetized with one using the reversing, diminishing direct current technique and special circuits. Demagnetization may be improved by the use of the transient method that applies a high-energy pulse of current in the opposite direction for a short period of time.

7.15 SELECTION OF EQUIPMENT

The selection of equipment for magnetic particle testing depends on several factors. What are the size, shape, and magnetic characteristics of the part to be tested? What types of flaws or defects are to be expected from the manufacturing process? Will circular, longitudinal, or simultaneous application of both magnetic fields be required for satisfactory test results? How should the magnetizing current be applied? Should test sensitivity be high, moderate, or low? How flexible is the equipment and where will the testing be done? Will wet, dry, or fluorescent magnetic particles be used? Will separate or built-in demagnetizing equipment be required? What will it cost to set up magnetic particle inspection and what will be the inspection cost per unit produced? Is equipment rental more cost effective than outright purchase? These are a few of the questions that deserve careful consideration before purchasing equipment.

Because of their relatively small size, light weight, and versatility, flexible magnetic yokes, such as shown in Figure 7.31, are very popular. Electromagnetic yokes produce longitudinal magnetic fields. The probe shown weighs

SELECTION OF EQUIPMENT

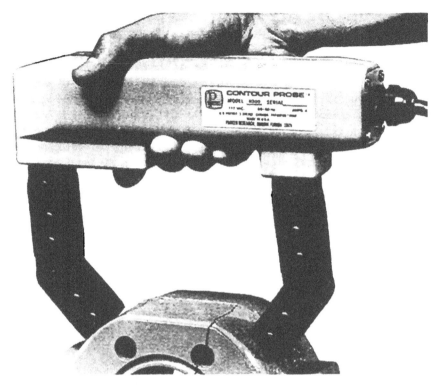

Figure 7.31 Flexible yoke magnetizer. Courtesy of Parker Research Corporation.

about 7 lb and can be adjusted to almost any contour. Probe spacing is adjustable from 0 to 12 in. and continuous 115 VAC or pulsed 300 VDC is available for magnetization. Magnetic yoke standards, such as MI1-STD-271E, specify minimum detectable defect size and minimum lifting power. In this case, the reference standard specifies that a magnetic yoke must have the proven ability to detect a notch 1/16 in. long × 0.006 in. wide × 0.02 in. deep. Lifting power must be 10 lb with continuous AC or 40 lb with DC or a permanent magnet yoke.

A summary of features of portable and mobile MT equipment is listed in Table 7.1. A typical portable magnetic particle test unit is shown in Figure 7.32 and two mobile units are shown in Figures 7.33 and 7.34. Ammeters, similar to the one shown in Figure 7.35, are used to determine the magnetizing current being used. Periodically, the ammeters are checked against a calibration shunt in accordance with established ASTM procedures. Stationary, horizontal machines for wet particle inspection (Figure 7.36) are used by the aircraft and automotive industries in equipment overhaul and maintenance programs.

TABLE 7.1. Portable and Mobile Equipment Features[a]

Magnetic Current	Voltage	Contacts (ft length of cable with prods)	Weight (lb)	Demagnetization
450 A-HWDC	115 VAC at 15 A[b]	12	24	None
1–750 A-HWDC	115 VAC at 35 A	12	33	Manual adjustment
1–1000 A-HWDC or AC	230 VAC at 35 A	12	40	Manual adjustment
1–1000 A-HWDC or AC	440 VAC at 20 A	12	42	Manual adjustment
Mobile Units				
1–2000 A-HWDC or AC	230/460 VAC 100/50 A	12	Cart mounted	Manual adjustment
1–4000 A-HWDC or AC	230/460 VAC 300/150 A	20 (remote control)	410 (mobile)	Automatic
1–6000 A-HWDC or AC	230/460 VAC 465/235 A	20 (remote control)	420 (mobile)	Automatic

[a] Optional equipment includes remote control stations, prod assemblies, magnetizing cable assemblies, wraparound cables, test meters, magnetism detectors, magnetic penetrameters, and various connectors.
[b] Input current requirements depend in part on required duty cycle.

Figure 7.32 Field application of portable unit with prods. Courtesy of ESC/Econospect Corp.

SELECTION OF EQUIPMENT 283

Figure 7.33 Mobile two-wheel unit for AC or DC magnetization. Courtesy of ESC/Econospect Corp.

Figure 7.34 Heavy-duty mobile unit with automatic demagnetization. Courtesy of ESC/Econospect Corp.

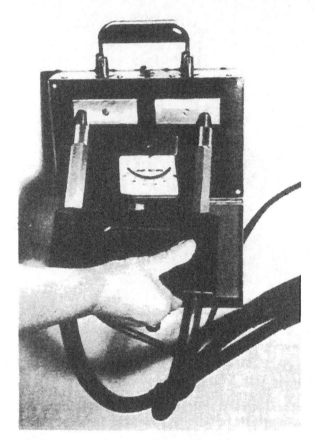

Figure 7.35 Meter for measurement of prod magnetizing current. Courtesy of ESC/Econospect Corp.

Figure 7.36 Fixed AC coil magnetization unit for wet magnetic particle testing. Courtesy of ESC/Econospect Corp.

7.16 ADVANTAGES AND DISADVANTAGES OF THE METHOD

The major advantages of the magnetic particle method are that it is simple, inexpensive, and rapid. The method is used extensively to detect surface and near-surface subsurface flaws in ferromagnetic parts and materials. However, critical parameters for reliable application of the technique depend on:

- The nature of the particles themselves
- Particle concentration
- Illumination and visibility
- Current density
- Field direction
- Human factors

The probability of (defect) detection (POD) can consistently reach the 90% level for threshold-type flaws with the fluorescent particles, through the use of adequate black-light illumination, closely controlled particle concentration, procedures verified with quantitative artificial flaws or shims, qualified personnel, and adequate process control tools. The major disadvantage of the method is that the POD is not 100%. There is still room for improvement and continued improvements will be made.

7.17 MAGNETIC RUBBER INSPECTION

7.17.1 Introduction

Magnetic rubber inspection (MRI) is a highly reliable, nondestructive inspection technique for detecting cracks and flaws in ferromagnetic parts. The MagRubber™ system combines the principles of magnetic particle inspection with a rubber replicating system that vividly displays cracks and other flaws on a cast impression of the inspected parts.

MagRubber was developed to inspect aircraft parts for cracks and to extend inspection capabilities to areas with limited visual or mechanical access such as bolt holes, blind holes, thread or gear root areas and tubular components. Other useful applications include inspection of areas with complex shapes, coated surfaces, or classifying defects requiring magnification for interpretation. In addition to the aerospace industry, MagRubber is widely used in power plant inspections and other general industry applications.

7.17.2 Inspection Principles

MRI principles are shown in Figure 7.37. MagRubber consists of two components—a vulcanizing rubber material containing specialized magnetic

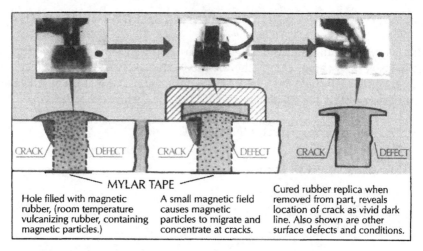

Figure 7.37 Dynamold MRI process. Courtesy of Dynamold, Inc.

particles and a hardener. MagRubber cures at room temperature—no special equipment is required. For inspection of a hole in a metal component, the rubber material is mixed with the hardener according to instructions and poured into the hole to be inspected. A magnetic field is applied that causes the magnetic particles to migrate and concentrate at the flaw area. After allowing as little as 10 minutes to cure, the rubber replica is removed from the part and examined for defects. Defects in the part will be displayed as vivid dark lines on the replica and other surface defects and conditions will also be accurately reproduced. After examination, the replica can be identified and stored—it becomes a permanent record of the inspection.

7.17.3 Advantages of MRI

- Excellent sensitivity to flaws as small as $\frac{1}{2}$ micron in width
- Permits in-situ inspections of portions of large parts in the magnetic field
- Can inspect areas where visual inspection is impossible
- Replicas are easily examined at a time and place of inspector choosing
- Works through paint or plating, on rough surfaces and on parts of any shape
- Only easy method for inspecting very small or threaded holes accurately
- Replicas also display machining quality, surface conditions, and physical dimensions
- Indisputable physical evidence—a permanent record
- Minimum shrinkage of replicas

MAGNETIC RUBBER INSPECTION

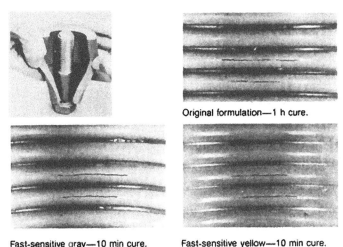

Original formulation—1 h cure.

Fast-sensitive gray—10 min cure.   Fast-sensitive yellow—10 min cure.

Figure 7.38 Comparison of MRI formulations—threaded bolt inspections. Courtesy of Dynamold, Inc.

- Relatively fast procedure, requiring as little as 10 minutes for complete test

Figure 7.38 shows the excellent results obtained for finding small cracks in thread roots. The three thread photos correspond to raised thread areas in the replica.

7.17.4 Formulations

MR-502: Regular—gray color with cure time of 45 minutes to one hour

MR-502K: Kwik Cure—gray color with cure time of 8–10 minutes

MR-502Y: Yellow Kwik Cure—high-contrast yellow base with cure time of 8–10 minutes

Notes:

- Cure times can be varied to sensitivity requirements of the inspection.
- Each formulation of MagRubber is offered in three sizes of complete inspection kits that contain all materials needed to perform any inspection.
- Each formulation is offered in Quart Kits and Gallon Kits.
- MRI complies with the following specifications:
  - ASTM E 709-80, p 686
  - Mil Spec Mil-l-83387

- AMS 2308
- AMS 2309
- Air Force//Army/Navy NDI Manual; TO-33B-1-1

7.18 UNDERWATER MRI

For several decades, the number of giant offshore oil drilling platforms has increased dramatically. Underwater inspection of these structures is required to assure their integrity for the safety of personnel working and living on them. Magnetic particle testing (MT) is often selected because it is a surface-sensitive method ideally suited for detecting large fatigue cracks. For underwater magnetic particle testing, flexible yoke, continuous alternating current, and fluorescent magnetic particles were originally used. The amount of equipment required, poor visibility, washing away of particles by underwater currents, and lack of permanent record were some of the disadvantages encountered. Because of these problems, special Dynamold MR-502Y rubber bags with isolated catalyst were developed for underwater use.

7.18.1 Technique

Underwater inspection is most efficiently done with a yoke that applies a small holding current attached to the structure until the diver is ready to use it. The diver/inspector takes a container full of inspection bags to the test area. At the inspection site, he breaks the barrier separating the rubber base and catalyst, thoroughly kneads the material, and applies the bag intact to the inspection site. He then magnetizes the site with a permanent or electromagnet, waits about 2 to 3 minutes, removes the bag from the inspection site, and examines the rubber mold for flaws. The bag is identified and retained for a permanent record of the inspection. This technique prevents currents from washing away particles and eliminates the need for making hasty interpretations when limited visibility is a problem.

7.18.2 Disadvantages

Some potential disadvantages of MRI are as follows:

- Rubber castings of small threaded parts are somewhat fragile and must be removed with care. Release agents may be required.
- As with all two-part systems, the catalyst must be thoroughly mixed with the base material to assure uniform distribution and consistent setup times. Results are not instantaneous.
- Leakage flux can be a source of false indications if magnet positioning is not correct.

AUTOMATIC AND SEMIAUTOMATIC INSPECTION

Note: Manufacturer's recommendations and established inspection procedures should always be followed—there are no satisfactory shortcuts!

7.19 MAGNETIC PENETRAMETERS

Magnetic penetrameters can be used to determine the direction of the magnetic field and test efficiency. One penetrameter consists of a remanence-free (free of residual magnetism when the applied magnetizing force is removed) shielding ring into which an iron cylinder is placed. The iron cylinder is notched in the form of a cross, thereby creating an artificial flaw. The iron cylinder is covered with a thin copper plate that can be varied in distance from the artificial flaw.

To use the penetrameter, it is placed on a magnetized part and dusted with dry or sprayed with wet magnetic particles. By slowly turning the penetrameter, the maximum indication of the artificial flaw indicates precisely the magnetic field direction. To determine test efficiency, the outside ring of the penetrameter is turned slowly, increasing the distance of the copper plate above the artificial flaw. The amount of lift-off at the point where the flaw indication just disappears is a measure of magnetic particle test efficiency. The amount of lift-off is read to $\pm\frac{1}{4}$ mm on a shielding ring.

7.20 AUTOMATIC AND SEMIAUTOMATIC INSPECTION

The Magwerks Corporation specializes in building *magnetic particle inspection (MPI)* equipment for inspecting ferromagnetic materials (iron, cobalt, nickel, and their alloys) and locating flaws and discontinuities in castings, forgings, and machined parts. MT provides a qualitative method for the inspector to accept or reject material based on visually observed indications that are matched to written quality control criteria for specific parts. MPI is faster and more cost effective than X-ray inspection, and when properly applied to ferromagnetic parts, is more versatile than other NDT methods used today.

Magwerks manufactures a full line of MPI machines, from small portables to complete automated systems. Unique equipment features include stainless-steel pumps, stainless-steel tanks, hinged and gasketed doors with twist latches, linear power controls, and contactless power switching even between magnetic heads and coils. Cabinet heat exchangers and closed-loop current controllers also can be provided.

The company also offers a complete line of free-standing demagnetizers, calibration instruments, and high-output black (UV) lighting. It also provides Level I, II, and III operator and inspection training at its 10,000 sq. ft training facility. The company also maintains its own calibration lab to guarantee that instruments are NIST traceable. Please contact the company for additional details.

7.21 MAGWERKS INTEGRATED SYSTEM TRACKING TECHNOLOGY

This state-of-the-art system provides inspection traceability when ISO 9000, QS 9000, aerospace, and military audits are required. Information that can be provided includes:

- The day and shift when parts in question were tested
- Identification of the inspector who processed the parts in question
- Identification of the machine that was used to inspect the parts
- Whether the machine was calibrated and, if so, when
- Both the good (acceptable) and bad (rejected) part count
- If good parts were reworked, the total number of reworked parts

Before the inspection begins, the operator can determine if the following checks were made and documented:

1. Bath concentration
2. Bath contamination (solids)
3. Bath clarity (liquids)
4. Ketos ring test at proper amperages and waveform output
5. UV lighting intensity
6. White light infiltration

The integrated system can also verify that the correct procedure was used for the parts being inspected. Variables that can be verified are:

- Correct contact current
- Correct coil current
- Correct output waveform (AC, FWDC, and HWDC)
- Correct sequence of operation
- Correct number of vectors used at a given time (multivector vs. single-vector processing)

7.21.1 Basic Operation

The Magwerks Integrated System (IS) is shown in Figure 7.39. When powered up, the operator is prompted to select either the "automatic" or "manual"

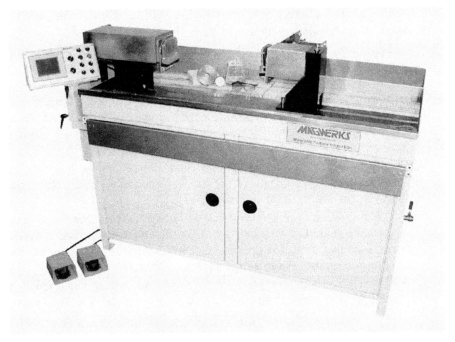

Figure 7.39 Magwerks Integrated System with twenty-first-century tracking technology. Courtesy of Magwerks Corporation.

operating mode. In the manual mode the machine operates like any other standard MPI machine. In the automatic mode there are many special features that make this system unique. To begin with, there are three selectable levels of security—first, the "operator" level, second, the "supervisory" level, and finally, a restricted confidential "factory" level.

The operator level is used for the daily operation of the machine, where the daily part selection and processing is performed. Machine functions available to the operator include selecting a preprogrammed part from the lookup table, entering the lot number of the part, entering the results of the daily process control data, an override for the autobath, preview of part parameters, enabling of automatic special features, a demag override, and printing of some reports. The supervisory level is used when part name and procedure data are set up along with operator accounts and some machine functions. The supervisory and factory levels of access are not available to the operator.

7.21.1.1 Basic Operation—Automatic Mode The basic sequence of operation in the automatic mode is:

1. The operator logs in from the main menu and enters his or her unique pin number, preventing false identity. The machine's real-time clocks

time stamp this action and tie it to the machine's serial number. The operator's console for entering data is shown in Figure 7.40.
2. If desired by the supervisor, the machine will not process any parts until the following Ketos test cycle has been performed and time stamped by the system.
3. The Ketos test cycle can be automatically run at three predetermined settings by the supervisor. The automatic sequence of operation is started by a single depression of the automatic foot pedal and is described as follows:
    - The central conductor with ring is clamped and demaged at a higher setting than it is magnetized (usually 2800 amps). This setting and waveform is supervisor selectable.
    - The mag waveform is automatically set to DC. The machine pulls the predetermined first current value from the lookup table and mags the ring with two shots. The operator is then prompted to input the number of lines seen into data storage via the touch screen.
    - The machine pulls the predetermined second current value from the lookup table and mags the ring with two shots. The operator is prompted to input the number of lines seen into data storage via the touch screen.
    - The machine pulls the predetermined third current value from the lookup table and mags the ring with two shots. The operator is prompted to input the number of lines seen into data storage via the touch screen.

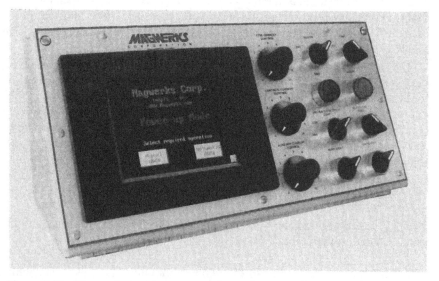

Figure 7.40 Magwerks instrument consolette. Courtesy of Magwerks Corporation.

- All of the above results and the associated magnetizing current levels are logged into memory and tied to the machine's serial number, operator, and time/date.
- The foot pedal is depressed again and the ring is demagnetized before it is put away.
- The Ketos operation is then complete, with the entire process taking about 2 minutes.

4. Particle and light data are then logged into the system and tied to the operator, machine, and date.
   - Concentration test values are logged into the system. Because of setting time required, the pop-up reminder will be displayed to the operator after a predetermined time set by the supervisor.
   - Pass/Fail conditions for contamination tests of both solid particulate and carrier contamination are logged into the system.
   - Black-light intensity values are logged into the system.
   - White-light infiltration values are logged into the system.
   - White-light source values are logged into the system.
   - All bath adjustments are logged into the system, such as additions of carrier, particle, antifoaming agent, wetting agent, and rust inhibitor.
   - The supervisor sets the accept/reject limits of these variables and has the ability to deactivate the machine in either, or both, automatic and manual mode until the problem is corrected.
5. The operator then returns to the main menu to select the part to be processed. Using the "Part Menu" button, the operator makes a selection from the list of parts whose parameters are already in the database.
6. The operator is given the option to further identify the part batch by heat code, lot number, die number, rework number, or any other alphanumeric delimiter set up by the supervisor.
7. All of the process control data are now tied to the operator, machine, date/time, and now the part.
8. The operator is returned to the main screen for part processing.

Two extra inputs have also been programmed into the computer for optional electric eyes that can be installed on conveyor systems, one on the good-parts conveyor and one on the reject-parts conveyor or chute. These two conveyors count the passing parts, totaling each type and inserting the value into the database that is tied to the part number, heat code (if desired), date, operator, number of current alarms, system performance tests, and machine ID.

At the end of the shift the operator logs out and the time is recorded. At this time the data can be either viewed or printed for insertion into a permanent log. Up to 15 operators' worth of data may be stored and printed by the system if desired. A sample operator's log is shown in Figure 7.41.

Operator

Operator Login	Logout	Session	Machine	Total	Alarm PPM	Good PPM	Control Data
Andrew Kennedy							
8/15/2001 15:42	8/16/2001 13:19	22 Hrs 37 mins	8675309	11	0 0.0	0.0	Ketos Test Failed, Black Light Low, White Light Source Low, Particle Concentration Bad, White Light Infiltration High, Carrier Contamination Failed
Mike Simpson							
8/8/2001 9:55	8/8/2001 9:57	2 mins	8675309	3	0 1.5	1.5	Particle Concentration Bad, Particle Contamination High, Carrier Contamination Failed
8/14/2001 13:23	8/14/2001 13:26	3 mins	8675309	8	1 2.7	2.3	Particle Contamination High
Greg Bartlett							
8/4/2001 13:01	8/4/2001 13:03	2 mins	8675309	8	5 4.0	1.5	Particle Contamination High, Carrier Contamination Failed
8/9/2001 13:17	8/9/2001 13:20	3 mins	8675309	6	1 2.0	1.7	Ketos Test Failed, Black Light Low, Particle Contamination High
Anita Ortez							
8/10/2001 12:44	8/10/2001 12:48	4 mins	8675309	7	2 1.8	1.3	Good
8/10/2001 12:48	8/10/2001 12:56	8 mins	8675309	12	4 1.5	1.0	Good
Jeff Smith							
8/10/2001 10:42	8/10/2001 10:54	12 mins	8675309	16	10 1.3	0.5	Ketos Test Failed, Black Light Low, Particle Concentration Bad, Particle Contamination High

Wednesday, November 14, 2001

Figure 7.41 Typical integrated system operator's log. Courtesy of Magwerks Corporation.

If the remote data-retrieval option is installed, these data are polled on a predetermined time period and logged into a Microsoft Access® database on the company's local area network server. In the event that defects are passed, anyone with network privileges can instantly find out how many of a given part were processed, who processed them, on what machine or machines they were checked, if the machine was in calibration, if the operator performed the daily system checks, and if the parts passed inspection. Also, the part can be further delineated by heat code, steel supplier code, or other relevant descriptor that was used. If multiple locations are processing the same parts and the data are being stored to a common database, the test location will also be shown.

Since all variables are stored in the database, operator sessions can be analyzed for the number of parts per minute processed to help gauge productivity. Samples of both the "Part Run Analysis" and the "Operator Analysis" are shown along with session control data. Therefore, the Magwerks Integrated System can enhance the quality control program while reducing both operator error and paperwork. This can result in substantial cost savings.

7.21.1.2 Applications Magwerks IS applications include the following:

- Crankshaft inspections
- Vehicle suspension component inspections
- Naval vessel component inspections

Other Magwerks products include light meters, black lights, MPI accessories, yokes, ammeters, shunt meters, and analog and digital gaussmeters. Gaussmeters are typically used to determine the strength and direction of the earth's magnetic field, measure the leakage flux from various flaws and defects, determine if there is any residual flux remaining after demagnetization, and measure the strength of permanent magnets.

Figure 7.42 shows a Magwerks Series 1200 gaussmeter. For ranges up to 25 gauss, this precision digital gaussmeter is calibrated using a Helmholtz uniform magnetic field standard coupled with a precision DC power source. The field generated by the Helmholtz coil is verified by a precision laboratory-style, $5\frac{3}{4}$-digit gaussmeter traceable directly to an NBS nuclear magnetic resonance magnetometer. The DC power source is calibrated by a $6\frac{1}{2}$-digit HP multimeter and crosschecked with a Fluke 5100B master calibrator.

Meters under test have their sensors mounted in the horizontal position and centered in the Helmholtz coil, normal to within $\frac{1}{2}$ degree to the axis of the field source. Calibrations are done in the Magwerks Corporation's Metrology Laboratory. For ranges over 100 gauss, instruments are calibrated using a laboratory-style electromagnet having a standard uniform magnetic field coupled with a precision DC power source. The electromagnet and power

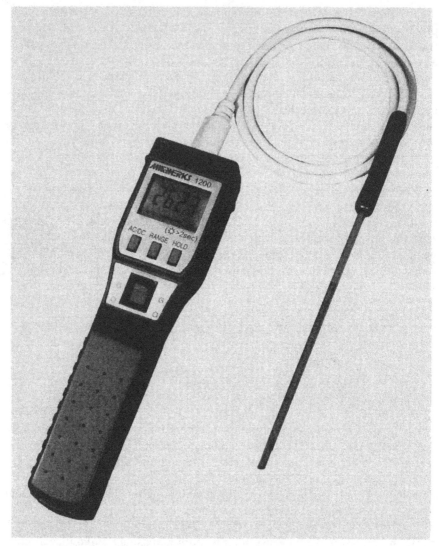

Figure 7.42 Magwerks hand-held gaussmeter. Courtesy of Magwerks Corporation.

source are calibrated as previously described. Calibration procedure MWC-CAL-200 is utilized for calibration of the Series 1200 guassmeters.

7.22 DISCONTINUITIES AND THEIR APPEARANCES

Discontinuities detected by MT are not necessarily defects. A discontinuity is a change in the normal physical structure or configuration of a part that pro-

NONRELEVANT INDICATIONS 297

duced an MT indication. Defects may be thought of as flaws, irregularities, or discontinuities that reduce normal service life, resulting in premature failure of the part. One definition of a flaw is that when the depth of the defect is five times its width at the surface it is a flaw. In cases where premature failure of a part can be hazardous to human health or life, a 100% inspection of parts is warranted. Some discontinuities that can appear on or near the surface of a part and can be detected by MT are listed in Table 7.2.

Nonmagnetic inclusions are usually hard fragments found near the surface. They produce subsurface indications at various angles to the metal grain flow in forging.

Cooling cracks in rolled bars produce indications somewhat similar to seams. They are long, straight, and parallel to the direction of rolling. They may vary in width and usually vary considerably in depth.

Laminations in plate or sheet are similar to seams, usually subsurface, and they cannot be detected by MT, except at part edges or where they intersect with hole surfaces.

Forging laps are irregular in shape, at right angles to the direction of metal flow, and at an acute angle to the metal surface. They produce weak, poorly defined MT indications.

Forging bursts are ruptures, which can be internal or external. They are sharp defects running in all directions. The internal ruptures are called "flakes," which cannot be detected. The external ruptures are irregular, of varying depth, and of somewhat similar appearance to radial cracks.

Large pockets of gas porosity or shrinkage porosity near the surface of a casting can be detected by MT. Indications would show large void areas with poorly defined edges.

Weld shrinkage cracks tend to run parallel to the weld near the center and turn off slightly at the ends.

7.23 NONRELEVANT INDICATIONS

Nonrelevant indications are MT indications produced by leakage flux fields that are independent of the service life of the object. For example, longitudinal magnetization at high amperage will produce leakage flux fields that accumulate magnetic particles along sharp edges, where part thickness changes, and at the ends of parts. These indications relate to the geometry of the part, but have nothing to do with its useful life.

Parts that have internal holes, splines, or keyways near the surface of the part will produce fuzzy, shadowy indications of the internal machining patterns. Identical parts would have identical nonrelevant MT indications.

Unmagnetized parts that come in contact with magnetized parts and parts that vary in hardness or heat treatment may produced random, loosely held, irregular indications that are nonrelevant. These indications are known as "magnetic writing."

TABLE 7.2. Detectable Discontinuities

Casting Process

Blowholes	Surface discontinuity caused by gas porosity
Cold laps	Caused by rapid surface cooling
Cracks	May be caused by thermal or mechanical stresses
Gas porosity	Caused by shrinkage or trapped gases in the casting
Hot tears	Surface ruptures caused by failure of metal to fill mold
Misruns	Caused by failure of metal to fill mold
Sand inclusions	From sand molds, generally subsurface, not always serious

Forging Process

Forging bursts	Material forged at too low a temperature
Forging flakes	Internal ruptures caused by rapid cooling and released gases
Forging laps	Irregular cracks caused by mechanical stress
Inclusions	Caused by inclusions in casting

Rolling Process

Cooling cracks	Caused by material cooling too rapidly during rolling
Laminations	Caused by inclusion, pipe, or blowholes in ingots
Nonmetallic inclusions	Caused by inclusions in billets
Stringers	Another name for nometallic inclusions
Surface seams	Caused by surface cracks or defects in billets

Welding Process

Incomplete fusion	Caused by lack of sufficient heat during welding
Lack of penetration	Improper amount of heat during welding
Porosity	Caused by gas evolution or moisture during welding
Shrinkage cracks	Caused by rapid cooling of weld metal
Surface seams	Caused by nonmetallic impurities; part contamination

Final Machining

Etching cracks	Appear after etching a highly stressed surface
Grinding cracks	Appear after grinding of a highly stressed surface
Plating cracks	Appear after plating a highly stressed surface
Quench cracks	Caused by rapid cooling during heat treatment

Parts that Have Been in Service

Fatigue cracks	Usually originate at high stress areas such as notches, seams, sharp corners. Likely to appear after thermal or mechanical cycling.

Other nonrelevant indications may be formed at the edge of cold-worked areas or at the edge of weld lines. Cold-working indications can be eliminated by stress relieving the part. Joints between dissimilar metals, brazed joints, and roughing tool surface cuts can all cause nonrelevant indications. Nonrelevant indications can be identified by consistent patterns on identical parts, through knowledge of known internal features of construction, or by lack of a true defect appearance to trained operators. If there is any question as to whether an indication is relevant or nonrelevant, the part should be reexamined to determine if a defect is present.

8

NEUTRON RADIOGRAPHIC TESTING

8.1 INTRODUCTION

A neutron is an uncharged particle with a mass slightly greater than a proton. The neutron's energy is expressed in electron volts (eV). One of its unique properties in neutron radiography is that it is electrically neutral, resulting in negligible electrostatic interaction with the atom's electrons.

In many ways, gamma radiation and neutron radiation complement each other. Heavy, dense objects absorb gamma radiation and neutron radiation is absorbed by lighter rare earth elements. The rare earth elements are those with consecutive atomic numbers from 57 to 71 inclusive. *Neutron cross section* is the term used to describe the neutron absorbing power of a material. It is expressed in area units or capture cross sections, such as a barn (b), which is equal to 10^{-24} cm^2/nucleus. The absorption of the neutrons in an object depends on its neutron cross section, the quantity of nuclides it contains, and its thickness. A comparison of neutron and X-ray mass attenuation coefficients for the elements as a function of atomic number is shown in Figure 8.1.

Most hydrocarbon foams, formed in petrochemical reactors, are transparent to gamma radiation, but readily absorb thermal neutrons. The same is true of most plastic materials. Gamma ray penetration is limited in steel and lead, but neutron radiography can inspect large, thick heavy metals and discriminate between neighboring elements such as boron (B) and carbon. Figure 8.2 shows the variation of element thickness required to provide a 50%

Introduction to Nondestructive Testing: A Training Guide, Second Edition, by Paul E. Mix
Copyright © 2005 John Wiley & Sons, Inc.

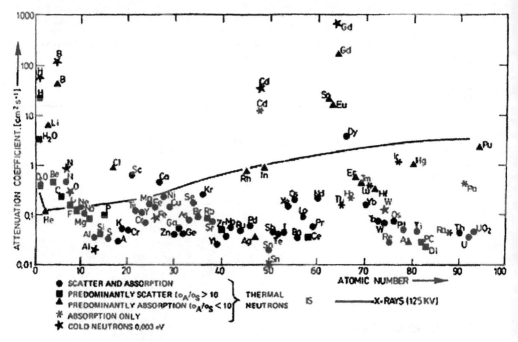

Figure 8.1 Neutron and X-ray mass attenuation coefficients for the elements. Courtesy of Cornell University.

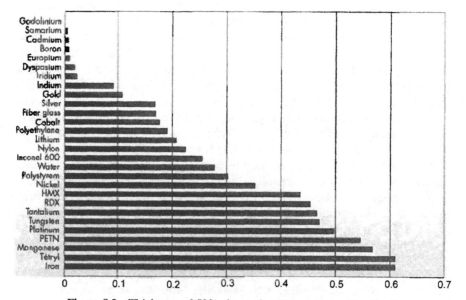

Figure 8.2 Thickness of 50% thermal neutron attenuation (cm).

attenuation of thermal neutrons or half value layer. In this example, boron is one of the better thermal neutron attenuators and iron is one of the worst.

8.2 PHYSICAL PRINCIPLES

Material containing hydrogen atoms (hydrogenous material) has a high macroscopic scattering cross section. These materials are commonly used in fast neutron detectors. The term *high macroscopic cross section* means that the material has a high linear attenuation coefficient for neutrons. The linear coefficient for neutrons (μ) is defined by Eq. 8.1:

$$\mu = N\sigma_t \tag{8.1}$$

where N = number of nuclei per cubic centimeter
σ_t = total cross section in square centimeters (the sum of absorption and scattering cross sections)

Accelerators, radioactive sources, and reactors produce fast neutrons with energies above 10,000 eV. Boral shutters are used to turn the neutron beam of a nuclear reactor on and off. When a fast neutron strikes a nucleus, inelastic scattering or capture of the radiated neutron occurs. Because of the collision, the neutron is slowed down, loses energy, and changes direction. Fast neutrons can be slowed down (moderated) or captured by paraffin, water, graphite, and most plastics (moderators). Water is a good moderator and is used in light water moderated reactors to slow down neutrons. Plastics produce high attenuation of thermal neutrons by scattering. Thermal neutrons, relatively free of high-energy components, are produced by moderation with water, paraffin, or graphite. A thermal neutron is a neutron whose energy is the same as a gas molecule at room temperature or about 0.026 eV; the thermal neutron energy range is 0–0.3 eV. The energy range of thermal and epithermal neutrons is 0.3–10,000 eV.

The penetrating ability of a thermal neutron beam is dependent on the attenuation characteristics of the material being penetrated. Neutron beams contain ionizing (gamma) and neutron particle radiation. Increasing neutron energy increases neutron penetration, but the energy of thermal neutrons decreases as their temperature decreases.

There is no relationship between the thermal neutron mass absorption coefficient of a material and its atomic number. Neutrons can easily penetrate lead, but are heavily moderated by paraffin. Boron has the largest mass absorption coefficient for thermal neutrons and lead is a poor neutron absorber. Rare earths are frequently used in neutron radiography because they have large neutron absorption cross sections. The rare earths are chemically similar, make up about one-sixth of all naturally occurring elements, and occupy one position (lanthanides) in the periodic chart. The rare earth elements are listed in Table 8.1.

TABLE 8.1. The Rare Earth Elements

Atomic Number	Symbol	Name
57	La	Lanthanum
58	Ce	Cerium
59	Pr	Praseodymium
60	Nd	Neodymium
61	Pm	Promethium
62	Sm	Samarium
63	Eu	Europium
64	Gd	Gadolinium
65	Tb	Terbium
66	Dy	Dysprosium
67	Ho	Holmium
68	Er	Erbium
69	Tm	Thulium
70	Yb	Ytterbium
71	Lu	Lutetium

8.3 NEUTRON RADIATION SOURCES

As with other radioactive isotopes, the half-life of a neutron source is the time required for one half of the atoms in the isotope to disintegrate. As the atoms disintegrate, short-wavelength gamma radiation is produced. Americium-241/Be and Californium-252 are popular source materials because of their portability. Californium-252 is a spontaneous fission source with a half-life of 2.65 yr. Table 8.2 lists some neutron sources that are currently available.

The best source of high-intensity thermal neutrons is a nuclear reactor because high-quality neutron radiographs are produced with core fluxes ranging from 1×10^{12} to 5×10^{14} nv. The intensity of neutron radiation is measured in nv (neutrons/cm$^2 \cdot$s). Material that has been exposed to thermal neutron beams may continue to be radioactive after the neutron bombardment has ceased; however, the level of radioactivity will be very low.

8.4 NEUTRON ACTIVATION ANALYSIS

A periodic chart similar to the one shown in Figure 8.3 helps classify the elements according to their chemical activity, atomic weight, and neutron activation analysis sensitivities. Since neutrons activate the nucleus of an atom, neutron activation analysis (NAA) is sensitive to the total elemental content of a sample regardless of its state of oxidation, chemical form, or physical location. High-flux neutron activation analysis has a median interference-free-sensitivity of 0.001 µg for about 70 elements as shown in Figure 8.4, making it

TABLE 8.2. Neutron Source Activity and Emission

A-241/Be

Nominal Activity (mCi)	Emission (±10%) neutrons/s
1	2.2×10^3
3	6.6×10^3
10	2.2×10^4
30	6.6×10^4
100	2.2×10^5
300	6.6×10^5

Ca-252

Nominal Content	Nominal Activity	Emission (neutrons/s)
0.01 μg	5 μCi	2.3×10^4
0.1 μg	54 μCi	2.3×10^5
0.5 μg	268 μCi	1.15×10^6
1.0 μg	536 μCi	2.3×10^6
2.0 μg	1.07 mCi	4.6×10^6
5.0 μg	2.7 mCi	1.15×10^7
10.0 μg	5.4 mCi	2.3×10^7
20.0 μg	10.7 mCi	4.6×10^7
50.0 μg	27.0 mCi	1.15×10^8
100.0 μg	54.0 mCi	2.3×10^8
200.0 μg	107.0 mCi	4.6×10^8
500.0 μg	268.0 mCi	1.15×10^9
1.0 mg	536.0 mCi	2.3×10^9
2.0 mg	1.07 Ci	4.6×10^9
3.0 mg	1.61 Ci	6.9×10^9

Neutron oil Well Logging Sources—Am-241/Be

Nominal Activity (Ci)	Emission (neutrons/s)
3	6.6×10^6
5	1.1×10^7
10	2.0×10^7
20	4.0×10^7

Source: Adapted from information supplied by Amersham Corporation.

one of the most sensitive tools today for elemental analysis. This is also a recognized identification of materials (IM) method.

Neutron activation analysis is accomplished by placing from a few tenths of a gram of sample to several grams of a sample in a nuclear reactor for both short and long periods of time. Gamma ray spectrum is then measured using Ge(Li) and Na(Ti) detectors. Gamma-ray spectra are recorded after decay times of 1 min, 1 h, 1 day, 1 week, and 3 weeks. Data obtained are fed into a

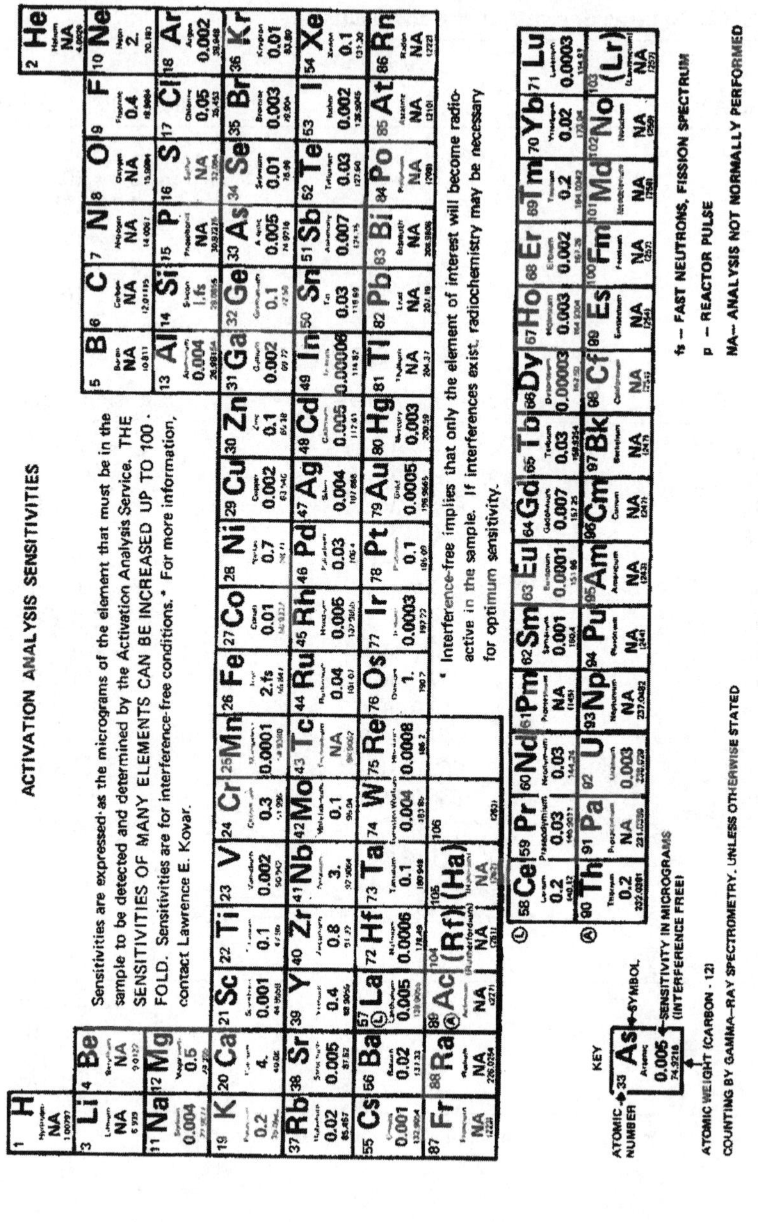

Figure 8.3 Periodic chart.

Symbol	Element	NAA sensitivities in micrograms	Atomic Number
Ac	Actinium	NA	89
Ag	Silver	0.004	47
Al	Aluminum	0.004	13
Am	Americium	NA	95
Ar	Argon	0.002	18
As	Arsenic	0.005	33
At	Astatine	NA	85
Au	Gold	0.0005	79
B	Boron	NA	5
Ba	Barium	0.02	56
Be	Beryllium	NA	4
Bi	Bismuth	NA	83
Bk	Berkelium	NA	97
Br	Bromine	0.003	35
C	Carbon	NA	6
Ca	Calcium	4	20
Cd	Cadmium	0.005	48
Ce	Cerium	0.2	58
Cf	Californium	NA	98
Cl	Chlorine	0.05	17
Cm	Curium	NA	96
Co	Cobalt	0.01	27
Cr	Chromium	0.3	24
Cs	Cesium	0.001	55
Cu	Copper	0.002	29
Dy	Dysprosium	0.00003	66
Er	Erbium	0.002	68
Es	Einsteinium	NA	99
Eu	Europium	0.0001	63
F	Fluorine	0.4	9
Fe	Iron	2. fs	26
Fm	Fermium	NA	100
Fr	Francium	NA	87
Ga	Gallium	0.002	31
Gd	Gadolinium	0.007	64
Ge	Germanium	0.1	32
H	Hydrogen	NA	1
(Ha)	(Hahnium)	NA	105
He	Helium	NA	2
Hf	Hafnium	0.0006	72
Hg	Mercury	0.003	80
Ho	Holmium	0.003	67
I	Iodine	0.002	53
In	Indium	0.00006	49
Ir	Iridium	0.0003	77
K	Potassium	0.2	19
Kr	Krypton	0.01	36
La	Lanthanum	0.005	57
Li	Lithium	NA	3
Lu	Lutetium	0.0003	71
(Lr)	(Lawrencium)	NA	103
Md	Mendelevium	NA	101
Mg	Magnesium	0.5	12
Mn	Manganese	0.0001	25
Mo	Molybdenum	0.1	42
N	Nitrogen	NA	7
Na	Sodium	0.004	11
Nb	Niobium	3.	41
Nd	Neodymium	0.03	60
Ne	Neon	2.	10
Ni	Nickel	0.7	28
No	Nobelium	NA	102
Np	Neptunium	NA	93
O	Oxygen	NA	8
Os	Osmium	1.	76
P	Phosphorus	NA	15
Pa	Protactinium	NA	91
Pb	Lead	NA	82
Pd	Palladium	0.03	46
Pm	Promethium	NA	61
Po	Polonium	NA	84
Pr	Praseodymium	0.03	59
Pt	Platinum	0.1	78
Pu	Plutonium	NA	94
Ra	Radium	NA	88
Rb	Rubidium	0.02	37
Re	Rhenium	0.0008	75
(Rf)	(Rutherfordium)	NA	104
Rh	Rhodium	0.005	45
Rn	Radon	NA	86
Ru	Ruthenium	0.04	44
S	Sulfur	NA	16
Sb	Antimony	0.007	51
Sc	Scandium	0.001	21
Se	Selenium	0.01	34
Si	Silicon	1. fs	14
Sm	Samarium	0.001	62
Sn	Tin	0.03	50
Sr	Strontium	0.005	38
Ta	Tantalum	0.1	73
Tb	Terbium	0.03	65
Tc	Technetium	NA	43
Te	Tellurium	0.03	52
Th	Thorium	0.2	90
Ti	Titanium	0.1	22
Tl	Thallium	NA	81
Tm	Thulium	0.2	69
U	Uranium	0.003	92
V	Vanadium	0.002	23
W	Wolfram-Tungsten	0.004	74
Xe	Xenon	0.1	54
Y	Yttrium	0.4	39
Yb	Ytterbium	0.02	70
Zn	Zinc	0.1	30
Zr	Zirconium	0.8	40

Figure 8.4 As shown, average high flux NAA interference-free sensitivity is about 0.001 microgram.

computer and analyzed. Short-life, highly activated isotopes are detected first with long-life, less activated isotopes detected later.

A faster and less expensive nondestructive approach to NAA is accomplished by irradiating the sample in a reactor, then counting the gamma emission with a Ge(Li) detector and γ-ray spectrometer after a suitable decay period.

8.5 WARD CENTER TRIGA REACTOR

The neutron radiography facility at Ward Center consists of a filtered and collimated beam of neutrons emanating from the through-port of the TRIGA reactor as shown in Figure 8.5. The L/D ratio for a 1″ aperture ranges from

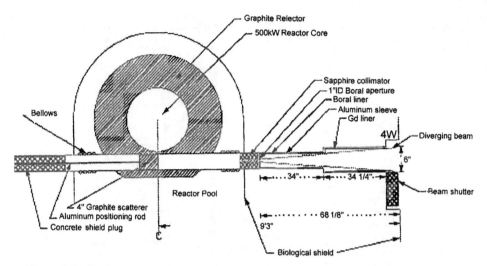

Figure 8.5 Ward Center TRIGA Reactor 4W tangential-port. Thermal Neutron Radiography Facility. Courtesy of Cornell University.

~116 for real-time radiography (10″-diameter beam) utilizing the Thomson tube camera, to ~151 (14″-diameter beam) using film. The introduction of a smaller 0.5″ aperture in the collimation system results in L/D ratios twice the above values with a fourfold reduction in beam intensity. With the usual 1″ aperture, the typical film radiograph can be completed in 15 minutes, resulting in an average film density of 2. High-speed pulse radiography with exposure times of ~30 milliseconds is also available. The gamma dose to the object for the typical radiograph is ~1 RAD. The gamma dose ratio and uniformity make it a Class I radiography facility.

The Cornell Teaching, Research, Isotope, General Atomics (TRIGA) reactor, was manufactured by General Atomics and went critical in January 1962. The design by Mark Nelkin was noted for its inherent safety. The fuel is alloyed in a matrix containing zirconium hydride whose prompt coefficient of reactivity is negative and large, assuring that the neutron multiplication rate decreases as temperature rises. Power excursions are limited by the special nature of the fuel. Therefore, human, electronic, and mechanical intervention is not required to guarantee reactor safety during operation.

In 2002 and 2003 the reactor had to be either relicensed by the Nuclear Regulatory Commission (NRC) or decommissioned under NRC approval. Having no Department of Nuclear Science and Engineering since 1994, the university had no further need for the facility and the decision was made by Cornell University to decommission the reactor pending NRC approval. In its 40 years of operation, the TRIGA reactor has served the university, community, and industry well. However, rather than rebuilding and expanding the facility, it is now time to say goodbye to an old friend and convert the 16,000-square-foot facility to other academic and research activities. Other NAA (neutron activation analysis) facilities are available for continued NAA research. Applications previously evaluated at Cornell's Ward Center neutron radiography facility, prior to its shutdown, include:

- Evaluation of microcracking in structural concrete
- Imaging of corn rootworm larvae in situ with live corn roots and real-time imaging of larval movement in soil
- Ionization chamber defect analysis
- Examination of fossilized dinosaur bones
- Quality assurance testing of a ventilation unit for outer space applications
- Study of two-phase flow of liquid and vapor
- Imaging plant root growth in contaminated or compacted soils in situ; studying fluid flow in porous media

Note: Additional information is still posted on the Ward Center website.

$T(d,n)^4He$ accelerators are among the few types of neutron generators that have a significant neutron yield with low bombardment energies. The designation $T(d,n)^4He$ means that a tritium (T) impregnated lithium target is used, and that neutrons (n) are generated by the bombardment of the target with deuterons (d). Helium^{-4} (^4He) is a harmless by-product. This type of accelerator produces neutron energy of about 14.3 MeV. Many of these generators were built in the 1960s. Some are still in existence, but few are currently being built.

8.6 RADIATION HAZARDS AND PERSONAL PROTECTION

The rem (R) as used by the U.S. Nuclear Regulatory Commission is defined as a measure of the dose of ionizing radiation corresponding to the absorption of 84 ergs of energy per gram of body tissue. Radioactivity is measured in curies or the number of disintegrations per unit time. One curie or 1000 mCi = 3.7×10^{10} disintegrations. A 1-curie source will produce a dose rate of several rems per hour, depending on its composition, energy intensity, and spectrum. The maximum whole-body dose that a radiation industry worker is permitted to receive in any one calendar quarter is 3 R.

When working at a neutron radiographic facility, workers should be aware that beta, gamma, and neutron radiation will be present. Because of their high intensity and high permeability, neutron radiographic sources are best handled with remote handling equipment. Neutron exposure can be due both to the direct beam from the neutron source and scatter radiation from objects in the direct beam. As a general rule, personnel should not be exposed to excessive radiation—any unnecessary exposure to radiation is considered excessive. Geiger counters are frequently used to monitor for both low levels of radiation and small radioactive leaks in radiation barriers. Radiation levels near biological shield areas should not exceed 200 mrem/h at the surface or 2 mrem/h at a distance of 1 m.

Note: rad = material absorption = 0.01 joule/kg = 1 Gray; rem = personnel absorption = 84 ergs of energy per gram of body tissue (rems = rads multiplied by the quality factor, Q, which equals 1 for X-, gamma, or beta radiation). Neutron monitoring outside a radiographic exposure area may be done with a BF_3 proportional counter.

For purposes of radiation safety, the time and distance relationships to exposure should be remembered. Keeping exposure time to a minimum and keeping as far away as possible from the source during exposure periods can minimize a person's exposure to radiation. The dose rate of gamma (γ) rays is measured in units of rems (R/h) or millirems per hour (mR/h). Time is a direct relationship; distance is an inverse-square relationship. With regard to time, if you are exposed to a dose rate of 1200 mR/hr at 6 feet, you will receive a radiation dose of 600 mR in 30 minutes and a radiation dose of 1200 mR in one hour. However, if your dose rate at 6 ft is 1200 mR/h, your dose rate at 12 feet would be reduced to 1200/4 or 300 mR/h. Divergent neutron beams follow the same inverse-square law of distance as X-rays and γ-rays.

X-ray and γ-ray exposure can be determined by film badges, dosimeters, and radiation survey meters. One advantage of monitoring personnel exposure with direct reading pocket dosimeters is that it provides an immediate reading of beta and gamma dosage. The main disadvantage of the direct reading pocket dosimeter (small ion chamber) is that it does not provide a permanent record of the exposure. Care must be taken when making surveys with some hand-held Geiger-Müller (G-M) survey meters because saturation can occur. If the meter needle rises during a survey then drops suddenly, it should be presumed that the meter has saturated because too much radiation is present. Special survey meters, such as the Victoreen® Model 190N portable survey meter with BF3 proportional counter, can be used to making low-level neutron radiation surveys.

Neutron radiation exposure is the product of neutron flux for a given area and time. The average thermal neutron flux that produces a dose of 100 mrem in 40 h (maximum permissible exposure) is 700 neutron/$cm^2 \cdot s$. Special film dosimeters, sensitive to beta, gamma, and neutron radiation, should be worn by neutron radiographers. These dosimeters use screen combinations similar to those used in radiography. For monitoring neutron energies <0.5 eV, personnel monitoring films are constructed using cadmium- or rhodium-covered film badge films. Li_6F thermoluminescent detectors are also available. Thermal neutron sensitive radiation counters usually contain boron. For fast neutron monitoring, a thermal neutron sensitive detector may be used in combination with a moderator. The personnel-monitoring detector should respond primarily to the type of radiation to be monitored; its response should be minimal to other types of radiation that are simultaneously produced.

Small amounts of exposure to neutrons or γ-rays may have a cumulative effect on personnel working with radiation. This cumulative effect should be taken into account when radiation levels are high and the radiation safety officer (RSO) is monitoring for maximum permissible doses. However, the calculations of *occupational doses* do not take into account radiation doses that an individual receives as the result of medical diagnosis or therapy. Overexposure to neutrons or γ-rays may cause damage to blood, skin, and internal organs. A radiation dose of 200 to 800 rems would prove harmful or fatal if it were received as a whole-body dose in a short period of time. With identical beam intensities, higher-energy fast neutrons produce larger biological doses than thermal neutrons.

8.7 RADIATION DETECTION IMAGING

When neutrons bombard objects, they can become radioactive and emit gamma radiation. In neutron radiography, some imaging screens can retain their radioactivity and indirectly transfer the test object image to the radiographic film. This method is known as the "transfer," "indirect," or "indirect transfer" method of radiation detection imaging. This method has the advantage that it is not sensitive to other sources of gamma radiation. Disadvantages of the indirect transfer method of imaging are that it is time consuming, requires many imaging foils, provides lower resolution than the direct method, and is more costly.

Direct conversion screens immediately emit γ-rays and conversion electrons to produce the test object image on the radiographic film. Direct conversion screens do not retain their radioactivity for any significant period of time.

Five important neutron radiographic parameters that can be controlled to optimize results are:

1. Neutron energy
2. Exposure time
3. Film type
4. *L/D* ratio (source-to-target distance divided by beam diameter)
5. Type of conversion screen

Desirable features for neutron radiography include low gamma radiation intensity, relatively low fast neutron intensity, and low angular divergence of the neutron beam. Low angular beam divergence improves the resolution capabilities of the systems when radiographing thicker test objects. Real-time imaging of thermal neutrons can be achieved using zinc sulfide and lithium fluoride detectors.

8.7.1 Conversion Screens

Conversion screens convert neutron energy into ionizing radiation (conversion electrons), which can be captured as an image on radiographic films. The choice of conversion screen is dependent on the energy of the neutrons in the neutron beam. As the effective energy of radiation from the conversion screen increases, radiographic definition decreases. Gadolinium and cadmium have high thermal neutron cross-sections that drop off rapidly at high energies. Common conversion screen materials are dysprosium (Dy), gadolinium (Gd), and indium (In). One characteristic of a good converter screen, such as boron (B), lithium (Li), or Gd is that they do not become very radioactive, but they emit ionizing radiation as soon as they absorb neutrons. Gadolinium is the most commonly used converter material.

8.7.2 Indirect Transfer Method

Dysprosium and indium screens are used for the indirect transfer method because these screens have reasonably long radioactive half-lives. These screens are used in highly radioactive areas to reduce film fogging. With fast films, the minimum thermal neutron flux that can be used by Dy screens is 10^4 nv. Medium-energy epithermal neutrons and In screens are used to obtain improved penetration and resolution when radiographing highly radioactive 6-mm-diameter, 28% plutonium/enriched uranium fuel pellets. When long exposures with low-energy beams are acceptable, the *track-etch neutron imaging method* is used.

Dysprosium conversion screens emit high-energy betas, low-energy gammas, and internal conversion electrons. After being exposed to a neutron beam, a Dy conversion screen will retain about one-eighth of its original activation after 6.9 h.

8.7.3 Direct Transfer Method

Gadolinium conversion screens emit γ-rays and conversion electrons. High-quality direct neutron radiographs are obtained with Gd imaging screens. These screens provide the best resolution and greatest contrast when used with direct neutron radiography. With direct neutron radiography using Gd screens, electrons cause X-ray film darkening. Filtering reduces unwanted γ-ray background. Gadolinium has high neutron absorption per unit thickness; Gd screens can resolve high-contrast images separated by as little as 0.0004 in. Gadolinium, B, and Li conversion screens can be used in combination with fluoroscopic screens for TV and optical imaging systems.

The main disadvantage of Gd screens is that they are expensive because they are produced by vacuum vapor deposition techniques. Gadolinium screens are frequently mounted in rigid vacuum cassettes to assure positive film–to–conversion foil contact. Identification labels of neutron absorbing cadmium or gadolinium are attached to the cassettes for film identification purposes. Neutron converter screens should be visibly inspected for flaws and dirt every time they are used. Dust or lint between the film and Gd screen will appear as a flaw on the radiograph. For best results, test objects should be mounted on aluminum backing material.

8.7.4 Fluorescent Screens

When fluorescent screens are used in neutron radiographic imaging, reciprocity law failure often occurs because less energy is deposited in the film grain. Failure of the reciprocity law means that the efficiency of the emulsion is dependent on light intensity. The reciprocity law failure can be partially alleviated by using scintillator converters at low temperature. Scintillator screens can be used when 0.002-in. resolution is acceptable. Neutron sensitive scintillators provide lower quality radiographs with shorter exposure times. LiF, ZnS(Ag) neutron scintillators emit blue light when used with fluorescent screens and provide a good match with the spectral response of most industrial X-ray films.

8.8 ELECTRONIC IMAGING

Electronic imaging devices provide amplification of both signal and noise. Higher electronic gains are used with electronic imaging when broadband Gd-oxysulfide (Gd_2O_2S) converter screens are used in place of ZrS-Li_6F converter screens. Gd-oxysulfide screens generate a combination of light and electrons over 10 different energy levels.

8.9 NONIMAGING DETECTORS

Nonimaging types of neutron detectors include gaseous ionization detectors, which detect alpha particles produced by neutron collisions, boron trifluoride (BF_3) proportional counters, plastic scintillation counters, and gold foil activation detectors. With proportional counters, the charge collected for a given amount of radiation is larger than, but proportional to, the amount of original ionization. With the gold foil activation technique, the degree of activation of the gold foil is an indication of neutron intensity. The output of most nonimaging neutron detectors is an electrical signal, often a number of pulses, whose count or count rate can be determined.

8.10 NEUTRON RADIOGRAPHIC PROCESS

Radiographs are photographic images produced when neutrons pass through a test object into the film. The transmission of neutrons by a material varies exponentially with material thickness. Exposure time is a linear function of thermal neutron flux intensity. In order to minimize shadow formation, the axis of the neutron beam should be perpendicular to the film. Filters such as bismuth may be used to improve the neutron-to-gamma ratio. Gamma radiation is an unwanted by-product of neutron generation.

The ability to detect a small discontinuity or flaw is a measure of radiographic sensitivity. Graininess of the film, unsharpness of the flaw image, and contrast of the flaw image affect radiographic sensitivity. ASTM standard E-545-99 defines what constitutes an acceptable neutron radiographic sensitivity and specifies the number of sensitivity indicators.

Radiographic definition is the sharpness of the outline of the image of the radiograph. Large focal spot, poor film–screen contact, or bad geometry can cause poor definition or radiographic unsharpness. The accidental movement of the test part or film during exposure, or the use of too small a source-film distance, will result in radiographic unsharpness. Flaw images close to the source side of the specimen become less clearly defined as the thickness of the part increases. Unsharpness of a radiograph is directly proportional to the size of the source and inversely proportional to the source-to-test part distance. In order to decrease geometric unsharpness, the neutron source should be as small as possible.

In direct neutron radiography, higher resolution can be achieved by increasing the L/D ratio of the collimation system. The L/D ratio is the distance from the source to the test piece (L) divided by the beam diameter (D). If the L/D ratio for a given neutron beam is 250 and the test part is 0.250 in. thick, the geometric unsharpness of the object will be 0.001 in. In cases where it is necessary to increase neutron beam intensity, the L/D ratio must be lowered.

Beam collimation is used to focus the neutron energy for improving radiographic resolution and image sharpness. One type of neutron beam collimator is the divergent beam collimator. With this type of collimator, image magnification occurs as the distance between the test piece and detector increases. If L is the source-to-test piece distance, and d is the test piece–to–detector distance, the magnification (M) is determined by Eq. 8.2:

$$M = (L+d)/L \qquad (8.2)$$

The advantages of the divergent beam collimator are that it is simple to manufacture, it has minimum neutron reflections, and there are no dividing slats that could possibly cause lines to show on a radiograph. Its disadvantage is that a very large collimator is required to achieve satisfactory L/D ratios.

Radiographic contrast, the difference between the densities of two areas of a radiograph, is dependent on the thickness differences of the specimen, neutron energy, and the intensity and distribution of scattered radiation. In neutron radiography, radiographic contrast is a function of the gradient of the characteristic curve of the film and neutron energy. The contrast observed for a radiographic image is greater when the film density gradient is maximized.

When hydrogenous objects are being radiographed, placing a neutron absorbing collimator between the object and film can reduce scattering. Radiographic contrast is only slightly affected by beam collimation.

Test parts of uniform thickness and compression have low subject contrast. Films having wide exposure latitude also have low contrast. Contrast enhancement can be achieved by using photographic techniques with X-ray duplicating film.

A graph showing the relationship between optical film density and the log of exposure or relative exposure is called the characteristic curve. Optical film density is a quantitative measure of film blackening or film density. The uniformity of a radiograph can be determined by measuring the density of the radiograph at several locations with no parts in place.

The density of a film increases with increasing exposure up to the film reversal point; beyond this point, further increases in exposure can cause film density to decrease. A film with a density of 2.0 will transmit only 1% of incident light falling on it. Photographic density (D) is defined by Eq. 8.3:

$$D = \log I_o / I_t \tag{8.3}$$

where I_o = incident light intensity
I_t = transmitted light intensity

As film development time increases, the characteristic curve grows steeper and moves to the left. The shape of the characteristic curve is independent of the energy of the neutron beam. The slope or steepness of the characteristic curve is a measure of film contrast. The slope of a straight line joining two points of specified density on the characteristic curve of a film is a measure of average gradient.

Real-time fluoroscopic imaging has several advantages over film radiography, including low-cost and high-speed processing. However, its main disadvantage is that its high image brightness on the screen can quickly lead to operator fatigue.

8.11 INTERPRETATION OF RESULTS

The main reason for using neutron radiography instead of X radiography is the ability to make images of objects and materials that are transparent to X-rays. Placing image-quality indicators (IQIs) or penetrameters on the source side of the cassette determines the adequacy or quality of the neutron radiographic technique. Image-quality indicators are considered standard test pieces; they are made of the same material as the test piece being radiographed.

ASTM beam-purity indicators indicate contrast. Some causes of poor definition of flaws are poor geometry, improper contact between the film and conversion screen, and film graininess. The lower limit of detectability for microporosity in material is dependent on the L/D ratio of the neutron beam and the graininess of the film.

The best method for determining how resolution affects the ability to interpret radiographs of parts is by comparing radiographs of parts having known defects. Comparison standards are best for establishing image-object relationships. The range of densities that are satisfactory for interpretation of results is a measure of the latitude of a radiograph.

Neutron radiographs can be used to detect the presence of an explosive in a metal container, detect oils and lubricants in metal systems, and determine the hydrogen content of metals. It also can be used to determine the integrity of thin plastic parts inside lead housings.

One specific application of neutron radiography is the detection of cracks in small plutonium pins. Dark, intermittent, or continuous lines appear on the radiograph of a part that contains a crack. Unwanted inclusions in a part can appear as either light or dark spots depending on relative neutron absorption of the part and defect.

Cold neutrons, neutrons that have an energy level less than that of thermal neutrons, are used for the detection of small amounts of explosives in steel. One characteristic of cold neutrons is that they are scattered less and absorbed more than thermal neutrons. This characteristic tends to produce a radiograph with good contrast in this application.

Scattered radiation from a wall or floor undercuts the specimen and adversely affects film quality. When it is noted that an image of the back of the cassette is superimposed on the image of the part, the most likely cause of the double exposure is *backscatter* radiation. A mask of cadmium, cut to the shape of the part, can be used to minimize the effects of scattered neutron radiation.

8.12 OTHER NEUTRON SOURCE APPLICATIONS

One popular application using a small nuclear reactor is neutron activation analysis (NAA). Because neutrons activate the nucleus of an atom, NAA is sensitive to the total elemental content of a sample regardless of its state of oxidation, chemical form, or physical location. High-flux neutron activation analysis has a median interference-free sensitivity of 0.001 µg for about 70 elements, making it one of the most sensitive tools today for elemental analysis. The technique is also recognized as an identification of materials (IM) method.

The small-angle neutron scattering instrument (ANS) at Oak Ridge National Laboratory (ORNL) provides an inexpensive source for neutron scattering research. Research at this facility has led to the development of tough polymer plastics used in automobile safety glass, unbreakable plastic toys, bullet-proof police vests, and tough, difficult-to-open food packages. In addition to new polymers, research conducted at ANS includes:

- Developing new colloidal suspension systems and studying the shear effects on oils
- Detecting and analyzing microcracking in concrete structures
- Determining the composition of paints and pigments for authentication of works of art
- Detecting and analyzing ion chamber defects
- Studying two-phase flow of various liquid and vapor systems
- Detecting early signs of corrosion in fighter aircraft wings
- Studying plant growth in porous or compacted soil systems
- Developing improved magnetic structures for high-density recording media
- Measuring stresses and stains in manufacturing and welding applications
- Developing lightweight magnetic materials using neodymium, iron, and boron

A less sensitive, less expensive instrument uses americium-241/beryllium sources for NAA. With this instrument, process material flows through coils around both the source and detector. A sample of material is held in the source coil for a few minutes where it is activated. Then the material passes on to the detector coil where it is also held for a few minutes. Sensitive gamma detectors measure the gamma spectrum and supply information to the microprocessor that analyzes the data and provides readout for the components of interest.

Another important use for neutron sources is in oil well logging applications. Am-241/Be sources of 1 to 20 Ci or californium sources of 536 µCi to 27 mCi are used for neutron porosity logging. Figures 8.6 and 8.7 show the geometry of systems used for determining the moisture and porosity content of various well formations. In both these applications, fast neutrons are moderated by hydrogenous material in the formation to produce thermal and epithermal neutrons that can be detected by BF_3 or helium-3 detectors, respectively. The detected neutron flux is proportional to the hydrogen in the formation, which exists within the porosity of the formation, and instrumentation can be calibrated in terms of either moisture or porosity.

There are several online or rapid analysis batch analyzer systems currently on the market that incorporate the use of neutron sources. Some disadvantages of these systems are:

- Neutron sources are expensive.
- Neutron moderation and shielding (with polyethylene, water, oil, and boron-filled materials, etc.) tends to be bulky and expensive.

Moisture gauging

Technique

Fast neutrons emitted by the source are moderated by collision with hydrogen atoms in moisture contained in the material. These moderated or thermal neutrons are detected by a neutron detector (usually a boron trifluoride (BF_3) proportional counter) to give a measure of the concentration of hydrogen atoms.

Geometry

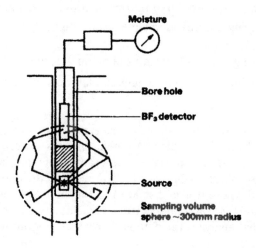

Applications

- Soil moisture content for agricultural and construction use.
- Moisture content of materials in silos.
- Continuous moisture content gauging in raw material supplies e.g. gravel, wood chips etc.

Sources

Nuclide	Typical activity range
Americium-241/Beryllium	30–250mCi 1.11GBq – 9.25GBq
Californium-252	0.1 µg (54 µCi 2MBq)

Figure 8.6 Neutron moisture gauging. Courtesy of Amersham Corp.

Neutron porosity logging technique

Fast neutrons emitted by a neutron source are slowed down by the formation and may undergo three interactions: 1) inelastic scatter, 2) elastic scatter, 3) absorption. Therefore, by collision with hydrogen atoms in the formation the neutron will be moderated to thermal or epithermal energies where it is soon captured by hydrogen nuclei and emits a secondary gamma ray. The detection of these three interactions by using different types of neutron detectors (BF_3 (thermal), ^3He (epithermal)) can be used to determine the hydrogen content of the formation. Since the majority of hydrogen in a formation generally exists within the pore space, the neutron flux will then be related to the porosity.

Geometry

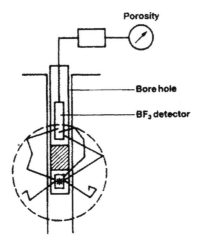

Applications

- Determination of formation hydrogen content
- Formation porosity for oil and mineral logging

Sources

Nuclide	Typical activity
Americium-241/Beryllium	1Ci – 20Ci 37GBq – 740GBq
Californium-252	536µCi – 27mCi 20MBq – 1GBq

Figure 8.7 Neutron porosity logging. Courtesy of Amersham Corp.

- Almost every instrument developed is unique in its design and application; fabrication is labor intensive and expensive.

Instruments currently in production that use neutron sources are neutron level gauges and neutron moisture analyzers, bench-top and online versions. The online version measures the moisture content of coke over the range of 0 to 20% and combines this measurement with a density measurement to calculate the mass flow of dry coke using a microprocessor.

8.13 NEUTRON LEVEL GAUGES

Neutron level gauges are used to measure the level of hydrocarbons and hydrocarbon foam. The neutron level gauge is sufficiently sensitive to both detect and suppress hydrocarbon foaming in stages, until the hydrocarbon material reaches the desired control level.

Light, portable hand-held probes can be used to measure hydrogen in foams, liquids, and solids in most insulated and noninsulated tanks and vessels. The neutron probe head consists of a sealed fast neutron source mounted next to a slow helium-3 neutron detector. The fast neutrons penetrate the walls of the vessel and react with hydrogen atoms in the material, where they lose their energy and are backscattered. These slow reflected neutrons are then counted with the neutron count rate meter, and converted to an analog scale reading. An audio signal is also produced that can be monitored with earphones in noisy production areas.

The foam or vapor/liquid interface in the tank is located by moving the probe head vertically on the outer surface of the tank. When the foam interface is detected, the neutron count rate will drop off significantly. Liquid level measurement usually can be detected to within ±0.5″.

Typical specifications for the portable neutron backscatter unit are:

- Source: Am-Be 241, 200 mCi
- Radiation at operator: 0.5 mrem/hr
- Linearity: ±5% of full scale
- Power supply: 2 D cells; ~100 hr operation
- Response time: 3 or 11 s
- Max. wall thickness: 3″
- Min. process material thickness: 2″
- Tank wall temperature limits: −112 to +500°F
- Ambient temperature limits: −112 to +221°F

Fixed neutron backscatter systems may be permanently mounted on a tank to precisely control interface levels, density, and/or percent solids in polyethylene-Freon slurries. Alarm functions can also be provided.

8.14 CALIFORNIUM-252 SOURCES

Californium-252 sources are used with prompt neutron analysis to determine the impurities in coal or cement. In this application, epithermal neutrons from the californium source cause prompt gammas to be emitted (immediately) from the coal or cement as it is transported past the source. The detected gamma spectrum then can be analyzed by a microprocessor to determine the amount of alumina, silica, calcium oxide, magnesium oxide, sulfur, potassium, or other impurities in the sample.

NUREG/CR-3110 (Vol. 1, Part 2, Chapter 6) states that neutron radiography has limited but specific applications to nuclear reactor systems. Where large differences in attenuation coefficients between neutrons and protons can be used to advantage, potential applications include:

1. Radiography of highly radioactive objects such as nuclear fuel and reactor vessels and piping.
2. Imaging of hydrogenous materials such as seals, gaskets, explosive charges, aluminum corrosion, oil, and grease.
3. Radiography of materials that have high-neutron absorption markers deliberately added. An example might be the addition of gadolinium to the core material of aircraft turbine blades.
4. Radiography of materials such as those listed in (2) and (3) that may be embedded or hidden by large, thick heavy metals (copper, lead, brass, and steel).
5. Testing of reactor control and target elements for the correct balance of neutron absorbers or radioactive isotopes.

8.15 NEUTRON RADIOSCOPIC SYSTEMS

Sensitive neutron radioscopic systems developed by Industrial Quality, Inc., are capable of detecting minute amounts of aluminum corrosion in aircraft parts, thereby providing early warning of an impending problem.

8.15.1 Introduction

Many nonreactor thermal neutron sources can be used in inspections for aircraft maintenance, explosive devices, and cast turbine blades, and for evaluation of electronic and/or electromechanical modules. Important considerations are neutron output, portability, ease of use, and cost. From an economic point of view, neutron sources should be also usable with electronic neutron radioscopic cameras for real-time inspection and computer analysis. There are many types of electronic neutron imaging systems, image intensifiers, flat panels, and scintillator-camera systems, as examples. These latter systems typically consist

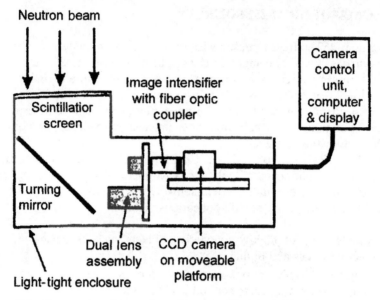

Figure 8.8 Neutron imaging system schematic. Courtesy of Industrial Quality, Inc.

of a scintillator screen, mirror and lens arrangement, image intensifier with fiber optic coupler, CCD camera, camera control unit, computer, and display as shown in Figure 8.8.

8.15.2 Neutron Imaging System Components

The major neutron imaging system components are described below:

- Thermal neutron screen: Li-6 based ZnS(Ag); MTF = up to 6 lp/mm with an image size of about 3″
- Fast neutron screen: 2.4-mm-thick ZnS(Ag) with polypropylene binder, Type S-25; MTF = 1 lp/mm at image sizes up to 12″
- Intensifier: Gen II with SuperGen photocathode; MTF = 10 lp/mm with 7.6 mm field of view
- Camera: cooled CCD with Texas Instruments TC-217 2/3-inch frame transfer CCD; MTF not measured
- Turning mirror: front surfaced; MTF not measured
- Lens: 25 mm focal length $f/0.95$; MTF not measured
- Lens: 85 mm focal length $f/1.2$; MTF not measured

Note: MTF (modulation transfer function) is the spatial frequency response of an imaging system or component; it is the contrast at a given spatial frequency relative to the contrast at low frequencies. Spatial frequency is typically measured in line pairs per millimeter (lp/mm), which is analogous to cycles per second (Hertz) in audio systems. High spatial frequencies correspond to fine image detail. The more extended the response, the finer the detail and the sharper the image.

8.15.3 Online Inspection Systems

Online inspection systems for detecting and characterizing corrosion in honeycomb aircraft structures are of prime importance. Neutron imaging is very sensitive to hydrogenous aluminum corrosion product and can be used to provide an early warning system for aluminum corrosion. Aluminum metal loss as small as 25 µm has been shown to be detectable. For thermal neutron scans to be economically practical for large-area inspection, as needed for on-aircraft inspection, the systems should be able to inspect 500 sqft/day of component surface with the system showing contrast sensitivity of at least 5% and image detail in the order of 4 mm for parts 10 cm thick. For three-shift operation, the selected source must be able to provide an incident thermal neutron flux of $5.6 \times 10^4 \, \text{n cm}^{-2} \text{s}^{-1}$ with an L/D ratio of 30. A stop-and-go inspection approach, with improved resolution, would require a source with similar characteristics.

Recently, a TRIGA reactor has been used for film and electronic neutron inspection of aging parts that can be removed from aircraft. The fuel for the TRIGA reactor is typically alloyed in a matrix containing zirconium hydride whose prompt coefficient of reactivity is negative and large, assuring that the neutron multiplication rate decreases as temperature rises. As mentioned earlier, power excursions are limited by the special nature of the fuel and may not require human intervention in all cases to guarantee safe reactor operation.

A robotic neutron system with 30 mg ^{252}Cf radioactive neutron source (initially installed) and thermal neutron imaging system also has been used for scanning aircraft surfaces with parts installed.

8.15.4 Characteristics of Aluminum Corrosion

Aluminum corrosion consists of both aluminum hydroxide and hydrated aluminum oxide. Neutron inspection development has shown that aluminum corrosion can be detected after test surfaces are subjected to a 6.5-hour supersaturated saltwater spray. Electronic imaging techniques can detect hydrogen content as low as 0.18 mg H/cm. This is equivalent to a typical hydrogen density of 30 mg/cc in natural aluminum corrosion products or a metal loss in aluminum as small as 25 µm. This sensitivity has been demonstrated using a reactor source and calculations indicate success with a transportable accelerator source as described earlier. Further tests have shown that small amounts of aluminum corrosion in aircraft structures may also be detectable in the presence of other hydrogenous materials such as sealants or adhesive bonding.

Despite its relatively high cost, neutron inspection of aging aircraft is considered cost effective based on its high sensitivity factor.

8.15.5 Thermal Neutron Inspection System Requirements

The following requirements assume that aircraft parts to be inspected are about 10 cm thick so that 10 cm is the separation between the farthest inspection part surface and the detector. The neutron beam L/D ratio can be increased to improve resolution at the expense of neutron flux and vice versa.

- Practical neutron beam L/D ratio = 30
- Calculated geometric unsharpness = 3.3 mm
- Motion unsharpness assumed = 3.3 mm
- Screen unsharpness = fractional mm
- Total unsharpness of imaging system ~4.15 mm
- Inspection rate = 500 sqft/day (3-shift, 21-hour day)
- Electronic detector image height = 1 foot
- Required linear scan rate = 23.8 ft/hr or 2 mm/s
- Calculated TV frame time = 1.65 seconds or 0.6 frames/s
- Calculated required neutron flux ~5.6×10^4 n cm^{-2}s^{-1} (considering statistical variables)
- Defined minimum incident neutron beam = 5.6×10^4 n cm^{-2}s^{-1} at L/D of 30

8.15.6 Conclusions

A number of neutron thermal sources and thermal neutron imaging systems have been evaluated. There are several accelerator-camera systems available that meet the needs for a practical neutron inspection system for early detection of aircraft part corrosion. The relatively low weight and portability of the (d-T) accelerator makes it a logical source for robotic, on-aircraft inspection applications. Conceptual designs for transportable cyclotron and RFQ accelerator systems have been proposed. All source systems can be used as fixed systems. Higher output sources offer choices between time and quality or sensitivity of inspection.

High neutron output, high L/D sources also can be considered for high-resolution inspection of details closer to the camera input. For example, for details 10 mm from the camera input, the geometric unsharpness can be in the order of 0.1 mm. Moving camera operation could be stopped for closer examination. Neutron cameras are available with spatial resolution of 4 lp/mm or better, so unsharpness in the order of 0.125 mm is now possible. These figures indicate the camera would display about 3 lp/mm. Suspicious indications near the camera could be examined further to show details in the order of 0.2 mm or better if necessary.

9
RADIOGRAPHIC TESTING METHOD

9.1 INDUSTRIAL RADIOGRAPHY

9.1.1 Personnel Monitoring

According to the U.S. Nuclear Regulatory Commission (USNRC), most radiographic accidents happen because proper procedures for working with radiation are not followed. Reasons and excuses most frequently given for failure to follow the proper procedures include rushing to get the job done, fatigue, boredom, illness, personal problems, poor communications, and inadequate training.

In almost every case involving radiography accidents, the radiographer was guilty of making three separate identifiable mistakes. These were:

1. Leaving the radiographic source out of the camera
2. Making the required radiation survey incorrectly or not making it at all
3. Failing to lock the radiographic source in its retracted, safe, shielded position

Licensed radiographers are responsible for supplying appropriate personnel-monitoring devices and making sure these devices are properly used. All persons entering a restricted, high-radiation area or using any source of radiation must wear one or more personnel-monitoring devices.

Introduction to Nondestructive Testing: A Training Guide, Second Edition, by Paul E. Mix
Copyright © 2005 John Wiley & Sons, Inc.

All radiographers and radiographic assistants must wear an individual direct reading pocket dosimeter and either a film badge or thermoluminescent dosimeter while doing their work. These devices must be assigned to and used by only one individual during the course of the radiographic work.

The pocket dosimeter must be read and have its exposure value recorded daily. If an individual's pocket dosimeter is discharged or "off scale," that person's film badge or luminescent dosimeter must be processed immediately! These records and readings must be maintained for state or federal agency inspection. In case of overexposures, the readings can be used to help determine what medical treatment, if any, is needed and the length of time the exposed individual should be barred from future radiographic work, based on the absorbed radiation dose.

9.1.2 Selected Definitions

Radiation, Absorbed Dose, and Dose Equivalent

- *Absorbed dose*—the energy imparted by ionizing radiation per unit mass of irradiated material. The units of absorbed dose are the rad and the gray (Gy).
- *Activity*—the rate of disintegration or decay of radioactive material. The unit of activity is the curie (Ci) or SI unit of activity Becquerel (Bq).
- *ALARA*—acronym for "as low as reasonably achievable." Means making every reasonable effort to maintain exposures to radiation as far below the dose limits of 10 Code of Federal Regulations (CFR) 20 as is practical for the general public and occupational workers.
- *ALI*—annual limit on intake. A derived limit for the amount of radioactive material taken into the body of an adult worker by inhalation or ingestion in a year by the reference man.
- *Fission or nuclear fission*—the result of absorbing an additional neutron. Uranium-235 is a fissile isotope because it has this property. This isotope is used in commercial nuclear reactors. When nuclear fission occurs, radiation is released, two or three neutrons are usually released, and two new nuclei (fission products) are formed.
- *Gray (Gy)*—the SI unit of absorbed dose. One gray is equal to an absorbed dose of 1 joule/kilogram (100 rads).
- *Ionizing radiation*—Ionization removes electrons from atoms, leaving two electrically charged particles (ions) behind.
- *Man-made radiation*—includes tobacco, television, medical X-rays, smoke detectors, lantern mantles, nuclear medicine, and some building materials.
- *Natural background*—includes small quantities of cosmic, terrestrial, and internal radiation.

INDUSTRIAL RADIOGRAPHY

- *Nonionizing radiation*—pure energy. Examples are light, microwaves, and radio waves, which do not have enough energy to remove electrons.
- *Quality factor (Q)*—the modifying factor that is used to derive *dose equivalent* from absorbed dose. $Q = 1.0$ for beta, gamma, and X-radiation, 20 for heavy particles of unknown charge, and 10 for neutrons of unknown energy and high-energy protons. See 10 CFR 20.1004 for additional information.
- *Rad*—special unit of absorbed dose. One rad is equal to an absorbed dose of 100 ergs/gram or 0.01 joule/kilogram (0.01 gray).
- *Radioactive decay*—when large unstable atoms from radioactive elements or compounds become more stable by emitting radiation in the form of a positively charged alpha particle, a negatively charged beta particle, or gamma rays.
- *Rem*—the special unit of any of the quantities expressed as *dose equivalent*. The dose equivalent in rems is equal to the absorbed dose in rads multiplied by the quality factor Q (1 rem = 0.01 sievert).
- *Roentgen*—a unit of radiation exposure; the dose of ionizing radiation that will produce 1 electrostatic unit of electricity in 1 cc of dry air. Original calibration unit for direct-reading pocket dosimeters.
- *Sievert*—the SI unit of any of the quantities expressed as *dose equivalent*. The dose equivalent in sieverts is equal to the absorbed dose in grays multiplied by the quality factor (1 Sv = 100 rems).

Note: rad = material absorption = 0.01 joule/kg = 1 Gray; rem = personnel absorption = 84 ergs of energy per gram of body tissue. (Rems = rads multiplied by the quality factor, Q, which equals 1 for X-, gamma, or beta radiation.) Alpha, beta, and gamma survey meters measure instantaneous absorbed dose rate in mR/hr or Sv/hr. The total absorbed dose received = integrated dose rate during the exposure period.

9.1.3 Survey Instruments

The radiographer's best friend may well be his or her portable, hand-held radiation survey meter. The front panel meter of the radiation survey meter indicates the dose rate in millirems per hour (mR/h) at that particular moment and place. The radiation dose received by exposed personnel is the integrated dose rate for the exposure time. Film badges and pocket dosimeters provide an indication of total dose received during the exposure period.

Survey meters typically use gas-filled sensors. There are two main types, those using ionization chambers, sometimes called "Cutie Pies," and those using Geiger-Müller (G-M) tubes. Ion chambers are primarily used for measuring low-energy X-rays. G-M survey meters similar to the Victoreen® Advanced Survey Meter (ASM-990) shown in Figure 9.1 with G-M Pancake Probe Model 489-11(OD) are frequently preferred because they are sensitive

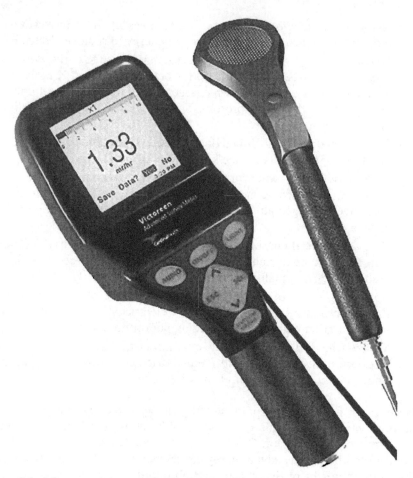

Figure 9.1 Victoreen Advanced Survey Meter ASM-990 Series with Model 489–110D Pancake Probe. Courtesy of Cardinal Health, Inc.

to low levels of gamma radiation and extremely rugged. With the proper probe combination, this meter can be used as a general survey meter, an area monitor, a wipe test counter, and a contamination monitor.

The ASM-990 series can detect alpha, beta, gamma, neutron, or X-ray radiation within an operating range of 1 µR/hr to 1 R/hr or 1 to 5,000,000 counts per minute (CPM), depending on probe selection.

Series ASM-990 features include:

- Advanced survey meter for hospital/environmental applications
- Multifunction key for easy menu navigation
- Backlit analog/digital display with error-free visual indication
- Autoranging, designed for multiple probe use

- Full-range audio output capability
- Survey mode data logging
- Timed peak hold feature
- Auto-power-down (extends battery life)

Note: Analog display may be preferred when excessive noise is encountered.

The Victoreen Model 489-110D G-M pancake probe is a hand-held, thin-window detector capable of measuring alpha, beta, and gamma radiation. It can also be used with other manufactured G-M survey meters. The probe is designed for ease of operation when making tabletop, floor, equipment, and personnel surveys. For handling and carrying ease, the probe can be clipped into the ASM 990 series handle.

Victoreen Model 489-110D G-M pancake probe features include:

- High detection efficiency
- Lightweight (2.1 lb) ergonomic design
- Detachable probe cable with BNC or MHV connector
- Durable housing that is easy to decontaminate

This G-M general-purpose pancake probe is suitable for a wide range of applications where alpha, beta, gamma, and X-ray detection is required. Prime applications include nuclear medicine countertops and frisker stations, leakage detection for low-energy diagnostic X-ray machines, geological and environmental surveys, or surveys of any place where there exists the suspicion that some form of radiation is present. Figure 9.2 shows G-M pancake probe efficiency for various isotopes.

Survey meters must be calibrated every 6 months and after instrument servicing. Meters with "past due" or expired calibration dates should not be used, but must be sent to a calibration laboratory or service company that has the appropriate radioactive source standards. For linear scale instruments, each survey meter is calibrated independently, typically at two points, representing 1/3 and 2/3 of full scale on each scale. For logarithmic scale instruments, meters should be calibrated at midrange of each decade, and at two points of at least one decade. Digital instruments should be calibrated at 3 points between 0.02 and 10 millisieverts (2 and 100 millirems) per hour, and so that an accuracy within plus or minus 20% of calibration source can be demonstrated at each point checked. The licensee must maintain records of the results of instrument calibration in accordance with Section 34.65 of federal regulation 10 CFR.

9.1.4 Leak Testing of Sealed Sources

When working with radioactive materials, the main areas of concern are radiation and contamination. Leak tests of sealed radioactive sources are performed to assure that smearable contamination is not present. Industrial

Isotope	%Efficiency
^{14}C	5
^{99}Tc	12
^{137}Cs	24
^{90}Sr	59
^{36}Cl	26
^{241}Am	8
^{129}I	2
^{230}Th	15
^{239}Pu	12

NOTE: The efficiency formula used to calculate the % Efficiency is:
Eff. % = (CPM x 100) / (μCi x 3.7 x 10^4 x 60)

Figure 9.2 Typical Victoreen G-M pancake probe efficiency for various isotopes. Courtesy of Cardinal Health, Inc.

sources such as Co-60, Cs-137, and Ir-192 are typically double encapsulated and sealed in small stainless-steel capsules. The source material is placed in the first stainless-steel capsule, welded shut, leak tested, placed in a second stainless-steel container, welded shut, and leak tested again. Subsequent leak tests by the radiographer, or industrial user's radiation safety officer (RSO) assures that the integrity of the initial capsulation has not been compromised by either physical damage or corrosive chemical attack.

Leak tests on sealed radiographic sources must be performed every 6 months as required by Nuclear Regulatory Commission (NRC) Regulation 10 CFR, Section 34.25(b). Purchased sources that are initially leak tested by the manufacturer must be supplied with a dated leak test certificate. Subsequent users must retest the source every 6 months and obtain a new leak test certificate from an approved service facility.

Gamma survey meters and leak test kits are used when testing radioactive sources for leaks. The leak test kit typically contains two cotton swabs, two patches (cloth or absorbent paper), and a small quantity of soap powder (wetting agent). Each test kit has either a blank or preprinted label for identifying the serial number of the source to be tested. If the same sources are repetitively tested every 6 months by the same service company, the service company may supply preprinted labels and deliver the test kits a couple of months in advance of the test date.

The person performing the leak tests makes up a 250-ml sample of soap solution and equips him or herself with a pocket dosimeter, film badge, disposable gloves, and gamma survey meter before beginning work. The survey meter is turned on as the source and source holder is approached to verify that the surveyed radiation dose rate is about the same as observed on previous surveys. A dry cotton swab and dry absorbent patch are used to wipe the external and internal surfaces of the source holder. These items are then returned to the test kit and the next cotton swab and patch are dampened with the soap solution and the external and internal surfaces of the source holder are wiped again. These items are returned to the test kit and the gamma survey meter is used again to verify that high readings are not obtained for the test kit materials. If high readings are observed, the RSO should be notified immediately.

If high readings are not observed, the leak test kits can be returned to the leak test service company for final evaluation. Specially trained personnel must evaluate the used test kits. If radioactive contamination on a swab or patch exceeds $0.005\,\mu Ci$ (microCurie), the contaminated equipment must be removed from service and decontaminated in accordance with established procedures. A report must be filed with the NRC if a leaking source is discovered.

Personnel conducting leak or wipe tests should not smoke cigarettes while performing tests. They should thoroughly wash their hands in warm, soapy water for at least two minutes after completing the tests to decontaminate themselves in case minute amounts of radioactive material are present.

9.1.5 Survey Reports

Radiographic operations shall not be conducted unless operable, calibrated survey instruments are available at each site where radiographic exposures are made. A radiation survey must be made after each radiographic exposure to verify that the radiographic exposure device or sealed source has been returned to its shielded position. In like manner, a radiation survey must also be made after each X-ray exposure to verify that the production of X-rays has been terminated. Radiographic exposure devices and storage containers must be locked and surveyed to assure that the sealed source is in the shielded position prior to movement to a new location and prior to storage at a given location. The equipment user must maintain records of all radiation surveys for possible future use by the NRC or appropriate state radiation agency.

9.2 WORK PRACTICES

A "restricted area" is defined as an area that the licensee restricts general personnel access to for the purpose of radiation protection. Restricted areas are roped off and posted to protect the general public from radiation exposure. The barricade for a restricted area is placed where the measured dose rate is 2 mrem/h or less. Persons entering a restricted area should have knowledge of the type of radiation present and methods of avoiding or controlling their radiation exposure. People who ignore the barricades and warning signs and enter the restricted area should be told that a radiation source is in use and be asked to leave the area immediately.

A "radiation area" is an area where radiation exists in which a person could receive a radiation dose rate to a major portion of the body in excess of 5 mrem/h or a dose of 100 mrem in five consecutive days. Radiation areas should be posted with "Caution Radiation Area" or "Danger, Radiation Area" signs displaying the yellow and magenta radiation symbol. A barricade for a radiation area should be set up where the dose rate is 5 mrem/h or less.

A "high radiation area" is an area where the dose rate to anyone could exceed 100 mrem/h. The area must be posted with "Caution, High Radiation Area" or "Danger, High Radiation Area" signs. Radiographers must place a barricade at the restricted area boundary and appropriate signs at the radiation area and high radiation area boundaries.

Finally, a "very high radiation area" is an area where additional measures must be taken to assure that an individual is not able to gain unauthorized or inadvertent access. In these areas, radiation levels of 500 rads (5 grays) or more in 1 hour at 1 meter from a radiation source (or surface through which the radiation penetrates) could be encountered.

9.3 TIME—DISTANCE—SHIELDING—CONTAINMENT

There are three basic ways to lower your accumulated dose when working with radiographic sources. Minimize the time you spend near a radiographic source or camera. Any unnecessary exposure should be avoided! Accumulated dose is a function of dose rate and time as shown in Eq. 9.1:

$$\text{Accumulated dose (rem)} = \text{dose rate} \times \text{time} \tag{9.1}$$

Some ways to reduce your exposure time include:

- Crank the source out of or into the camera as rapidly as possible to shorten your exposure during these operations.
- Carry the radiographic camera rapidly to its intended location for use or storage.
- Do not sit unnecessarily near a radiographic camera. Whenever possible, position yourself outside the restricted area when the source is exposed.

Keep as much distance as possible between you and the source at all times. The amount of radiation received decreases rapidly with distance from a point source. Radiation follows a straight line and spreads out rapidly as it moves away from the source. As distance from the source doubles, radiation intensity decreases by a factor of four. This is known as the inverse square law, which is expressed by Eq. 9.2:

$$D = D_o(R_o/R)^2 \qquad (9.2)$$

where D = calculated dose rate
D_o = original dose rate at known location
R_o = distance where dose rate is known
R = distance where you want to calculate dose rate

In like manner, if the distance from the source triples, the radiation intensity decreases by a factor of 9; if the distance quadruples, radiation intensity decreases by 16, and so on.

Finally, reduce the radiation dose by shielding. Shielding refers to the practice of placing a radiation-absorbing material between you and the source. Figure 9.3 shows the shielding effect of steel, lead, and uranium on Co-60 and

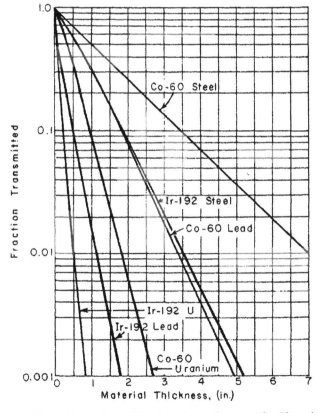

Figure 9.3 Shielding effect of steel, lead, and uranium on Co-60 and Ir-192 radiographic sources.

Ir-192 sources. Generally, heavy, dense materials tend to absorb X-rays and γ-rays more effectively than lighter, less dense materials. For this reason, uranium, tungsten, and lead are good shields, steel is fairly good, and concrete is less effective. However, concrete is still a popular shielding material because of its relatively low cost, even for thick-walled sections. Figure 9.4 shows the shielding effect of concrete on Co-60 and Ir-192 sources.

One of the most practical uses for shielding is for the fabrication of collimators. Generally, collimators are small pieces of lead, tungsten, or uranium that are used to shape or collimate the radiation beam in a specific direction and absorb radiation in other directions. The radiation beam can thus be shaped and directed by the collimator somewhat analogous to the shaping and direction of a flashlight beam by a reflector. Collimators reduce the dose from an Ir-192 source by a factor of 20 to 10,000. They reduce the dose from a Co-60 source by a factor of 3 to 10. Usually, collimators have two holes. The first hole directs the radiation beam; the second hole is where the source enters

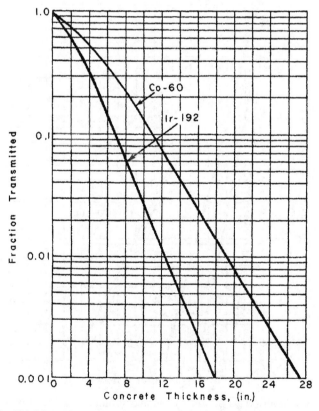

Figure 9.4 Shielding effect of concrete on Co-60 and Ir-192 radiographic sources.

the collimator. The beam from the second hole must also be taken into account when making surveys and setting up barricades to protect personnel.

A term commonly used with regard to shielding is half-value thickness or half-value layer. The half-value thickness is the thickness of a material that is required to reduce the radiation dose by a factor of two. For example, the half-value thickness of lead for an Ir-192 source is 0.19 in.; the half-value thickness of lead for a Co-60 source is 0.49 in., more than double. Graphics showing the attenuation of radiation from various sources as a function of material thickness are frequently used for shielding studies.

9.4 REGULATORY REQUIREMENTS

Radiation surveys are required to protect the radiographer and nearby personnel from potential overexposures. Surveys must be taken with a suitable survey meter after each radiographic exposure for two reasons:

1. To make sure the source is in its shielded position
2. To define the perimeter of the restricted area barricade or other area boundaries

Although USNRC regulations do not specifically require rope barricades, radiation signs are posted at the restricted area boundary and at the high-radiation area boundary where personnel could be exposed to a dose rate of 100 mR/hr. All signs must be conspicuously posted so that anyone approaching the area will see them.

The radiographer is responsible to see that no one enters a high-radiation area while a radiographic source is exposed. Direct visual surveillance of the area is required if it is possible for others to enter the area of high radiation without an alarm being sounded. Direct surveillance is not required if others can be locked out of the area or if automatic retraction of the source can be provided when it is approached.

9.5 EXPOSURE DEVICES

The daily maintenance and inspection of radiographic cameras is essential to assure that they are in good operating condition. A daily check of a crank-type camera should include the following:

1. Make a radiation survey of the camera. Expected survey values can be determined from previous surveys of the same camera.
2. Examine the camera for visual damage. Determine if damage is superficial or substantial. If substantial, do not use.

3. With the camera still locked, examine the locking mechanism and any portion of the pigtail that is visible with the cap removed.
4. Check the pigtail connector and drive cable connector for wear, dents, and other physical damage.
5. With the camera still locked, connect the drive cable. Remove the safety plug from the front of the camera and check to see that the source hole is round and smooth so that the source will not stick when it is cranked out or retracted.
6. Examine the source guide tube for any physical defect that might cause the source to get stuck.
7. Check the source locking mechanism to make sure it is clean. Lubricate as necessary.
8. Check for sticking or binding as the source is positioned and retracted during the first exposure.

Note: Do not use the camera if any problems are noted. Report all problems to your supervisor and the radiation safety officer (RSO).

A similar daily maintenance procedure can be used with beam-type cameras by omitting steps pertaining only to crank-type cameras. Radiographic cameras must also be inspected quarterly by specially trained personnel. Each radiographic source, source holder, or logging tool containing a radiographic source shall have a durable, legible label having the standard radiation caution symbol and the following words:

DANGER

RADIOACTIVE—DO NOT HANDLE

NOTIFY CIVIL AUTHORITIES (OR NAME OF COMPANY)

This label is to be placed on the smallest component that is transported as a separate piece of equipment. Each transport container shall also have a permanently attached label with the standard radiation caution symbol and the following words:

DANGER (OR CAUTION)

RADIOACTIVE

NOTIFY CIVIL AUTHORITIES (OR NAME OF COMPANY)

In addition, each licensee's operating and emergency procedures shall include instructions in the following:

- Handling and use of sources such that the operator will not be likely to exceed permissible radiation doses
- Methods and reasons for conducting radiation surveys

- Methods for controlling access to radiographic areas
- Methods and reasons for locking and securing radioactive sources
- Personnel monitoring including instructions on the use of monitoring equipment
- Control of sources during transportation to field locations including packing for transportation, posting of vehicles, and so on
- Plans for minimizing the exposure of people in the event of an accident
- Procedures for notifying the proper authorities in the event of an accident
- Procedures dealing with the required record keeping of leak test results, radiation surveys, and personnel monitoring records
- Inspection and maintenance of all radiographic exposure devices, storage containers, and machines

9.6 STATE AND FEDERAL REGULATIONS

The responsibility for regulating the use of radiographic materials used in gamma radiography has been given to the U.S. Nuclear Regulatory Commission by Congress. The responsibility for regulating the use of X-ray machines, accelerators, and Ra-226 remains with the individual states. However, the NRC may at its option relinquish the regulation of radioactive materials to the various states if the state wants the authority and has adequate resources to ensure that radioactive materials will be safely handled and used.

Thus far, 33 states have agreements to regulate the use of their radioactive materials and are known as *agreement states*. Three states have letters of intent to become agreement states and are known as *letter of intent—AS*. These states are MN, PA, and WI. All other states are called *nonagreement states*. The nonagreement states are AK, CT, DE, HI, ID, IN, MO, MT, NJ, SD, VA, VT, WV, WY, and the District of Columbia (DC).

Whether a state is an agreement state or nonagreement state, private companies must operate under a license to do radiographic work using radioactive materials. Companies must operate under an NRC license in nonagreement states and under various state licenses in agreement states. Agreement state regulations must be compatible with, and in some cases, are the same as NRC regulations. It a company performs industrial radiography using X-rays only, then the radiography is regulated by the state in which it is performed, whether an agreement state or a nonagreement state.

Radiography licenses fall under similar rules of reciprocity as automobile driver's licenses. For example, a new license is not needed to perform radiography work in another state, but the rules of that state also must be followed. Therefore, work could be done under an NRC license in an agreement state or work could be done under a state license in a nonagreement state. State regulations must not conflict with or contradict NRC regulations under any circumstance.

Radiographers must have training in the fundamentals of radiation safety, be knowledgeable in the use of radiation detection and personnel-monitoring equipment, and be familiar with the type of radiographic equipment to be used. In addition, radiographers must know pertinent state and federal regulations as well as the licensee's written operating and emergency procedures.

9.7 BASIC RADIOGRAPHIC PHYSICS

9.7.1 Introduction—Isotope Production

The use of X-rays for radiography dates back over 100 years. William Roentgen discovered X-rays while experimenting with high-voltage vacuum tubes in December 1895. He noted that the X-rays he produced caused a fluorescent material to glow. Later, he X-rayed various pieces of metal and his own shotgun to reveal variations and flaws.

Shortly after his discovery of X-rays, Roentgen X-rayed his wife's hand and, in January 1896, mailed pictures of his X-rays to scientists around the world. Within two months, hospitals around the world were using X-ray pictures to aid in surgery. By this time, excellent resolution had also been obtained with the early X-rays.

Roentgen's work quickly led French scientist Henry Becquerel to discover similar radiation in a uranium-bearing material. However, it was not until a year and a half later that Marie and Pierre Curie were successful in separating the new element, radium, from literally tons of uranium ore. Another 30 years would pass before enough radium could be accumulated for use in industrial radiography.

In this country, gamma radiography started when the Naval Research Laboratory was searching for a way to test steel castings thicker than 3 in. At that time, X-ray sources were too weak, but small radium sources, typically 0.1 Ci in strength, worked when exposure periods of several days were used. Shortly after World War II, new, stronger man-made sources of cobalt and iridium became available. After that point, the use of gamma radiography in industry grew by leaps and bounds.

Industrial radiography is defined as the use of penetrating radiation, such as X-rays, γ-rays, or neutrons, to make internal pictures of objects such as metal castings or welds to detect material variations, defects, or flaws. Medical radiography is considered a separate subject.

Industrial radiography is a valuable tool that carries a substantial risk of accidental exposure to harmful radiation to the radiographer. For this reason, NRC regulation 10 CFR Part 34 now requires a review of case histories of overexposures as part of the licensee's training program. A review of these case histories is beyond the scope of this book, but it should be pointed out

9.8 FUNDAMENTAL PROPERTIES OF MATTER

The atom is the basic building block of the elements. An atom consists of a nucleus made up of protons and neutrons with orbiting electrons. The atom is mostly empty space and there are large spaces between atoms. For this reason, the analogy is sometimes made between the atom and our solar system. Two or more atoms bound together in defined proportions form molecules.

Atoms can be found in nature or be "man-made." They can be stable or unstable; radioactive atoms are examples of unstable atoms. Combinations of atoms form elements and numerous compounds can be formed from the elements by various chemical reactions.

The electrons that circle the atom have minute mass and negative charge. Removal of one or more electrons from an atom or molecule forms ions. Atoms, molecules, and subatomic particles carrying a positive or negative charge are also called ions. As electrons are moved away from the nucleus, their energy increases and they are known as being "excited." Ionizing radiation has enough energy to remove electrons from atoms; nonionizing radiation, such as microwaves, can only excite electrons in place or cause them to vibrate and produce heat. Generally, chemical reactions affect only the outermost electron of an atom.

The nucleus of the atom is made up of protons and neutrons also called nucleons. Protons are positively charged particles with a mass of 1.007277 amu (atomic mass units) or 1.67×10^{-24} g; most of the atom's mass is in the nucleus. Nuclides are characterized by the atomic (or proton) number Z, and the atomic mass (or nucleon number), A. Standard notation is shown as:

$$^A_Z Pu$$

where Pu = element symbol for plutonium

Isotopes have the same value for Z but different values of A. Chemical properties may be similar, but nuclear properties may be quite different. For example, both U-238 and U-235 have 92 electrons and 92 protons; U-235 has three less neutrons in its nucleus. Therefore, the standard notation for U-235 would be:

$$^{92}_{235} U$$

A radioisotope or radioactive isotope is an unstable isotope. The instability is caused by the fact that the isotope has too many or too few neutrons in the nucleus. Radioisotopes generally attempt to reach a stable form by alpha (helium ion) emission, beta (electron) emission, or orbiting electron capture (K capture). Elements with more than 82 protons are considered unstable or ever changing.

9.9 RADIOACTIVE MATERIALS

Most artificial isotopes, made in large quantities and in widespread use today, are produced in fission reactors. For these reactors, the nuclear fuel cycle starts with the mining and milling of ores or the recovery of uranium from phosphates. Yellow cake, U_3O_8, is produced from sedimentary rock deposits in various parts of the world including the United States. There is also about 1 lb of U_3O_8 for every ton of 100% phosphoric acid. The yellow cake in phosphoric acid is recovered by solvent extraction.

Refining and chemical conversion remove the chemical impurities from yellow cake and uranium hexafluoride, UF_6, is converted to uranium dioxide, UO_2, for use in nuclear reactors. Natural uranium also contains about 0.7% U-235 and isotope separation plants increase the concentration of this element, known as enriched uranium-235.

Some nuclear fuel elements are made by loading UO_2 fuel pellets into aluminum tubes to form a fuel rod. Other fuel elements are made using natural uranium fuel cores clad with aluminum and vertically stacked in aluminum tubes. Enriched U-235 can also be alloyed with aluminum, formed in a billet, clad with aluminum, and coextruded into a fuel tube. Fuel rods, control rods, target rods, and a moderating medium form the core of a nuclear reactor. The moderating medium can be carbon, light water, or liquid sodium. The purpose of the moderating medium is to slow down the neutrons produced by fission, and help cool the reactor. The amount of U-235 in the reactor is sufficient to cause spontaneous fission if the reactor's control rods were to be removed. Cadmium control rods absorb neutrons and prevent spontaneous fission from occurring. They are slowly withdrawn partially from the reactor core to control the rate of reaction and the amount of power generated.

In nuclear fission, an atom of U-235 captures a neutron and splits into two highly unstable fragments, which in turn immediately release one or more neutrons. These neutrons then split more atoms, causing a rapidly accelerating chain reaction that converts mass to energy, mainly in the form of heat. In some cases, the heat can be used to make steam for operating conventional steam generators that produce electricity. Unreacted or "unburned" uranium, plutonium, and other isotopes can be recovered from "spent" fuel rods after a "cooldown" period. Target rods containing various isotopes can be inserted in a reactor for the specific purpose of producing a multitude of radioactive isotopes by neutron activation. The recovered radioactive isotopes can be used as a fuel in breeder reactors or as radioactive sources for many other appli-

RADIOACTIVE MATERIALS

cations. Radioactive sources are used in radiography, oil exploration, medicine, space, food processing, and industrial instruments.

The waste disposal of radioactive materials must be carefully considered at each step of isotope production. The disposal problem becomes more serious as radioactive materials are concentrated and their activity is increased.

9.9.1 Stability and Decay

Both proton-rich and neutron-rich atoms are unstable or radioactive. Proton-rich nuclei decay by positron emission or electron capture. Neutron-rich nuclei decay by beta emission or more rarely by neutron emission.

Elements with more than 83 protons are unstable and tend to follow their decay chains until stable end products are reached. As these elements decay, there may be alpha or positively charged helium nuclei emission, beta or electron emission, and spontaneous fission caused by neutrons released by decay. Photons usually accompany decay, gamma photons from nuclear rearrangement, and X-ray photons from electron rearrangement or *bremstrahling*, the sudden stopping of electrons.

Charts of the nuclides are plots of protons (Z) versus neutrons (N). These charts are used to identify stable and unstable nuclei, show the type of decay, and follow the decay chains to the stable end product. Other nuclear parameters may also be given.

With alpha emission, the nucleus of the helium atom is emitted; the atomic mass (A) decreases by four and Z decreases by two. With beta emission, an electron is emitted; A does not change and Z increases by one. With positron decay, a positive electron is emitted from the nucleus of a proton-rich nuclide; A does not change and Z increases by one. With electron capture, an inner electron is captured by the nucleus. This action tends to compete with positron decay in neutron-rich nuclides. A does not change and Z decreases by one. Neutron decay, proton decay, and spontaneous fission are very rare and seldom-observed forms of decay.

9.9.2 Activity

Source strength is referred to as activity. Activity is defined as the number of radioactive disintegrations per second. The unit of activity is the curie, abbreviated as Ci. One curie is defined as 3.7×10^{10} disintegrations per second (dps) or 2.2×10^{12} disintegrations per minute (dpm). Activity is often expressed in millicuries (mCi), Ci $\times 10^{-3}$, 37 million dps or microcuries (μCi), Ci $\times 10^{-6}$, or 37,000 dps.

In radiographic work, the energy, penetrating power, and dose rate at 1 ft vary widely for different source materials even though they have the same source strength or number of dpms. Table 9.1 shows the dose rate per curie at 1 ft for various radiographic source materials.

TABLE 9.1. Dose Rates for Radiographic Source Materials

Source Material	Dose Rate/curie at 1 ft (R/h)	Half-Life	Gamma Energy (MeV)
Ir-192	5.9	75 days	0.10–0.60
Co-60	14.4	5.3 yr	1.17 and 1.33
Cs-137	3.3	33 yr	0.66
Ra-226	9.0	1620 yr	0.20–2.2

9.9.3 Half-Life

Half-life is the time required for one-half of the initial number of unstable atoms to decay or disintegrate. Each radioactive isotope has a very specific half-life $T^{1/2}$ that can range from a few seconds to over 1000 yr. The half-life of a radioactive isotope never changes. After 1 half-life, the activity or total strength of the source will be also reduced by one-half. After 2 half-lives, the activity will be reduced by one-fourth of its original value. After 3 half-lives, the activity will be reduced to one-eighth the original value, and so on. After 10 half-lives, source activity will be less than one-thousandth of its original value.

The half-life of a small, low-activity, short-lived source, such as Ba-137m, can be easily determined in a few minutes using a stopwatch and scintillation counter. Data shown in Figure 9.5 and Table 9.2 were obtained from an actual laboratory experiment using a small Ba-137m source.

These data clearly show that the half-life of the Ba-137m source is somewhere between 2.5 and 3.0 min. Short-lived sources are frequently used as tracers in medical and industrial studies. Knowing the half-lives of radioactive sources can be useful in verifying source strength after long storage periods. It can also help provide a check on survey instruments when surface dose rate measurements are periodically made. For example, an unexpectedly low reading might indicate low battery voltage for the survey meter.

Graphs are frequently used to make rough approximations of source strength as a function of elapsed time. For more accurate source activity data, the source manufacturer provides a graph with each radiographic source giving the activity of the source in curies at various dates. The purpose of the dated decay curve is to enable the user to determine source strength at any time. A study of source decay charts quickly shows that a large source loses curies faster than a small source. In the case of the Ba-137m source, the counts decreased by 2810 in the first 30 s and only 699 in the last 30 s of the experiment. This same decay relationship exists from time zero regardless of the size, activity, or composition of the original source.

Source decay charts are generally plotted on semilog paper to obtain a straight-line plot of source activity as a function of elapsed time or date. Specific activity is the term used to describe the number of curies of activity per unit weight of source material.

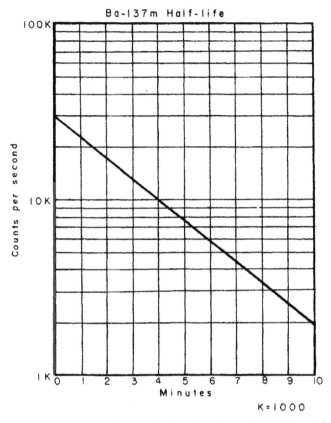

Figure 9.5 Half-life experiment showing the decay of Ba-137m with time.

9.10 TYPES OF RADIATION

There are two basic types of radiation, particle radiation and electromagnetic radiation. The velocity of electromagnetic radiation is 186,000 miles/s. The relationship between frequency, velocity, and wavelength, shown in Figure 9.6, is expressed by Eq. 9.3:

$$c = fL \qquad (9.3)$$

where c = speed of light
f = frequency
L = wavelength

Charged particle radiation is the result of ion or electron emissions from nuclear reactors or unstable isotopes. Alpha particles are positively charged

TABLE 9.2. Radioactive Decay of 10 μCi Ba-137m Source

Elapsed Time (min)	Counts Obtained[a]
0.0	29,535
0.5	26,725
1.0	22,411
1.5	19,787
2.0	17,478
2.5	15,177
3.0	13,326
3.5	11,807
4.0	10,030
4.5	8,932
5.0	7,982
5.5	7,051
6.0	6,111
6.5	5,255
7.0	4,556

[a] Background counts = 41.

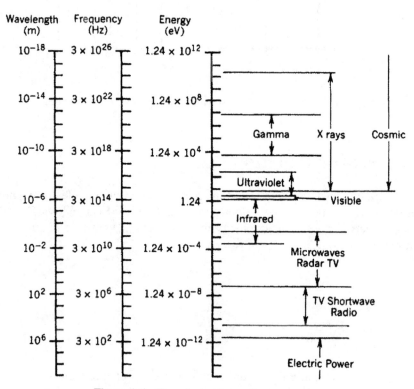

Figure 9.6 The electromagnetic spectrum.

TYPES OF RADIATION

helium nuclei. Fission fragments are composed of highly positively charged hydrogen nuclei. Heavy ions have a positive charge and Z more than two. Positrons are positively charged electrons. Meson particles may be positive, negative, or neutral. Beta particles are orbital or nuclei electrons. High-energy accelerators or cosmic rays can produce other particles.

Neutral particles tend to be more penetrating than charged particles. They consist of neutrons from reactions or decay and weakly interacting neutrinos.

Photons are pure energy, have no mass, and form a part of the electromagnetic spectrum. Gamma energy is released as unstable radioactive isotope decay and their nuclei are rearranged. X-ray energy can be produced by electron rearrangement in the atom or by bremsstrahlung, stopping electrons. Generally, gamma energy is derived from natural sources and X-ray energy is man-made. Typically, aiming a high-speed electron gun at a tungsten target causes X-rays to be produced as the electrons expend all their energy in the target.

The ability of radiation to penetrate an object depends on mass, charge, and energy. Alpha particles are considered nonpenetrating and only an internal hazard. If swallowed, they would expend all their energy in nearby tissue causing severe damage to that tissue. Beta particles are considered somewhat penetrating and large skin doses would be of concern. Positrons interact like beta particles and produce 8.5 MeV γ-rays. Gamma energy is very penetrating and shielding is usually required to protect personnel from this type of radiation. Neutrons are also very penetrating, require shielding, and can produce secondary gammas when they collide with a nucleus. Naturally occurring X-rays are not very penetrating, but man-made X-rays from machines can be very penetrating and require the same shielding as γ-rays.

The penetrating ability of γ-rays and X-rays is directly related to the intensity of the radiation from the radioisotope or X-ray generator. Because of its penetrating power, Co-60 is used for X-raying medium-density metals over a thickness range of 1.5 to 9 inches. Previously it was shown that the dose rate from a 1 Ci radioisotope can vary from 3 to 14 R/h at 1 ft. While 20–50 Ci sources may be encountered in radiography, sources of several thousand curies are commonly used in medical radiography. With a 1000 Ci Co-60 source, a dose rate of 14,400 R/h could be obtained at 1 ft. Remote-handling equipment should be used for handling high-intensity emitters.

X-ray generators, operating at energies of 60 to 2000 kV peak power, can generate equivalent dose rates of 200 to over 80,000 R/h/mA at 1 ft. Betatrons with energies of 10,000 to 100,000 kV peak can generate equivalent dose rates of 48,000 to 1.6 million R/h at 1 ft. For this reason, heavy shielding is required for X-ray machines, betatrons, and other particle accelerators. Another important factor to remember is that radioisotopes continuously emit radiation and X-ray machines are usually turned on for only a few seconds or minutes, depending on the target and its thickness. However, the potential danger of an unattended X-ray machine should never be underestimated.

TABLE 9.3. Energy Releases from a Nuclear Reactor

Source	Energy (MeV)	Percent
Fission fragment kinetic energy	168	84.0
Neutron kinetic energy	5	2.5
Prompt fission gammas	5	2.5
Fission product decay gammas	6	3.0
Fission product decay betas	6	3.0
Kinetic energy of neutrinos	10	5.0
Total	200	100.0%

In nuclear reactors, millions of electron volts (MeV) are released in fission. The amount of energy released from different sources and their respective percentage of the total energy released are shown in Table 9.3.

9.11 INTERACTION OF RADIATION WITH MATTER

Ionization occurs when an atom loses an electron. Two charged particles or ions are the result of ionization. The electron that has been removed from the atom is negatively charged and the remaining atom is positively charged.

Radiation survey meters respond to charged particles in their ion chamber or G-M tube detectors. Alpha particles, beta particles, and fast electrons can collide with an atom, knocking an electron out of orbit, causing ionization. Photons, γ-rays, and X-rays can also cause ionization when they expend their energy in an atom. When people talk about the benefits or dangers of radiation, they are generally talking about ionizing radiation.

The photoelectric effect refers to the process in which a photon transfers its total energy to an electron in the outer shell of an atom. When this happens, there may be enough energy to move the electron from one outer shell to another or even completely remove the electron from its orbit, causing ionization. When photon energy exceeds the binding energy of the electron, ionization occurs.

As photon energy increases, many photons collide with electrons and ricochet off in another direction colliding with even more electrons in the process. Normally, photons travel in straight lines. Thus, one photon may knock several photons out of orbit because the binding energy of the electron is small compared to the initial photon energy. This is known as the *Compton effect*. Based on the principles of energy and momentum conservation, the Compton scattering process leads to the conclusion that the energy of the scattered photon is always less than that of the primary photon. Furthermore, the decrease in energy can be calculated based solely on the angle of scattering.

When matter absorbs very high-energy photons, the photon is converted into an electron and positron, which have a mass equivalent to the energy of

INTERACTION OF RADIATION WITH MATTER

the photon. This process is known as pair production and it cannot occur below energies of 1.02 MeV. At higher energies, kinetic energy is merely imparted to the pair particles. The intensity of X-rays or γ-rays can be measured by the amount of ionization they produce in air. This ionization is measured in exposure units named after William Roentgen, who discovered X-rays, and is abbreviated as R. Previous sections have already discussed radiation exposures in terms of R/h or mR/h.

Radiation shielding is an effective way to decrease our external exposure to harmful radiation. Shielding is accomplished by placing a radiation-attenuating material between personnel in the area and the radioactive sources. However, since radiation shields are merely safety devices, quality assurance in design and periodic testing are essential.

The material used for a shield depends on the type of radiation being attenuated. For X-rays and γ-rays, high-density, high-Z materials are most effective. Lead, tungsten, and uranium are excellent. Concrete and steel are also used. Figure 9.7 shows the absorption characteristics of lead for low-activity Cs-137

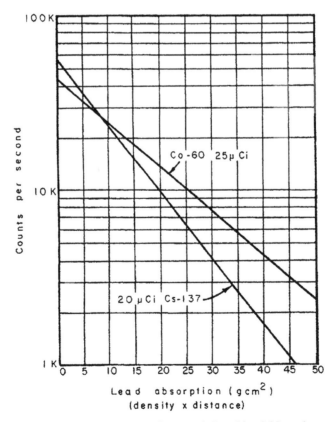

Figure 9.7 Gamma energy absorption characteristics of lead. Note the greater penetrating power of the Co-60 source.

TABLE 9.4. Relaxation Lengths of Various Shielding Materials (*l*)

Material	Density (g/cm³)	Relaxation Lengths (cm)		
		Fast Neutrons	1-MeV Photons	8-MeV Photons
Water	1.0	10	13.0	40.0
Concrete	2.3	12	6.7	18.0
Iron	7.5	6	2.2	4.4
Lead	11.2	9	1.2	1.9

and Co-60 gamma sources. For neutrons, materials with high hydrogen content are best. Water, plastic, and concrete are effective. Other elements that absorb neutrons, such as boron, paraffin, cadmium, and Li-6, are effective in reducing secondary gamma radiation. Shielding materials are also used for containment and as structural members in radioactive handling facilities.

One of the most important uses of shielding is for the containment and/or collimation of radioactive sources. Lead, uranium, and tungsten are frequently used for these purposes. Small, high-energy radiographic point sources may be safely handled and transported due to the effectiveness of this type of shielding. The shielding material may also be machined or formed to produce a precisely shaped beam of radiation for radiographic or other industrial work.

Photon and neutron shielding is often experimentally determined. Some of the shielding terms frequently encountered are the half-value layer (HVL), where the radiation intensity is decreased by a factor of 2, the tenth-value layer (TVL), where the radiation intensity is decreased by a factor of 10, and the relaxation length *l*, where the intensity is decreased by a factor of e. HVL = $0.693\,l$; TLV = $2.303\,l$. The relaxation lengths for some shielding material are listed in Table 9.4.

Most students are probably surprised to learn that water, with a density of $1.0\,g/cm^3$, attenuates neutrons almost as well as lead with a density of $11.2\,gm/cm^3$.

One of the most effective ways to reduce your personal exposure to radiation is to put as much distance between you and the source in the shortest possible time. By doing so, you reduce the number of particles or the amount of energy passing through your body per unit time. As previously discussed, most radioactive sources are point sources and the radiation intensity decreases with the square of distance. This is known as the *inverse square law*.

9.12 BIOLOGICAL EFFECTS

Nature's radioactive materials that surround us constantly emit radiation; this radiation is known as natural background radiation. How much background radiation we receive depends on where we live. If we live at high altitude or near naturally occurring uranium ore deposits, we would receive a much higher annual background radiation dose. Our eating and social habits can also affect the annual radiation dose we receive. For example, Brazil nuts are high in thorium, and Po-210 can be found in tobacco. In addition, the human body itself contains small amounts of K-40.

Man-made sources, such as nuclear bomb fallout, can also contribute to our current annual background radiation dose. There are many other man-made sources of radiation that can contribute to the annual radiation dose received by large segments of the population in the United States. Some sources of naturally occurring and man-made radiation are listed in Table 9.5. Since the radiation exposure numbers represent an average annual dose, they can vary widely from one individual to another.

According to the NRC publication, "Working Safely in Gamma Radiation," a person exposed to 1 R loses 1 day of life expectancy on the average. Similarly, a person exposed to 1 mR loses about 1 min of life expectancy on the average due to the increased risk of early death by cancer.

Based on the previous estimates, the normal exposure level for a one-pack-a-day cigarette smoker would be 8.7 R/yr as compared to about 0.7 R/yr for a nonsmoker. The average occupational annual radiation dose for people who work with or near radiation sources can add 270–670 mR/yr to the average annual radiation dose already discussed. The occupational dose limit for workers in the industry is 5 R/yr.

When we study the effects of radiation on people, we are concerned about the biological damage that could occur to living tissue if placed in a radiation beam. The measure of this biological effect is the radiation dose in rem or roentgen-equivalent man. When dealing with X-rays and γ-rays, as is the case

TABLE 9.5. Estimated Radiation Doses to U.S. Population

Natural Sources		Man-Made Sources	
Source	Dose (mR/yr)[a]	Source	Dose (mR/yr)[a]
Cosmic (sun and space)	28	Diagnostic X-rays	90
Building materials	7	Nuclear fallout	7
Human body	25	Color TV	1
Earth	26	Homes	75
		Natural gas	22
		Ambient radon	450
		Cigarettes[a]	8000
Total (less than)	100		8645

[a] 1000 mR = 1 R.

Source: "Radiation Exposure from Consumer Products and Miscellaneous Sources," NCRP Report No. 56, 1977.

in this book, 1 rem equals 1 roentgen and will be expressed in R to conform to the calibration of various gamma survey meters.

Another important concept is dose rate, which is a measure of how rapidly the radiation dose is being accumulated. Dose rates are typically expressed in terms of roentgen per hour (R/hr) or milliroentgen per hour (mR/hr). They could also be expressed in roentgen per minute (R/min) or milliroentgen per minute (mR/min). If exposure to radiation resulted in a dose rate of 5 R/h, a radiographer would accumulate his or her total allowable annual exposure in just 1 hour. Extreme care should be taken when using a survey meter calibrated in roentgen per minute or milliroentgen per minute since these meters are calibrated for radiation intensities that are 60 times stronger than meters calibrated in terms of R/hr or mR/hr.

Many people confuse radiation with contamination and ask: "Can radioactive sources make other objects radioactive?" The answer is no! Radioactive contamination refers to the spreading of minute, even invisible, radioactive particles. If we could imagine ordinary house dust was radioactive, we would have some insight as to the problems that could be caused by radioactive contamination. However, most radiography and industrial sources are double encapsulated and sealed to prevent contamination. The radioactive material is bound in a ceramic matrix, encapsulated in a stainless-steel cylinder, welded, and tested. Then, it is placed in a second stainless-steel container, welded, and retested again. Operations using radiographic sources are to be suspended upon detection of 0.005 μCi or more of removable radioactive material with leak test kits. Contamination, which is seldom a problem in radiographic work, is much more likely to occur as the result of a nuclear explosion or careless handling of radioactive waste. Therefore, some of the harmful effects of radiation on people will be considered in this section.

Pioneer radiographers often worked with radium until radiation burns were noted. In mild cases, the burns were comparable to sunburn with no apparent long-term harmful effects. In severe cases, finger and hand amputation and death by cancer was the result. Much of what is known today about the harmful effects of large doses of radiation was learned from early radiography pioneers, therapeutic radiation, and the Hiroshima and Nagasaki bombings.

We learned that young, growing, active basal cells, well-oxygenated cells, and the eyes are more sensitive to radiation than old, dormant, inactive, mature cells, low in oxygen content. For these reasons, persons under 18 years of age should not actively engage in radiography work. For the same reasons, cancer may spread more slowly in elderly persons. The relative radiosensitivity of human tissues is listed here with most sensitive tissue starting at the top.

1. Lymphocytes
2. White blood cells
3. Immature red blood cells (bone marrow)

4. Blood platelets
5. Gonadal cells (sperm and eggs)
6. Active skin (basal) cells
7. Blood vessels and body linings
8. Various glands, liver
9. Connective tissue and bone
10. Muscle
11. Nerve tissue, brain

Geneticists agree that there is no safe threshold or background level of radiation. However, there is also no way to avoid background radiation and the risks from background levels are so small that may other factors would mask the risk if it could be assessed. With short-term exposures below 50 R, there are no clinical effects that can be seen or measured. Also, we can only project the effects of long-term exposure to small doses of radiation based on population averages. If damage is done at these lower levels, it is believed that the body may heal itself to some extent. Some scientists even believe that exposures to background levels of radiation may have some beneficial effects. There is no doubt that acute doses of penetrating radiation are harmful to life as illustrated in Table 9.6.

TABLE 9.6. Acute Doses of Penetrating Radiation[a]

Dose (R)	Symptoms	Delay Time	Incidence of Death	Death Within	Effects
50	None	—	None	—	Blood profile
150	Nausea 25%	3 h	Nil	—	Blood changes
300	Nausea 100%	2 h	20%	2 months	Hemorrhage, infection
450	Nausea 100%	2 h	50%	2 months	Hemorrhage, infection
600	Nausea 100%	1 h	100%	2 months	Hemorrhage, infection
3000	Nausea, diarrhea	30 min	100%	2 weeks	Circulatory collapse
5000	Nausea, convulsions	30 min	100%	2 days	Respiratory failure

[a] The numbers shown can vary greatly from one individual to another based on incident circumstances, age, body exposure, and many other factors.

9.13 RADIATION DETECTION

According to the U.S. Nuclear Regulatory Commission, all persons performing radiographic work must wear a direct reading pocket dosimeter and either a film badge or thermoluminescent dosimeter (TLD). Pocket dosimeters must have a range of 0 to at least 20 mR and be recharged at the start of each work shift. Each film badge or TLD must be assigned and worn by only one person.

Pocket dosimeters shall be read and their exposure recorded daily. The operation of pocket dosimeters must be checked at least yearly and they must respond to within ±30% of true radiation exposure.

If an individual's pocket dosimeter is discharged beyond its range, their film badge of TDL must be processed immediately. Reports dealing with film badge or TDL exposures and daily pocket dosimeter information must be kept by the NRC or state inspection until the license holder is authorized to dispose of the information.

The self-reading pocket dosimeter, shown in Figure 9.8, is an air-filled ion chamber with a quartz fiber and metal wire attached to a charging electrode. Electrons are placed on the fiber and wire with a small charging unit. The quartz fiber is repelled from the wire having the same charge. Ionizing radiation neutralizes the charge and the quartz fiber moves closer to the wire. The position of the fiber, which is related to the received dose, is viewed through a miniature built-in microscope. Direct reading γ- and X-ray dosimeters are available in the ranges shown in Table 9.7.

Other equipment usually made and sold for use with direct reading dosimeters (DRDs) are chargers, calibrators, and viewers. The latest state-of-the-art dosimetry monitoring is the computer interfacing dosimetry system (CID), shown in Figure 9.9, which features a dosimeter viewing console, dosimeter

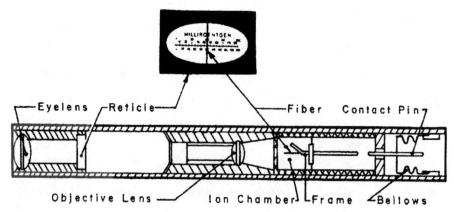

Figure 9.8 Cross-sectional diagram of a direct-reading pocket dosimeter. Courtesy of Dosimeter Corporation, Cincinnati, Ohio.

RADIATION DETECTION

TABLE 9.7. Ranges of Commercially Available Direct Reading Dosimeters

0–200 mR	0–20 R
0–500 mR	0–50 R
0–1000 mR	0–100 R
0–5 R	0–200 R
0–10 R	0–600 R

Source: Based on information provided by Dosimeter Corporation, Cincinnati, OH.

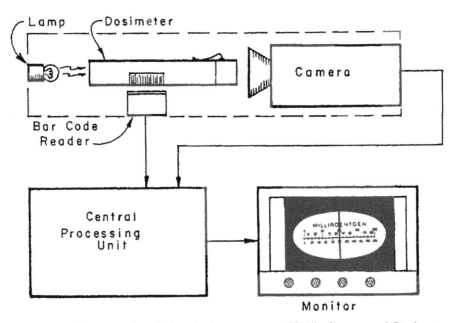

Figure 9.9 Computer interfacing dosimeter system (CID). Courtesy of Dosimeter Corporation, Cincinnati, Ohio.

holder, and an interface module. The computer interfacing dosimetry system reads the bar-code identification on the dosimeter and sends the information to the customer's computer or printer if provided.

Film badge dosimeters or thermoluminescent dosimeters are the second type of dosimeter that must be worn by radiographers. Film badges measure total radiation dose by measuring the darkness of an exposed, developed radiation film. The darker the film is, the higher the radiation dose.

Film badge dosimeters consist of a plastic film packet holder, film packet, small filter shims, and a spring-loaded clip. Unexposed film is sealed in a light-tight paper packet and placed in the film holder. The film packet is sensitive to light, heat, and moisture, as well as radiation. The various small metal and plastic filters that are built into the film badge holder are used to help determine the intensity and dose of radiation received by the wearer. Plastic filters shield the film from alpha and beta particles; metal filters shield the film from low-energy γ- and X-rays.

The TLD dosimeter is similar to the film badge in appearance, but contains a crystal chip that can store energy induced by radiation. To read the amount of stored energy, the dosimeter is heated and photons are released. A special reader measures the amount of light emitted, which is a measure of the accumulated radiation dose. The thermoluminescent dosimeters are more expensive than film badges but their chips can be reused.

9.13.1 Survey Instruments

Survey instruments use Geiger-Müller tubes, ionization chambers, scintillation chambers, or counters as detectors. Both the G-M tube and ionization chamber are gas-filled radiation detectors. They measure dose rate typically in terms of mR/h. With ionization chambers (Figure 9.10) radiation passes through the gas-filled chamber, ionizes the gas, and produces a number of pulses proportional to the radiation intensity. Depending on ion chamber voltage, some recombination of ions may occur. Output current pulses are very weak, requiring the use of a high-impedance, high-gain electrometer amplifier. Ion chambers tend to be more accurate at higher dose rates and more delicate than G-M tubes.

Geiger-Müller detectors operate at higher voltages and discharge pulses tend to spread out along the entire length of the detector's tube anode, producing large, constant amplitude discharge pulses. Little or no additional

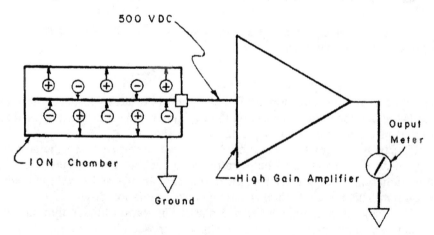

Figure 9.10 Ion chamber with high-impedance, high-gain amplifier.

RADIATION DETECTION

amplification is required because a single ionizing event is sufficient to cause a count or pulse to be produced. Radiographers usually prefer Geiger-Müller-type survey meters because they are rugged and sensitive when used in their calibration range.

A scintillation detector or counter counts the small flashes of light produced when radiation strikes a radiation-sensitive crystal such as sodium iodide. As the crystal absorbs energy, photons are produced. The crystal conversion efficiency is about 10%. The crystal has an integrating or averaging effect. Scintillation counters are relatively stable devices consisting of a sensitive, high-gain probe and moderate-gain electronics unit. The scintillation probe (Figure 9.11) consists of the radiation-sensitive crystal optically coupled to the top of a photomultiplier tube. The probe is usually wrapped in aluminum foil, to reflect the photons produced, and then in a magnetic shield to minimize electromagnetic pickup by the photomultiplier tube. The entire assembly must be kept light-tight to prevent ambient light from entering the assembly and saturating the electronics. Scintillation counters are often calibrated in terms of counts per minute.

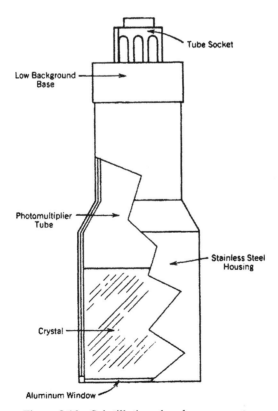

Figure 9.11 Scintillation chamber or counter.

Survey meters used by radiographers must be calibrated every 6 months. Each meter must have a label showing the last calibration date. Survey meters are calibrated with known sources at various distances. The meter is placed at one or more points where the radiation dose rate is known and an internal adjustment is made to make the analog or digital meter read correctly. If the survey meter has more than one range, each range is calibrated independently. Typically, a two-point calibration at 1/3 and 2/3 of range are provided.

9.14 RADIATION SOURCES

X-ray generators fall into three main categories. These are:

1. Industrial units (40–400 kV)
2. High-resolution systems (30–150 kV)
3. High-energy systems (>400 kV)

One of the most popular applications for low-energy X-ray tube generators is X-ray fluorescence. With X-ray fluorescence, a sample is placed in the instrument and a computer-controlled beam of X-rays is focused on the sample. The beam excites atoms in the sample, causing them to emit X-rays characteristic of the elements present. The characteristic X-rays are detected by a liquid nitrogen–cooled, solid-state, lithium-drifted silicon Si(Li) detector and analyzed by a microcomputer program.

X-ray tube generators, used for X-ray fluorescence, have dual modes, either molybdenum/tungsten (Mo/W) or rhodium/tungsten (Rh/W), a 0.25-mm-thick beryllium (Be) window, voltage range of 5 to 50 kV, and a tube current of 1 to 500 µA. Maximum tube power is 10 W and an incident angle of 50° between the X-ray tube and detector is common.

Shielded-cabinet X-ray units (Figure 9.12) are finding increasing use in X-raying everything from small castings to microelectronics. The X-ray cabinets hold samples up to 14 × 17 in., but 8 × 10-in. sample areas are more common. Beam size is typically 13 in. in diameter at the floor of the cabinet.

Maximum X-ray tube voltage is 10–30 kV with 3 mA of continuous current. Exposure times can be varied from 1 to 59 s or 1 to 59 min. Safe fluoroscopic viewing is available at reduced sensitivity.

There are several high-resolution real-time X-ray systems (Figure 9.13) currently on the market. These machines provide safe viewing by utilizing a computer-controlled 160 kV tubehead, sample manipulator, optical zoom system, image intensifier, TV camera, and monitor to present the magnified X-ray image.

9.14.1 Isotope Sources

Small, lightweight portable X-ray fluorescent analyzers, using 1 to 100 mCi radioactive sources, are used to prevent the catastrophic consequences of a

RADIATION SOURCES

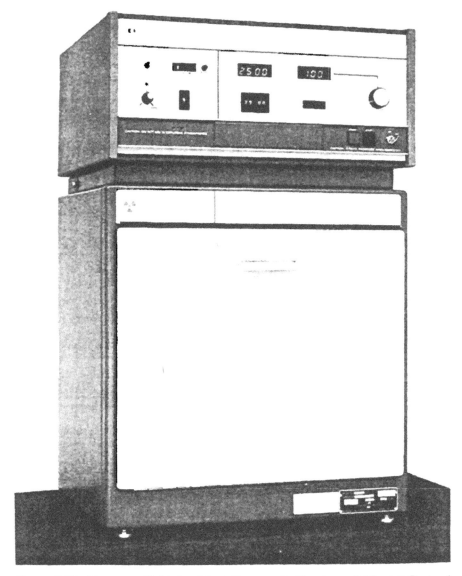

Figure 9.12 Compact shielded cabinet X-ray unit. Courtesy of Hewlett-Packard Company.

metal material mixup in critical metal equipment by nondestructively analyzing and identifying the metal alloy of construction and its elemental composition.

Figure 9.14 shows the Thermo Electron TN Alloy Pro with pistol-grip and top-mounted PDA with bright TFT display (65,000 colors) for clear viewing indoors or out. For alloy determination, the front of the pistol-shaped section is placed against the unknown sample and a short exposure is made. The PDA

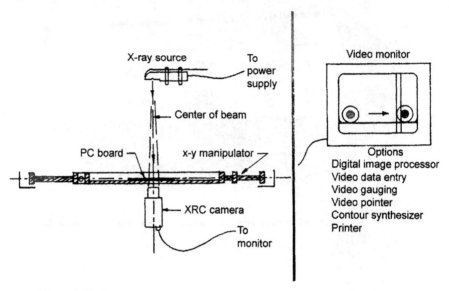

Figure 9.13 Real-time X-ray viewing unit. Courtesy of TFI Corporation.

can be pivoted to any viewing angle exclusive of the probe angle. The PDA can also be connected with a separate cable for TN Alloy Pro operation in tight spots.

Partial List of TN Alloy Pro Specifications

- X-ray excitation: 20 mCi Fe 55 and 4 mCi Cd 109 radioisotope sources
- X-ray detection: SiPIN detector with 250 eV resolution
- Measurement window: 6-micron-thick (0.00024-inch) Mylar (replaceable)
- Window aperture: 12 mm × 25 mm (1/2 × 1 inch)
- Dimensions: 21.6 cm × 20.3 cm × 3.8 cm (8.5 in. × 8 in. × 1.5 in.)
- Weight: 0.8 kgs (1.7 lb)
- Main carrying case: Hard-sided IATA (airport) approved; floats in water with 18.1 kg (40 lb) load
- HP iPAC processor: 400 MHz Intel® Xscale processor, 32 MB Flash ROM, 64 MB SDRAM, iPAQ backup.

X-ray tubes produce a continuous energy spectrum over their electromagnetic range with higher-intensity characteristic K and L bands formed when electrons strike target atoms, producing photons. For X-ray fluorescence work, the K X-ray energy bands are preferred for analysis of elements whose atomic numbers are close together.

TN Alloy Pro

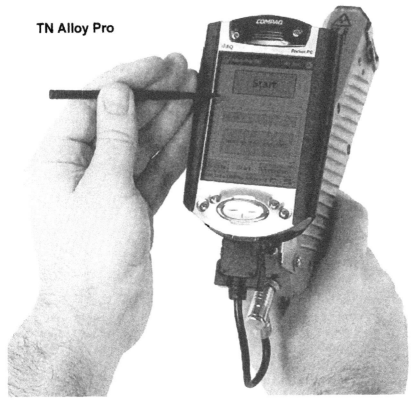

Figure 9.14 TN Alloy Pro. Courtesy of Thermo Electron Corp.

9.15 PORTABLE LINEAR ACCELERATORS

In 1957, high-energy, light-weight, portable linear accelerators were developed primarily for research and the in-service inspection of nuclear generating plants. High-energy X-rays are required to inspect steel sections thicker than 8 in. in high-temperature environments having an ionizing radiation background.

Portable accelerators inject high-energy electrons into a tuned waveguide structure where they are accelerated by microwave fields. Electrons strike a target and are decelerated, producing a spectrum of high-energy X-rays. Characteristics of the spectrum are dependent on target material and design, and the energy spectrum of the impinging electrons. These factors, along with electron beam current and X-ray efficiency of the target, determine the X-ray intensity. Major components of the portable accelerator are the control unit, power supply, RF modulator unit, and accelerator guide.

Commercial units are now available in 1.5, 4, and 6 MeV electron energy ranges. For these ranges, output radiation is 2, 120, and 300 R/min at a 1 m

TABLE 9.8. Half-Value Layers of Shielding Materials

Material	Density (g/cm³)	HVL (in.)
Tungsten	18.00	0.43
Lead	11.00	0.70
Steel	7.85	0.97
Aluminum	2.70	2.86
Concrete	2.35	3.30
Lucite	1.20	6.40
Water	1.00	7.70

target distance, respectively. With the midrange unit, some typical half-value layers are given in Table 9.8.

9.16 SPECIAL RADIOGRAPHIC TECHNIQUES

Flash radiography is a technique using exposure times of 20 to 60 ns duration to stop the action of high-velocity objects, for the purpose of investigating both internal and external characteristics of objects. Recording of these data is done with various film and intensifier screens to produce a stopped-in-motion negative of the object.

Flash radiography has found extensive use in the study of shaped charges, in-flight dynamics, ballistics (internal, transitional, external, and terminal), structural phase formations, and rocket motor studies. Many of these uses provide challenging problems with regard to equipment protection.

For ballistic applications, sealed high-vacuum tubes with a 3 to 5 min target provide the simplest, most reliable, cost-effective combinations.

Flash X-ray systems operate with peak voltages ranging from 150 kV to 2.3 MV at pulse duration of 20 to 60 ns. Penetration typically ranges from 1 to 8 cm in steel and 2 to 24 cm in aluminum at 2 m film-to-source distance (FTSD). X-ray dose at 1 m varies from 1.6 mR to 2 R depending on pulse duration and peak voltage.

Low-voltage flash X-ray systems have the advantage of lower cost, smaller size, portability, wider choice of objects, and better contrast. Splitting the pulser output and coaxially coupling it to dual remote tubeheads can be done to make stereoradiographs.

In addition to stop-motion studies, high-speed videography has recently been developed for optical and X-ray imaging. With this technique, up to 12,000 partial frames/s can be recorded and played back at 60 frames/s, resulting in a slowdown factor of 200. Two cameras can be used with the system and frame identification (time, date, and elapsed time) can be provided.

Cineradiography has also been developed where single flash systems are modified to flash several times a second. Medical-type film-changers have also been modified to synchronize the film-changing operation with the X-ray tube flashing. This technique has been used to study internal surface changes during online testing of solid propellant devices such as rocket motors and gas generators.

Some experimental cineradiographic work has also been done in the field of ballistics using six sequenced flash tubes in a single source head. Images are captured using six corresponding microchannel plate image intensifier tubes and polaroid film. This system results in one high-speed sequence of six pictures without film changing.

9.17 STANDARD RADIOGRAPHIC TECHNIQUES

9.17.1 Introduction

The goal of radiographic practice is to obtain optimum sharpness and quality consistent with reasonably short exposure times. Film quality and sensitivity depend on sharpness and contrast. Poor-quality or unsharpness can be defined as the blurring of image edges, resulting in a loss of definition of small cracks, defects, inclusions, or penetrating details. The quality of the radiograph also depends on the penetrating ability of the X- or γ-rays. Poor quality radiographs will result if the penetrating ability is insufficient or too great. Ideally, the radiographer would like to be able to identify changes in density of 1 or 2%.

Industrial X-ray tubes produce a continuous spectrum of X-rays because the electrons have such a wide range of velocities prior to striking the X-ray tube target material. Thin metallic filters of copper, brass, aluminum, and so on are placed at the source to reduce the effects of softer, undesirable radiation. In contrast, most radioisotopes have one or more very narrow, specific energy bands that are characteristic of the isotope. The successful use of X-rays depends on the intensity of the X-rays, their wavelength, the dimensions of the area from which they are emitted, and the duration of the emission. X-ray efficiency is improved when the target has a high atomic number.

X-rays emanate from the X-ray tube focal spot. Figure 9.15 shows an X-ray tube, test object, and typical radiographic setup. The penetrating ability of X-rays is determined by kilovoltage and waveform applied to the X-ray tube by the high-voltage transformer. A rectifier is used to change alternating current (AC) to direct current (DC) to further increase X-ray machine output. As the kilovoltage of the X-ray tube is raised, X-rays of shorter wavelength and greater penetrating power are produced. The total amount of radiation emitted from an X-ray tube is dependent on tube current (mA), kilovoltage (kV), and exposure time (T). If kilovoltage is held constant, the intensity of the X-rays will increase proportionally with tube current. However, the bandwidth or wavelengths of X-rays produced remain the same; therefore, X-ray quality or penetrating power does not change with the changes in X-ray tube

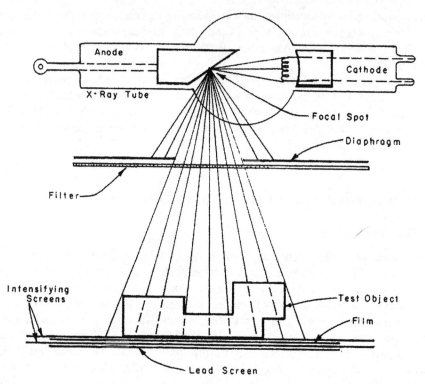

Figure 9.15 Typical X-ray setup with X-ray tube, test object, film, and intensifying screens.

current. On the other hand, increasing the kilovoltage of the X-ray tube increases both the intensity and bandwidth of the X-rays produced and does affect the quality or penetrating ability of the X-rays.

The output of exposure (E) of an X-ray tube is directly proportional to the product of current (I) and exposure time (T) and can be expressed by Eq. 9.4:

$$E = IT \qquad (9.4)$$

where E = exposure in milliampere minutes (mA-min) or milliampere seconds (mA-s)
I = milliamperes (mA)
T = minutes (min) or seconds (s)

The exposure factor involves the square of the source-to-object distance as shown in Eq. 9.5:

$$EF = (IT)/(d)^2 \qquad (9.5)$$

It should be pointed out that two identical machines operating under identical exposure conditions can produce different intensities and qualities of radiation.

9.17.2 Basic Principles

The same principles of shadow formation that apply to visible light transmission are applicable to radiographic shadow formation. In radiography the formation of shadows or the unsharpness of images on the film are related to the physical size of the source, the distance between the source and sample, and the distance between the sample and film. In addition, geometric distortion or a change in sample shape on film will result if the film is not positioned properly with regard to the energy beam. X-ray scattering can also cause shadows or unsharpness to be recorded on the film.

The quality and sharpness of radiographs will be greatly improved as the following ideal conditions are approached.

- The focal spot should be as small as possible; ideally it should be a point source.
- The radiographic source should be as far away from the sample as possible while still obtaining adequate penetration.
- The X-ray film should be as close as possible to the sample being radiographed.
- Source rays should be perpendicular to the film surface.
- The plane of the sample and the plane of the film should be parallel.

Some degree of radiographic distortion will always be present because some parts of the sample will always be further from the film than other parts. The shape of the object may also make it impossible to have the plane of the film parallel to all surfaces or planes of the sample being radiographed. Geometric unsharpness (Figure 9.16) is also referred to as the "penumbral shadow." Geometric unsharpness can be determined by Eq. 9.6:

$$U_g = F(t/d) \tag{9.6}$$

where U_g = geometric unsharpness in millimeters (mm)
F = source size in millimeters (mm)
t = object-to-film distance in millimeters (mm)
d = source-to-object distance (mm)

The size, shape, and location of a flaw or defect in a sample may be determined by radiographing the sample from different angles, perpendicular to its external surfaces. Shifting a source head equal distances in two directions from the

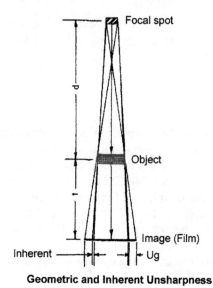

Figure 9.16 Geometric unsharpness is dependent on local spot size and t and d dimensions. Inherent unsharpness is dependent on object thickness.

original position is known as the "parallax method for triangulation" of flaws and defects.

9.17.3 Screens

Lead foil screens are commonly used in contact with X-ray film. Their effect is to enhance primary radiation in providing a sharp image while reducing the scatter radiation that contributes to image shadows. Lead screens can intensify the primary radiation by a factor of two to five while absorbing the secondary radiation, thus reducing the required film exposure time. Primary radiation striking a film holder, after passing through a thin section of a sample, will cause scattering into the shadows of thicker sections of the sample, producing an effect known as undercut. Lead masks, cut to the shape of the sample, can be used to decrease the scatter radiation that undercuts the specimen.

Fluorescent intensifying screens are made of a white cardboard base and coated with an 8-mil layer of luminescent chemicals. Since the X-ray films are sandwiched between the screens, there are a number of layers of material with various protective coatings. New bonded screens incorporate the use of special fluorescent material. The bonded screen assembly consists of lead foil, fluorescent layer, X-ray film, fluorescent layer, and lead film. Bonded screens are fabricated for specific applications such as for exposure by X-rays, γ-rays, and supervoltages. Lead foil thickness increases from 1 to 10 mils depending on

TABLE 9.9. Industrial Radiographic Film Intensifying Factors

				Intensity Factors	
Feature	Lead Thickness (in.)	Energy Level	Sample Thickness (in.)	Nonscreen Films	Screen Films
Hi definition	0.001	80–300 kV	steel 2	5–9	45–65
Hi speed	0.001	80–300 kV	steel 2	16–24	105–165
Hi definition	0.004	200–1 MeV	steel 4	4–6	30–50
Hi speed	0.004	200–1 MeV	steel 4	8–12	50–85
Hi definition	0.010	1–35 MeV	steel 8	3–5	20–30
Hi speed	0.010	1–35 MeV	steel 8	5–9	35–55

the expected energy level range. Industrial X-ray film intensifies sensitivity 3 to 24 times. Special screen-type films provide intensifications of 20 to 165 times as shown in Table 9.9.

For optimum performance, screens should be used only in high-quality film cassettes that assure good film-screen contact. Film-screen contact must be maintained over the entire surface area to provide good film image definition. Cassettes must be perfectly tight. Care should be taken to make sure screens and cassettes are stored in a clean, moisture-free area. Foreign matter will scratch screen surfaces and substantially reduce the life of all intensifying screens. Experienced personnel should load films into cassettes, because the improper mounting can be the cause of many radiographic problems. Some manufacturers make screen cleaners that are safe for all screen materials.

9.17.4 Film Composition

X-ray film has a thin emulsion coating on both sides of a clear plastic base support film. The emulsion is typically a 1-mil-thick suspension of microscopic silver sulfide crystals. Photofluorescent films differ from X-ray films in that they are coated with an emulsion on one side only. The latent image on the film is formed when light, γ-, or X-rays strike the film's emulsion, reacting with the silver sulfide crystals to form black metallic silver.

9.18 THE RADIOGRAPH

In photography, films are exposed to light (visible energy) and latent images are formed on a film's surface as silver sulfide crystals in an emulsion turn into black metallic silver. The more light that reaches the film, the darker the negatives. These same films would darken if exposed to γ- or X-rays and the procedures used in handling, developing, and caring for these films is almost identical to that of processing X-ray film.

In radiography, γ- or X-ray energy causes latent images to form. The contrast, darkness, or optical density of the film depends on the amount of energy reaching the film. As objects are X-rayed, thicker or denser objects absorb more energy and less energy reaches the film, causing lighter areas to be formed. As thinner or less dense sections are X-rayed, more X-rays reach the film and darker areas are formed on the film.

Inherent unsharpness occurs because objects being radiographed are not uniform in thickness or density. If they were, radiography would be pointless because we would have a uniform exposure that would show nothing but uniformity. Inherent unsharpness results because we do not have point sources, infinite distance between the source and test object, infinitesimal distance between the test object and film, single-plane test objects, and uniform perpendicular energy rays. All of this results in the scattering of secondary electrons in the film emulsion.

The relationship between current (mA), distance, and time for X-ray exposures is similar to that of source strength or activity, distance, and time for radioisotopes. The relationship between current and time is direct. For example, if X-ray tube current is reduced by a factor of three, the required exposure time will be three times longer to obtain the same quality exposure. If milliamperage is doubled, the required exposure time will be cut in half to obtain the same results. If an excellent radiograph was obtained with a tube current of 5 mA and an exposure time of 12 min, and all other conditions remain the same, a 6 min exposure would require a 10 mA tube current.

The relationship between tube current and distance follows the inverse square law; so does the relationship between exposure time and distance. Increasing the source to sample distance can compensate for a high-intensity source size. As exposure distance is doubled, milliamperage must be quadrupled to obtain the same-quality radiograph. If a 60 s exposure were required with a 4 ft source-to-film distance (STFD), a 15 s exposure would be required with a 2 ft STFD. If an excellent radiograph were obtained for a source-to-sample distance of 40 in., an exposure time one-fourth as long would be required for a 20 in. source-to-sample distance. These statements imply that the sample is in direct contact with the film for all practical purposes.

Photographic density is defined as the logarithm of the ratio of incident light intensity to transmitted light intensity, or

$$D = \log_{10}(I_i/I_t) \tag{9.7}$$

where I_i = incident light intensity
I_t = transmitted light intensity

With X-ray exposures, density is linear to the relative X-ray exposure (E) over a limited range, depending on film characteristics and the degree of development. Beyond a certain exposure level, the density curve flattens out. The characteristic curve for radiographic film is a plot of film density as a function of the logarithm of relative exposure. The characteristic curve (Figure 9.17) is also known as an H and D curve, in honor of Hurter and Driffield, who first used it for photographic density calculations near the turn of the twentieth century.

To use the characteristic curve, the density and log E is noted for a particular exposure. The desired density and its corresponding log E are then noted. The antilog of the log E difference is then calculated and the original exposure (X-ray tube mA current × exposure time minutes) is then multiplied or divided by the antilog number to achieve the new desired density. One advantage of using the H and D curve is that pairs of exposures with the same ratio will have the same interval on the log exposure scale even though their density changes may vary dramatically. The greater the slope of the characteristic

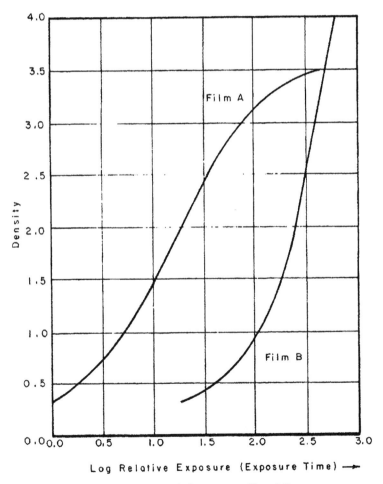

Figure 9.17 Characteristic curve or H and D curve.

curve, the greater the difference in density for a given log exposure interval. The greater the difference in density, the greater the contrast, definition, or visibility of details. As film development time increases, the slope of the characteristic curve will tend to grow steeper and move to the left. The slope of the characteristic curve is also called the gamma or gradient of the film.

The Bunsen-Roscoe reciprocity law states that radiographic film density is dependent only on the product of radiation intensity and exposure time and is independent of the absolute value of either considered separately. However, this law does not apply to fluoroscopic screens because of their construction and visible-light-emitting characteristics.

Other factors affecting image density are:

- Total amount of radiation emitted from the source
- Amount of radiation reaching the sample or test object
- Amount of radiation passing through the sample
- Screen intensifying action

Table 9.10 summarizes some causes of poor radiographs and lists corrective action that can be taken.

Exposure or technique charts (Figures 9.18 and 9.19) are plotted with exposure time on a vertical log scale as a function of material thickness on a horizontal linear scale. The curves are based on a particular piece of equipment,

TABLE 9.10. Radiographic Problems and Corrective Action

Symptom	Cause	Corrective Action
High film density	Overexposure	Decrease exposure—check meters and timers
High film density	Overdevelopment	Check darkroom timer and developer temperature
Low film density	Underexposure	Increase exposure factors
Low film density	Underdevelopment	Check development time and solution temperature
High film contrast	Sample too thick	Increase kilovoltage, use X-ray tube filter, or mask sample
Low film contrast	Sample too thin, wrong film speed, underdevelopment	Decrease kilovoltage, use higher speed film, check developer time and strength
Poor definition	Poor geometric exposure factors	Check all factors affecting shadows and unsharpness
Graininess	Wrong film for application	Use fine-grained film
Mottled appearance	Fluorescent intensifying screen used	Try lead foil screen
Foggy film	Exposure to light	Check for light leaks

THE RADIOGRAPH

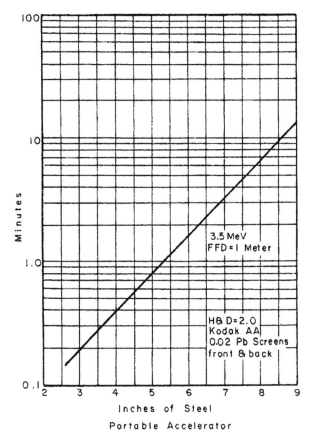

Figure 9.18 Technique chart for portable accelerator. Courtesy of Schonberg Radiation Corp.

operating at a specific energy with a specified source-to-film distance. X-ray exposure charts show material thickness, kilovoltage, and exposure time. Gamma ray exposure charts show material thickness, gamma energy, and exposure time.

Gamma ray exposure is calculated as follows:

$$E = AT \qquad (9.8)$$

where A = source strength in curies (Ci) or millicuries (mCi)
T = exposure time in seconds (s), minutes (min), or hours (h)

The radiation quality of a γ-ray source is determined by isotope selection. In order to determine the proper radiographic exposure, during the initial setup, the radiographic film manufacturer's recommendations should be fol-

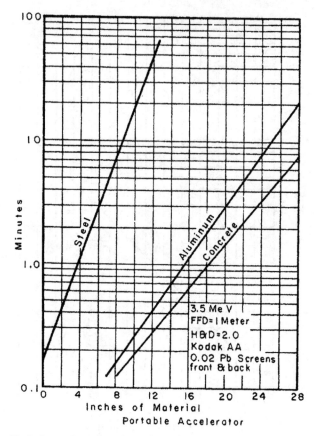

Figure 9.19 Technique chart for portable linear accelerator. Courtesy of Schonberg Radiation Corp.

lowed. If the film appears too light, the exposure should be increased by 30%; if the film appears too dark, the exposure should be decreased by 30%. A convenient, low-cost gamma radiation exposure calculator, available from JEM Penetrameter, calculates exposure time for Co-60 and Ir-192 sources and contains data on DuPont, AGFA-Gevaert, and Kodak films. Industrial X-ray films are classified as shown in Table 9.11.

9.18.1 Image Quality

The radiographer's goal is to obtain the best image quality and greatest radiographic sensitivity consistent with the current state of the art. To achieve this, the radiographer must answer the following questions. How can small flaws or discontinuities be reliably detected within the constraints of an existing or available system? What are some of the factors affecting sensitivity? Radiographic sensitivity is the ability to detect small flaws or discontinuities.

TABLE 9.11. Classification of Industrial X-Ray Films

Classification	Description and Uses
Class I	Extra fine grain, high-contrast film. Used with low-density metals. For direct use or use with lead screens.
Class II	Fine grain, high-contrast film. For low-density metals and lower kilovoltage. More widely used because of higher film speed. Used directly or with lead screens.
Class III	Highest speed film. For use with γ-rays and higher kilovoltages. Used directly or with lead screens.
Class IV	Highest speed and greatest contrast when used with fluorescent intensifying screens. Lower contrast when used directly or with lead screens. Used for examination of thicker or denser materials when limited kilovoltage ranges are available.

Three major factors affecting sensitivity are:

1. Film graininess
2. Image unsharpness
3. Flaw image contrast

Film graininess occurs during the development process when the developer used tends to clump many silver grains together into a mass that can be seen by the human eye. The degree of graininess is a function of both film type and processing.

Unsharpness is defined as the width of the visible density band change observed for the image of a sharp edge or knife-edge. Unsharpness is the cumulative effect of movement unsharpness, fluorescent screen unsharpness, and inherent unsharpness. Movement of the film, object, or radiation source causes movement unsharpness. Geometric unsharpness is caused by the finite size of the radiation source and relative distance of the source, object, and film. The light scattering of the screen materials causes fluorescent screen unsharpness. The scattering of secondary electrons in the film emulsion causes inherent unsharpness.

The thickness of the sample, the penetration characteristics of the source, and the degree of secondary scattering affect flaw image contrast. Table 9.12 summarizes the factors affecting radiographic sensitivity.

Contrast can refer to radiographic contrast, film contrast, or sample contrast. Radiographic contrast, which is dependent on sample contrast and film contrast, refers to the differences in density, for various radiographic areas. Film contrast depends on the type of film used and film processing. It is also dependent on film density over its useful range and limited to some extent by the ability to illuminate the film. Sample contrast is a function of the thickness or density of various sections of the sample. It depends on the radiographic

TABLE 9.12. Radiographic Sensitivity Controlling Factors[a]

Subject Contrast	Film Contrast	Geometric	Graininess
1. Sample thickness	1. Film type	1. Source size	1. Film type
2. Source quality	2. Processing	2. Source-sample distance	2. Screen type
3. Secondary radiation	3. Density	3. Sample-film distance	3. Radiation quality
	4. Developer activity	4. Thickness variations	4. Developing

[a] These factors affect image definition or sharpness, the opposite of unsharpness. Film selection depends on the thickness of the object to be radiographed, object material, and voltage range of the X-ray machine.

quality (kilovoltage and wavelength) and the amount of scattered radiation. The lead symbol "B" is attached to the back of the film holder to determine if excessive backscatter is present.

Penetrameters or image quality indicators (IQIs) provide a check on the adequacy of the radiographic technique. They help determine optimum radiographic sensitivity to an ideal flaw or defect. Penetrameters may consist of step changes in material thickness—machined grooves, sharp-edged holes, wires of known diameter, or a combination of these man-made variables. Step thickness, groove width, hole, or wire diameter are usually some multiple of the thickness (T) of the sample. Actual flaws or discontinuities may not have sharp-edged holes; therefore, it may be more difficult to detect their changes in density. In this case, actual flaws or simulated flaws may aid in contrast sensitivity studies.

Penetrameters with a 2% thickness of the object to be radiographed and $1T, 2T$, and $4T$ hole diameters are common. These are identified as 2-$1T$, 2-$2T$, and 2-$4T$ penetrameters. Usually, the penetrameter is made of the same material as the object to be radiographed and is placed on the source side of the object. Penetrameters for stainless steel are considered Group 1 materials and usually do not have an identification notch. Ideally, the penetrameter is placed in the area of interest where the flaw or defect is most likely to occur. The image of the required penetrameter and hole on the radiograph is an indication that the radiograph has the required sensitivity. In cases where the penetrameter cannot be placed on the actual object, such as weld reinforcement, a shim can be used to simulate the weld reinforcement and the penetrameter can be placed on top of the shim.

In the past, the American or ASTM standards, European or DIN standards, and military standards have varied somewhat in physical design. The ASTM has done much to correct plaque penetrameter size and tolerance ambiguities, establish a manner of indicating sample thickness, and set up a method for classifying similar radiographic materials. Four sets of wire penetrameter sizes, specified in ASTM Standard E747-80, are shown in Table 9.13.

THE RADIOGRAPH

TABLE 9.13. Wire Penetrameter Sizes

	Wire Diameter (in.)		
Set A[a]	Set B[a]	Set C[b]	Set D[b]
0.0032	0.0100	0.0320	0.1000
0.0040	0.0130	0.0400	0.1260
0.0050	0.1600	0.0500	0.1600
0.0063	0.2000	0.0630	0.2000
0.0080	0.2500	0.0800	0.2500
0.0100	0.3200	0.1000	0.3200

[a] Length of wire in sets A and B = 1 in.
[b] Length of wire in sets C and D = 2 in.

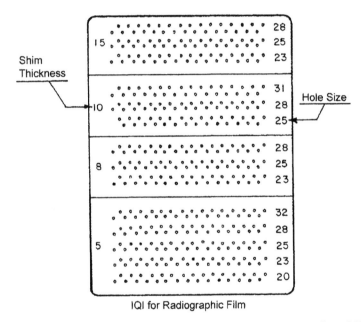

Figure 9.20 Shim penetrameters for determining the image quality of films.

For determining the image quality response of various radiographic films, ASTM Standard Method E746-80 recommends a four-step shim penetrameter having step thickness of 0.005, 0.008, 0.010, and 0.015 in. The shims (Figure 9.20) contain multiple rows of holes with 15 holes/row. The holes must be 0.2 in. apart vertically. Hole diameters range from 0.02 to 0.032 in. See ASTM E747-97 for the most recent standard.

Three readers independently reading three test films of any one film type do film evaluations. Each reader records the number of holes visible on each step of the IQI. Readers may use a 3× magnifier in an appropriately darkened room. The holes visible on each step are added and equivalent penetrameter sensitivity (EPS) is calculated in accordance with Eq. 9.9:

$$\text{EPS } (\%) = (100/X)(Th/2)^{1/2} \qquad (9.9)$$

where h = hole diameter in inches (in. or mm)
T = step thickness of IQI in inches (in. or mm)
X = thickness of the test object in inches (in. or mm)

9.18.2 Film Handling, Loading, and Processing

Good darkroom practices for handling, loading, and processing X-ray film are similar to those used for photograph negative handling and processing. Since long-term exposures to safelights will result in film fogging, exposed film should be kept at least 4 ft away from safelights. The weakest or most heavily filtered safelights should be used near the loading and unloading bench and where Class IV films are used. Medium-intensity safelights should be used in areas where films are developed and fixed, with the highest-intensity safelights reserved for use in areas where the films are washed and dried.

Load bench operations include handling of the unexposed and exposed films, loading and unloading of film cassettes, and hanging of exposed films for subsequent wet processing. The layout of the darkroom should provide a smooth flow of processing operations. Film holders and cassettes should be located near the loading bench. Unexposed boxes of X-ray film should be stored vertically on their ends or edges. A divided wall box with sliding doors should be provided for entering the exposed film for processing and passing out processed radiographs in their protective envelopes. Cleanliness is essential in all darkroom operations because a single fingerprint can destroy an otherwise acceptable radiograph.

Care should be exercised when opening film boxes and packets to assure that the films remain clean and dry. During cassette loading, the films should be handled primarily by the edges to prevent fingerprints and smudges. Crimp marks and black static marks on the radiograph, which appear treelike, bird-foot shaped, or circular in pattern, are caused by improper film handling. Films that have been left between lead screens for too long may become fogged if temperature and humidity are high.

The steps in both automatic and manual film processing are:

1. Developing the exposed film
2. Stopping the development
3. Fixing the film
4. Water washing
5. Rinsing in a wetting agent and drying

THE RADIOGRAPH

Automatic film processing is preferred in large operations because manual processing is too time-consuming and labor intensive. With automatic film processing, a dry, ready-to-read radiograph can be produced in about 12 min. One disadvantage of automatic processing is that a slowdown of the film transport system, film overlap, sticking, or wrapping can cause the film emulsion to become swollen, soft, and sticky.

The three major steps are developing, fixing, and washing. X-ray processing solutions can be mixed and used in stainless-steel, plastic, or enameled-steel vessels. The first step in film processing consists of immersing the exposed X-ray film in a developer solution to develop the latent image. The film processing solutions should be maintained at 65 to 75°F. Temperatures below 60°F result in underdevelopment. During manual film development, film hangers are sharply tapped after lowering the film into the developer to remove air bubbles clinging to the emulsion. Replenisher is added to developer as needed to maintain good developer activity. However, the developer should be discarded in three months or when the volume of replenisher added equals two or three times the volume of the original developer, whichever is shorter.

Areas on the film that have received the greatest exposure will be darker and have higher density. It is essential that films be agitated in the developer to achieve uniformity in development. Agitating the X-ray film during development renews the developer at the surface of the film. However, care should be taken when using mechanical agitators because they can create undesirable developer flow patterns. It is generally accepted that finer quality radiographs can be obtained by manual processing. Normal development time for manually processed X-ray film is 5 to 8 min at 68°F.

After development, the films are placed in a glacial acetic stop bath for 30s and agitated to neutralize the alkaline developer and stop the developing process. If a stopper solution is not available, the films can be thoroughly washed in running water for at least 2 min before fixing. When diluting acid to make the stop bath, the acid should be slowly added to the water to prevent splattering caused by the exothermic reaction of dilution. After the stop bath, the films are placed in a fixer or sodium thiosulfate "hypo" solution for 3 to 6 min. The hypo solution is a strong oxidizing agent that removes the undissolved silver salts, leaves the developed black silver metal as a permanent image, and hardens the film gelatin. Fixer activity diminishes in time due to contamination by soluble silver salts. Fixer solution should be disposed of when it can no longer do its job within 10 min.

Thorough film washing for 10 to 30 min is required to remove all traces of hypo from the emulsion surfaces. Failure to do so can result in excessive hardening and subsequent cracking of the film surface in a relatively short time. The hourly flow of cascading water in the wash tank should be four to eight times the volume of the tank. Before drying, the films should be dipped in a mild wetting agent for 30s to minimize waterspot formation and deter reticulation. After drying, radiographs should be stored in an area with a relative humidity of 30 to 50%.

Figure 9.21 Pickup truck with portable X-ray lab unit. Courtesy of Jem Penetrameter Manufacturing Company.

Processing tanks should be periodically cleaned and sterilized using a commercially available sodium hypochlorite cleaning agent, such as Clorox. Tanks should then be thoroughly rinsed with water to prevent interaction of trace amounts of hypochlorite with processing solutions that will be added later.

Improper film processing can result in reticulation and frilling or loosening of the emulsion from the film base. In most cases, reticulation, a puckered lacelike film surface, is caused by sudden extreme temperature changes during film processing. The use of warm or exhausted fixer solution usually causes frilling.

Portable X-ray labs (Figure 9.21) have been designed to provide radiographic services at remote construction sites, thus eliminating transportation time to and potential backlog problems at larger centrally located laboratories.

9.18.3 High-Intensity Illuminators

High-intensity illuminators are used by readers to help interpret radiographic images. These illuminators should be located where they will not interfere with film loading or developing operations. The film viewing room should be arranged so that background lighting does not reflect on the films being viewed. Wall colors should be selected so that their brightness is about the same as the images of interest for optimum human eye-contrast sensitivity.

High-intensity illuminators permit viewing of films with densities up to 4.0 H and D. They should have good masking to prevent glare from the edges of the radiograph. The ideal illuminator provides continuously adjustable, evenly distributed diffused light. Heat filters should be used with high-intensity illuminators to prevent damaging films while they are being viewed.

9.19 FLUOROSCOPY TECHNIQUES

Fluoroscopy involves the real-time viewing of X-rayed parts. The quality of the inspection technique depends on the resolution of the system and skill of the operator/interpreter. The operator must have good eye sensitivity, good eye adaptation to darkroom lighting conditions, and not tire easily when viewing rapidly moving parts. To minimize operator fatigue, test object scanning speed should not exceed 3 in./s and operators should be periodically rotated to other duties.

Operators should also receive extensive training on the characteristics of defects expected. The X-ray fluoroscope is a fixed machine operating under typical darkroom lighting conditions. To obtain satisfactory results, geometry, scatter radiation, source kilovoltage, and X-ray tube amperage must be carefully controlled. Care must also be taken to prevent material vibration or unwanted motion that will blur the image during inspection.

Because the small focal spot size of a fluoroscope approaches that of a point source, image enlargements can be made without loss of definition. The optimum magnification for best image clarity is determined by Eq. 9.10:

$$M = 1 + (U/F)^{3/2} \tag{9.10}$$

where M = optimum magnification
U = screen unsharpness in millimeters (mm)
F = focal spot in millimeters (mm)

Even though relatively low X-ray tube voltages are used in fluoroscopy, operator safety must be considered because of the operator's close proximity to the instrument. Operators should be well acquainted with the design of the machines they are using, know whether they are direct or indirect viewing, and know what background radiation levels to expect in the general work area.

Advantages of fluoroscope inspection include immediate viewing of the subject from a multitude of angles, the illusion of a three-dimensional presentation, and the ability to study moving parts in action. In addition, fluoroscopy does not have the added expense of the film and film processing costs when compared to standard radiography. The fluoroscopic image is a positive image. This means that holes in dense objects show up as lighter areas and denser inclusions in objects show up as darker areas.

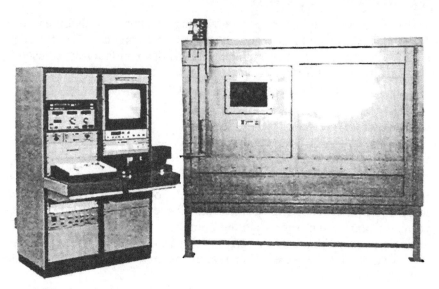

Figure 9.22 Real-time X-ray viewing unit. Courtesy of TFI Corporation.

The major disadvantage of traditional fluoroscopy is that image contrast does not compare favorably to that obtained with conventional film radiography. Fluoroscopic screens consist of minute crystals that glow when bombarded with X-ray energy. The large grain size of the screen crystals and the limit as to how brightly they can glow determine the degree of contrast and sensitivity that can be achieved. Fluoroscopic screens are easily damaged and dulled by exposure to direct sunlight or ultraviolet (UV) radiation.

Penetrameters can be used to determine when the operator's dark adaptation and visual acuity are satisfactory. Penetrameters also can be used to determine inspection sensitivity. Production samples with known flaws or defects are invaluable aids in establishing test standards.

Direct viewing of a fluorescent screen can be accomplished when an adequate transparent shield is provided to protect the operator. Other systems use lead glass mirrors to reflect the image while reducing the need for heavy radiation shielding. Both of these methods are becoming less popular as remote viewing systems incorporating the use of image-intensifier tubes and TV monitoring systems (Figure 9.22) become more available. With these systems, the visible light emission of the screen must be matched to the wavelength sensitivity of the image detector system for optimum results.

9.20 FLAT PANEL DIGITAL IMAGING SYSTEMS

Amorphous selenium, amorphous silicon, and CMOS panels are referred to as digital flat panel systems. Digital flat panels are rapidly growing in popularity for many reasons. However, even these newest panels are not perfect because they can have dead pixels that are not obvious. Each pixel is not independently wired. To compensate for image information lost by dead pixels, imaging information from surrounding pixels helps supply data. Flat panel pixels are typically wired in a series of rows or columns, and it is possible to lose an entire row or column of pixels. Pixels in CMOS panels are wired independently, but they can also have dead clusters that hide image data and compensate for data lost by using data from surrounding pixels. When flat panels are recalibrated, pixel compensation occurs and the dead pixels disappear from view, hiding minute defects.

Single dead pixels are analogous to minute film pickoff in film radiography. This happens when film emulsion is actually picked off the film base of single-emulsion film, providing no imaging information. Likewise, dead pixels provide no imaging information.

According to the Fuji Film Corporation, the only reliable digital replacement for film is Fuji Dynamix Computed Radiography. With this system, all of the pixels are alive all of the time in the imaging plate. However, if you scratch or damage an imaging plate, the scratch would show up on the radiograph the same as it would on a film radiograph. The damage to the imaging plate is permanent and visible. The plate is still usable and can be repositioned to minimize the plate defect before reshooting the test object.

9.21 FLAT PANEL SYSTEMS VS. FUJI DYNAMIX CR IMAGING SYSTEM

Recalibration of flat panels uses information from neighboring pixels to compensate for their loss. This may or may not be a problem depending on where the dead pixels are and the size and location of the smallest defect to be detected. Fuji Dynamix Computed Radiography (CR) direct film replacement imaging panels have an active surface where all pixels are active, but subject to physical damage. With scratched imaging plates, the scratches are obvious and can often be repositioned away from regions of interest (ROI) being X-rayed.

9.21.1 Resolution

Silicon panels have a resolution of 2 to 2.5 lp/mm and are capable of displaying a moving image, which make them suitable for real-time imaging applications. Selenium panel resolution is 3 to 3.6 lp/mm but they are unsuitable for real-time imaging. Flat panels usually compensate for lower resolutions by using geometric enlargement and sharply focused X-ray sources. The Fuji imaging plate has a standard resolution of 5 lp/mm without geometric magnification. This resolution is probably suitable for most midrange Class 1 radiographic film applications without special equipment needs. Figure 9.23 shows the Fuji Dynamix Series 4 fully automatic reader with throughput of 56 to 90 IPs (image plates) an hour.

Figure 9.23 Fuji Dynamix Series 4 CR high-speed reader. Courtesy of Fujifilm N.D.T. Systems, a Division of Fuji Medical Systems.

9.21.2 Ghost Images

Flat panels can hold a ghost image, known as "burn in," with image-intensifier systems when a test object is overexposed to radiation. When this happens, high-intensity external light sources can be used to help get rid of a ghost image. Recalibrating the panel can help, but this is also very time-consuming. When ghosting is encountered with a flat panel system, the system is down until the condition is corrected. Fuji imaging panels are also subject to "ghosting," but in this case the image plate can be erased or taken out of service. After a few days, the ghost will be gone and the imaging plate can be used again.

9.21.3 Image Lag

Image lag is a phenomenon that lasts for a few hours or days with a flat panel system. It is the panel's inability to normalize and get rid of a previous image. Some flat panels require several flat field exposures with readout checks to verify image lag has been eliminated. Other flat panels use an internal light exposure and readout to verify removal of the last image. Image lag is not a problem with the Fuji imaging panel because quick erasure removes latent images.

9.21.4 Dark Current Noise

Flat panels are subject to dark current noise when thermal electrons fill up the photodiodes in the panel. This can be a problem when long exposure times are required. Dark current may provide more signal intensity than conventional X-rays. Therefore, exposure times may be severely limited in these applications. CR imaging plates do not have dark current electrons.

9.21.5 Portability

Flat panel digital panels are connected to the computer and are therefore more limiting than CR imaging plates, which replace film. The maximum source-to-test object distance is determined by the length of the interconnecting cable, which is also limiting. CR imaging panels and readers are completely portable.

9.21.6 Temperature Sensitivity

Flat panel systems share the same temperature sensitivity as their silicon, selenium, and CMOS counterparts. In some cases, temperature compensation may be applied. CR imaging panels are not sensitive to normal daily temperature changes encountered during radiography work.

9.21.7 Flexibility

Flat panel systems are flat and limited by their physical size. They may be placed in test objects or behind test objects, but they cannot be bent to conform to the shape of test objects. CR imaging plates can be rolled into a 3-inch diameter. They are about as flexible as X-ray film and can be reused thousands of times.

9.21.8 Fragility

Flat panel systems are costly and fragile by nature. Care should be taken in their handling and use. Fabricators should design flat panel systems to withstand normal radiographic handling and use. CR imaging panels are less costly and sturdier. They should be handled with the same care as a film cassette.

9.21.9 Advantages

Flat panel systems appear to have many advantages for in-process inspections and real-time imaging applications. CR image plates and readers appear to have many advantages over most standard X-ray film processing systems.

9.22 INDUSTRIAL COMPUTED TOMOGRAPHY

Within the range of nondestructive methods, only ultrasound and X-ray are able to investigate the internal structure of an object. While ultrasound has its advantages when searching for plain defects (cracks, debonding) in homogeneous material, X-rays are used for detecting volume defects even in complex objects.

One consequence of the steadily increasing complexity of technical modules is the need for improved NDT methods, providing more detailed but easy-to-interpret information. Besides defect analysis—as done with conventional X-ray techniques—more and more spatial geometric measurements on internal structures have to be performed with high accuracy. Up to now, *computed tomography* is the only nondestructive way to do this.

9.22.1 Scan Procedure

When X-rays pass through an object, they are attenuated by absorption and scatter. The amount of attenuation depends on the kind of material and the beam path length within. The portion of radiation that penetrates the object forms a projective image. All inner and outer structures of the object are shown overlapping (radiography). Defects such as gas inclusions or less dense material give rise to local increases in intensity as fewer X-rays are absorbed. They are seen as gray-level deviations in the radiographic image in Figure 9.24.

These two-dimensional X-ray images have inherent limitations, such as:

- Details shown as overlapping are hard to distinguish—only an experienced operator, familiar with the test part, may be able to reliably interpret test results.
- No information about the position along the beam path (depth) is available from the projective image, so the exact location of flaws and anomalies is difficult to determine.

Both limitations can be avoided by the use of computed tomography. Instead of generating only one image in one position, a series of projections is taken under different angles when rotating the object in the radiation beam. A tomogram, which contains the entire volume information, is calculated out of these raw images (Figure 9.25). The tomogram comprehends the spatial composition of the object; materials can be distinguished regarding their density (more precise: absorption). Depending on the application and objective, the follow-

Figure 9.24 Gas inclusions and less dense material are seen as gray-level deviations in conventional radiography. Courtesy of XYLON International.

ing evaluation steps are done based on volume data (visualization, testing) or slicewise (geometrical measurement).

9.22.2 Applications of Industrial Computed Tomography

Compared to medical diagnostics, industrial applications deal with a broad range of objects, materials, and sizes. Therefore, different types of CT scanners are designed to fulfill the specific requirements of various applications. Within an R&D environment, CT has to be flexible to cover a wide range of object sizes and materials. Besides geometry (magnification), the imaging parameters have to be set individually to get optimum results. Accuracy and flexibility are more important than throughput and short scan times.

In terms of quality control during serial production, only a few, well-defined images have to be generated on a large number of similar objects within a short time. CT systems for that kind of application are fast and easy-to-operate, but less flexible for use for different purposes. With increasing throughput, the potential to use CT for quality assurance gets more and more attractive.

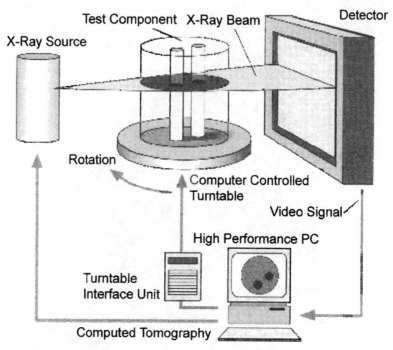

Figure 9.25 A tomogram containing volumetric information is calculated from new image data. Courtesy of XYLON International.

9.22.3 CT System Components

Radiographic inspection is usually done by looking through the flat side of an object in order to minimize beam path length. Computed tomography, however, requires sufficient penetration in each projective image to achieve sophisticated image quality. Therefore, high-energy sources with 450 kV or more are commonly used whenever microfocus X-ray sets do not offer enough penetration power. Once the energy range is defined, all the other components within the CT system have to follow this guideline. So, three major groups cover the range of applications: μ-CT, 450 kV-CT, and Linac-based CT.

Spatial resolution, contrast sensitivity, and dynamic range are the main parameters to describe the usability of detector systems or a specific application. For visualization of internal structures in terms of NDT, a full-volume image has to be acquired within a minimum of time. Detail visibility as the indicator in practical use is a mixture of spatial resolution and contrast discrimination. As these investigations are more qualitative than quantitative, both requirements are in a medium range. Industrialized versions of ASD (amorphous silicon detector, flat panels) fulfill these requirements in most cases.

INDUSTRIAL COMPUTED TOMOGRAPHY

Performing CT-based geometrical measurements is an increasing requirement within the range of applications. Sharp edges and a precisely reproduced image of the real contours form the basis for an accurate evaluation. Scattered radiation forms a "shining cloud" around the object. The shape and intensity of this cloud depends on radiation energy, material, structure, and its position in the image. Its effect to smooth and shift the edges in the scanned image is critical for any measurement. This effect is visible at low energies, but it gets critical when approaching 450 kV.

The light-sensitive photodiode array (amorphous silicon), shown in Figure 9.26, is coupled to a scintillation screen (GOS or CsI) that fluoresces during X-ray bombardment. When X-rays strike the scintillation material, they are converted to visible light that is detected by the photodiode array and converted into an electrical current, which in turn is converted into a digital image.

The linear diode array (LDA), Figure 9.27, mainly consists of a set of photodiodes laminated with a scintillator screen or individual crystals. The incoming X-ray photon is first converted into visible light, and then the photodiode produces a current proportional to the light flash intensity. This current is digitized and forwarded to the computer. Images taken with a linear diode array do not suffer from this effect. A deep slit collimator masks the sensitive area and eliminates almost the whole scattered (indirect) radiation. So, 2D CT images (slices) are best for geometrical evaluation at high energy. Moreover the spatial resolution (pixel size) can be adapted to the needs for the specific application.

One major trade-off has been made when using LDA—each scan is done within seconds, but a full-volume scan may last an hour or more, compared with only minutes when using an ASD. On the other hand, a high-resolution scan of an entire object is not the typical way to perform geometrical analy-

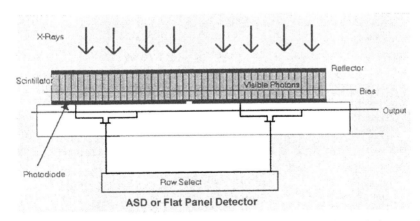

Figure 9.26 Light-sensitive photodiode array (amorphous silicon) coupled to a scintillator screen. Courtesy of XYLON International.

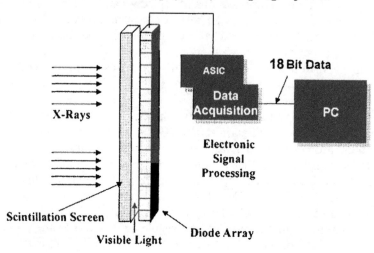

Figure 9.27 Linear diode array with photodiodes laminated to a scintillator screen of individual crystals. Courtesy of XYLON International.

sis, which focuses on well-defined positions within the objects. Whenever both high-speed imaging and precise measurements are needed, a system having both types of detectors guarantees optimal performance. Degradation and aging are long-term effects that decrease imaging quality with ASDs during usage. This decrease is hardly seen on LDAs of high quality.

As development goes on, technical data for resolution and dynamic range change rapidly. Today, 1000–5000 pixels in a line detector are common; AD-conversion is done at 18 bits. ASDs take 1000^2–3000^2 pixels of some hundred microns each, typically covering 400 × 400 mm of sensitive area.

Time was when computing power was a serious challenge to handling incoming data. But today, even big series of high-resolution images are reconstructed in parallel with the measurement, so reconstruction is done shortly after the scan is finished.

Medical scanners are built with a rotating gantry for source and detector in order to keep the patient in a stable, horizontal position. Technical objects usually are stiff enough to tolerate a smooth rotation. Therefore, a turntable is used to rotate the object instead of the imaging device. In order to fit to various object sizes, the object may be shifted along the beam path (focus-object-distance—FOD). Scan time and overall geometry may be optimized by moving the detector (focus-detector-distance—FDD) while the area of interest along the object is irradiated by moving either the object or the imaging device vertically.

AUTOMATIC DEFECT RECOGNITION

"One-fifth of the resolution" is a thumbnail rule to define the accuracy of the manipulation system, which leads to single-μm accuracy in micro-CT systems and a 10μm range for 450kV systems.

A high-end CT system capable of inspecting medium-sized castings would typically consist of the following:

- A 450kV X-ray source with a primary collimator to reduce scattering (Figure 9.28)
- An ASD to create volume data sets for NDT purposes (Figure 9.29)
- An LDA to perform highly accurate cross sections for geometrical evaluation (Figure 9.29)
- A manipulation system that allows adjustment to optimum parameters for various objects and tasks (Figure 9.30)

As mentioned before, penetration is a must under all directions. Looking on the steel wires assembly in the bead of a tire, it is obvious that truck tires have to be scanned with a linear accelerator (Figure 9.31). ASDs do tolerate such high energy; however, inspection of tires is based on radial scans. So an LDA is used to generate slice images. The overall size of the system is determined by the space needed to rotate the tire ex-centrically.

9.23 AUTOMATIC DEFECT RECOGNITION

Real-time radiographic applications lend themselves to *automatic defect recognition (ADR)*. Successful ADR application results in faster inspection times, better repeatability, and additional cost savings. Components that can be used with ADR include conventional, mini-, or micro-X-ray sources, analog or digital detectors, and the ADR platform (neural network, golden image comparison, or rule base using specific algorithms for a specific region of interest—ROI).

Advantages of ADR are increased quality through repeatable, objective inspection and improved manufacturing, testing, and quality control processes. The resultant increase in productivity automatically decreases labor costs. ADR integrates the X-ray inspection process with machine vision. X-rays penetrate the product, the imaging system generates images for evaluation, and machine vision automates the analysis of the image and makes pass/fail decisions.

9.23.1 Imaging Improvements

Radioscopic inspection systems have been in operation for over a decade. The first systems used an image intensifier, solid-state CCD camera, and an operator to make the final pass/fail decision. Because of design limitations of

Figure 9.28

Figure 9.29

Figure 9.30

Figure 9.31

Figures 9.28–9.31 All figures courtesy of XYLON International.

the image-intensifier tubes, these systems did not produce an image of high enough intensity and contrast for ADR.

Linear diode arrays (LDAs) were introduced about 10 years ago. These systems were improved, but still had specific hardware problems that had to be solved to increase reliability and decrease installation and routine maintenance times. Engineers and designers met the challenge and solved the problems and produced today's technically advanced LDA digital imaging systems.

AUTOMATIC DEFECT RECOGNITION 389

The advanced linear diode array imaging system features higher resolution, longer life, greater reliability, lower maintenance costs, and utilizes an 18-bit microprocessor to accommodate a wide dynamic range. Current LDA systems feature a PC with IBM architecture, Microsoft Windows NT® operating system, SVGA monitor, and 1536-element diode array.

9.23.2 LDA Design and Operation

The LDA consists of 64 diodes laminated with a scintillation screen to create X-ray-sensitive diodes. X-rays penetrating the test part are received by a single row of diodes. The scintillation screen converts the photon energy emitted by the X-ray tube into visible light on the diodes. The diodes produce an output voltage proportional to the light energy received. This voltage is amplified, multiplexed, and converted to an 18-bit digital signal by the interface board and sent to the computer. Then the diodes are reset and ready to receive the X-rays for the next line of the part to be inspected. Figure 9.27 shows a typical LDA imaging system. With the LDA imaging system, the part moves at a controlled rate past the LDA, or vice versa.

9.23.3 ADR Techniques

ADR takes the data provided by the detector, analyzes it, and returns a pass/fail decision. See the ADR schematic representation in Figure 9.32.

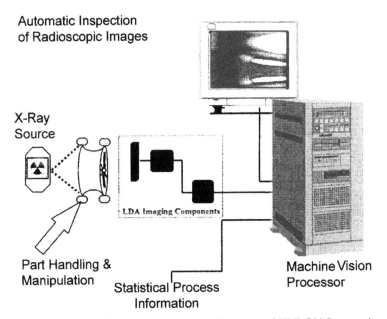

Figure 9.32 Automatic defect recognition. Courtesy of XYLON International.

Golden Image Technique. Golden Image Software's final pass/fail decision depends on the results of a comparison of an acquired production image to a stored golden image of the same part. The golden image database is established by acquiring and using the images of acceptable production parts.

The following steps outline ADR operation using the golden image technique:

1. A product code that defines a set of scan parameters for a specific product is entered by an operator or determined by an automatic product identification system.
2. The system receives and acknowledges the scan parameters, then requests and receives an ID to identify the image.
3. The system sends the scan parameters to the panel/LDA and starts the acquisition.
4. The panel/LDA scans the image into the system's image buffer.
5. The acquired image is compared to the image stored in the *golden image archive*.
6. The system requests and receives image results for alignment, feature extract, and feature classification.
7. The image results are returned to the operator.

Acquiring a Golden Image. Acquiring a golden image is referred to as *training a part*. The process usually requires only one revolution of an acceptable part. Simple point-and-click drawing tools, such as box and paintbrush, are used in the training process along with an "undo" function. The operator acquires the image of a good part and defines the start and stop locations on the image with a box tool. Then the operator defines the regions of noninterest with paintbrush tools. Finally, the ROIs are defined and the reject level for each defect type is set. Reject levels for each defect type are indicated by different colors on the processor display. Selectable defects include cavity, gas, inclusion, and sponge. After the variables have been set, the golden image is saved in the golden image database.

The Inspection Process. Proprietary software algorithms capture an image of the part and align the image to a golden image. The part is compared to the golden image at each ROI for each inspection type and level. Differences are extracted and the accept/reject decision is based on the extraction results. Flawed parts are identified, and the ADR system provides a pass or fail signal. Common defects are detected as manufacturers are able to define various levels of defects. Quantitative measurements of defect size and density can be used to provide process feedback.

9.23.4 Neural Network Artificial Intelligence (AI)

The AI ADR system is designed based on a qualitative image model. The model assumes that the image is defect free and is composed from areas of relative smooth density separated by narrow regions of rapid variations (edges). Edges are not sharp structures in X-ray imagery because of the penetration and divergent rays of the central projection. In the edge regions, noise is high. Defects are described as local deviations from the smooth intensity given by the continuity of the surface. Gas inclusions or less dense material give rise to local increases in intensity as fewer X-rays are absorbed. Prior to the development of digital detectors, which have a high dynamic range, defects could not be repeatably detected near the edge regions due to the high relative noise.

The analysis involved in the identification of defects in complex X-ray imagery requires the following:

1. *Edge detection.* The actual location of edges provides information about the orientation of the object, and is used for the trained neural network to estimate the location of the ROIs that the given geometry contains. Further, a segmented edge image is used to mask the edge regions in order to eliminate false detections.
2. *Regions-of-interest network.* The polygonal ROIs are determined with the edge image as an input; the model is imposing and trained by example. Examples are derived by operators drawing typical configurations in the teach-in phase. The ROI network outputs a mask so that areas outside the ROIs are masked out along with areas where an edge may enter an ROI.
3. *Adoptive reference subtraction.* Within each ROI a fixed-weight neural network estimates a smooth average intensity surface and subtracts the surface model from the actual noisy surface. The network operates with a single sensitivity parameter that is calibrated in the teach mode. The parameter determines the flexibility of the adoptive surface.
4. *Region-of-interest masking.* The ROI originating from the neural network is used to mask out noncritical areas. Regions of noninterest (RON) can be specified in ROIs to minimize false reject sources.
5. *Defect identification.* The final imaging process step is to convert the noisy residual reading from the subtraction in step 3 into a two-level image, which is high on the defects and low on the noisy background. A novel nonlinear filter based on the so-called Hopfield-Tank neural network is employed to perform this critical task. The algorithm is unique in combining both local smoothing as a low-pass filter and sensitivity to small defects as a high-pass filter.
6. *Quality estimation.* The ROI specific configuration of defects is quantified and fed to the final classification system. The classification is based on look-up tables or user standards.

Note: Prior to the development of digital detectors, edges produced unwanted noise due to low dynamic range that is inherent in image intensifiers.

9.23.5 Rule Base Using Specific Algorithms

Rule Base/Specific Algorithms Method. The rule base software system is designed to accept/reject from known standards or algorithms. Test programs are loaded either automatically by the test sample recognition module or by the operator using a standard computer and mouse to point and click. The advantage of the automated test sample recognition module is that sorting of different test samples is not required. This allows the system to be fed with all production samples in random order.

9.23.5.1 Operating Sequence
1. Enter or recall a test routine.
2. Adjust the test position.
3. Adjust the proper X-ray technique and ADR parameters. Integration, contrast, sensitivity, and image segmentation is adjusted for every X-ray image so that reference defects are recognized true to size and location.
4. Enter the maximum allowable defect size and allowable cluster parameters.
5. Regular image structures are self-taught by the system by running multiple test samples through the system.

9.23.6 ADR Advances of a PC Platform Over Proprietary Hardware

The open architecture of the PC platform makes many ADR packages affordable, easy to maintain, and easy to upgrade. With Microsoft Windows® NT as the operating system and point-and-click graphical user interface (GUI), the operator interface is user friendly and setup time is lower. Images can be stored to the system hard drive, CDROM, or network drive.

9.23.7 ADR Techniques

On the average, false accepts are <0.1% while false rejects are <3 percent. However, normal process variations may sometimes appear to be defects as shown in Figure 9.33. It is statistically improbable to find very tiny or marginal flaws, but never reject good product. This is true for human as well machine inspection.

9.23.8 SADR

Semiautomatic defect recognition (SADR) fills the gap between visual inspection and ADR. SADR tests a smaller quantity of parts, so it is more economical when compared to ADR. It's also faster than visual inspection. Flaws and anomalies are indicated to the user, which speeds up the testing process. Due

THE DIGITOME® PROCESS

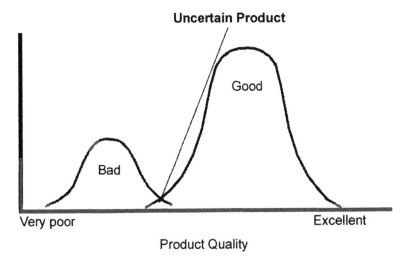

Figure 9.33 Uncertainties in product testing. Courtesy of XYLON International.

to the nonlinear diffusion algorithm used by the system, teaching or training the system is not necessary. Most of the regular structures are recognized. An additional use of white or black lists will minimize pseudos.

9.23.9 Conclusions

ADR techniques offer manufacturers a significant competitive edge through labor savings, reduced liability, higher-quality product, and improved process control. ADR is currently being used in semiprecious metals, cast aluminum wheel, general-purpose castings, and tire industries with excellent results. As technology advances, so will ADR.

9.24 THE DIGITOME® PROCESS

The *Digitome® X-ray imaging process* combines the data from multiple, flat X-ray images of an object to reconstruct volumetric representations of the object (digital tomosynthesis). Both horizontal and vertical image planes can be reconstructed to analyze or measure internal details. The equipment allows the user to examine and precisely measure internal features throughout an inspection volume. The information made available by a Digitome exam provides a full volumetric examination after the collection of only a small number of input images, features that separate Digitome from computed tomography (CT). In addition the Digitome image collection and reconstruction are vastly

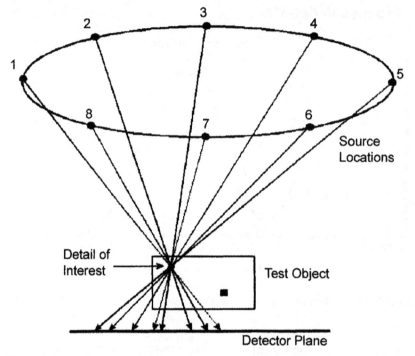

Figure 9.34 Schematic showing exposure geometry for Digitome system. Courtesy of Digitome Corporation.

different and provide many advantages over CT. The Digitome imaging process is protected by U.S. and Japanese patents. The recent development of high-resolution digital X-ray detectors has enabled the Digitome process to collect and analyze the necessary data for a volumetric analysis quickly and efficiently. The complete examination of test objects is often completed in only a few minutes.

9.24.1 Examination Concept

Many forms of inspection geometry can provide the multiangle views used in the Digitome process. Figure 9.34 illustrates the mapping of a detail of interest on the detector plane with differences in the source position. The variation in exposure geometry can be achieved by using multiple sources at different locations, moving a single source in 45° increments (or any angle) as illustrated, or rotating the test object to be examined.

A prototype fixture used to rotate the test object in a typical X-ray laboratory is shown in Figure 9.35. Examination of the test object is achieved by placing the test object on the turntable and positioning an X-ray source at an

THE DIGITOME® PROCESS 395

Figure 9.35 Prototype rotational device for Digitome system. Courtesy of Digitome Corporation.

angle. After each exposure the turntable is rotated 45° to position the test object for the next exposure. This produces an exposure geometry equivalent to that shown in the schematic. Methods for Digitome field inspections are also available.

9.24.2 Digital Flat Panel Detector

There are several types of flat electronic X-ray detectors currently available, including amorphous silicon (a-Si), amorphous selenium (a-Se), and complimentary metal oxide silicon (CMOS) arrays. These detectors are commonly offered in pixel sizes ranging from 80 to 400 microns and in sizes that can range up to several feet. As an example, the a-Si panels offer:

- 127 micron pitch capable of >3.8 lp/mm when measured at the panel
- Range of sizes from 4″ × 4″ to 12″ × 16″

The digital detectors typically have the following advantages:

- Wide dynamic range, typically 4096 grayscale shades generation (12-bit data)
- Little or no blooming—edges can be inspected with the same resolution as the rest of the part
- No geometric distortion
- Faster image acquisition times

9.24.3 Image Acquisition

Digitome® software is an easy to use, fully automated image acquisition system that collects multiview X-ray images. The digital flat panel detector assures short exposure times for each image and fast downloads of images to the operator console for processing. Powerful image correction methods are used to prepare the highest-quality images for reconstruction of planar views. The Digitome software puts the power of this unique process at the operator's fingertips.

9.24.4 Flaw Location and Measurement

Any horizontal or vertical layer of the inspected part can be selected and viewed with the click of a mouse. Inspectors can continuously scan through a section of the part and selectively reconstruct any layer for detailed inspection. Image enhancement methods can be applied and the processed image can be archived or printed for subsequent comparison or review.

Real-time scans are very useful for locating the position of flaws; then a horizontal view of the flaw can be reconstructed. Vertical and horizontal views can be used to determine flaw (or detail) dimension with high accuracy. In one application, small weld pores and other internal details have been clearly shown in TIG welded aluminum plates, with greater contrast than is possible with conventional radiographs.

9.24.5 Other Applications

In addition to precise reconstruction and location of internal flaws, as needed in most X-ray inspections, the Digitome system can be used to reliably inspect printed circuit ball grid arrays (BGA) on double-sided boards to determine solder ball height and quality. In like manner, the planar thickness of most multilayered electronic structures and mechanical devices can be measured for quality control purposes.

9.25 MANUFACTURING PROCESSES AND DISCONTINUITIES

The three major metallurgical processes of concern to the manufacturer with regard to quality control are the casting process, wrought processes, and welding processes. Flaws and discontinuities can start out in the casting process, progress through different manufacturing stages such as extrusion or machining, and end up as a defect in the final product. Manufacturers would like to detect flaws or discontinuities as early as possible so that defective parts can be rejected before additional manufacturing costs are incurred. Table 9.14 lists some defects that may be encountered by the radiographer. Burnthrough, root-pass, aligned porosity, and tungsten inclusion defects are shown in Figure 9.36.

9.26 OTHER ISOTOPE APPLICATIONS

9.26.1 Electron Capture Detection

With this technique, a low-energy beta source in an ion chamber establishes equilibrium current with a stream of pure argon gas. This current is amplified and displayed on an output meter. When a gas with a high infinity for electrons enters the ion chamber, ion chamber current decreases. The amount of current decrease is proportional to electron affinity. Beta sources for electron capture detection include 10 mCi Ni-63, 500 mCi tritium, and 5 mCi Fe-55.

Gas eluded from a gas chromatograph column can be fed into the electron capture detector for qualitative as well as quantitative analysis. The electron capture detector is unaffected by the presence of other gases that do not have an electron affinity.

Applications include the determination of sulfur hexafluoride in accelerators, hydrogen in air, carbon tetrachloride in air, solvent fume analysis in degreasing operations, gas leaks, and explosive detection. The electron capture detector is capable of detecting 0.00001 ppm sulfur hexafluoride in air and 0.001 ppm nitro compounds in air.

9.26.2 Moisture Gauging

Fast neutrons emitted by a neutron source are moderated by collision with hydrogen atoms in the moisture contained in the material of interest. A boron trifluoride (BF_3) proportional counter detects the moderated (deflected and slowed down) neutrons. The number of moderated neutrons is dependent on the concentration of hydrogen ions or moisture content of the sample. Fast neutron sources include 30–250 mCi combination sources of Am-241/Be and 54 µCi of Cf-252.

TABLE 9.14. Defects and Appearances

Defect	Cause or Appearance	Radiographic Appearance
Casting Defects		
Core shift	Mold movement during casting operation	Uneven internal wall thickness
Gas porosity	Gas released during solidification or moisture	Smooth dark round or oval spots
Cold shuts	Imperfect fusion between two molten streams	Distinct dark line or band with smooth outline and variable length and width
Inclusions	Sand, slag, and other foreign objects	Irregular light or dark spots in castings depending on density. Light or dark streaks in forgings or extrusions.
Misrun	Failure to fill mold cavity	Appearance of large voids.
Shrinkage porosity	Improper pouring temperature or wrong alloy composition	Same as gas porosity
Shrinkage cavities	Inadequate crucible feeding during casting	Moderately large irregularly shaped voids
Hot cracks	Fracture of metal in a hot plastic state	Parallel fissures
Hot tears	Restricted contraction prior to solidification	Dark ragged lines of variable width with numerous branches
Welding Defects		
Mismatch (hi–lo) SMAW[a]	Misalignment of pieces to be welded	Abrupt change in film density across weld width image
Offset w/lack of penetration (LOP) SMAW[a]	Misalignment and insufficient filling at bottom of weld	Abrupt density change across weld width. Straight dark line at center near density change.
External concavity or insufficient fill. SMAW[a]	Insufficient filling	Weld density darker than pieces across full width of weld
Excessive penetration (icicles, drop-thru) SMAW[a]	Hot weld with extra metal at root of weld	High density in center of weld image; extended along weld or in isolated circular drops
External undercut SMAW[a]	Gouging out of top edge of piece to be welded	Irregular dark density along edge of weld image
Internal (root) undercut SMAW[a]	Gouging out of parent metal along bottom edge of weld	Irregular dark density streak near center of weld width

TABLE 9.14. Continued

Defect	Cause or Appearance	Radiographic Appearance
Welding Defects		
Internal concavity (suck back) GTAW[c]–SMAW[a]	A depression in the center of the root pass	Elongated darker density area with fuzzy edges
Burnthrough SMAW[a]	A deep depression at the bottom of the weld	Localized darker density with fuzzy edges at center of weld
Lack of penetration (LOP) SMAW[a]	Bottom edges not welded together	Dark density band with very straight parallel edges at center of weld image
Slag inclusions SMAW[a]	Nonmetallic impurities solidified on weld surfaces	Irregular dark density spots
Elongated slag lines (wagon tracks) SMAW[a]	Solidified impurities on bottom surfaces	Elongated parallel dark density lines of varying width in longitudinal direction
Lack of side wall fusion (LOF) GMAW[c]	Elongated voids between weld beads	Elongated single or parallel dark density lines with dispersed darker density spots
Interpass cold lap GMAW[c]	Lack of fusion along top surface and edge of lower passes	Smaller spots of darker density, some with elongated tails
Scattered porosity SMAW[a]	Rounded voids of random size and location	Rounded spots of darker density, random size and location
Cluster Porosity SMAW[a]	Rounded or elongated closely grouped voids	Darker density spots in randomly spaced clusters
Root pass aligned porosity GMAW[c]	Rounded or elongated voids in bottom center of weld	Round or elongated dark density spots in a straight line at bottom center of weld image
Transverse crack GMAW[c]–GTAW[b]	Weld metal fracture across the weld	Feathery twisting line of darker density running across weld image
Longitudinal crack GMAW[c]–SMAW[a]	Longitudinal fracture of the weld metal	Feathery twisting line of darker density running length-wise along weld image
Longitudinal root crack SMAW[a]	Fracture in weld metal at edge of root pass	Feathery twisting line at edge of root pass
Tungsten inclusions GTAW[b]	Random bits of tungsten fused into weld metal	Irregularly shaped light spots randomly located in weld image

[a] SMAW = shielded metal arc welding.
[b] GTAW = gas tungsten arc welding.
[c] GMAW = gas metal arc welding.
Source: Based on information supplied by E. I. du Pont NDT Systems.

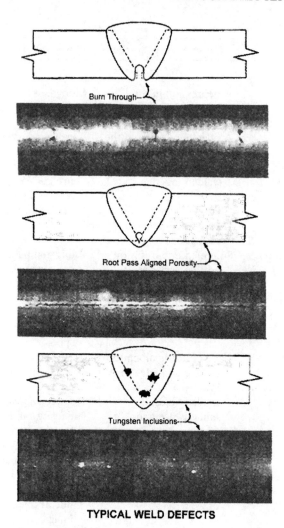

Figure 9.36 Three common weld defects. Courtesy of DuPont NTD Systems.

Applications for this type of moisture gauge include the measurement of moisture in soil, materials stored in silos, and the online monitoring of moisture in coke, cement powders, gravel, wood chips, and other building materials.

9.26.3 Bone Density

This technique uses direct transmission of a collimated beam of low-energy gamma radiation from a point source. The beam is transmitted through body tissue and bone where it is attenuated by the mineral content of the bone. The amount of radiation absorption determines the bone mineral content of the part of the body being examined. Simultaneous measurement at two different photo energies permits evaluation of complex bone structures such as those found in the hip or spine areas. Suitable single sources are Ie-125, Gd-153, and Am-241. Combination sources are Ie-125/Am-241 and Am-241/Cs-137.

9.26.4 Gamma and Beta Thickness Gauging

For transmission thickness gauging, the source and detector are placed on the opposite sides of the sample to be measured. As long as the density of the material is uniform, the amount of radiation transmitted will be directly related to the thickness of the sample. With gamma thickness gauging, a fan-shaped radiation beam is detected by a long ion chamber. Volumetric flow is determined by keeping material on the belt at a constant depth and cross-sectional area and measuring belt speed with a digital tachometer. Accuracy of ±1% is practical.

Other applications for gamma gauging include the thickness measurement of sheet metal, glass, plastic, and rubber where the thickness is too great for beta testing. Steel thickness up to 100 mm can be measured depending on source selection and source activity. Sources for gamma transmission thickness measurement include Am-241, Ba-133, Cs-137, and Co-60 with activities ranging from 1 to 1000 mCi depending on the application.

Beta gauging is used to measure the thickness of thin materials, textiles and paper, weighing of cigarettes, and the measurement of dust and other pollutants on filter paper samples. The useful measurement range is 1–1000 mg/cm^2. Sources for beta transmission measurements include Pm-147, Kr-85, Tl-204 and combination sources Sr-90/Y-90 and Ru-106/Rh-106 with activities ranging from 1 to 1000 mCi depending on the application.

9.26.5 Gamma and Beta Backscatter Thickness Gauging

With gamma backscatter, the source and detector are located on the same side of samples having a low mean atomic number (Z). The intensity of the backscattered radiation from the sample is a measure of sample thickness or mean atomic number. In cases where the sample thickness is known, the technique is called "Z gauging." Gamma backscatter is used in applications where transmission measurements are not adequately sensitive. A sodium iodide (NaI) crystal or proportional detectors may be used.

Gamma backscatter is used to measure the thickness of light alloys, glass, plastics, rubbers, and tube walls in applications where beta sources are not suitable. The thickness range varies from 1 to 30 mm depending on source activity and material properties. Beta backscatter, typically detected by G-M counters, is used to measure the thickness of paper and thin coatings of plastic and rubber on steel rolls. Sources of Am-241 and Cs-137, with activities of 50 to 100 mCi, are used for gamma backscatter techniques.

Beta backscatter can be used to measure almost any thin coating as long as there are significant differences in the atomic numbers of the coating and substrate. Coating thickness ranges of 1 to 100 μm are possible depending on source activity and material characteristics. Promethium-147, Kr-85, Tl-204, and combination sources of Sr-90/Y-90 and Ru-106/Rh-106 are used for beta backscatter measurements. Source activities of 1 to 5 mCi are typical depending on the application.

9.26.6 Gamma Level Gauging

The transmission of gamma energy through a tank, vessel, or container is dependent on the level of the material inside the container. Containers can be as small as food container cans or as large as power company fly ash hoppers. The energy transmitted is detected with G-M tubes in point level systems and long ion chambers in continuous level applications. With fly ash hopper systems, one source can be used with two hoppers and two detectors (Figure 9.37) to indicate a high level in either hopper. In this case, the source has two beams of radiation, 180° apart, directed through two different hoppers. Geiger-Müller tubes are used to activate alarms, SCRs, relays, switches, pumps, or other devices when a preset count rate is reached.

With the source beam collimated in a fan shape and directed through a vessel, one source and two detectors can be used for both high-level and low-level indication and alarm. The useful range of measurements is 2 in. to 40 ft, depending on source activity and vessel contents. For small can content monitoring, a 100 mCi Am-241 source is used. For larger containers, Ba-133, Cs-137, and Co-60 sources with activities to several curies are used.

9.26.7 Gamma Density Measurement

When a pipe section is kept full of liquid during normal operation, the density of material in the pipe can be easily determined by gamma energy absorption. The TN 3680 Smart Density Transmitter is shown in Figure 9.38. This instrument combines an improved scintillation detector with the power of a "smart" transmitter, operating on Hart® Protocol, in an explosion-proof housing.

The system simply mounts around an unmodified process pipe having adequate clearance. When mounted, the centerlines of the source, process pipe, and detector are automatically aligned. The amount of gamma energy that passes through the pipe is inversely proportional to the density of the material within the pipe. The scintillation-based detector produces photons of light when exposed to the gamma beam. The photons are amplified by a photomultiplier tube and the number of pulses produced is directly related to the intensity of the gamma energy received. The pulses are conditioned, counted, and scaled by the transmitter's on-board microprocessor and converted to a process density reading.

OTHER ISOTOPE APPLICATIONS

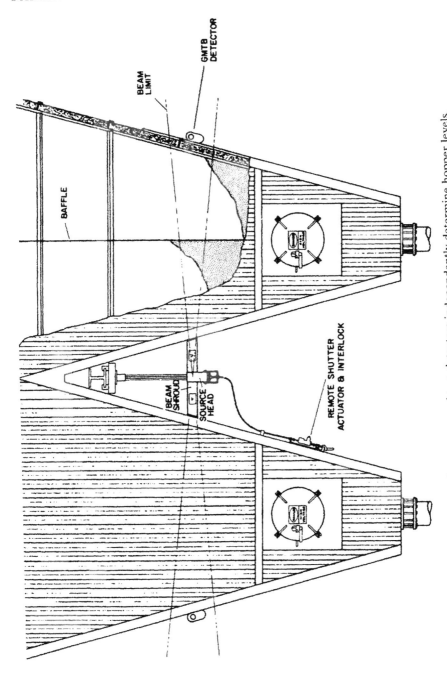

Figure 9.37 One source and two detectors independently determine hopper levels.

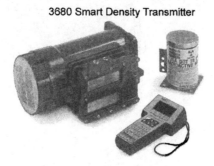

Figure 9.38 Courtesy of Thermo Electron Corporation.

Figure 9.39 Courtesy of Thermal Electron Corporation.

Communication with a Hart communication interface is accomplished by a separate, intrinsically safe terminal or via a 4–20 mA current loop. The Hart protocol uses 1200 baud frequency shift keying (FSK) based on the Bell 202 standard to superimpose digital information on a conventional 4–20 mA current loop. The overall effect is to combine calibration and communication functions into a single Hart documenting process calibrator, which simplifies technician calibration.

9.26.8 Point Level Switch

The TN LevelPro, shown in Figure 9.39, is a cost-effective point level switch for on/off control of circuits. It is mounted external to tanks, bins, hoppers, pipes, and other vessels to sense the levels of liquids, slurries, and solids. A relay contact output is used to control high or low level or operate alarms that signal level changes.

OTHER ISOTOPE APPLICATIONS

The LevelPro consists of a steel-jacketed detector containing the electronics unit and a shielded gamma source that emits a narrow beam of radiation, which passes through the vessel walls to the detector. When process material is in the beam path, the gamma energy is attenuated. When the process material is not in the beam path, the maximum energy is detected. Each energy pulse received by the detector is counted. Typically, a low count rate indicates a high level and a high count rate indicates a low level.

9.26.8.1 Features and Benefits

Features and benefits of the gamma point level switch include:

- Noncontacting, noninvasive high, low, or normal level indication.
- Highly reliable digital technology in a rugged package
- Simple installation with easy setup and adjustable response time
- Unaffected by pressure, temperature, viscosity, and density
- FM approval for hazardous areas
- Source shutter lockout for safe installation, shutdowns, shipping, etc.

9.26.9 Oil Well Logging

This technique is based on the principle that γ-ray scattering is a function of the bulk density of the ground formation being irradiated and studied. The number of γ-rays deflected from the source back to the shielded detector in the well logging tool is a measure of bulk density or porosity of the formation being studied. Oil filling the ground formation porosity, would show an increase in bulk density. Other minerals of interest would also increase or decrease the bulk density of the formation based on their specific gravity. Cesium-137 sources of 150 mCi to 2 Ci are used for γ-ray density logging.

10

THERMAL/INFRARED TESTING METHOD

10.1 BASIC MODES OF HEAT TRANSFER

Heat is a form of energy that can be converted to work. The most common units of heat are the calorie and British thermal unit (Btu). The three basic modes of heat transfer are *conduction, convection*, and *radiation*. Conduction refers to the transfer of energy between adjacent molecules in solids or liquids at rest. It is usually thought of as a very slow process in the case of gases. However, highly conductive metals transfer heat by molecular contact very rapidly. Electrical heating elements are good examples. The rate of heat flow in a rod of material is proportional to the cross-sectional area of the rod and to the temperature difference between the ends and inversely proportional to the length as shown in Eq. 10.1.

$$H = -k(A/l)(T_2 - T_1) \qquad (10.1)$$

where H equals heat flow
A is the cross-sectional area of the rod
l is the length of the rod
$(T_2 - T_1)$ is the temperature difference
k is the thermal conductivity of the metal

Introduction to Nondestructive Testing: A Training Guide, Second Edition, by Paul E. Mix
Copyright © 2005 John Wiley & Sons, Inc.

The minus sign indicates that heat flows from the higher to lower temperature. Materials with high thermal conductivities are high heat conductors.

The transfer of heat by convection for liquids or gases in a state of motion is thought of as a rapid process since it combines thermal conduction with fluid flow. A heat exchanger would be a good example of convection heat transfer. Finally, radiation transfers heat in the form of electromagnetic waves or photons. We can both see a campfire and feel its heat on a cold winter day. Infrared imaging systems detect only radiated heat in the form of electromagnetic energy or photons.

10.2 THE NATURE OF HEAT FLOW

10.2.1 Exothermic and Endothermic Reactions

Large amounts of heat are emitted by exothermic reactions and absorbed by endothermic reactions. Examples of exothermic and endothermic reactions are given below.

10.2.1.1 Exothermic Reactions   *Steel Mills.* Deoxidation carried out in a steel mill ladle is exothermic and raises the temperature of the liquid steel, but the liquid steel loses heat by radiation at the top surface of the ladle and by heat transfer through the ladle lining and shell. Just holding the steel in the ladle can also lower the temperature of the steel as much as 2°C per minute.

Molybdenum Processing. Multiple hearth furnaces are used to convert MoS_2 into molybdic oxide for commercial uses. Molybdenite concentrate is fed into the top of the first furnace against a current of heater air and gases blown from the bottom. Each hearth has four air-cooled arms rotated by an air-cooled shaft. The arms rake the material from the sides to the center of the roaster where the material drops to the next hearth. In the first hearth the concentrate is preheated and flotation reagents ignite, initiating the exothermic reaction that transforms MoS_2 to MoO_3. The exothermic reaction continues and intensifies in the following hearths. Temperature is controlled by the adjustment of oxygen and water sprays to cool the furnaces as necessary.

Thorium Reduction. The Spedding process is used to reduce powdered thorium tetrafluoride (ThF_4) to thorium metal. Powdered ThF_4, finely divided calcium, and a zinc halide are placed in a sealed refractory-lined "bomb." After heating to 650°C, an exothermic reaction occurs that reduces the thorium and zinc to metal and produces a slag of calcium halide. After solidification, the zinc thorium alloy is heated to the point where the zinc evaporates, leaving a highly purified sponge of thorium that can be melted and cast into ingots.

Alkali Metals and Water Reactions. Alkali metals all react violently with water. A small cube of sodium reacts violently with water, forming sodium hydroxide and releasing large amounts of heat and hydrogen gas that can react with oxygen, further increasing the heat. Alkali metals can be burned in atmospheres of various halogens to form corresponding halides. These reactions are also highly exothermic, producing up to 235 kcal/mole for lithium fluoride.

Industrial Polymerization. Addition polymerization and bulk polymerization reactions are also exothermic. On a small scale they seldom present a problem, but on a large scale they can cause autoacceleration, which causes the polymerization reaction to accelerate at explosive rates unless there are efficient means of heat dissipation and cooling.

Bulk Polymerization. Bulk polymerization occurs without solvent or dispersant and is the simplest reaction in terms of formulation. It is used for most step-grown polymers and many types of chain-growth polymers. In the case of chain-growth reactions, which are usually exothermic, the heat evolved may cause the reaction to become too vigorous and difficult to control unless efficient cooling coils are installed in the reactor vessel. Bulk polymerizations are difficult to stir because of the high viscosity of high-molecular-weight polymers.

10.2.1.2 Endothermic Reactions Many endothermic reactions are relatively easy to control. However, condensation polymerization requires careful temperature control and heat dissipation.

Condensation Polymerization. Condensation polymerization is an endothermic process that requires an addition of significant heat from an external source. The chemical reactor used must supply the necessary heat to maintain a satisfactory production rate of polymer. Reactor design also needs to take into account the need for removal or recycling of solvents and catalysts.

Mixing of Solutions. Most simple molecules mix with a small endothermic heat of solution. Endothermic heats of mixing typically show that the volume of the mixed liquids increases slightly after mixing, whereas exothermic heats of mixing tend to contract the volume of the mixed liquids slightly.

Calcination. Calcination, in which solid material is heated to drive off either carbon dioxide or chemically combined water, is an endothermic reaction.

Bottom-Blown Furnaces. Another, less common, oxygen steelmaking system is a bottom-blown process known as the Q-Q BOP (quick-quiet BOP). Using this steelmaking system, oxygen is injected with lime through nozzles, or tuyeres, located in the bottom of the vessel. The tuyeres consist of two concentric tubes: Oxygen and lime are introduced through the inner tube, and a hydrocarbon such as natural gas is injected through the outer annulus. The endothermic (heat-absorbing) decomposition of the hydrocarbon near the molten bath cools the tuyeres and protects the adjacent refractory.

10.3 TEMPERATURE MEASUREMENT

Temperature is one of the most important measured properties found in science and engineering. It affects all aspects of our life including health, sports, and medicine as well as every service, manufacturing, chemical, petrochemical, and process control industry.

The thermodynamic or Kelvin scale of temperature used in SI has its zero point at absolute zero and has a fixed point at the triple point of water. The triple point of water is the temperature and pressure where ice, liquid water, and water vapor are in equilibrium. This temperature is defined as 273.16°Kelvin. The triple point is sometimes defined as 0°centigrade, 32°Fahrenheit, and 273.15°Kelvin.

The centigrade or Celsius temperature scale is derived from the Kelvin scale. The worldwide scientific community has adopted the Celsius scale of temperature measurement. Equation 10.2 is used to convert Celsius temperature to Fahrenheit temperature.

$$°F = (9/5 \times °C) + 32 \qquad (10.2)$$

English-speaking countries favored the Fahrenheit temperature scale until about 1970. Daniel Gabriel Fahrenheit determined the freezing point of water as the temperature of an equal ice-salt mixture. His original scale was modified a couple of times, resulting in the final determination of freezing point of water as 32°F, and normal human body temperature as 98.6°F. Equation 10.3 is used to convert Fahrenheit temperature to Celsius temperature.

$$°C = (5/9 \times °F) - 32 \qquad (10.3)$$

10.4 COMMON TEMPERATURE MEASUREMENTS

10.4.1 Melting Point Indicators

Melting point indicators are available as pellets, crayons, lacquers, and labels.

Pellets. Pellets were one of the first forms of melting point indicators. They are still used and preferred in applications where a lot of indicating material is needed because observations must be made from a distance or heating periods are long and oxidation can mask other temperature indicators. Pellets are frequently used to determine the temperature of air spaces or heat zones in industrial furnaces and ovens.

Flat pellets are typically 7/16 in. in diameter and 1/8-in. thick. Smaller pellets, measuring 1/8 in. diameter by 1/8 in. thick, can be used for thermal switches or circuit breakers when they melt. When the pellet melts it releases a spring, allowing the circuit breaker to open, thereby cutting off the electrical supply. Pellets come in an extended range of 100°F to 3000°F for temperature measurement in hydrogen, carbon monoxide, and reducing atmospheres. Check manufacturer for special series temperature ranges.

Crayons. These temperature-indicating devices are heat-sensitive fusible materials that consist of crystalline solids that turn to a liquid when the melting point is reached. They are relatively independent of ambient temperature and a good indication of surface temperature. They are also economical and easy to use.

Crayons are temperature-sensitive sticks that have calibrated melting points. They are typically available in 100 different temperature ratings, ranging from 100°F to 2500°F with an accuracy of ±1% of its temperature rating.

The workpiece to be tested is marked with the crayon and heated until the crayon melts, indicating the proper temperature has been reached. In cases where the workpiece takes a long time to reach the desired temperature or the surface is highly polished and does not mark well or the surface absorbs the liquid crayon, a different technique must be used. In these cases the inspector must repeatedly stroke the workpiece with the crayon, observing when the liquid state is reached, or use another technique.

Lacquers. A dull lacquer-type liquid that turns glossy and transparent at a predetermined temperature is another highly effective low-tech temperature-indicating device. This phase-changing liquid can be used on more surfaces than the crayon technique. This liquid material contains a solid material that has a calibrated melting point. The liquid that suspends the solids is volatile, inert, and nonflammable. When the solids melting point temperature is reached, the dull lacquer mark liquefies. Subsequent cooling leaves a shiny crystalline coating indicating the proper temperature was reached.

Lacquer-type liquid applications include glass, smooth plastic surfaces, and even paper and cloth surfaces. Response time is a fraction of a second or even milliseconds when thin coatings are applied. The number of melting point temperatures available, overall melting point temperature range, and accuracy of the liquid system is the same as for temperature melting crayons. Temperature-sensitive lacquers are supplied in the proper consistency for surface brushing. If spraying or dipping of the lacquer is advantageous to the user, special thinners are added to alter the viscosity. These thinners do not alter the melting point temperature of the solids.

Labels. Adhesive-backed temperature labels are also available. In this case, one or more temperature sensors are sealed in a transparent heat-resistant window. The centers of the indicators turn from white to black at the temperature shown on the label face. The color change is irreversible and is caused by the temperature-sensitive material being absorbed by the backing material. After recording the temperature change, the label can be removed and attached to the item's service report as part of its permanent record.

10.5 COLOR CHANGE THERMOMETRY

10.5.1 Irreversible Color Change Indicators

Self-Adhesive Strips and Indicators. Thermographic Measurements Co. (TMC) Ltd. of England is the world leader in color change thermometry and they offer a wide range of irreversible, reversible, and quasireversible color changing strips and indicators for the measurement of surface temperatures. Liquid Crystal Resources, Inc. distributes their products.

Self-adhesive standard indicators and strips are available in the following combinations:

- Ten-level strips in 4 temperature ranges
- Eight-level strips in 5 temperature ranges
- Six-level ministrips in 8 temperature ranges
- Five-level strips in 10 temperature ranges
- Four-level microlabels in 9 temperature ranges
- Three-level strips in 16 temperature ranges
- Single-level encapsulated indicators in 49 spot temperatures

Color Change Paints. Irreversible paint packs provide a nonreversible, accurate method of measuring temperature at 41 standard change points for temperatures of 104–1270°C (219–2318°F). The advantages of these irreversible paints are:

- Primer is not required for most surfaces.
- Reaction time is very fast.
- Excellent performance in normal and hostile environments.
- Ideal for conducting thermal surveys.
- Strong and clear color contrasts.
- Resistant to oil, steam, and water.
- Require no additives.

Single Color Change Paints. Single change paints are colorfast in direct sunlight and resistant to chemical abuse. These paints provide useful aids for the industrial, chemical, and petrochemical industries. Applications include monitoring lined-vessel paint and detecting rises in surface temperature when refractory linings have broken down. These paints cover the range of 135–630°C (275–1160°F) with an accuracy of ±5% and are available in 250 ml, 500 ml, 1 L, and 2.5 L containers. Quasireversible paints are available in 48°C and 80°C temperatures only. These paints revert back to their original color when water is applied.

Multichange Paints. Multichange thermal paints withstand hostile environments and are used in gas turbine, jet engine, rocketry, and high-speed flight applications. These are the only paints known to retain their surface adhesion at these temperatures, thereby offering a high degree of protection from erosion. These paints cover the range of 104–1270°C (219–2318°F) with an accuracy of ±5%. These paints are also available in 250 ml, 500 ml, 1 L, and 2.5 L containers.

Autoclave Ink. This ink has a very distinct color change under steam autoclave conditions. The ink can be airbrushed onto paper, cloth labels, plastic, glass, or metal. Applications include the processing of fruit, meat, and vegetables. The color change occurs after 15 to 30 minutes' exposure to saturated steam at 116–127°C (241–261°F).

Color Change Crayons. When applied to a preheated surface, these crayons change color within 1 or 2 seconds when the temperature is within ±5% of the crayon's temperature rating. An instantaneous color change indicates the object's temperature is above the crayon's temperature rating. A color change after 2 seconds indicates the object's temperature is below the crayon's temperature rating. Color change crayons are used in the rubber, textile, plastics, chemical, and electrical industries. They are also used for zinc foundries, aluminum processing, glass industry, iron and steel industry, enameling applications, welding, and fabricating.

- Light-gray crayons turn violet blue at 120°C.
- Pink crayons turn blue-violet at 195°C and gray at 295°C.
- Pale-blue crayons turn light green at 215°C and buff white at 305°C.
- Light-purple crayons turn bright blue at 225°C and gray at 320°C.
- Orange-brown crayons turn black at 245°C and light gray at 335°C.
- Yellow-brown crayons turn red-brown at 300°C.
- Dark-violet crayons turn light violet at 320°C and buff white at 460°C.
- Aqua-green crayons turn buff white at 360°C.
- Red crayons turn white at 470°C.
- Apple-green crayons turn white at 600°C.

Special Color Change Applications. These applications include *firewatch glove indicators* that can be attached to a fireman's glove. The indicator color changes to black when the temperature exceeds recommended firefighter safety levels. There are also *firewatch indicators* that can be placed on hotel doors or spyholes in doors, *fire ladder indicators* that show when aluminum ladders have been exposed to temperatures above 300°F, and battery indicators that ensure the casing is bonded to the internal components. Battery temperature can be measured at 210, 219, or 300°F. Firewatch indicators are shown in Figure 10.1.

10.5.2 Thermochromic Liquid Crystal Indicators

Reversible Color Change Indicators. Thermochromic liquid crystals (TLC) were developed for space and defense applications in the 1960s. This in turn led to the development of color change thermometers and many other products. Accuracy of these devices is ±1°C on any standard product and ±0.5°C on any medical product. Standard temperature ranges are −30 to +90°C (−22 to +194°F) with color bandwidths of 1–10°C (2–18°F).

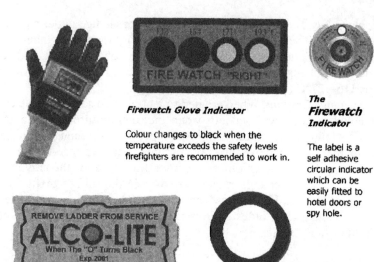

Figure 10.1 Temperature-sensitive indicators. Courtesy of Liquid Crystal Resources, Inc.

Applications for TLC liquid crystal thermometers include medical, educational, industrial, promotional, and domestic sectors. For example, TLCs can be used to measure and evaluate electronic chip and component temperatures to determine when they are operating within their design temperature range or overheating. Procedures involve careful selection of TLC temperature range, preparing the surface of the device, providing good illumination, applying power to the device, photographing color changes with a digital camera for documentation, and making a permanent record for future use.

Labels. Reversible predetermined color change labels include room, refrigerator, and freezer strips. Labels are available in a 16-level range, 7- or 13-level horizontal, and 7- to 9-level vertical strips.

Tapes. A number of predefined tapes are also available. These include reversible twin tape, hand hot indicators, reversible indicators, reversible tape-strip, reversible strips, and humidity/temperature indicators.

Liquid Crystal Inks. These inks are available as water-based screen inks and water-based sprayable inks that are usable from −30 to +120°. Screen Printers use screen ink formulations for promotional products. Inks are ready to use and cost effective for large-area applications. The liquid temperature profile can be produced to suit the customer's needs. *Material safety data sheets (MSDSs)* are available upon request.

Chromazone$^{(tm)}$ Heat Reactive Ink. These inks in printed form are colored below a specific temperature and change to colorless or another lighter color as they are heated through a defined temperature range. The activation temperature is defined as being the temperature above which the ink has completely changed to its final clear or light color end point. The color will start to fade at approximately 4°C below the activation temperature and will be in between colors within the activation temperature range. The ink is reversible and will return to its original color upon cooling.

Thermochromic flex ink is available in water-based and UV formulations. *Thermochromic screen ink* is available in solvent-based, plastisol-based, UV, and epoxy formulations. *Thermographic offset ink* is available in lithographic, dry offset, letterpress, and UV cure wet and dry offset processes.

Applications for heat reactive inks include document security to deter counterfeiting and its use with promotional items such as interactive advertising in brochures or flyers or for use on coffee mugs. Other applications include temperature indicating labels for foods and beverages to ensure that the product is fit for consumption and hasn't been overheated, and novelty items used for mouse pads, pens, T-shirts, etc. typically aimed at children.

Note: Water-based inks must not come in contact with any solvents.

SolarZone$^{(tm)}$ Ink—Photochromic Plastisol. Photochromic inks change color when exposed to UV light. These inks are colorless indoors and turn into vibrant colors outdoors. The ink becomes intensely colored after 15 seconds in direct sunlight and returns to clear after about 5 minutes indoors. On overcast days about 80% of the UV light still gets through the clouds, changing the ink color significantly.

Thermochromic ink colors are black, blue, red, orange, green, magenta, purple, and burgundy. Photographic ink colors are charcoal, red, blue, green, purple, yellow, orange, magenta, brown and teal. Standard available temperature ranges for both types of ink are –10 to 69°C, 29°C, 31°C and 47°C.

10.5.3 Liquid in Glass Thermometers

Liquid in a sealed capillary glass tube expands and contracts with temperature. The capillary space above the liquid is filled with nitrogen to prevent the liquid column separation during temperature fluctuations and high temperatures. The expansion of liquid contained in the thermometer bulb raises the liquid in the stem. The temperature is read on a scale that is an integral part of the thermometer or mounted next to the thermometer. Bulb and stem temperature differences can be a source of error, particularly in the case of total immersion thermometers. Modern LIGTs feature high accuracy and precision. Most industrial thermometers are graduated in 1 or 2° increments and range in length from 305 to 405 mm (12 to 15.94 in.).

For most home and garden use, shorter red liquid (alcohol) filled thermometers, are commonly used. The American Society for Testing and Materials (ASTM) standardized high-accuracy precision thermometers, measuring to 0.1°C or 0.2°F, are available for some limited-range applications. For most scientific and educational uses, mercury thermometers are preferred because of mercury's lower freezing point (−38.9°C).

Design criteria for LIGTs include low vapor pressure, wide temperature range, linear temperature expansion, large expansion coefficient, and non-wetting properties. Factors affecting the accuracy of LIGTs are the uniformity, concentricity, and smoothness of bore diameter and the temperature coefficient of expansion of glass.

Bimetal dial thermometers utilize the fact that two different metals can be selected with well-known but widely differing thermal expansion and contraction characteristics. The dial gauge itself is usually 1 or 2 in. in diameter and its stem length is typically 5 or 8 in. long. Dial divisions can be 1, 2, or 5°F depending on range. Rated accuracy is typically ±1% of total scale range. Industrial dial thermometers usually have 304 stainless-steel construction and water proof glass or plastic windows. There may be a small hex nut at the base of the head for minor calibration adjustments on some models. Varying stem lengths are available as specials.

10.6 TEMPERATURE SENSORS WITH EXTERNAL READOUTS

10.6.1 Thermocouple Sensors

In 1921, Thomas Seebeck discovered that when two different metals were joined at both ends, an electrical current was generated when one of the junctions was held at a constant temperature and the other junction was heated or cooled. The effect became known as the *Seebeck effect* and the bimetal thermocouple was born. A modern-day alloy sorter operating on the Seebeck effect is described later in this chapter.

Four thermocouples discussed below are the Type T copper/constantan thermocouple, Type J iron/constantan thermocouple, Type K chromel/alumel thermocouple, and high-temperature Type C tungsten–5% rhenium/tungsten–26% rhenium thermocouple. In each case the first wire named is the positive wire and the second wire is the negative wire. The alloys and metals used in the thermocouple are chosen to produce large Seebeck coefficients that result in high sensitivity for the intended operating temperature range. Thermocouple extension wires are usually made of the same material as the thermocouple to avoid the creation of additional undesirable thermocouple junctions. Cold junction compensation circuits are incorporated within the TC instrumentation to provide a 0°C reference junction.

The composition of the thermocouple wires and their insulation materials determine the maximum temperatures that can be measured. The maximum temperature for a copper/constantan thermocouple is 400°C compared to 750°C for iron/constantan, 1370°C for chromel/alumel, and 2315°C for a tungsten–5% rhenium/tungsten–26% rhenium thermocouple.

For a temperature of 100°C (0°C Ref.), the absolute output of a copper/constantan thermocouple is 4.28 mV compared to 5.27 mV for an iron/constantan thermocouple and 4.10 mV for chromel/alumel thermocouple. The sensitivity of all thermocouples is nonlinear. Calibration tables are provided in the *CRC Handbook of Chemistry and Physics* and manufacturer's publications, such as Pyromation, Inc.'s *Publication EMF-96*. The National Voluntary Laboratory Accreditation Program (NLAP) certifies calibration laboratories whose criteria meet the requirements of *ISO/IEC Guide 25* and relevant requirements of ISO 9002 (ANSI/ASQC Q92-1987).

While thermocouples themselves are not electronic measuring devices, they are frequently connected to instruments that provide a cold junction reference, voltage amplification, and digital readouts. Figure 10.2 shows a protected thermocouple in threaded fitting, thermocouple connector, thermocouple extension cable, and digital readout with cold junction temperature compensation.

The main advantages of thermocouples are that they can be used at higher temperatures than RTDs and are inexpensive to manufacture. The main disadvantages of thermocouples are that extension wires must be made of the same material as the thermocouple, they are not as accurate as RTDs or interchangeable, and a cold junction temperature reference must be provided. The *limit of error* for a Type C thermocouple is ±4.5°C at 400°C and ±1.0% for higher temperatures.

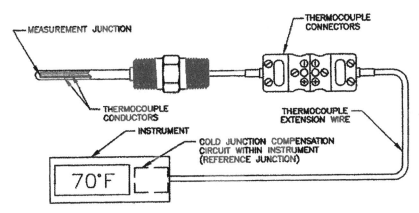

Figure 10.2 Thermocouple sensor construction. Courtesy of Pyromation, Inc.

10.6.2 Special Thermocouple Products

Specialty products for some thermocouple applications include:

- Protection tubes for molten aluminum
- Abrasion-resistant thermocouples and RTDs for rubber compound mixing applications
- Wear-resistant thermocouples for asphalt mixing and abrasive granular mixing processes

10.6.3 Resistance Temperature Devices (RTDs)

RTDs are electrical resistors whose resistance changes as a function of temperature. The sensing element is usually a coil of wire or a grid of conductive film that has a conductor pattern to it. Figure 10.3 shows both types of construction. Platinum is the metal of choice because of its purity, stability, and reproducibility. The resistance of an industrial platinum RTD is a 100 ohm resistor at 0°C. Extension wires are connected to the RTD sensor so that the temperature can be measured at some distance away. RTDs and thermocouples are typically insulated, enclosed in metallic sheaths, and isolated from the process they are measuring. A threaded metallic thermowell reaches the same temperature as the process at the measuring point. Figure 10.4 shows a protected RTD sensor, connector, instrument cable, and readout device.

10.6.3.1 RTD Sensing Elements and Typical Temperature Ranges
- Copper -100 to $300°F$ (-73.3 to $148.9°C$)
- Nickel/iron 32 to $400°F$ (0 to $204.4°C$)
- Nickel -150 to $600°F$ (-101.1 to $315.6°C$)
- Platinum -450 to $1200°F$ (-267.8 to $648.9°C$)

The temperature limits for RTDs are generally lower than those of thermocouples and the temperature limits for connecting wires for both RTDs and thermocouples must be as high or higher than the temperature limit of the measuring sensor.

The typical RTD temperature transmitter is a "two wire" loop powered resistance to current transducer. The transmitter produces a linearized 4- to 20-mA DC output current proportional to the temperature of the RTD temperature sensor. The transmitter is small and can be panel mounted or mounted in standard or explosion-proof thermostat housings. Ambient temperature range is -30 to $65°C$ (-22 to $149°F$).

RTDs are used in applications where repeatability and accuracy are the most important factors. Platinum RTDs have a very repeatable resistance versus temperature curve that is not significantly affected by time. Because of their high repeatability RTDs are interchangeable and do not require

TEMPERATURE SENSORS WITH EXTERNAL READOUTS

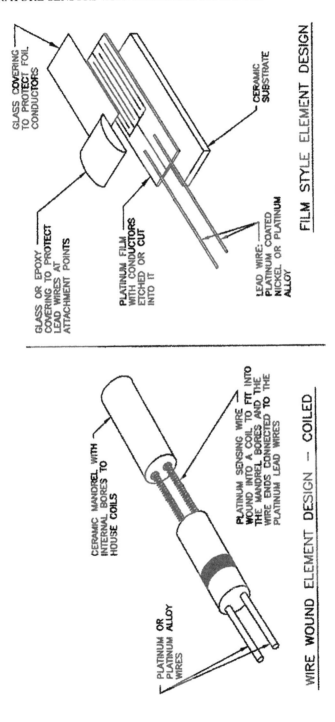

Figure 10.3 RTD element construction. Courtesy of Pyromation, Inc.

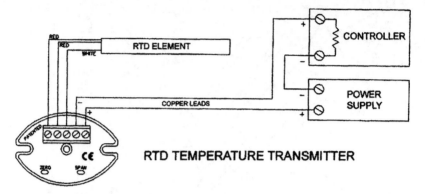

Figure 10.4 RTD temperature transmitter. Threaded Thermowell not shown. Courtesy of Pyromation, Inc.

individual calibration for more precise temperature measurements. Unlike thermocouples, RTDs have a specified resistance tolerance at a specified temperature. Another advantage of RTDs is that the connecting leads between the RTD and its instrumentation can be made from standard instrumentation cable, as long as the operating temperature requirement is met.

One disadvantage of RTDs is that they are expensive, costing 4 to 10 times more than a base metal thermocouple. RTD manufacturing costs are higher in order to achieve the accuracy, repeatability, and interchangeability. Another disadvantage of RTDs is that they are more susceptible to damage from high vibration and mechanical shock that can cause open circuits or short circuits. Thermocouples are more rugged and fabrication is simple—twisting two wires together and fusing the junction forms the basic thermocouple.

10.6.4 Resistance Temperature Elements (RTEs)

These platinum resistance elements are found in gas turbine propulsion plants. As temperatures increase, the resistance of the RTE increases. All RTEs have a platinum element and a resistance of 100 ohms at 0°C (32°F). Probe temperature ranges from −60 to +1000°F (−51.1 to +537.8°C). Probe length range from 2 to 10 inches. Most RTE values are read directly as ohmic values by propulsion electronics. Some RTEs are mounted in thermowells and read by remote modules that produce a 4- to 20-mA current output.

10.7 INFRARED IMAGING ENERGY

According to *Kirchoff's law,* energy radiated by an object equals the energy absorbed when the object is in equilibrium. Therefore, when infrared energy (heat) strikes an object, the total amount of infrared energy is transmitted, reflected, or absorbed by the object.

Emissivity is defined as the relative power of a surface to emit heat by radiation; it is the ratio of the radiant energy emitted by a surface of a body to that emitted by a blackbody at the same temperature. *Absorptivity* is the fraction of total incident radiation that is absorbed by the body. *Reflectivity* is the fraction of the total incident radiant energy that is reflected from a body. *Transmissivity* is the fraction of the total incident radiant energy striking the surface of a body that is transmitted from the body.

10.8 HEAT AND LIGHT CONCEPTS

While visible light (photons) and infrared (radiated heat) are neighbors on the electromagnetic spectrum as shown in Figure 10.5, there are significant differences in their heat absorption and reflection characteristics. For example, visible light transmission through the atmosphere is excellent. On a clear day we can see for miles and on a clear night we can see the moon, stars, and planets, which appear to be stars that don't blink. However, heat or IR transmission is significantly attenuated or effectively blocked by carbon dioxide (CO_2) and water (H_2O) at more than one wavelength. Therefore, IR analyzers are very effective in detecting and quantifying the amount of these gases in so-called anhydrous gases at specific wavelengths where these molecules are absorbed.

Clear plastics and plastics that appear to be semitransparent to the eye are also heavy absorbers of IR at several wavelengths. Smooth glass plate that appears transparent to the eye may appear as an IR reflector to IR imaging devices. Materials, such as metals, which are good electrical conductors, are usually good thermal conductors as well. Materials such as glass and quartz are poor electrical and thermal conductors, but once heated may tend to hold heat for a relatively long time. Glass has about 60 to 70% transmission for a near-IR wavelength range of 1.8 to 2.7 µm. Man-made sapphire windows are frequently used with sample cells for IR process analyzers because of their improved IR transmission characteristics.

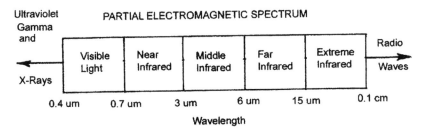

Figure 10.5 Partial electromagnetic spectrum covering VIS–IR range.

10.9 PYROMETERS

Pyrometer (fire measuring) devices read temperatures by measuring the intensity of IR radiation emitted from an object. All objects above absolute zero temperature (0°K) radiate and absorb thermal energy. Narrow-band pyrometers typically operate in accordance with Planck's law and broad-band pyrometers operate in accordance with the Stefan-Boltzmann law. Noncontacting pyrometers have a broad price range with models incorporating blackbody correction commanding the highest prices.

10.9.1 Error Correction

The radiation intensity of a perfect broad-band blackbody irradiator can be summarized by Eq. (10.4):

$$W_b = \sigma \cdot T^4 \tag{10.4}$$

where σ = Stefan-Boltzmann constant, 5.6697×10^{-8} W/m²·K
T = absolute temperature (K)

Note: Transmissivity = 0, reflectivity = 0, and emissivity = 1.0 or 100% and based solely on temperature.

Some common materials with an emissivity of 0.9 or greater are listed below:

- Asbestos board at 20°C = 0.96
- Asphalt paving at 4°C = 0.97
- Brick, rough red at 21°C = 0.93
- Cloth, black at 20°C = 0.98
- Formica at 27°C = 0.94
- Glass, smooth at 22°C = 0.94
- Ice, smooth at 0°C = 0.97
- Lampblack at 20 to 400°C = 0.95 to 0.97
- Marble, gray, polished at 20°C = 0.93
- Paper, dull black at 40 to 100°C = 0.96 to 0.98
- Steel, rough plane surface at 50°C = 0.95 to 0.98
- Tar paper at 20°C = 0.91 to 0.93

Different objects at the same temperature have different heat absorption, reflection, and emission characteristics. An object's heat emission characteristics may change in time due to oxidation, surface roughness, dirt, and other reasons. Radiant power intensity of a graybody can be calculated by the fol-

PYROMETERS

lowing equations, which apply to narrow-band and broad-band pyrometers as noted:

Planck's law:

$$W_a(\lambda, T_m) = \varepsilon\lambda \cdot 2\pi c^2 h/\lambda^5 \cdot [e^{(hc/k\lambda T)} - 1] \qquad (10.5)$$

(Used with narrow-band pyrometers)

Wein's radiation law:

$$W_a(\lambda, T_m) \approx \varepsilon\lambda \cdot 2\pi^2 h/\lambda^5 \cdot e^{(hc/k\lambda T)} \qquad (10.6)$$

(Simplification of Eq. 10.5)

Stefan-Boltzmann law:

$$W_a(T_m) = \varepsilon \cdot \sigma \cdot T^4 \qquad (10.7)$$

(Used with broad-band pyrometers)

where W_a = radiant intensity of the body
λ = wavelength in meters (m)
T_m = measured temperature (K)
$\varepsilon = \varepsilon\lambda$ = emittance = constant
h = Planck's constant, 6.625×10^{-34} J·s
c = the speed of light, 3×10^8 km
k = Boltzmann's constant, 1.380×10^{-23} J/K

Note: The geometry of the probe tip and detection angle of the sensor may also affect accuracy of the measurement and need to be included as an additional factor, f_g. Manufacturers of IR instruments and material suppliers have tabulated tables of emissivity factors.

10.9.2 Principles of Operation

10.9.2.1 Narrow-Band Optical Pyrometers Portable Laser IR Thermometer. The Mikron Quantum I Portable Laser IR thermometer is used for measuring, computing, and displaying target emissivity and temperature. This advanced noncontact IR thermometer uses patented laser technology to actually measure emissivity and determine true target temperature. A powerful built-in microcomputer and specialized software corrects and calculates the effects of reflected ambient radiance within a fraction of a second. Precision optics provides a large, clear view of the target area but with a very small, well-defined target zone. Accuracy is independent of target distance errors.

The Quantum I Portable Laser IR thermometer overcomes three common problems that many pyrometers have. These are:

1. Determination of the emissivity factor, which is a characteristic of the surface of the target. Emissivity can be measured or manually set and fed into the computer. It is then displayed in the viewfinder and stored for subsequent use until reset.
2. Ability to measure the ambient radiance, compute the reflected component, and make the necessary correction. When furnace walls are scanned, the computer averages and stores wall temperature for subsequent use. Procedure is repeated only if updated values are desired.
3. Selection of operating wavelengths in a very narrow spectrum where gas flames and flue gases in a furnace have little or no IR emission and do not interfere with the temperature measurement. Locate target through the viewfinder, focus, press trigger, and read target temperature in the viewfinder.

These hand-held portable IR thermometers measure and display true target temperature in a bright LED digital display in the viewfinder. The units, shown in Figure 10.6, have a rugged cast aluminum housing and include the following features:

- Instrument range of 350–3000°C
- ±3°C accuracy
- Readout in 400 ms
- Accuracy independent of distance from the target and ambient temperature
- Instant warmup
- Lightweight—easy to hold and aim
- Wide-angle field of view for easy target location
- National Institute of Standards and Technology (NIST) traceable

A simple block diagram of the Quantum I Portable Laser IR thermometer is shown in Figure 10.7.

10.9.2.2 Broad-Band Optical Pyrometers

Optical pyrometers measure the intensity of radiation in eight optical bands in the visible and near-IR spectral regions at high sampling rates. First the instantaneous temperature of the radiating material is determined by calibration procedures that relate the emissivity of the material to its temperature under specific plant conditions. In the selected time interval, the temperature of the material and its related errors are statistically calculated from the large number of measurements taken.

PYROMETERS 425

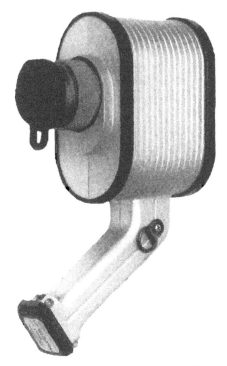

Figure 10.6 Quantum I Portable Laser thermometer. Courtesy of Mikron Instruments, Inc.

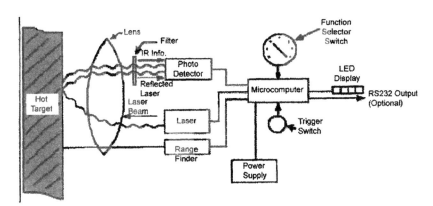

Figure 10.7 Simple block diagram of Quantum I Portable Laser thermometer. Courtesy of Mikron Instruments, Inc.

Information about radiation intensity in the eight spectral bands is used to eliminate disturbances that negatively effect the accuracy and reliability of the temperature measurement under industrial conditions. To provide this degree of compensation the internal computer of the pyrometer recognizes and analyzes the conditions that exist during the measurement, optimizes the measuring process, and makes temperature calculations based on actual plant conditions. The broad-band optical pyrometer circuitry compensates for:

- Dust or smoke between the pyrometer and the emitting surface
- Slag islands on molten surfaces
- Inadvertent obstacles in the field of view
- Actual environmental conditions at the time of measurement

It provides status codes to aid application evaluation when problems are noted.

10.9.3 Design and Operation of Optical Pyrometers

The optical pyrometer consists of an optical head and hermetically sealed case containing both optics and electronics. The optical head measures radiation from an emitting surface. The case is usually stainless steel with a sapphire window. Optical pyrometers have either a built-in optical system for visual positioning or a laser system for laser positioning of the pyrometer head for temperature measurement of the test object or surface. The optical head can also be connected to a pressurized source of air for instrument cooling and shielding of the sapphire window against dust. The optical head is connected to the main case by an optical cable that can be hermetically sealed and protected by an armored shield.

The main case of the pyrometer contains a special polychromator. The electronic unit is a programmable computer compatible processor with internal software for calculating the instantaneous temperature, the average temperature in the selected interval, and the experimental error. The algorithm for temperature calculation is optimized for specific plant conditions with regard to measured material characteristics, temperature range, time sampling, minimization of background disturbances, and formatting of output data.

Optical pyrometers have serial ports for connection to supervisory computers. In the standalone mode, the pyrometer can send the temperature measurement data stream to the main computer. The optical pyrometer also can be set up to interact with a supervisory computer, following commands coming from the main computer to make the final temperature evaluation and sending the status codes back to the main computer. Most optical pyrometers can be remotely calibrated.

10.9.4 Applications for Broad-Band Optical Pyrometers

Because of their broad-band approach, which enables optical pyrometers to minimize outside disturbances, they can provide reliable results in the following applications:

- Measurement of moving molten metals in steel mills, blast furnaces, and foundries
- Measurement of slag islands on the surface of molten metals
- Measurement of temperatures in hostile environments involving smoke, dust, obstacles, and intense light
- Measurement of temperatures inside plants having hostile environments such as heat emissions, dust, vibrations, and hot metal droppings

10.9.5 Installation of Optical Pyrometers

The installation of broad-band pyrometers must be optimized by modification of technical parameters for specific plant applications. Then a suitable location must be selected for installation. Where is the best place to mount the pyrometer based on the material being measured and the expected temperature range, taking into account the plant environment? Is special hardware required? Have all of the factors affecting calibration and optimization of the temperature measurement been considered? How does the customer's application and user demand affect time sequences? Will the optical pyrometer provide standalone information to a central computer, or will the central computer optimize the temperature measurement? Fine-tuning of the system is usually required and will change when there are application changes.

10.10 INFRARED IMAGING SYSTEMS

With infrared (IR) imaging, the camera or sensor's detector detects only radiated energy. However, conduction and convection are still important factors, which can alter or enhance the radiated energy emitted by the heat source.

Blackbodies are theoretical surfaces that absorb all incident heat radiation and have an emissivity factor of 1.0. The IR blackbody represents the perfect IR or thermal emitter. Its emissivity is 1 or 100 percent and its reflectivity and transmissivity is zero. Therefore blackbody sources are frequently used to calibrate IR sensors.

10.10.1 Blackbody Calibration Sources

There are a number of blackbody calibration sources ranging from an economical, portable single point temperature-controlled hot plate design to more sophisticated units with built-in RTD reference probe and 3-point NIST traceable calibration certificate. These units feature portable rugged designs.

The radiation reference target temperature range for the hot plate design is 212 to 752°F. Some target temperature ranges for other infrared blackbody calibrators are:

- Ambient +10 to 215°C (+20 to 420°F)
- Ambient +10 to 400°C (+20 to 752°F)
- 100 to 398°C (212 to 750°F)
- −19 to 149°C (0 to 300°F)
- 100 to 982°C (212 to 1800°F)

An object that is not a perfect emitter but has an emissivity that is independent of wavelength is known as a *graybody*. The emissivity of a graybody is the ratio of graybody photons to blackbody photons.

10.11 SPACIAL RESOLUTION CONCEPTS

10.11.1 FOV, IFOV, MIFOV, and GIFOV

- *Field of view (FOV)* is the angular extent of the full image.
- *Individual (detector) field of view (IFOV)*, sometimes referred to as pixel resolution, is the angular extent of an individual detector (a hardware specification).
- *Measured individual field of view (MIFOV)* is the projection of the detector IFOV at the specified measurement distance.
- *Ground individual field of view (GIFOV)* is the projection of the detector IFOV on the ground or "ground spot footprint" (geographic systems).

10.11.2 Angular Resolving Power

The human eye is similar to optical systems whose inherent resolving power is determined by their optics. The angular resolving power of an optical detector in radians is a function of receptor cell size (pixel size) and image distance. The instantaneous field of view (IFOV) is a relative measure because it is an angle and not a length. The IFOV for optical systems is calculated by Eq. 10.8.

$$\text{IFOV (rad or radian)} = L/r \tag{10.8}$$

where L = the receptor cell size
r = image distance

For the average human eye, the receptor cell (L) is 4 μm and the image distance within the eye (r) is 20,000 mm. Therefore IFOV of the eye is 4/20,000 or 0.0002 rad or 0.2 mrad.

10.11.3 Error Potential in Radiant Measurements

The greatest potential error in radiant measurement systems appears to be in the determination of the emissivity factor. To help minimize these errors, material fabricators and manufacturers of IR imaging systems have generated tables of emissivity factors to aid in calibration. Reflected radiation and radiation from secondary sources also provide additional sources of error. In these cases camera or detector positioning can help minimize these potential sources of error. Sources contributing to the conduction or convection of heat also need to be taken into account.

10.12 INFRARED TESTING METHOD

10.12.1 Preventive and Predictive Maintenance Programs

Exhaustive record-keeping programs on individual mechanical and electrical parts were formerly maintained in an effort to determine the practical life of individual parts as well as their associated assemblies. Under preventive maintenance programs parts were often replaced before any signs of failure were observed. Unexpected failures still occurred, some with very costly consequences. In some cases preventive maintenance programs also called for complete overhaul of equipment before failures occurred. This was very costly, but often less costly than a failure that could shut the plant or process down for a substantial period of time. Still, there was no way to predict when equipment failure might occur.

Most mechanical and electrical failures occur when excessive heat is generated by the component of interest for an undetermined period of time. The development of lightweight, hand-held, portable infrared imaging systems allows trained maintenance technicians to view complete assemblies and zoom in on suspected trouble areas. It is now possible to determine when excess heat is being generated, and in many cases the part or component can be changed before catastrophic failure occurs.

10.12.2 Electrical PdM Applications

Electrical predictive maintenance applications include:

- Hot transformers—possible winding problem or low transformer oil level
- Shorted or high-resistance motor windings
- Poor or corroded bolted bus bar connection

- Hot fuse connection caused by faulty fuse or corroded connection
- Hot circuit switch bolted connection—poor mechanical connection/corrosion
- Internal circuit breaker heating—possible faulty breaker
- Hot line jumper connection—poor mechanical connection
- High-resistance circuit board connection

10.12.3 Mechanical PdM Applications

- Hot motor or fan bearing
- Hot pulley drive belt
- Excessive tire temperature
- Excessive brake friction

10.13 HIGH-PERFORMANCE THERMAL IMAGER FOR PREDICTIVE MAINTENANCE

The newest entry in high-performance imaging for predictive maintenance programs is the Raytek® noncontact ThermoView™ Ti30 thermal imager. A front view of the instrument is shown in Figure 10.8. Shown in this view are, the top gain control switch (image contrast) on the left and level control switch (image brightness) on the right. The recessed switch-bay, which controls °F or °C temperature, LCD illumination, measurement mode, and laser on/off selections, is located immediately in front of these switches. The lens cover, immediately below the lens, slides down to turn on the unit and up to turn off the unit.

A rear view to the Ti30 instrument is shown in Figure 10.9. Shown in this view is the instrument's color display readout, and button controls: down button (parameter values) bottom left, mode button (cycle between operations) bottom center, and up button (parameter values) bottom right.

Not shown in either view are the docking station (recharging stand with data communications port), USB field connection port, and battery compartment (in handle). The following items are included with the ThermoView Ti30 thermal imager:

- Interactive user's manual (CDROM)
- InsideIR™ companion software
- Docking station/charger with universal power adapter
- Hardshell carrying case
- USB computer cable
- Rechargeable battery pack
- Nonrechargeable battery pack (batteries not included)

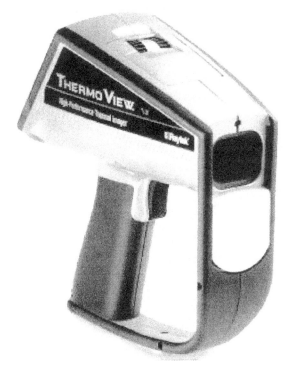

Figure 10.8 ThermoView Ti30 thermal imager. Courtesy of Raytek Corporation.

- Multilanguage training materials (CDROM)
- Carrying pouch
- Quick Reference Card

10.13.1 Predictive Maintenance Program

The goal of both preventive and predictive maintenance (PPM) programs is to minimize repair and labor costs, reduce parts inventory, determine product variation, and minimize production losses. The ThermoView Ti30 thermal imager displays clear, clean thermal images while automatically recording radiometric readings for establishing complete maintenance records. The built-in laser functions as a sighting aid to help pinpoint target details. A cross-hatched indication at the center of the image display is used to locate and identify high-temperature spots for recording purposes. The captured radiometric readings and related thermal images shown on the LCD screen can be used for future quantitative or qualitative reporting.

For record purposes, unique names and locations are assigned to equipment when making the initial inspection. Thermal images and data are then downloaded using InsideIR software to establish an initial equipment database and

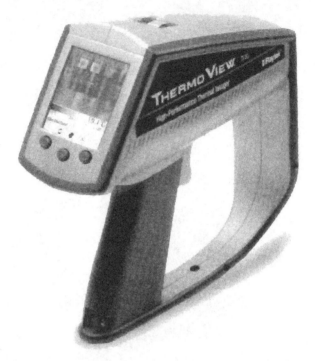

Figure 10.9 Rear view of ThermoView Ti30 thermal imager showing display area. Courtesy of Raytek Corporation.

inspection report. Operating temperatures can be compared to maximum recommended operating temperatures if known. Subsequent thermal imaging records can then be downloaded for comparison with previous records and established guidelines. When maximum operating temperatures are exceeded, corrective action can be taken.

10.13.2 Specifications

10.13.2.1 Thermal

Measurement range: 0 to 250°C (32 to 482°F)

Accuracy: ±2% or ±2°C, whichever is greater at calibration geometry and 25°C

Repeatability: ±1% or ±1°C, whichever is greater

Noise equivalent temperature difference (NETD): 250 mK

Temperature indication resolution: 0.1 (°F or °C)

Optical/IR

Spectral range: 7 to 14 microns

Target sighting: single laser dot (meets IEC Class 2 and FDA Class II requirements)

Optical resolution: 90:1

Minimum diameter of measurement spot: 7mm (0.27") at 60cm (24")

Image frame rate: 20 Hz

Field of view (FOV): 17° horizontal × 12.8° vertical

Instantaneous field of view (IFOV): 1.9 mrad

10.13.2.2 Controls

Focus: focusable from 61cm (24") to infinity

Temperature scale: °C or °F—selectable

Palettes: gray, ironbow, or rainbow—selectable

Measurement modes: automatic, semiautomatic, or manual—selectable

Laser on/off, gain, and level controls: same as above

LCD backlight: bright, dim, or off—selectable

10.13.2.3 Optional Features

Emissivity, adjustable: 0.10 to 1.00 in 0.01 steps (contact Raytek for details)

Liquid crystal display: thin film transistor (TFT) technology—optimized for indoor and outdoor use

Reflected background temperature: −50 to 460°C (−58 to 860°F)

Ambient operating temperature: 0 to 50°C (32 to 122°F)

Relative humidity: 10 to 90% noncondensing

Storage temperature: −25 to 70°C (−13 to 158°F) (without batteries)

Stage capacity: 100 images

Laser on, low-battery, palette, and measurement mode icons: available

Thermal analysis software: InsideIR (included)

PC software operating systems: Microsoft®, Windows® 98/2000/XP

10.13.2.4 Other

Weight (including batteries): 1 kg (2.2 lb)

10.14 HIGH-PERFORMANCE RADIOMETRIC IR SYSTEMS

10.14.1 Introduction

High-performance, real-time digital thermal video imaging systems are being applied in new applications almost daily. With the proper lens, applications range from microscopic to telescopic in nature. Typical applications include circuit design and testing, microcircuit analysis, arterial restriction research, engine design and development, detection of minute thermal changes, gas detection, ignition process development, HVAC control analysis and recording of rapidly changing thermal events, nondestructive testing (NDT) of composites and other advanced aerospace materials, and numerous military applications.

10.14.2 Applications

Examples of NDT using such powerful and sensitive InSb (indium antimonide) cooled midwave IR detection as embodied in the TVS-8500 are seen in aerospace applications ranging from imaging subsurface delamination of cracks in multilayer aerospace composites to detecting disbonded or corroded rivets along an airplane skin.

Some automotive applications are seen in developing more effective safety systems such as the thermal pattern on an air bag after inflation, or the extreme temperature variations across the surface of a brake rotor after multiple panic-stop testing.

In the process and manufacturing industries, applications include monitoring multicavity injection-molded temperature differences due to plugged cooling channels and imaging of bonded surface junctions for consistency. Thermal imaging can also determine the existence of excess moisture created in a process, which is especially useful when out-of-control processes result in excessive final drying operations and thousands of dollars being wasted.

Note: In a paper mill process, the first section or wet end of the process is pressed out by the second section rollers. If enough water is not pressed out in the roller section, then thousands of dollars in excess energy are wasted in the drying process to counteract the excess moisture left going into the dryers at the third section of the process.

Finally, in the research-and-development areas, the system provides unequalled performance when it comes to designing more effective cooling for laptop computers, comparing product design performance during environmental testing, and noncontacting thermal evaluation of component items within the whole design.

With regard to arterial research, the system is so sensitive that a thermal image of a bare arm is able to differentiate the blood coming from the heart via arteries (warmer) versus the blood returning to the heart via veins (cooler). The high-performance CMC Electronics Cincinnati TVS-8500 can even see changes in individual sweat pore temperature of the fingertips as they cool the human body by convection.

10.14.3 Theory of Operation

The heart of CMC Electronics Cincinnati's Model TVS-8500 thermal video system (TVS), Figure 10.10, starts with the closed-system Sterling-cycle cooled

Figure 10.10 CMC Electronics TVS-8500 thermal video system. Courtesy of CMC Electronics.

indium antimonide (InSb) photodiode focal plane array (FPA). The photodiode detector provides over 60,000 noncontacting individual IR detectors or pixels on a single silicon wafer, which corresponds to individual pixels viewed on the integral video screen or external display. Add one of four top-quality lenses to this thermal imaging system, and you have a combination that's hard to beat. Finally, user-friendly PE Professional thermographic analysis and reporting software tops off a great system.

The InSb focal plane array provides the highest level of performance available today in the midwave infrared (MWIR) range of 3–5 micrometers (μm). CMC Electronics Cincinnati utilizes a patented process focal plane array construction method that yields superior discrimination and far less cross talk than competitive systems. Photons from the object being inspected are detected by the individual photocells on the focal plane array to provide real-time full-view thermal images up to 120 hertz. The performance of the focal plane array drops off rapidly if the cryogenic temperature is not properly maintained. While liquid nitrogen is the most cost-effective way to cool these detectors, it is not practical for small, highly portable equipment.

The Sterling-cycle cooling engine provides a practical, if somewhat expensive, alternative. The Sterling cooler keeps the detector at 77°K; the temperature of liquid helium, where the semiconductor array is nonconductive. At this temperature, each photon that strikes the detector produces a measurable output. Costs are expected to come down as applications and sales increase. As listed below, the minimum temperature resolution for an individual photodiode is 0.020°C taken at 30°C, NEDT blackbody. However, an absolute change of this order of magnitude cannot be detected due to the unknown variance in emissivity of the object being inspected. Therefore, the accuracy range of ±2°C or ±2%, whichever is greater, defines the accuracy of the system.

As inspection surface temperature increases, the number of photons striking each detector increases, producing digital counts. As temperature goes up, digital counts increase. For each of the six available temperature ranges up to 900°C (2000°C option), 256 colors are possible for each detector, resulting in a full-view display of more than 16,000 levels (14 bits) of color possibilities.

Additional flexibility of the system is obtained with four precise lenses as described in Figure 10.11. The lenses are identified as the standard lens, wide-angle lens, closeup lens, and microscope lens. The focus distance for these focus lenses are 30 cm to infinity, 30 cm to infinity, approximately 77 mm, and approximately 27 mm, respectively. The field of view is 14.6°H × 13.5°V, 44.0°H × 40.5°V, 25.6 mm (H) × 23.6 mm (V), and 2.56 mm (H) × 2.36 mm (V), respectively. The TVS-8500 camera recognizes the lens being used and provides automatic calibration for the lenses via built-in look-up tables. When pointed at an object to be inspected, the camera automatically can be set to select the best temperature range for the detector based on the range of temperatures within the FOV. Figure 10.12, not to scale, shows the FOV as a function of distance for the standard lens. The individual photocell spatial resolution or pixel IFOV within the FOV is represented by the square in the center. Therefore, at the distance of 1.0 meter, the camera images an area 25.4 cm wide by 24 cm high. Each photocell in the focal plane array averages the temperatures contained within the IFOV, in this case a 1 mm square.

10.14.4 Operating Technique

This high-tech thermal imaging camera is about as automatic as it can be, thereby placing the burden of satisfactory operation on the nondestructive inspection (NDI) technician. It would be just as foolish to point the camera at a stationary airplane reflecting bright sunlight as it would be to point the camera at the sun and saturate the detector. Simple radiant heat from the ground or concrete on hot days may cause irrelevant moving bands of color and heat. The NDI technician must place the camera in the correct position, at the correct distance, to observe the expected defects of interest.

In most cases, the best procedure is to place the imager at a distance where the maximum desired IFOV size is small enough to detect the anticipated size of the defect. The defect could be a subsurface delamination within a composite; a crack under the surface finish of a fiberglass material, or voids within injection molded parts. Different lenses are available to achieve imaging of the object so that it fills the field of view and still achieves satisfactory imaging with good spatial resolution. The simple heating of the surface by a halogen lamp for a short period, or warm air for a longer period, will produce a thermal differential where such cracks, bubbles, or voids within the subsurface will appear.

Applicable Lens List for TVS-8500

Lens Type	NACL Part Number	Lens Name	Focal Length	Spatial Resolution	Field of View	Focus Distance	Depth of Field	Measurement Temp. Range	NEDT at 30C blackbody	Dimension (mm)
Standard Lens (f=30mm)	7402000-100	TVL-8530A	30mm	1mrad (0.27mm at 30cm)	(H)14.5°x(V)13.5°	30cm to infinity	±12mm at 30cm	-40 to 900C	20mK	φ75x70
Wide Angle Lens(x3) (f=10mm)	7000279-011	TVL-8510A	10mm	3mrad (0.87mm at 30cm)	(H)44.0°x(V)40.5°	30cm to infinity	±70mm at 30cm	-40 to 900C	approx.30mK	φ75x70
Close-up Lens	7000279-009	TVL-8510OU	25mm	100µm	(H)25.6mmx(V)23.6mm	approx.77mm	±2.2mm	-20 to 300C	approx.30mK	φ75x40
Microscope Lens	7000279-010	TVL-8510U	30mm	10µm	(H)2.56mmx(V)2.36mm	approx.27mm	±0.2mm	Room Temp. to 300C	approx.50mK	φ75x120

Notes : 1. The field of view indicated in the above list is correspondent with the effective pixels(H256xV236) of TVS-8500.
2. The focus distance indicated in the above list shows the distance between the front of lens and object.

Figure 10.11 Lens options for CMC TVS-8500 thermal video system. Courtesy of CMC Electronics.

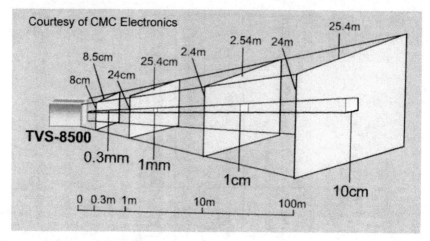

Figure 10.12 Field of view vs. distance with TVL-8530A standard lens. Courtesy of CMC Electronics Cincinnati.

10.14.5 Typical Specifications

The following list itemizes the TVS-8500 system specifications:

- Detector: InSb focal plane array (FPA), 256 × 256 pixels.
- Detector cooling method: Sterling-cycle cooling engine.
- Temperature measurement range: −40°C to 900°C (2000°C optional).
- Minimum temperature resolution: 0.020°C or less (30°C blackbody).
- Accuracy: ±2°C or ±2%, whichever is greater. Greatest variance is knowing the object's emissivity.
- Scanning speed: up to 1/120 fps image acquisition.
- Measuring wavelength: 3.4 to 4.1 μm and 4.5 μm to 5.1 μm (broad-band 2.5 μm to 5.0 μm available first Q 2003).
- FOV: 14.6 (H) × 13.5 (V) when used with standard lens.
- Measured distance: 30 cm to infinity.
- A/D conversion: 14 bit.
- Correction function: emissivity/reflectivity correction, auto-room-temperature correction.
- Display: 5-inch-high gain color matrix liquid crystal display (LCD).
- Functions: Multipoint temperature measurement, memory, zoom, and isotherm. Highest temperature display, display color selection, and averaging. Auto/manual temperature range sensing and switching.
- Fourteen-bit thermographic recording/play back: real-time recording at 120, 30, 10, 5, 2, or 1 frame per second. Compact flash card 100 frames using the supplied 16 MB card, 300 frames with a 48 MB card.

- Image output: National Television System Committee (NTSC), phase alternating line (PAL).
- External interface: RS-232C, specialized Institute of Electrical and Electronics Engineers (IEEE) 1394 Firewire® PC interface, remote trigger and synch terminals via 4 integral strip terminals with TTL inputs.
- Input power: 120 VAC/240 VAC, 50/60 Hz.
- Dimensions: 200 mm wide × 250 mm deep × 120 mm high (excluding protrusions).
- Weight: approximately 5.0 kg.

10.15 MIKRON INSTRUMENT COMPANY, INC.

The Mikron Instrument Company manufactures a complete line of IR imaging and temperature-measuring instruments. Included in their product line are lightweight, hand-held IR and VIS/IR cameras, fixed-mount IR thermal imagers, a single spot radiometric IR camera, High-Speed IR-line cameras, and software for managing images, analyzing images, and generating reports. The Mikron TJ-200 IR Man is a data storage and display system capable of producing a matrix of temperature distribution with 64 distinct areas superimposed on a visible image. The company makes blackbody radiation calibration sources covering the range from −20°C to 3000°C, traceable to NIST.

One unique product is the SpyGlass lens for the MikroScan 7200/7515 IR cameras. Figure 10.13 shows the camera/lens system. The SpyGlass lens permits thermal inspection of electrical switchgear without opening the enclosure door or disconnecting circuits. The camera can view the field of interest through a 5/8-in.-diameter hole in the cabinet. Other products include advanced portable IR thermometers with through-the-lens sighting, an IR thermometer with precision laser sighting, and process control instruments

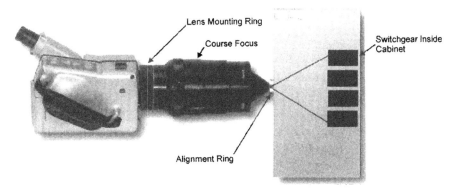

Figure 10.13 Mikron IR camera with SpyGlass lens. Courtesy of Mikron Instruments.

such as IR temperature sensors with TC or linear voltage output and IR temperature transmitters with 2-wire linear 4–20-mA outputs. One model of this transmitter has a laser sighting for pinpoint aiming accuracy.

Process instruments include a fiber-optic IR temperature transmitter for inaccessible or severe environmental applications, a 2-color Infraducer with 4–20-mA linear output that is independent of emissivity and unaffected by dust and other contaminants. Other process control instruments include a high-accuracy, high-resolution IR thermometer with exceptional software capabilities, miniature IR temperature sensors, and an IR thermocouple. Additional process control instruments are a smart IR fiber-optic temperature transmitter with wide temperature span and focusing optics, and a multi-channel fiber-optic Infraducer with wide temperature range and absolute accuracy of 0.2% of reading.

10.16 MIKRON 7200V THERMAL IMAGER AND VISIBLE LIGHT CAMERA

10.16.1 General Features

This camera is designed for comfortable one-hand point-and-shoot operation and combines on-board digital voice recording with 14-bit thermal and digital visual imaging. The camera is housed in a self-contained splash-proof metal case and is battery operated. Images and other data can be stored on PCMCIA cards. Images can also be viewed in real-time via the video output or through an optional built-in IEEE 1394 Firewire®. Figure 10.14 shows the MikroScan 7200V thermal imager/visible light camera.

10.16.2 Technical Data

10.16.2.1 Performance
- Temperature ranges: –40 to 120°C, 0°C to 500°C, and 200 to 2000°C
- Measurement accuracy: ±2% or 2°C of reading
- Field of view: 29°(H) × 22°(V)
- Focus range: 50 cm to infinity
- IFOV/spatial resolution: 320 × 240 uncooled focal plane array (UFPA) vanadium oxide (VOX) microbolometer
- Spectral band: 8.0 to 14.0 µm
- Atmospheric transmission correction: input correction by outside temperature, humidity, and measuring distance
- Emissivity setting: automatic based on operator input
- Alarm: upper or lower
- Image freeze: provided

MIKRON 7200V THERMAL IMAGER AND VISIBLE LIGHT CAMERA

Figure 10.14 MicroScan 7200V thermal imager/visible light camera. Courtesy of Mikron Instruments.

10.16.2.2 *Presentation*
- File format: 14 bit
- Digital visual recording: on-board
- Digital voice recording: on-board
- B&W/color: several palettes available
- Automatic gain control (AGC): automatic level, gain and focus
- Viewfinder: standard (color LCD optional)
- Video output: NTSC/PAL, S-video
- Image zoom: 2:1 (4:1 with spatial filtering)

10.16.2.3 *Measurement*
- NUC: flag correction by specifying the interval time. (Manual/auto selectable. Interval time setting available at auto.)
- ΔT Display: display temperature difference between points A and B
- Region-of-interest setting: display max/min temperature in operator-defined box
- Peak temperature hold: keep max/min temperature in an operator-defined box

- Isotherm: variable bandwidth. Multicolor for regions available
- Temperature span: automatic
- Temperature range setting: auto and manual
- Multispot temperature measurement: 10-point max with EMISS setting

10.16.2.4 Interface
- Communication: RS-232/C (computer control available)
- Memory card: provided—PCMCIA 16 mb
- Remote control operation: GPIB, RS-232C or LCD remote panel, IEEE 1394 Firewall® interface (optional)

Note: The MikroScan 7302 fixed-installation thermal imager and MikroScan 7515 single spot IR camera with thermal imaging software have image update rates of 30 frames/sec or 60 frames/sec.

10.17 HIGH-SPEED IR LINE CAMERAS

10.17.1 General Information—MikroLine Series 2128

These cameras are housed in rugged industrial housings and have air purges for lenses and water-cooling. They can be operated independently or with a PC and feature fiber-optic data transmission. There are no optomechanical scanners. These cameras feature uncooled IR linear arrays and 16-bit A/D conversion. There are 128 data points per line and a measuring rate of 128 lines per second. These units have a large dynamic range and provide for triggered inputs and alarmed outputs. Data recording and application-specific hardware and software are provided. There are seven models available in this series, which are well suited for fast noncontacting temperature measurements of production line processes.

10.17.2 High-Speed Temperature Measurement of Tires

The MikroLine 2250 is specially designed for the high-speed temperature measurement of tires. The camera has a rugged industrial housing, air purge for its lens system, and no operating elements. Data transfer is by fiber optics. The unit provides parallel measurement of 160 measuring points, using lead selenide (PbSe) sensors, and a measuring frequency of 18,000 lines/s.

The triggered measurement provides exact geometrical assignment of the individual measuring points and efficient online software controls thresholds and processes. Rapid temperature changes can be accumulated as a film, providing a 360° temperature image of a tire traveling at 100 mph. Figure 10.15 shows a line diagram of the temperature measurement and recording system.

HIGH-SPEED IR LINE CAMERAS

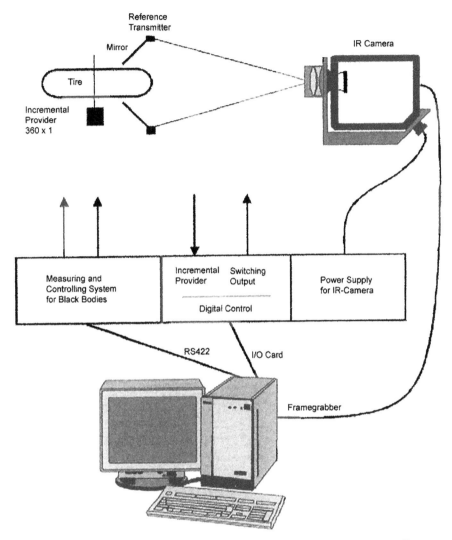

Figure 10.15 MicroLine 2250 camera for tire temperature measurements. Courtesy of Mikron Instruments.

10.17.2.1 Camera Specifications
- Spectral range: 3 to 5 µm
- Temperature range: 50 to 180°C
- Sensor: PbSe-160 element sensor with CMOS multiplexer
- Opening angle: 30° × 0.13°
- Measuring distance: 10 cm to infinity
- Spatial resolution: 3.3 mrad (50% modulation)

- Temperature resolution: 0.5 K at 50°C
- Temperature accuracy: ±2 K ±2% from measurement (°C)
- Line scan frequency: 2000 Hz up to 18 kHz
- Response time: approximately 1 second
- Warmup time: <30 minutes
- Interface: fiber-optic/PCI-PC card
- Camera housing: protective housing IP-65
- Operating temperature-camera: 0 to 40°C
- Storage temperature: −20 to 70°C

10.18 OTHER THERMAL TESTING METHODS

10.18.1 Fourier Transform Infrared Spectrometer

Design Features. The Shimadzu Prestige-21 Fourier transform infrared (FTIR) spectrophotometer features high sensitivity and accuracy with improved operability and expandability. This highly optimized instrument features a high-throughput optical system with gold-coated mirrors and temperature stabilized *deuterated L-alamine triglycine sulfate (DLATGS)* detector that results in a signal-to-noise (S/N) ratio of 40,000 to 1. The instrument can operate in the near-IR, mid-IR, and far-IR ranges with simple exchanges of the light source, beam splitter, and detector.

Other features include a patented precision control and stabilization of the optical bench that ensures reliable, reproducible results. Powerful IR Solution Software also provides a Windows®-based 32-bit control to enable operation to be performed quickly and easily using dedicated analysis screens.

Key Specifications. The Shimadzu Prestige 21 Fourier transform infrared spectrophotometer is shown in Figure 10.16. An IR spectrum is seen on the computer display. Some key specifications are shown below:

- Interferometer—30° incident angle Michelson interferometer, built-in *advanced dynamic alignment (ADA)* function, sealed with autodryer
- Optical system—single-beam optics
- Beam splitter—Ge-coated KBr (moisture-proof) for mid-IR (standard), Ge coated CsI (moisture-proof) for mid/far-IR (optional), Si coated CaF_2 for near-IR (optional)
- Light source—air-cooled high-intensity ceramic light source for mid/far-IR (standard), tungsten-iodine lamp for near-IR (optional)
- Detector—temperature-controlled DLATGS detector for mid/far-IR (standard), liquid-nitrogen-cooled mercury cadmium telluride (MCT) detector for mid-IR (optional), InGaAs detector for near-IR (optional)

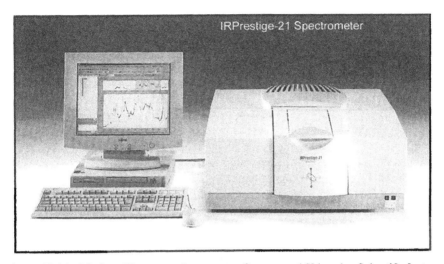

Figure 10.16 Modern IR spectrophotometer. Courtesy of Shimadzu Scientific Instruments, Inc.

- S/N—above 40,000:1 (4 cm^{-1} resolution, 1 minute accumulation, around 2100 cm^{-1}, P-P)
- Wave number range—7800 to 350 cm^{-1} for near- and far-IR ranges (standard)
- Resolution—0.5, 1, 2, 4, 8, 16 cm^{-1}
- Mirror speed—3 steps: 2.8 mm/s, 5 mm/s, 9 mm/s
- Data processing functions—arithmetic calculation, peakpick, spectrum subtraction, smoothing, baseline correction, data cut, data correction, normalization, Kubelka-Munk conversion, Kramers-Kronig analysis, ATR correction, Fourier transform, derivatives, transmittance/absorbance conversion, peak area integration, peak ratio calculation, emission correction, deconvolution, quantitation, multilinear regression quantitation, spectrum search, JCAMP conversion, ASCII conversion, logarithmic conversion, wavelength/wave number conversion, shift along X-axis, and optional programs
- Validation program—provided as standard (validation program conforming to the Japanese Pharmacopoeia/European Pharmacopoeia/ASTM), optional items for the Japanese/European Pharmacopoeia validation program
- Operating temperature—15 to 30°C

Note: JCAMP conversion refers to the JCAMP standard of spectroscopic data transfer. Existing JCAMP standards represent the first nonbinary approach, which is vendor independent (not owned by anyone), features printable characters only, and has reasonable compression rates. It is extendable and its open definitions provide for future improvements.

Horizontal ATR Operation. Thin film sampling techniques frequently use single reflection or *attenuated total reflection (ATR)* of the IR signal. With this technique, the sample can be in direct contact with a diamond prism bonded into a tungsten carbide support disc or germanium prism bonded into stainless-steel support disc version of a single reflection top plate. The sample is held in place with a quick-release bridge and self-leveling pressure anvil arrangement having a sapphire insert. The single reflection horizontal ATR system, shown in Figure 10.17, is well suited for the measurement of corrosive solutions, strongly absorbing polymers and rubbers, powders, thin films on semiconductors, macro and micro sample volumes, and forensic science samples, many of which are difficult to handle in a vertical cell.

Schematic Figure 10.18 shows how the FTIR beam is reflected by mirrors and passes through a beam-condensing lens into the ATR element fixed in the top plate where the sample is held in place. Beam-condensing optics is ZnSe or KRS-5, depending on application. The signal is reflected back from the sample through another condensing lens, mirrors, then on to the input of the FTIR spectrometer. Total reflection takes advantage of the refractive index of

Figure 10.17 Golden Gate single reflection diamond ATR System. Courtesy of Specac Limited.

OTHER THERMAL TESTING METHODS

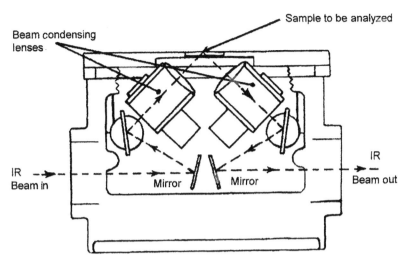

Figure 10.18 Internal view of Golden Gate single reflection ATR system. Courtesy of Specac Limited.

the prism to assure that all light entering the prism is reflected back, thereby providing 100% reflectance at the prism/sample surface. The ratio of total light reflected, when there is and is not a sample, provides a transmission-like ATR spectrum. With the ATR cell, flat samples can be placed on or painted on the top surface of the prism for measurement. Cameras may also be used with some ATR systems.

10.18.1.1 DLATGS Pyroelectric Detectors The main features of *d*euterated L-*a*lamine *t*riglycine *s*ulfate detectors include high detectivity, from 1 Hz to 4 kHz, and wide spectral range, from 0.1 to over 1000 microns. Applications for these detectors include gas analysis, radiometric measurements, high-speed spectroscopy (FTIR), laser modulation studies, and interferometry.

Window materials typically used with these detectors include:

- Sapphire (Al_2O_3)—0.15 to 6.0 microns
- Calcium fluoride, polycrystalline (CaF_2)—0.2 to 11.0 microns
- Coated germanium (Ge)—1.8 to 17.0 microns
- Potassium bromide (KBr)—0.25 to 25 microns
- Thallium bromide iodide (KRS-5)—0.55 to 40 microns
- Cesium iodide (CsI)—0.25 to 50 microns

10.18.1.2 FTIR Evaluation of Hard Disk Fluororesin Coating A fluororesin coating is applied to the surfaces of computer hard disks as a lubricant. *Reflectance absorbance spectroscopy (RAS)* is an effective means of measuring the thickness of fluororesin film coatings. A reflection IR device with a large angle of incidence is used to measure thin films on a metal plate. An RAS device with an angle of incidence of at least 70° is able to measure film thickness of 1.0 μm or less. Thinner films several nanometers thick can be measured using a combination RAS device with polarizer to measure the polarized component only.

A combination RAS device with polarizer was used to measure film coating thickness of 11.8, 25.4, 41.0, and 48.8 Å using the measurement conditions listed below:

- Resolution—8.0 cm^{-1}
- Accumulation—300 times
- Detector—DLATGS
- Mode—reflection

At a peak area near 1300 cm^{-1}, a linear calibration curve was obtained for the wavelength range of 1200 to 1350 cm^{-1} for a film coating thickness range of 10.0 to 50.0 Å.

10.18.1.3 Measurement of Film Thickness on a Silicon Wafer Film thickness can be calculated from the interference fringe spectrum obtained by FTIR using Eq. 10.9:

$$d = \Delta m / 2\sqrt{(n^2 - \sin^2 \theta)} \cdot 1/(1/v_1 - 1/v_2) \qquad (10.9)$$

where n = the refractive index
θ = the angle of incidence to the sample
Δm = the number of peaks or troughs in the calculated wavenumber region
v_1 and v_2 = are the maximum and minimum values of the wavenumber region

FTIR software calculates the equation in the quantitative analysis mode, displays the IR spectrum, and identifies wavelength range numbers with a vertical cursor. This FTIR nondestructive evaluation method provides a rapid and simple way to evaluate semiconductor materials.

10.18.2 Advanced Mercury Analyzer

10.18.2.1 Introduction While not a nondestructive test in the traditional sense of the word, LECO's AMA254 *advanced mercury analyzer* is a unique *identification of materials (IM)* thermal testing method (TTM). What makes it unique is its high-speed, high-accuracy determination of mercury in nominal sample sizes of 100 mg for solids and 100 μl for liquids without sample pretreatment for solids or preconcentration for liquids.

OTHER THERMAL TESTING METHODS

LECO's AMA254 advanced mercury analyzer (Figure 10.19) complies with EPA Method 7343 dealing with the measurement of *mercury in solids and solutions* by *thermal decomposition, amalgamation,* and *atomic absorption spectrometry*. It also complies with recently approved ASTM Method D-6722 as it pertains to efficient, automatic determination of *total mercury in coal and combustion residues*. Because it has been estimated that 60% of the total mercury deposited in the U.S. environment comes from airborne emissions, the EPA has proposed that coal-fired power plant emissions of mercury be released by 2003 and adopted as final rules by 2004. When fully implemented, the new regulations should reduce airborne mercury emissions by 50% compared with 1990. In the future, even more restrictive regulations are anticipated.

Mercury from power plants eventually makes its way into both fresh- and saltwater systems where it is converted by biological processes into methyl mercury. This highly toxic compound is readily absorbed by fish and eventually by human tissue when the fish are eaten. People who eat moderate or large amounts of fish and women of childbearing age are most at risk because mercury ingestion can cause both neurological and developmental damage.

The concentration of mercury in coal is very low, in the order of 0.1 part per million by weight (ppmw), but mercury is extremely volatile when coal is combusted. And after years of combusting millions of tons of coal, dangerous levels of mercury compounds have accumulated in some fish populations in the United States and elsewhere. Health officials frequently quarantine contaminated bodies of water, but the problem needs to be solved, not contained.

10.18.2.2 Theory of Operation Designed with a front-end combustion tube that is ideal for the decomposition of difficult matrices like coal, combustion residues, soils, and fish, the instrument's operation may be separated into three distinct phases for a given analysis.

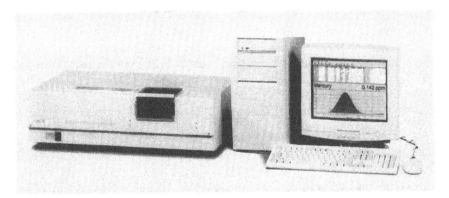

Figure 10.19 LECO AMA254 advance mercury analyzer. Courtesy of LECO Corp.

The first phase is the *decomposition* phase. During this phase, a sample container with a nominal amount of material is placed inside a prepacked combustion tube. The combustion tube is heated to 750°C through an external coil that provides the necessary decomposition of the sample into a gaseous form. The evolved gases are then transported via oxygen carrier gas to the other side of the combustion tube. This portion of the tube, which is prepacked with specific catalytic compounds represents the area in the instrument where all interfering impurities, such as ash, moisture, halogens, and minerals, are removed from the evolved gases.

Following decomposition, the cleaned, evolved gas is transported to the amalgamator for the *collection* phase of the system. The amalgamator, a small glass tube containing gold-plated ceramics, collects all the mercury in the vapor. With a strong affinity for mercury and a significantly lower temperature than the decomposition phase, the amalgamator traps all of the mercury for subsequent detection. When all the mercury has been collected from the evolved gases, the amalgamator is heated to 900°C, releasing the mercury vapor to the detection system.

The released mercury is transported to the final phase of analysis, the *detection* phase. During the detection phase, all vapor passes through two sections of an apparatus known as a curvette. The curvette is positioned in the path length of a standard atomic absorption spectrometer. This spectrometer uses an element-specific lamp that emits light at a wavelength of 253.7 nm, and a silicon UV diode detector for mercury quantification. The dual-path length of the curvette expands the dynamic range of the mercury from the subpart per billion (ppb) levels to the upper part per million (ppm) levels.

The advanced mercury analyzer houses an *autoloader*, which is a rotating carousel that holds up to 45 nickel-sample boats. The sample boats are automatically inserted in the combustion/catalyst tube. Each nickel sample boat can contain up to 500 milligrams (mg) of various samples.

10.18.2.3 Software Highly effective Windows-based operating software, known as Quicksilver, maximizes mercury analysis results, while streamlining the process. This user-friendly software provides easy-to-use icons, spreadsheet formats, and on-board manuals for improved user–machine interaction. The instrument and its standard were compared to *cold-vapor atomic absorption spectroscopy (CVAAS)*. Results obtained with the advanced mercury analyzer were more accurate and precise than those obtained with the CVAAS. In addition to the improved performance, the AMA254 reduced total analysis time to less than six minutes.

10.18.3 Identification of Materials

10.18.3.1 Thermoelectric Alloy Sorting Introduction. C. S. Taylor and J. S. Edwards appear to be the first Americans to report information on thermoelectric testing of aluminum-manganese and other alloys in the *Transactions*

OTHER THERMAL TESTING METHODS

of the American Electrochemical Society in 1930. Thirty years later, E. H. Greenberg developed and patented an improved instrument operating on the Seebeck effect.

Thermoelectric alloy sorting using the Seebeck effect has been popular for a long time, probably due to its low cost-to-benefit ratio. Applications for these systems are somewhat limited, but in applications where they work, they work very well. The Industrial Instruments, Inc. thermoelectric alloy sorter, known as the Identomet G-II, complies with ASTM Standard Practice E977-84.

Alloy sorting is of concern for many industries because improper alloy identification can result in pipe and equipment failures, plant shutdowns, and possibly injury or death to personnel. For example, pipe corrosion can be rapidly accelerated, causing sudden pipe failure and worse when the wrong alloy is used in the wrong service. Correct identification of materials is of prime importance in refineries, salvage yards, material storage areas, process areas, and manufacturing. In some cases, a plant shutdown can cost thousands of dollars a day.

Theory of Operation. Basically, a thermoelectric sorter consists of a power supply, sensitive voltmeter, temperature-controlled hot probe, and cold probe as shown schematically in Figure 10.20. Both the hot probe and cold probe are combined in one probe assembly. Probe tips can be made of copper, nickel, or gold, depending on application. Gold has the highest thermal and electrical conductivity. The difference in thermocouple voltage between the hot and cold probe junctions is a function of the thermoelectric properties of the alloy, which vary with alloy composition, but are independent of the size and shape of the test specimen. Automatic temperature compensation can be provided

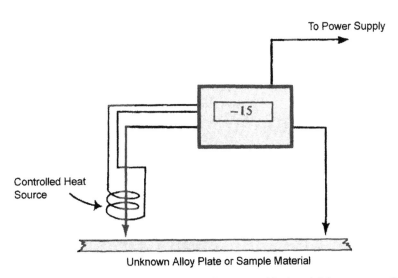

Figure 10.20 Thermoelectric alloy sorter. Courtesy of Industrial Instruments, Inc.

when test specimen temperatures vary. Instrument sensitivity is in the range of 50 to 100 µV. When utilizing the Seebeck effect, voltage readings may be positive or negative, depending on alloy composition.

Operation. A photograph of the Identomet G-II is shown in Figure 10.21. Instrument operation is straightforward as outlined in the following steps:

1. *Inspect the probe assembly.* The two fine wires brazed to the tip of one side identify the hot probe, which is the most vulnerable part of the instrument.
2. *An internal heavy-duty 6V rechargeable battery powers the Identomet.* The power switch is on the left side of the front panel. Turn the instrument on and note the three dots or decimals that appear on the display with a reading of zero.

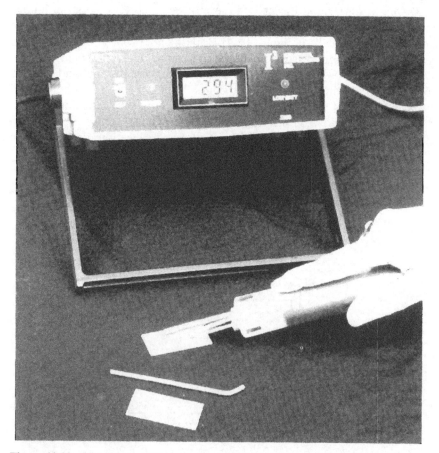

Figure 10.21 Identomet G-II instrument. Courtesy of Industrial Instruments, Inc.

OTHER THERMAL TESTING METHODS

3. *Allow about 2 minutes for probe warmup.* When the hot side of the probe is at operating temperature, the dots on the display will disappear and the green AT TEMP light on the right side of the display will turn on.
4. *Hold the probe handle the same way you would hold a dinner knife*, palm down and index finger extended. Always apply the cold side to the specimen before applying the hot side. Allow the cold tip to make contact first. Then make contact with the hot tip while keeping contact with the cold tip. Hold the probe assembly at an angle of about 30 degrees with the plane of a flat specimen.
5. *Watch the display closely.* It will go up quickly, settle down, and lock on a reading. At this point, note the reading, then remove the probe and allow the probe to reheat before reading the next specimen. The probe takes longer to recover when samples have high thermal conductivity.

Calibration. The Identomet G-II is calibrated before it leaves the shop, but recalibration is simple. Remove the top cover by unscrewing four bolts and locate the slide switch on the left front circuit board. Slide the switch to its other position. Connect the probe to the instrument and turn the power switch on. Allow the probe to reach operating temperature. When the AT TEMP light turns on, turning the trim pot, R9, near the center of the circuit board until the display reads 404 sets the probe temperature. After the probe temperature has been calibrated, return the slide switch to its original position and reinstall the cover.

10.18.3.2 Applications Successful applications include the thermoelectric sorting of refinery alloys, expensive alloys, electronic industry alloys, and tool steel saw blade alloys.

Refinery Alloy Comments

- 5 chrome springs out and produces a positive identification reading of −85 +3 for easy identification.
- Monel also springs out and provides rapid and easy identification.
- Carbon steels usually produce signals less than −51. Therefore, the Identomet G-II can be used to make sure there are no carbon streaks where specs call for $2\frac{1}{4}$, 5, or 9 chrome.
- With nickel probes signals from $2\frac{1}{4}$ and 9 chrome can be separated.

Steels that nominally contain $1\frac{1}{4}$% chrome may actually contain as little as 1% chrome and will produce somewhat lower chart readings.

Expensive Alloys. Within certain limitations, alloys such as hastelloy, inconel 800 and 825, monel, and copper-nickel can easily be sorted. Typical Identomet G-II values are shown in Figure 10.22.

Identification of Saw Blade Materials. Electric jigsaw and circular saw blades are fabricated by welding a wire of tool steel to a carbon steel base. The assem-

Identification of Nickel Alloys
Typical Values

5 Cr	-83	Hastelloy	+5-+9
Hastelloy B*	-66	Incoloy 825*	+13
9 Cr	-61	Incoloy 800*	+24
1010	-51	Monel 400	+202
Inconel 600*	-53	Copper-nickel*	+287

*Nonmagnetic

Figure 10.22 Identification of nickel alloys. Courtesy of Industrial Instruments, Inc.

bly is then notched, formed, and heat-treated to the proper hardness. The resulting blade has the flexibility and toughness of carbon steel with teeth having the hardness of tool steel. Figure 10.23 shows the Seebeck voltage chart for the saw blade application. Other charts for other applications are available from the manufacturer.

Many types of tool steel are used for different saw blades in large manufacturing plants. The welding, notching, and heat-treating operations obliterate alloy identification marks and products can easily get mixed up. Therefore, a rapid nondestructive method is needed for identification of tool steel teeth. In seconds, the thermoelectric sorter is able to identify M2, M42, and D6 tool steels. In this application the thermoelectric sorter is more feasible than most other methods because it does not depend on the size and shape of the specimen. It is also very cost effective in cases where a relatively small number of well-defined choices exist.

Note: Consult the manufacturer for additional application data.

10.18.4 Advantages and Disadvantages

10.18.4.1 Advantages

- Instrument operation and calibration is simple. Readings settle down and lock in as described in the operating section.
- The test is not sensitive to surface size or shape—it can measure fine wires or large girders.
- Portability and high-speed response—alloy identification can be made in minutes.
- For known alloys of 4340 and 17-4PH, the instrument reading may be helpful in determining the degree of heat treatment.
- The alloy composition of thick plating or cladding materials can be determined.
- The Identomet G-II instrument readily identifies alloys used for saw blade teeth.
- Diagnostic analysis is fast, predictable, and reproducible.

OTHER THERMAL TESTING METHODS

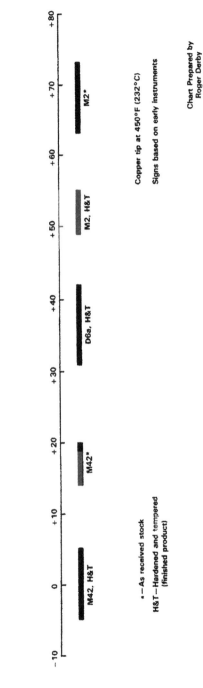

Figure 10.23 Thermoelectric sorting of saw blades. Courtesy of Industrial Instruments, Inc.

10.18.4.2 Disadvantages
- The probe assembly is somewhat position sensitive.
- The cold side of the probe assembly must be applied to specimen first to minimize cooling of the hot side.
- One tip of the probe assembly is hot, capable of reaching a maximum temperature of 450°F.
- The probe assembly should be shielded from air currents, which can limit its use.
- The test is sensitive to the heat treatment of some alloys as well as alloy composition.
- Good readings are difficult to obtain on thinly plated alloys.
- The instrument cannot tell the difference between 304 and 316 stainless steel.

11

ULTRASONIC TESTING

11.1 INTRODUCTION

Modern ultrasonic transducers are used for the nondestructive analysis of solids, liquids, and gases. Recent improvements include super-high-sensitivity, high-frequency, high-resolution, high-power, dry coupling, and noncontacting transducers for a host of applications.

The advantages of *ultrasonic testing* include:

- It can be used to determine mechanical properties and microstructure.
- It can be used for imaging and microscopy.
- It is portable and cost effective.
- It can be used with all states of matter except plasma and vacuum.
- It is not affected by optical density.

Ultrasonic transducers can be used in the time, attenuation, frequency, and image domains. *Time domain transducers* measure the time of flight and the velocity of longitudinal, shear, and surface waves. Time domain transducers measure density and thickness, detect and locate defects, and measure elastic and mechanical properties of materials. These transducers are also used for interface and dimensional analysis, proximity detection, remote sensing, and robotics.

Introduction to Nondestructive Testing: A Training Guide, Second Edition, by Paul E. Mix
Copyright © 2005 John Wiley & Sons, Inc.

Attenuation domain transducers measure fluctuations of transmitted and reflected signals at a given frequency and beam size. These transducers are used for defect characterization and determining surface and internal microstructures. They also can be used for interface analysis.

Frequency domain transducers measure the frequency dependence of ultrasonic attenuation, thereby providing ultrasonic spectroscopy. These transducers are especially used for microstructure analysis, grain boundary studies, determining porosity and surface characterization, and phase analysis.

Image domain transducers measure the time of flight and are used for attenuation mapping as function of discrete point analysis by raster C-scanning or synthetic aperture techniques. These transducers can provide surface and internal imaging of defects, microstructure, density, velocity, or mechanical properties. True 2D or 3D imaging can be provided.

11.2 DEFINITION OF ACOUSTIC PARAMETERS OF A TRANSDUCER

Nominal frequency (F)—nominal operating frequency of the transducer (usually stamped on housing)

Peak frequency (PF)—the highest frequency response measured from the frequency spectrum

Bandwidth center frequency (BCF)—the average of the lowest and highest points at a −6 dB level of the frequency spectrum

Bandwidth (BW)—the difference between the highest and lowest frequencies at the −6 dB level of the frequency spectrum; also % of BCF or of PF

Pulse width (PW)—the time duration of the time domain envelope that is 20 dB above the rising and decaying cycles of a transducer response

11.3 NONCONTACTING ULTRASONIC TESTING

One of the most significant advances in recent years has been the development of noncontacting ultrasonic transducers with perfect air/gas impedance (Z) matching. Non-Contacting Ultrasound, NCU™, was made possible by the development of high-transduction piezoelectric transducers in 1997 (U.S. and international patents) and the creation of a dedicated noncontacting ultrasonic analyzer in 1998 to 2003. As a result of this work the Ultran Group formed Second Wave Systems.

One of the greatest advantages of NCU is its total freedom from touch and contamination, which is of great importance to many high-tech manufacturing, food, pharmaceutical, and biotechnical industries. NCU provides a cost-effective alternative to X-ray, γ-ray, neutron, infrared, laser, and nuclear magnetic resonance (NMR) methods in many applications.

NONCONTACTING ULTRASONIC TESTING

Conventional ultrasonic testing (UT) methods currently extend well beyond traditional flaw detection and material thickness measurements into material characterization and 2D and 3D imaging surface and internal applications where flaws, composition, homogeneity, structure, texture, and material properties can also be determined. These applications are of prime importance for:

- Manufacturers of electronic materials and components
- Manufacturers of plastics and composites
- Aircraft and aerospace industries
- Chemical and petrochemical industries
- Food and pharmaceutical industries

For medical diagnosis, UT can replace harmful X-rays for fetus visualization, cornea measurement, tissue characterization, imaging of plaque in the arteries, and gum disease. Other medical uses for UT include measurement of brain waves, detection of skin and breast cancer, and blood-flow monitoring. In general, the UT method is also portable and cost effective. See Figure 11.1.

The primary limitations of conventional UT (200 kHz to 5 MHz) include its high attenuation in air and its need for a physical coupling agent such as water

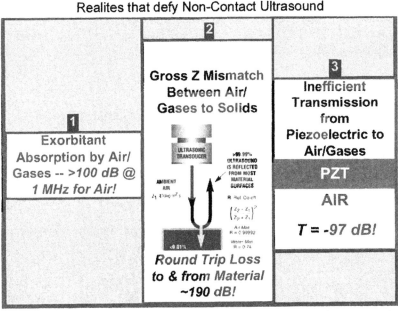

Figure 11.1 Problems that need to be overcome for Noncontacting ultrasound. Courtesy of Ultran Labs.

or grease between the transducer and test medium. Because of these limitations, NCU was once considered an impossible dream because of the acoustic impedance mismatch, which can be as high as six orders of magnitude for propagation from air to hard alloys and dense ceramics. For NCU to work, this acoustic barrier had to be broken.

The development of dry coupling for longitudinal and shear wave transducers operating at frequencies up to 25 MHz was the NCU transducer precursor. Since 1983, these transducers have been used to characterize thickness, velocity, elastic, and mechanical properties of green, porous, and dense materials. This research was followed by the development of planar and focused air/gas propagation transducers, which utilized a less than 1 Mrayl acoustic impedance matching layer of a nonrubber material on the piezoelectric material. These 250 kHz to 5 MHz air-coupled (AC) transducers with polymer acoustic impedance matched layers depended on high-energy or tone burst excitation, and high signal amplification, and were somewhat application and range limited.

In 1997, Mahesh C. Bhardwaj* produced and evaluated transducers with compressed fiber as the final acoustic impedance matching layer. These transducers produced unprecedented and phenomenal transduction in air. This work was instrumental in the development of current noncontacting transducers with perfect air acoustic impedance matching. As a result, current noncontacting transducers, covering the range of ≈ 50 kHz to >5 MHz, can now be propagated though practically any medium including very-high-acoustic-impedance materials such as steel, cermets, and dense ceramics.

Note: NCU™, NCT™, NCI™, GMP™, iPass™, iMove™, and iStand™ are trademarks of Ultran Labs/Second Wave Systems.

11.3.1 NCU Transducers

NCU™ transducer signal-to-noise ratio (SNR) is determined by Eq. 11.1:

$$\text{SNR} = 20 \log V_x / V_n \, [\text{dB}] \qquad (11.1)$$

where V_x is the received signal in volts
V_n is noise voltage

The SNR is determined without signal processing and includes the noise associated with measuring instruments, cables, etc.

NCU transducer sensitivity (S) is determined by Eq. 11.2:

$$S = 20 \log V_x / V_0 \, [\text{dB}] \qquad (11.2)$$

---

* Mahesh C. Bhardwaj, "Non-Destructive Evaluation: Introduction of Non-Contact Ultrasound," in *Encyclopedia of Smart Materials*, Editor Mel Schwartz, John Wiley & Sons, New York (2002).

where V_x is the received signal in volts
V_0 is the excitation voltage

Figure 11.2 shows a group of NCU transducers having a frequency range of ≈ 50 kHz to 10 MHz with active dimensions ranging from <1 mm to >75 mm. Since the invention of a gas matrix piezoelectric composite transducer by Mahesh C. Bardwaj (U.S. and international patents pending), transduction efficiency has been further improved, including the possibility of producing extremely large transducers with dimensions up to 1×1 m. Optimum transducer performance is a function of the piezoelectric material characteristics and its frequency constants. As an example, optimum dimensions for a 200 kHz GMP (gas matrix piezo) transducer are about 50×50 cm. To date Ultran Labs has successfully produced transducers as large as 25×25 cm between ≈ 50 kHz and 200 kHz.

Figure 11.3 shows the setup for determining the transmission characteristics of both AC and NC transducers. Frequency, bandwidth, and signal-to-noise ratio can be directly read from the oscilloscope.

NCU transducers generate immense acoustic pressure in air over their frequency range, making it possible to provide transmission in 25 mm of steel (6 orders of magnitude Z mismatch) using 2 MHz transmission with only one 16 V_{pp} burst and 64 dB amplification. Figure 11.4 shows acoustic pressure as a function of transducer-to–reflecting target distance in air (mm) for 50 mm transducers operating at frequencies between 100 KHz and 1 MHz. Note the significant spread in the first 50 mm of distance.

Figure 11.2 Selection of noncontacting transducers. Courtesy of Ultran Labs.

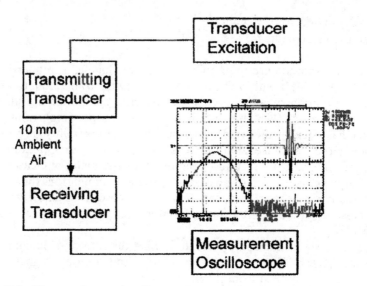

Figure 11.3 Theory of operation diagram for noncontacting transducers. Courtesy of Ultran Labs.

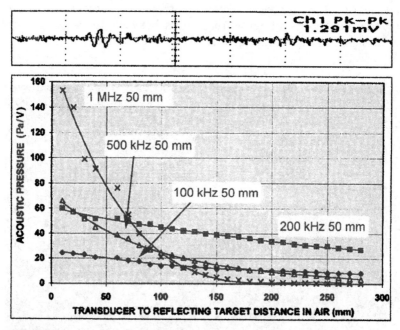

Figure 11.4 Acoustic pressure vs. target distance. Courtesy of Ultran Labs.

NONCONTACTING ULTRASONIC TESTING

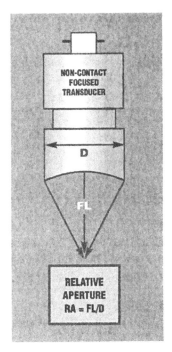

Figure 11.5 Ultran focused transducer. Courtesy of Ultran Labs.

NCU transducers can be compensated for velocity fluctuations in air caused by thermal currents, humidity variations, and turbulent air flow. Compensation is typically recommended for high-accuracy thickness, velocity, transmittance, reflectance, and density measurements.

NCU transducers may be cylindrically (line) or point focused. Figure 11.5 shows a focused transducer along with the equation for calculating relative aperture (RA). Active transducer diameters range from 6.3 to 50 mm, focal lengths in air range from 25 to 300 mm, and focal points in air range from about 0.28 to 10 mm, depending on transducer frequency and diameter. In special cases, transducers with an RA of 1.0 have been made.

11.3.2 Instant Picture Analysis System

iPass™ is a Microsoft Windows-based industrial pulser/receiver system designed for noncontacting transducers for analytical and imaging applications. It features square wave pulsed ultrasound and true 76 dB gain amplification for high resolution through complex high-attenuation materials. The system has a pulse repetition rate of greater than 1 kHz for extremely fast reading rates and sampling speeds. iPass measures defects, attenuation, time of flight, and velocity. Two transducers are used for the direct transmission or transmit-receive (T-R) pitch-catch reflection mode (Figure 11.6).

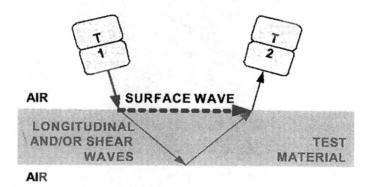

Figure 11.6 Ultran noncontacting pitch-catch mode. Courtesy of Ultran Labs.

iPass measures data in two modes. The first mode is a continuously rolling trend line that acquires data across the sample in one or two dimensions. The x-axis of the trend plot is the distance traveled by the transducer in one or both directions. The y-axis represents the transmittance value in dB through the material. Scanning directions are established using an iMove™ translation device. The ultrasonic picture created by iPass is analogous to an X-ray. iPass can provide A- and C-scans.

The iMove translation device provides standard travel speeds of 0.0125 mm/s to 510 mm/s. Custom travel speeds greater than 2.5 m/s can be provided. Resolution of the scanning device is 0.0125 mm and maximum thrust force is 91 kg or 200 lb. iMove devices are available in a multitude of sizes and degrees of freedom depending on customer application and needs.

AirTech 4000 is a Microsoft® Windows-based burst ultrasonic pulser/receiver system exclusively designed for non-contact (air/gas coupled) Ultrasound (NCU™) analytical and imaging applications. Again, rectangular pulses are used for measurement of materials with very high acoustic impedance.

Specifications for the transmitter/excitation section, preamplifier, and receiver/amplifier are the same as for the iPass system described above. The only exception is that in the ADC and software section of the specifications, *iPass™ for windows* is used for A- and C-scans for the iPass system and *Hillgus for Windows® NT®* is used for A- and C-scans for the AirTech 4000 system. Both systems use industrial computers suitable for operation in demanding environments. This equipment has been successfully applied to imaging and analysis of materials in all stages of formation. Ultrasound now can be applied to all critical process control functions, saving material, energy, and time costs while providing excellent quality control.

Materials successfully tested using the AirTech 4000 system and NCU transducers include:

- Uncured and cured polymers
- Prepregs

- Multilayered structures
- Green and sintered ceramics
- Powder metals
- Porous material
- Food and pharmaceutical products
- Tissue and bone
- Rubber and tire
- Automotive and aircraft components
- Wood and lumber
- Concrete and other construction materials

11.3.3 Limitations

- A single NCU transducer cannot be used in pulse-echo testing techniques at the present time.
- Similar to conventional ultrasound, NCU is limited by the complexity of material shape and size.
- Extremely high-acoustic-impedance materials have special requirements for successful NCU testing.
- It is nearly impossible, without special considerations, to transmit ultrasound through materials at or above 250°C.

Figure 11.7 shows degradation of hybrid rocket motor insulation. The lightest circles show about 67% of original insulation value, gray areas within the lighter areas show about 40% of original insulation value, and dark areas within the gray show about 10 to 20% of original insulation value.

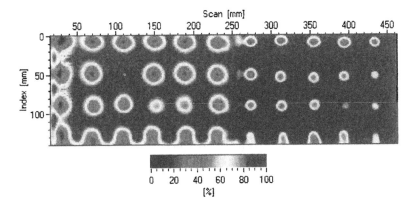

Figure 11.7 Light areas show nonbonding for 50 mm hybrid rocket motor insulation. Courtesy of Ultran Labs.

11.3.4 Bioterrorism

One timely and interesting original article by Kelli Hoover, Mahesh Bhardwaj, Nancy Ostiguy, and Owen Thompson, titled, "Destruction of Bacterial Spores by Phenomenally High Efficiency Non-Contact Ultrasonic Transducers," was published in *Materials Research Innovation* (volume 6, Springer-Verlag 2002). This article deals with the destruction of disease-causing microorganisms that are highly resistant to killing and exhibit high toxicity in low numbers, making it difficult to control human exposure through air-delivery systems. This problem was brought to the forefront by the recent contamination of government offices and postal inspection facilities through the anonymous mailing of anthrax spores. The contamination resulted in extraordinary confinement and destruction costs, and worst of all, the irreplaceable loss of human life.

Freeze-dried spores of Bt were used as a model for evaluating the destruction of anthrax spores because they are very closely related and are safe to work with outside of a biocontainment facility. The Bt spores were irradiated with 50 mm NCU planar transducers in the pulsed mode with a 50 dB power amplifier, generating about 10 MPa of acoustic pressure in ambient air. A frequency range of 70 kHz to 200 kHz was used and exposure times were varied from 10 to 180 seconds. Pulse repetition rate was kept at 50 ms for all tests.

As a result of these tests, 98.12% to 99.99% of the spores were destroyed at 161 and 93 kHz respectively. Exposure of 93 kHz destroyed 99.99% of the spores, representing a reduction in spore loads of 4900-fold and 6500-fold following 30 and 60 seconds of exposure, respectively. Treatments for longer than 60 seconds did not improve the destruction efficiency.

While 60 seconds seems like a long time for exposing a relatively small area, future NCU scanning arrays and advancements could conceivably keep pace with post office processing rates. Other opportunities for NCU sterilization can be easily visualized for the medical services and food processing industries (U.S. patent pending).

In light of significant implications of germicidal applications of noncontact ultrasonic transducers, a new company, SONIPURE, has been formed by the Ultran Group and the Penn State Research Foundation of Pennsylvania State University.

11.4 ULTRASONIC PULSERS/RECEIVERS

JSR Ultrasonics, a division of Imagilent, provides UT instruments, systems, and system components for NDT applications and research projects. The company makes a PRC35 pulser/receiver card for PCs, a DPR300 pulser/receiver, DPR500 dual pulser/receiver, and remote pulsers for use with the DPR500 unit. They also make the μP501 PELT® multi-layer ultrasonic thickness gauge and Robotic PELT automated coating thickness measurement system for rapid measurement of up to five coating layers of automotive finishes.

ULTRASONIC PULSERS/RECEIVERS

Remote pulsers and pulser/preamps provide improved performance and increased system reliability in cases where transducers must be located at relatively long distances from the receiver or when a transducer has a center frequency greater than 50 MHz. Figure 11.8 shows remote pulsers and preamps. Remote pulsers are selected based on their performance characteristics, namely fall time, pulse amplitude, and pulse width or energy, based on the specific application. Significant improvements in performance may result when cable lengths between the pulser and the transducer can be minimized.

The PRC35 pulser/receiver card is a 35 MHz computer-controlled ultrasonic pulser/receiver on a 2/3 length ISA card. Instrument controls include receiver gain, pulse repetition rate, pulse energy, pulse-echo or through-transmission mode select, pulse trigger source select, high- and low-pass filter cut-off frequency select, selection is correct and damping adjustment. A standby mode can be selected to reduce power consumption for use in power-sensitive applications.

The PRC35 card (Figure 11.9) is fully shielded from electromagnetic noise and interference while inside the PC to ensure a high signal-to-noise ratio. The fast recovery amplifier provides rapid recovery from the initial excitation pulse and large interference echoes.

A turnkey software front panel control program provides an unlimited number of instrument setups to be stored and retrieved through named setup files. The instrument base I/O port address is selectable allowing multiple cards to be installed in the same host computer. Windows® 95/98, NT and LabVIEW

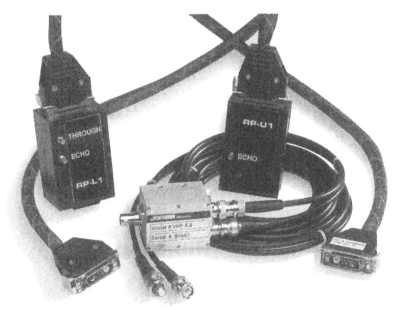

Figure 11.8 Remote pulsers and preamplifiers. Courtesy of JSR Ultrasonics.

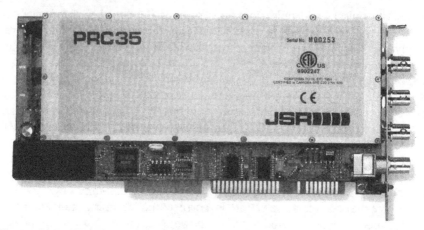

Figure 11.9 PRC35 computer-controlled pulser/receiver card. Courtesy of JSR Ultrasonics.

drivers and C-language source code are provided to enable rapid development of custom software.

Applications include:

- Computer-controlled imaging and measurement systems
- Materials analysis and inspection
- Transducer evaluation
- Portable NDE systems

The DPR500 pulser/receiver is a dual channel, modular instrument consisting of two complete pulsers/receivers integrated into one unit. Standard receiver modules are available in 500 MHz, 300 MHz and 50 MHz bandwidths. The unit can be configured as a single channel unit using any of the available modules. The DPR500 is shown in Figure 11.10.

Remote pulsers can be located in close proximity to UT transducers. The elimination of long cable lengths between the pulsers and transducers minimizes undesirable UT reflections and ringing. Interchangeable remote pulsers accommodate a wide range of transducer frequencies and energy requirements.

Instrument functions include adjustable damping, gain, pulse amplitude, pulse energy, pulse repetition rate, high-pass filters, low-pass filters, echo or through mode selectable operation (based on pulser selection), and pulser trigger source selection.

The receiver can be configured for one or two of the following frequencies—500 MHz, 300 MHz, or 50 MHz. Phase is 0° (noninverting). Input referred noise, bandwidth, and high- and low-pass filter frequencies are operating fre-

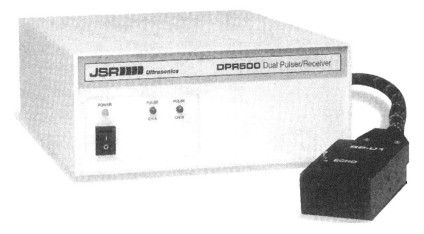

Figure 11.10 Dual pulser/receiver. Courtesy of JSR Ultrasonics.

quency dependent. Output impedance is 50 Ω in all cases and maximum output power is 5.5 to 5.2 dBm (approx. 0.58 to 0.60 V into 50 Ω, depending on frequency).

Applications include:

- Acoustic microscopy
- Thin material or coating thickness gauging
- Computer-controlled imaging and measurement systems
- Material analysis and characterization
- Transducer evaluations

The DPR300 is a computer-controlled ultrasonic pulser/receiver with an extremely low-noise receiver. Instrument controls include receiver gain, high- and low-pass filter cut-off frequency selection, pulse energy, pulse amplitude, pulser impedance, damping level, pulse-echo or through transmission mode select, pulse repetition rate, and pulser trigger source select. The DPR300 is shown in Figure 11.11.

The rapid recovery receiver is fully shielded from electromagnetic noise and interference to ensure a high signal-to-noise ratio. Pulser impedance, pulse energy, and pulse amplitude may be individually adjusted to optimize the excitation pulse for a specific application or transducer. A Windows-based software program is included for immediate usage in customer applications. Multiple DPR300's can be controlled from one computer using a hardware daisy-chain interconnection scheme. Windows® 98/95 NT and LabVIEW drivers are provided to enable rapid development of custom software.

Figure 11.11 DPR 300 pulser/receiver. Courtesy of JSR Ultrasonics.

Areas of application include:

- Computer-controlled imaging and measurement systems
- NDE systems
- Research and development
- Materials analysis and inspection
- Transducer evaluation
- Exacting low-noise measurement systems

11.5 MULTILAYER ULTRASONIC THICKNESS GAUGE

The µP501 PELT® (Pulse Echo Layer Thickness) gauge, shown in Figure 11.12, is a high-resolution multilayer thickness gauge that provides accurate and repeatable measurements on many types of coatings from ceramics to specialized elastomers. Up to five coating layers may be simultaneously measured on many different substrates. Individual layer thickness as thin as 8 microns (0.3 mil) may be measured regardless of how the coatings are applied, including "wet on wet."

The µP501 generates and displays an A-scan waveform for each measurement taken. Waveform analysis can be performed directly on the gauge and the resulting individual layer as well as total layer thickness is displayed. Larger data files can be transferred to a host PC for automated analysis. Figure 11.13 shows PELT Explorer (with Autogauge) host PC software.

CONVENTIONAL ULTRASOUND 471

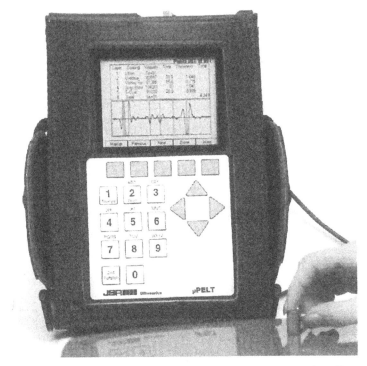

Figure 11.12 Multilayer ultrasonic thickness gauge. Courtesy of JSR Ultrasonics.

Instrument features include:

- Simultaneous measurement of up to five layers
- Unparalleled gauge repeatability and reproducibility
- Measurements can be made on virtually any substrate from polymer matrix composites to steel, plastic, wood, and glass
- Measurement accuracy is unaffected by varying thickness or composition of substrate
- Autogauge software provides easy data analysis
- Up to 8-hours of continuous operation on a single removable and rechargeable battery

11.6 CONVENTIONAL ULTRASOUND

Ultrasonic testing (UT) is widely used by industry for quality control and equipment integrity studies. Major uses include flaw detection and wall thick-

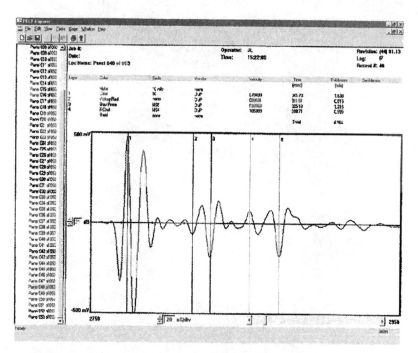

Figure 11.13 Explorer (with Autogauge) host PC software. Courtesy of JSR Ultrasonics.

ness measurements. Using ultrasonic techniques, it is possible to detect flaws and determine their size, shape, and location. It is also possible to measure the thickness of process pipes and vessels with ultrasonic transducers. Wall thickness measurements are especially important in corrosion studies where corrosion can cause a uniform reduction in wall thickness over a period of time.

Process pipes and vessels are designed to have a specific corrosion allowance or added amount of metal wall thickness based on design pressure, acceptable corrosion rates, and the expected life of the equipment. If unexpectedly high corrosion rates are encountered, wall thinning can occur at accelerated rates, resulting in dangerous conditions with regard to pipe and pressure vessel ratings. If pipes and vessels become too thin, they can burst, causing extensive plant damage as well as endangering the lives of workers. For these reasons, many chemical and petrochemical plants have active corrosion-monitoring programs where periodic wall thickness measurements are made on all process pipes and vessels.

Ultrasonics can also be used to determine differences in material structure and physical properties. Heat treatment, material grain size, and the modulus of elasticity or Young's modulus can affect the attenuation of ultrasound. Young's modulus is the ratio of stress to strain in a material within its elastic limit.

11.6.1 Flaw Detection

When a piezoelectric crystal is driven by high-voltage electrical pulses, the crystal "rings" at its resonant frequency and produces short bursts of high-frequency vibrations. These "sound wave trains" generated by the ultrasonic transducer or "search unit" are transmitted into the material being tested. When the search unit is in direct contact with the test material, the technique is known as "contact" testing.

If flaws or discontinuities are present, an acoustic mismatch occurs and some or all of the ultrasonic energy is reflected back to the search unit. The piezoelectric crystal in the search unit converts the reflected sound wave or "echo" back into electric pulses whose amplitudes are related to flaw characteristics and whose time of travel or *time of flight* through the material are proportional to the distance of the flaw from the entrance surface. Ultrasonic pulses are also reflected from the back surface of the material and this signal represents the total distance traveled. The pulse received from the back surface can also represent the width, length, or thickness of the material depending on its orientation. Ultrasonic thickness testing measures the wall thicknesses of pipes and vessels by measuring the total distance traveled by the ultrasonic pulses, which is represented by the distance from the initial pulse or front surface to the back reflection from the back surface. Ultrasonic flaw and thickness indications are frequently displayed on an instrument or computer display screen. Figure 11.14 illustrates the pulse-echo contact technique and shows the initial pulse, flaw indication, and back surface reflection display.

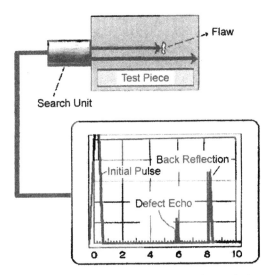

Figure 11.14 Pulse-echo contract testing technique.

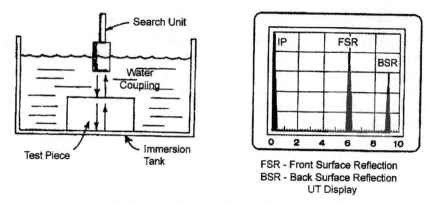

Figure 11.15 Pulse-echo immersion testing technique.

Immersion testing is another popular technique used for ultrasonic flaw detection. In immersion testing, both the ultrasonic search unit and the test piece are immersed in a liquid, usually water. The advantage of this technique is that the water excludes air and acts as a "couplant" or coupling agent for the transmission of ultrasonic energy. A couplant is used to couple (or complete the ultrasonic path) between the search unit and test surface. Figure 11.15 illustrates the immersion testing technique and shows a typical screen presentation.

11.6.2 Frequency

In theory, any frequency about 20,000 cycles per second (cps) (20 kHz) can be thought of as an ultrasonic frequency or frequency above the normal sound range. In practice, ultrasonic transducers and equipment operate in the frequency range of <50 kHz to 200 MHz. This frequency range extends well beyond the audio sound wave frequency range of 20 Hz to 20 kHz, but ultrasonic waves are still propagated as waves of particle vibrations. These waves travel with ease in uniform solids and low-viscosity liquids; voids or gases such as air quickly attenuate them.

Some important characteristics of ultrasonic vibrations are:

- They travel long distances in solid materials.
- They travel in well-defined sonic beams.
- Their velocity is constant in homogeneous materials.
- Circuitry is designed so that the energy from the first wave train is dissipated before the next wave train is introduced.
- Vibrational waves are reflected at interfaces where elastic and physical properties change; they are also refracted when elastic properties change.

CONVENTIONAL ULTRASOUND

- Vibrational waves may change their mode of vibration or be subject to mode conversion at material interfaces.

Ultrasonic pulses can be generated by radio frequency (RF) wave trains driving a crystal at a controlled frequency and precise time or they can be generated by a shock pulse that permits the crystal to resonate at its natural frequency thus establishing the vibrational frequency. For maximum sensitivity, the piezoelectric crystal should be driven at its fundamental resonant frequency.

In ultrasonic testing, a search unit may be thought of as an ultrasonic probe or transducer containing one or more piezoelectric crystals. The search unit is driven for 1 to 3 µs, producing a short burst of ultrasonic waves. The ultrasonic waves are transmitted through the material where it is reflected by the back surface. After this initial burst of pulses is transmitted, the transducer acts as a receiver, waiting to receive the reflected wave train or echo pulse. This transmitting–receiving cycle is repeated 60 to 1000 times or more based on transducer design and application requirements. To avoid confusion, sufficient time must be allowed to elapse between transmitted pulses to permit return of the echo pulse and provide for the decay of the initial pulse. Figure 11.16 shows UT pulse generation.

The relationship among frequency, wavelength, and velocity is given by Eqs. (11.3)–(11.5):

$$V = \lambda \times f \tag{11.3}$$

$$\lambda = V/f \tag{11.4}$$

$$f = V/\lambda \tag{11.5}$$

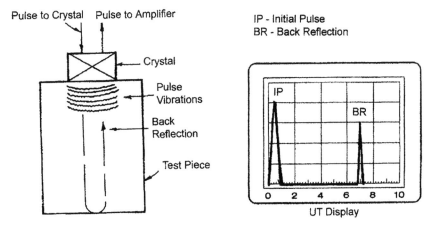

Figure 11.16 Ultrasonic pulse generator.

where V = velocity in cm/s
λ = wavelength in cm
and f = frequency in Hz

From these equations it can be seen that velocity varies directly with wavelength and frequency, but wavelength and frequency vary inversely with each other. Velocity is the speed at which ultrasonic vibrations pass through various materials; it is dependent on the elastic properties of the material and the mode of vibration. The elasticity and density of the material determines its sound velocity.

In practice, the test frequency selected depends on the sensitivity and sound penetration required. In general, high-frequency crystals are more sensitive to discontinuities and lower-frequency crystals provide greater depth penetration. All frequencies work equally well with fine-grained materials or in immersion testing. High frequencies usually are not used with coarse-grained materials because the material tends to scatter the energy. Also, frequencies above 10 MHz are seldom used with contact-type search units because of the fragile nature of the high-frequency crystals.

11.6.3 Ultrasonic Wave Propagation

The four fundamental modes of ultrasonic wave propagation are:

1. Longitudinal or compression waves
2. Shear or transverse waves
3. Surface or Rayleigh waves
4. Plate or Lamb waves

Longitudinal waves are similar to audible sound waves in that they are also compressional in nature. The alternate expansion and contraction of a piezoelectric crystal generates longitudinal waves. Particle displacement is in the direction of wave propagation as shown in Figure 11.17. Only longitudinal waves can travel through a liquid.

With shear waves, particle vibration is transverse (at a right angle) to the direction of wave propagation. Passing the ultrasonic beam through the material at an angle generates shear waves. Plastic wedges with angles of about 27.5° (1st critical angle) to 57° (2nd critical angle) are used with transducers to generate shear waves in steel at 33 to 90°. At 90° the transverse wave propagates along the surface of the object, becoming a surface wave. Figure 11.18 illustrates the shear wave relationship between the direction of particle vibration and direction of wave propagation.

Surface waves (Figure 11.19) travel with little attenuation in the direction of wave propagation. However, their energy decreases rapidly as the wave penetrates below the surface of the material. Surface waves do not exist in

CONVENTIONAL ULTRASOUND

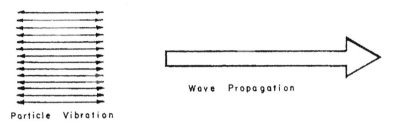

Figure 11.17 Particle motion and wave propagation with longitudinal waves.

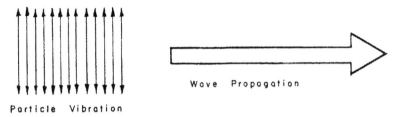

Figure 11.18 Particle motion and wave propagation with shear waves.

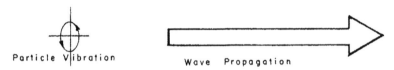

Figure 11.19 Particle motion and wave propagation with surface waves.

immersion testing. Particle displacement of the wave follows an elliptical orbit.

When ultrasonic energy is introduced into relatively thin plates, it is propagated by Lamb waves. Lamb waves have multiple or varying wave velocities. Lamb wave velocity is dependent on the thickness of the material and frequency. With Lamb waves, a number of modes of particle vibration are possible, but the two most common modes of vibrational motion are symmetrical and asymmetrical as shown in Figure 11.20. The complex particle motion is somewhat similar to the elliptical orbits of surface waves.

11.6.4 Acoustic Impedance

The characteristic impedance (Z) of a material is defined as the product of density (d) and longitudinal wave density (v). The equation for acoustic impedance is shown in Eq. (11.6):

$$Z(g/cm^2 \cdot s) = d(g/cm^3) \times v(cm/s) \qquad (11.6)$$

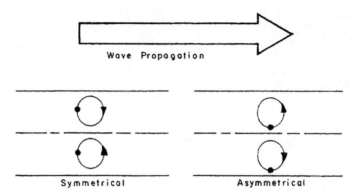

Figure 11.20 Particle motion and wave propagation with Lamb waves.

The acoustic impedance of a material determines its reflection and transmission characteristics. As the impedance ratio of two dissimilar metals increases, the amount of sound coupled through their interface decreases. Acoustic impedance values for some common materials are listed in Table 11.1.

11.6.5 Reflection and Refraction

Ultrasonic vibrations are reflected at the interface of two different materials when a mismatch occurs in acoustic impedance. Acoustic mismatches are likely to occur at the water–metal interface, metal–flaw surface interface, or metal–metal interface where the material properties are quite different. One important property of the reflected wave is that the angle of reflection is always equal to the angle of incidence. Figure 11.21 shows the angular relationship of the incident, reflected, and refracted waves.

When the ultrasonic beam passes at an angle from one material to another, refraction and mode conversion can occur. Refraction occurs when the ultrasonic wave changes direction and velocity as it crosses a boundary between different materials. Both reflection and refraction are analogous to what can be observed with light beams. The reflection of light beams with mirrors is similar to the reflection of sound waves when an acoustic mismatch occurs. The refraction of the sound wave is similar to viewing a stick half in and half out of the water. The stick appears to be broken or disjointed at the surface of the water or air–water interface and its direction appears to change. In contact testing, the search unit utilizes a plastic wedge with the ultrasonic crystal to introduce the ultrasonic beam at an angle to produce shear waves. In immersion testing a transducer manipulator is used to angulate (vary the angle of) the transducer to produce shear waves. Mode conversion occurs when the longitudinal waves in plastic or water are converted into shear waves in the material. The refraction angle of waves passing from water into metals,

TABLE 11.1. Acoustic Impedance of Various Materials

Material	Longitudinal Velocity (cm/s)[a]	Density (g/cm^3)	Acoustic Impedance (g/cm$^2 \cdot$s)[a]
Acrylic	2.67	1.18	3.15
Air	0.33	0.0001	0.00033
Aluminum 250	6.35	2.71	17.2
Aluminum 17ST	6.25	2.80	17.5
Beryllium	12.80	1.82	23.3
Brass	4.43	8.10	35.9
Bronze	3.53	8.86	31.3
Cadmium	2.80	8.57	24.0
Copper	4.66	8.90	41.5
Glass, crown	5.30	3.56	18.9
Glycerin	1.90	1.26	2.4
Gold	3.20	19.56	62.6
Ice	4.00	0.88	3.5
Inconel	7.82	8.25	64.5
Iron	5.90	7.69	45.4
Iron, cast	4.60	7.22	33.2
Lead	2.16	11.40	24.6
Magnesium	5.79	1.74	10.1
Mercury	1.40	14.00	19.6
Molybdenum	6.30	10.19	64.2
Monel	6.02	8.33	53.2
Neoprene	1.60	1.31	2.1
Nickel	5.63	8.80	49.5
Nylon, 66	2.60	1.11	2.9
Oil, SAE 30	1.70	0.88	1.5
Platinum	3.30	21.15	69.8
Plexiglass	2.70	1.15	3.1
Polyethylene	1.90	0.89	1.7
Polystyrene	2.40	1.04	2.5
Polyurethane	1.90	1.00	1.9
Quartz	5.80	2.62	15.2
Rubber, butyl	1.80	1.11	2.0
Silver	3.60	5.99	21.6
Stainless 302	5.66	8.03	45.4
Stainless 410	7.39	7.67	56.7
Steel	5.85	7.80	45.6
Teflon	1.40	2.14	3.0
Tin	3.30	7.33	24.2
Titanium	6.10	4.47	27.3
Tungsten	5.20	19.42	101.0
Uranium	3.40	18.53	63.0
Water	1.49	1.00	1.49
Zinc	4.20	7.05	29.6

[a] Times 100,000.

Source: Based on information supplied by the courtesy of Automation/Sperry, a unit of Qualcorp, and Krautkramer Branson, Incorporated, Manufacturers of Ultrasonic Nondestructive Testing Equipment.

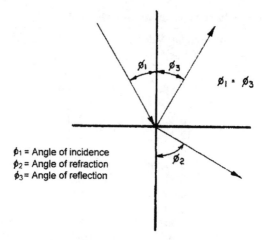

Figure 11.21 The relationship among the angle of incidence, angle of reflection, and angle of refraction.

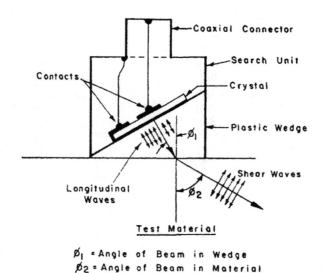

Figure 11.22 Angle beam construction and mode conversion.

at angles other than normal, varies as a function of the relative velocities of sound in water and metal. Figure 11.22 shows angle beam construction, refraction of the ultrasonic beam, and resulting particle motion. *Snell's law*, Eq. (11.7), defines both the angle of reflection and refraction as:

$$\sin\theta_1/V_1 = \sin\theta_2/V_2 \tag{11.7}$$

where θ_1 = angle of incidence = angle of reflection
θ_2 = angle of refraction
V_1 = velocity of sound in first material
V_2 = velocity of sound in second material

The ratio between wave speed in one material and wave speed in a second material is called the *index of refraction*.

11.6.6 Diffraction, Dispersion, and Attenuation

Diffraction, dispersion, and attenuation are considerations that are hard to quantify because of undefined transducer, flaw, and material variations. Wave interference or diffraction patterns are found in transmitted ultrasonic beams and in reflections received from small discontinuities. Diffraction refers to the breaking up of sound waves into high- and low-energy bands. Some of the factors that affect diffraction are transducer beam spread and beam profile or pattern. Transducer beam spread or divergence is dependent on transducer frequency and diameter as shown in Table 11.2.

From information in Table 11.2, it can be concluded that the higher-frequency crystals are more directional than lower-frequency crystals and larger-diameter crystals are more directional than smaller crystals operating at the same frequency. The angle of beam divergence of a quartz crystal is given by Eq. (11.8):

$$\sin \theta/2 = 1.22 \times \text{wavelength}/\text{diameter} \quad (11.8)$$

The term *dispersion* can have a definition similar to *diffraction* or be thought of as the dispersion of energy due to rough and irregular surfaces. These surfaces reflect the ultrasonic energy in directions that cannot be detected by the search unit. Dispersion is primarily affected by material or defect reflection factors.

TABLE 11.2. Divergence versus Frequency and Diameter in Steel

Crystal Frequency (MHz)	Diameter (in.)	Divergence (in./ft)	Angle of Divergence (deg)
5	1	0.62	6
2.25	1	1.50	14
1	1	3.00	30
0.5	2	3.00	30

Source: Adapted from information supplied through the courtesy of Automation/Sperry, a unit of Qualcorp.

It is difficult to measure the attenuation of ultrasonic energy in material because of the effects of beam divergence, scattering, field interference or diffraction, and possible coupling losses. The attenuation in test parts can be inferred by observing the amplitude of back reflections from distance amplitude blocks of similar materials. Most of the time, all losses in sonic energy are considered attenuation losses. They can be empirically determined by measuring the amplitude of multiple reflections and calculating the small incremental losses in consecutive pulses. Some material properties affecting attenuation are hardness, heat conduction, grain size, viscous friction, crystal structure changes, elastic hysteresis, and Young's modulus.

11.6.7 Fresnel and Fraunhofer Fields

Ultrasonic search units produce beam patterns composed of two zones, the Fresnel zone or near-field zone and the Fraunhofer zone or far-field zone. Fluctuations in sound pressure, also known as interference fields, occur in the Fresnel zone near the face of the transducer. In the far field of a uniform axial beam, the sound intensity is maximum at its center. The amplitude of an indication diminishes exponentially with distance in the far-field zone.

Flat-faced immersion transducers have a *natural focus* where the pressure peak in the axial beam is at its maximum relative intensity. This point of natural focus (Figure 11.23) is identified by the symbol (Y_0^+) and corresponds to the transition distance (N) from the near field to the far field. Y_0^+ occurs at the same point as N.

The approximate value of N can be calculated using Eq. (11.9):

$$N = (R/W)^2 \text{ or } N = 1/4(D/W)^2 \tag{11.9}$$

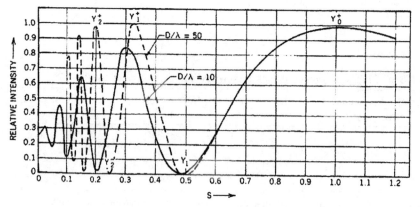

Figure 11.23 Beam intensity as a function of near-field focusing.

TABLE 11.3. Near Field Distances of Flat Transducers in Water

Frequency (MHz)	Diameter D (in.)	Near Field Distance N (in.)
0.50	1.00	2.131
1.00	1.00	4.262
2.25	1.00	9.589
3.50	1.00	14.916
5.00	1.00	21.309
7.50	1.00	31.963
10.00	0.50	10.654
10.00	0.25	2.664
15.00	0.50	15.982
15.00	0.25	3.995
20.00	0.50	21.309
20.00	0.25	5.327
25.00	0.25	6.659
25.00	0.12	1.665
30.00	0.12	1.998

where D = crystal diameter
R = crystal radius
W = wavelength of sound in water at the specified crystal frequency.

Some typical transition distances are shown in Table 11.3. As shown in the table, the near-field distance increases with increasing crystal frequency and decreases as a function of the square of the crystal diameter. With immersion testing, unwanted near-field effects can be eliminated by selecting an appropriate water path.

The transition distance, N, is of importance because a transducer cannot be focused beyond its near field. In practice, focusing factors >0.6N produce little or no increase in sensitivity, and therefore, they are not used. However, transducers can be focused at distances of 0.2N to 0.6N, which produces significant increases in sensitivity to small flaws without significant decreases in ultrasonic energy.

11.6.8 Generation of Ultrasonic Waves

The conversion of electrical pulses to mechanical vibrations and the conversion of returned mechanical vibrations back into electrical pulses are the basis of ultrasonic testing. The piezoelectric transducer that accomplishes this sending and receiving conversion is a plate of polarized ceramic or crystalline material with electrodes on the opposite surfaces. Some materials that exhibit piezoelectric properties are lithium niobate, barium titanate, lead metaniobate, lead zirconate titanate, lithium sulfate, tourmaline, quartz, and composite materials.

The three most common materials used in the manufacture of ultrasonic search units are quartz, lithium sulfate, and polarized ceramics. Quartz crystals are known for their electrical and thermal stability. They are insoluble in most liquids, resist wear and aging, and have high mechanical strength. One major limitation is that quartz crystals have relatively low electromechanical conversion efficiency (high-energy pulses produce weak vibrations). They are therefore the least effective ultrasonic generators. Lithium sulfate crystals have good dampening properties for best resolution, negligible mode interaction, and moderate conversion efficiency. For these reasons, lithium sulfate transducers are considered the best receivers. Their major limitation is that they cannot be used at temperatures above 165°F. Finally, polarized ceramic elements have the highest conversion efficiency and highest search unit sensitivity (ability to detect small flaws). Ceramic transducers, such as barium titanate, are considered to be the most efficient ultrasonic generators. Lead zirconate titanate is frequently supplied as spherical focused transducer crystals. Their disadvantages include lower mechanical strength, higher electrical capacitance, and greater vibration mode interaction. High electrical capacitance can limit the upper frequency of operation of a crystal by decreasing its impedance. To minimize some of these disadvantages, these crystals are frequently supplied in frequencies from 2.25 to 5.0 MHz.

Piezoelectric crystals expand and contract when they are subjected to a changing radio frequency (RF) field or socked by a high-voltage electrical pulse. Their thickness changes are a function of the applied voltage, typically 100 to 2000 V. Figure 11.24 illustrates this principle. Likewise, if a mechanical pressure is applied to the face of a crystal, a small electrical signal, typically 0.001–1.0 V, will be produced at the same frequency as the applied mechanical vibrations. The transducer or search unit thereby generates ultrasonic vibrations and acts as its own receiver when these bursts of energy are properly timed.

11.6.9 Search Unit Construction

An ultrasonic search unit consists of an appropriate housing with electrical connector, backing material for dampening, active piezoelectric material, lens

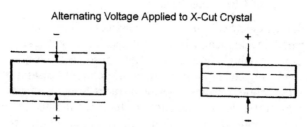

Figure 11.24 Effect of alternating voltage on crystal thickness and vibration.

for acoustic focusing, or time delay material if required. In addition, a thin coupling layer, protective layer, or wear plate may be used ahead of the active element.

The effectiveness of the search unit for a particular application depends on Q, bandwidth, frequency, sensitivity, acoustic impedance, and resolving power. The Q of a transducer is given by Eq. (11.10):

$$Q = F_r/F_2 - F_1 \qquad (11.10)$$

where $F_2 - F_1$ = bandwidth
F_r = resonant frequency of the element
F_2 = frequency above F_r where amplitude = 0.707 F_r amplitude
F_1 = frequency below F_r where amplitude = 0.707 F_r amplitude

Note: In most nondestructive testing applications, the Q of the search unit will vary from 1 to 10.

Sensitivity is the ability of the search unit to detect reflections or echoes from small defects or flaws. Search unit sensitivity is directly proportional to the product of its efficiency as a transmitter and its efficiency as a receiver.

The acoustic impedance of a transducer is the product of its density and the velocity of sound within it. Its resolving power includes the ability to separate reflections from two closely spaced flaws or reflectors. For example, in contact testing, the transducer must have good resolution to separate the front surface pulse, initial pulse, or "main bang" from the near-surface defect reflections. The ability of a transducer to resolve or detect near-surface defects is related to its pulse width or length. The transducer must stop vibrating or "ringing" after it is shocked in order to resolve or "see" small near-surface flaws. Resolving power is also related to crystal dampening and bandwidth. Ideally, we would like to reduce crystal ringing or bandwidth without adversely affecting sensitivity. In practice, compromises must be made to achieve test goals.

For most search units, the product of bandwidth and sensitivity is a constant. As bandwidth increases, sensitivity decreases; as bandwidth decreases, sensitivity increases. By knowing the sensitivity bandwidth product and bandwidth of the transducer, its sensitivity can be calculated.

Piezoelectric elements are usually coated with gold or silver on their front and back surfaces to form electrodes. Figure 11.25 shows a few of the available electrode configurations for flat crystals and some typical transducers. Small wire leads are then attached to the electrode surfaces by welding, soldering, or cementing them in place with conductive epoxy. Good electrical insulation must be used in search unit construction because high-voltage pulses (100–2000 VDC) are applied to the electrodes for 10 µs or less. The high-voltage pulses are transmitted from the pulse circuit of the electronics unit to the search unit through a coaxial cable to an ultrahigh frequency (UHF) or baby N connector (BNC)-type coaxial connector on the search unit.

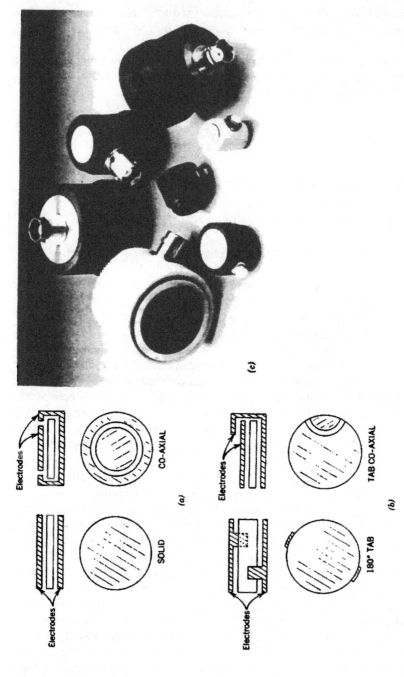

Figure 11.25 (a, b) Crystal electrode arrangements and (c) factory assembled transducers. Electrode diagrams courtesy of Valpey-Fisher Corp.; transducer photo courtesy of Krautkramer Branson, Inc., manufacturer of Ultrasonic Nondestructive Testing Equipment.

The backing material of the search unit serves two purposes. First, it controls crystal dampening and therefore bandwidth, and second, it attenuates the energy on the back side of the crystal so that unwanted reflections will not be received from the back side of the transducer. Transducers can be air backed or backed by fibrous or cellular plastic material that effectively attenuates sound on the back side of the crystal while reducing its undampened bandwidth from as much as 20,000 to a range of 1 to 10, depending on the application.

Some manufacturers embed a hardened surface ring and coat the front sides of their transducers with an aluminum oxide coating to reduce transducer wear. In contact testing, plastic wedges are used to position crystals at an angle for shear wave testing. Delay tips are also attached to contact-type transducers for testing thin materials. While increasing the versatility of the transducer, the Lucite wedges and shoes decrease transducer sensitivity. Lucite shoes are contoured wedges that are shaped to the contour of the test surface. In high-speed production lines, the Lucite shoes may actually ride on the test surface, become worn, and have to be periodically replaced.

Acoustic lenses can also be attached to the front surface of the transducer to act as a lens for focusing the sound beam. As the radius of curvature of a curved lens increases, the focal length of the lens increases. Ideally, the acoustic impedance of the focusing lens is between that of the transducer and material under test. Transducers can be spherically or "spot" focused and cylindrically focused. Spherically (spot or point) focused transducers are used when improved resolution to small flaws is required for the test material. Cylindrically focused transducers are typically used for pipe and tubing inspections.

Acoustic lenses with cylindrical curvatures focus sound energy into cylindrical surfaces normally or at right angles. Focusing lenses effectively shorten the Fresnel zone by shifting the transition distance N toward the transducer. Backing material variations, lens misalignment, or lens porosity can result in the propagation of nonsymmetrical beams when using focused transducers.

Focusing can increase the echo amplitude from small flaws near the focal point. This technique is used to obtain better near-surface resolution without increasing transducer frequency. The disadvantage is that the sensitivity to defects in the far field is greatly decreased. Cylindrical focusing is used to shape the sonic wave front to conform to part geometry. This produces a clean front surface reflection and better resolution of near-surface flaws in pipe and tubing. This type of focusing is also called "contoured focusing." The effects of crystal diameter, focal distance, and water path length are given in Table 11.4.

To calculate the focal distance (FD) under other conditions, Eq. (11.11) may be used:

$$\text{FD (water)} = \text{metal path (water equivalent)} + \text{water path} \quad (11.11)$$

TABLE 11.4. Focused Search Unit Characteristics

Crystal Diameter (in.)	Focus[b]	Beam FD-Water (in.)	Beam Diameter (in.)	Actual Water Path (in.)	Test Conditions[a] Beam FD-Aluminum (in.)	Beam FD-Steel (in.)
1.50	S	4.2	0.13	1.0	0.75	0.82
1.50	M	6.5	0.23	1.2	1.25	1.35
1.50	L	9.2	0.37	1.8	1.75	1.87
0.75	S	2.5	0.05	1.5	0.25	0.25
0.75	M	4.0	0.09	1.9	0.50	0.53
0.75	L	5.2	0.15	1.0	1.00	1.07
0.50	S	1.7	0.04	0.9	0.18	0.19
0.50	M	2.7	0.06	1.1	0.38	0.41
0.50	L	3.7	0.12	1.9	0.44	0.46
0.37	S	1.3	0.03	0.8	0.13	0.13
0.37	M	2.2	0.05	1.2	0.25	0.25
0.37	L	3.0	0.08	1.4	0.38	0.41
0.25	S	0.8	0.02	0.4	0.09	0.10
0.25	M	1.5	0.05	0.8	0.16	0.18
0.25	L	2.2	0.07	0.9	0.31	0.33
0.19	S	0.6	0.02	0.4	0.05	0.05
0.19	M	1.0	0.04	0.6	0.10	0.10

[a] (1) Water path may vary ±25% without adversely affecting test sensitivity. (2) Data applicable to frequencies of 10 MHz or above.
[b] S = short, M = medium, L = long.

where metal path (water equivalent) = metal path × (velocity of sound in metal/velocity of sound in water)

water path = distance from the transducer face to the part surface

The relationship between crystal frequency and thickness is given in Table 11.5.

Some special-purpose transducers are mosaic and paintbrush transducers. A multiple crystal transducer, whose crystals operate in the same plane and in phase with each other, is known as a crystal mosaic. A paintbrush transducer is a large-area, rectangular-shaped transducer with uniform-intensity beam pattern. Since both mosaic and paintbrush transducers are used for rapid scanning of relatively large areas, search units with smaller crystals are frequently used to accurately locate and further evaluate discontinuities found with these transducers.

Today's modern search units are quite rugged and designed to survive heavy use. However, some common sense should be used with regard to the care and storage of search units. They should not be dropped or mechanically

CONVENTIONAL ULTRASOUND

TABLE 11.5. Crystal Frequency versus Thickness[a]

Crystal Frequency (MHz)	Crystal Thickness (in.)
0.20	0.500
0.50	0.200
1.00	0.100
2.25	0.050
5.00	0.020
10.00	0.010
15.00	0.007
20.00	0.006
25.00	0.005

[a] Crystal diameters vary widely, typically in the range 0.5–1.5 in. for 1 MHz to 0.125–0.375 in. for 25 MHz.

deformed; their front surfaces should be protected to prevent scratches and minimize wear. Units should be stored at normal operating temperatures and not subjected to excessively harsh chemical environments. Tank-type immersion units should be immersed or loaded before they are pulsed, and the transmitter should be turned off before they are removed from the immersion tank.

11.6.10 Test Methods

The active area of the search unit affects beam divergence and determines the amount of energy that can be transmitted into the test material. In reality, the amplitude of crystal vibration is so small (in the micro-inch range) that a fluid couplant must be used between the search unit surface and the test piece surface. In immersion testing, the sound is transmitted into the material through a water path or liquid column. In contact testing, a thin film or oil, water, or glycerin is used as a coupling agent.

Immersion transducers include bubbler and squirter units, liquid-filled wheels, and any waterproof transducer that can be attached to a probe manipulator for use underwater.

Search units are designed for contact testing or immersion testing. The contact units can be designed as straight beam or angle beam units. The straight beam unit transmits longitudinal waves into the test surface when the sonic beam is at an angle to the entrant surface. The angle beam unit introduces sonic waves at an angle to the entrant surface. Most contact search units require a thin film solid or liquid couplant because air attenuates most of the ultrasonic energy. Noncontacting transducers are an exception. In contact testing, the entry surface indication is known as the initial pulse, transmitted pulse, or "main bang." In contact testing, defects near the surface cannot always be detected because of the dead zone.

Straight beam contact search units project a beam of ultrasonic vibrations perpendicular to the test surface. They can be used with the pulse-echo or reflection technique; they also can be used with the through-transmission technique that will be discussed later. Straight beam testing is effective in detecting parallel, laminar, or planar flaws in plates. Straight beam testing propagates longitudinal waves, which have the greatest velocity in a given material. Straight beam testing from the part end is also effective in detecting root cracks in threaded cylinders. The pulse echo or reflection technique can be used with either one or two search units. When one search unit is used, the technique is known as the "single-search-unit reflection technique." With this technique, the search unit acts as both the transmitter and receiver. It sends longitudinal waves through the material and receives reflections from the back surface and any flaws in the beam path as shown in Figure 11.26. With straight beam testing, nonparallel front and back surfaces can cause partial or total loss of the back reflection. Inadequate coupling, an angularly oriented flaw, or a near-surface defect that cannot be resolved from the transmitted pulse also could cause a reduction in the back surface reflection. The shape of the initial pulse determines the ability to separate the front pulse from a defect echo.

Two search units can be used when the test object is irregular or when its back surface is not parallel to the entrant surface. As shown in Figure 11.27, one search unit is used as the transmitter and the second search unit is used as the receiver. Sound vibrations are transmitted through the material and reflected in a number of directions. The second transducer receives reflections from parallel surfaces or flaws with parallel surfaces.

Angle beam contact units have the piezoelectric crystal mounted on a plastic wedge so that the ultrasonic beam enters the test material at a predetermined angle to the entrance surface. Angle beam search units can be used for testing flat sheet and plate or pipe and tubing as shown in Figure 11.28. Angle beam units are designed to induce particle vibrations in Lamb, longitudinal, and shear wave modes. A 45° shear angle transducer produces a wave

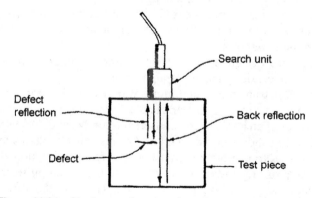

Figure 11.26 Single-search-unit pulse-echo reflection technique.

CONVENTIONAL ULTRASOUND

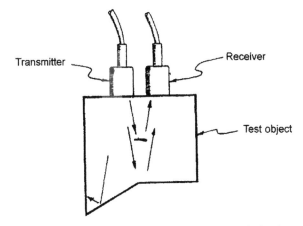

Figure 11.27 Two search units being used to inspect irregularly shaped test object.

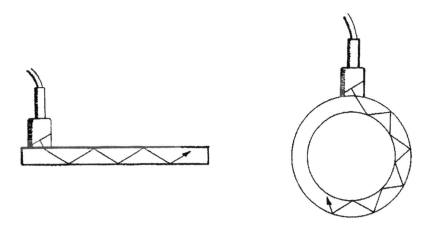

Figure 11.28 Angular beam search unit being used to induce shear waves in flat plate and tubular elements.

that is polarized perpendicular to the direction of propagation and 45° to the entry surface.

Some manufacturers, such as Jem Penetrameter, offer an "ultrasonic calibrator" that simplifies the calculation of flaw depth in flat plates. The calculator, a plastic rule with sliding crosshair, can be used to measure the defect depth, sound path length, or surface path length using 45, 60, or 70° angle beam contact probes. The calculator also lists the information shown in Table 11.6.

Figure 11.29 shows a typical defect in a 2-in.-thick plate. With this plate thickness, a 60° contact probe would be chosen in accordance with Table 11.6. Next, the sound path or probe-to–object distance (Y) would be determined

TABLE 11.6. Contact Angle Probe Data

Beam Angle (deg)	Plate Thickness T (in.)	Thickness Measurement Surface Path (P)[a]	Sine	Cosine	Tangent
45	2.5+	P = 2 T	0.707	0.707	1.000
60	1.5–2.5	P = 3.5 T	0.866	0.500	1.732
70	0.25–1.5	P = 5.5 T	0.940	0.342	2.748

[a] P = First full surface path; distance from entrance point to point of first front surface bounce.
Source: Adapted from information supplied by Jem Penetrameter Mfg. Corp.

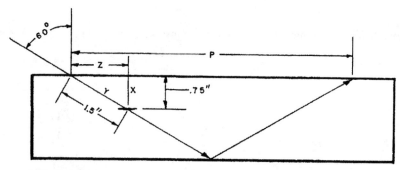

X = Defect Depth
Y = Defect Sound Path
Z = Defect Surface Path
P = Full Surface Path of a Reflection

Figure 11.29 Relationship between defect sound path and defect depth. Adapted from Jem Penetrameter Mfg. Corp.

from the oscilloscope trace. In this case, the UT display indicated a sound path length of 1.5 in. Finally, we would place the hairline of the slider on scale C for a 60° probe and read the defect depth on the standard inch ruler at the top. The indicated defect depth is 0.75 in., or the cosine of 60° times the sound path length. In angle beam shear wave testing, the skip distance will increase with material thickness. Shear waves are commonly used to detect discontinuities in welds, tube, and pipe. Shear waves are more sensitive to defects than longitudinal waves because of their shorter wavelength; shear waves are about half the velocity of longitudinal waves with particle motion normal to the direction of propagation. However, shear wave testing of a plate will often miss laminations that are parallel to the front surface.

Surface wave contact units are similar to angle beam units in that a plastic wedge is used to correctly position the crystal with respect to the test surface. Increasing the incident angle until the second critical angle is reached results in the generation of surface waves. The wedge angle is selected so that the refraction angle is 90°. Surface waves are generated by mode conversion and they travel along the test surface as shown in Figure 11.30. The velocity of surface waves is about nine-tenths the velocity of shear waves in the same material; they propagate both longitudinally and transversely. Rayleigh waves are influenced most by defects close to the surface; therefore, oil and dirt on the surface will also attenuate sound and produce display screen indications. Surface waves will travel around gradual curves with little or no reflection from the curve. Surface waves are reduced to one-twenty-fifth of their original power at a depth of one wavelength.

Lamb waves are used for detecting laminar defects near the surface of a thin material. They are also used for bond testing thin laminar structures. The velocity of the Lamb waves depends on the material characteristics, plate thickness, and frequency. Lamb waves are generated when the angle of incidence is adjusted so that the velocity of the longitudinal incident wave equals the velocity of the desired mode of Lamb waves. Since this is difficult if not impossible to determine mathematically, some manufacturers recommend that the angle of incidence be varied experimentally to determine the optimum Lamb wave mode for the material of interest. Low-frequency sound waves are not used on thin materials because of their poor near-surface resolution.

To experimentally determine the optimum Lamb wave mode, reference standards of the material of interest are an invaluable aid. When a single search unit is used, an echo pulse or reflection will be received from the reference discontinuity. Adjusting the search unit angle for maximum peak amplitude will determine the proper incident angle for that material at that test frequency.

When two transducers are used for Lamb wave testing, the receiving transducer can be angulated to receive the maximum signal when a flaw is not present or a minimum signal when a flaw is present, thus determining the optimum incident angle. Or, the second transducer can be set up at a right

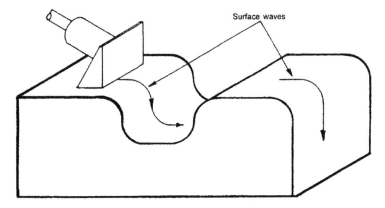

Figure 11.30 Illustration of surface waves following gradual contours.

angle to the plate surface and left there as the transmitting unit is angulated. The angle of incidence of the transmitting unit will be correct when the maximum signal is received from the simulated flaw because the flaw will reflect energy in many modes and many directions. Generally, the angle of incidence for Lamb wave testing is less than the angle of incidence for surface wave testing.

Through transmission testing uses two search units; one unit is used as a transmitter and the other unit is used as a receiver, as shown in Figure 11.31. With this technique, the ultrasonic beam passes through the test piece or is attenuated by one or more discontinuities. Total or partial attenuation of the signal is possible depending on the severity of the discontinuity. Both transducers must be properly coupled with a liquid coupling agent to obtain reliable results. As with other techniques using two search units, greater efficiency may be obtained by using a ceramic element in the transmitting search unit and a lithium sulfate element in the receiving unit.

Many of the same techniques used in contact testing can be used in immersion testing. One advantage of immersion testing is that water makes a very effective coupling agent. A wetting agent is often used with the water to reduce surface tension and minimize air bubble formation on probes and test parts. The main advantages of immersion testing are speed of inspection, ability to direct the sound at any desired angle, and the ease of incorporating automatic scanning techniques. With immersion testing, the time to send the beam through the water is usually greater than the time to send the beam through the test piece. All immersion search units are basically straight beam units that are directed to produce either longitudinal or shear waves in the test material.

With straight beam immersion testing, the ultrasonic beam is projected straight into the material perpendicular to the test surface. Reflected pulses are received from the entrant surface, any discontinuities present, and the rear or back surface. When sound travels from a liquid through a metal, it will con-

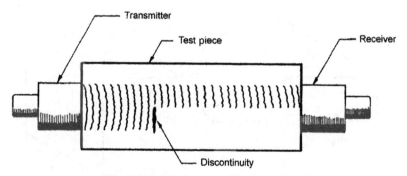

Figure 11.31 Through transmission testing using two search units. Sketch shows attenuation of sonic vibrations by discontinuity.

verge if the surface is concave or diverge if the surface is convex. With straight beam testing, the water path is usually adjusted so that its scan time is longer than the time of the scan in the test piece. By doing this, the first multiple of the front surface signal will be beyond the first back reflection signal. Since sound waves travel about four times faster in steel and aluminum than they do in water, a general rule of thumb is that the water distance should be $\frac{1}{4}$ the part thickness plus $\frac{1}{4}$ in. When immersion testing is used for tapered plates, there should be a uniform water path above the test surface. With immersion testing, false indications from contoured surfaces will result in broad-based noise echoes.

Angle beam immersion testing is accomplished by angulating the search unit to introduce shear waves into tubular or plate test pieces as shown in Figure 11.32. Angulation is needed during immersion testing to obtain the maximum signal response to discontinuities obtained at various angles to the entry surface of the object. Angulation is also used to more accurately determine the orientation of defects in the test piece.

Table 11.7 shows refraction and mode conversion data for three materials immersed in water. When the incident angle is between the first and second critical angles, shear waves are produced.

Through transmission immersion testing can be used with both tubular and plate-shaped parts as illustrated in Figure 11.33. When testing tubular parts, acoustic lenses are often used to normalize the beam so that it enters the test surface perpendicularly. The receiving transducer is generally positioned to receive the maximum amount of transmitted energy for best efficiency.

There are several modified immersion testing techniques such as squirter probes, bubbler probes, and liquid-filled tire or wheel-type probes. A single

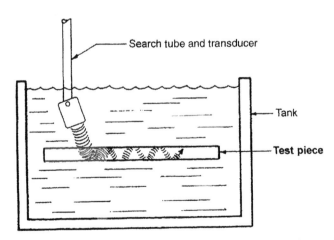

Figure 11.32 Introducing shear waves in a plate using angle beam immersion testing technique.

TABLE 11.7. Angle of Refraction and Mode Conversion—Immersion Testing

θ1 Water (deg)	θ2 Steel (deg)		θ2, Aluminum (deg)		θ2, Bronze (deg)	
Longitudinal	Longitudinal	Shear	Longitudinal	Shear	Longitudinal	Shear
1	4.0	2.1	4.2	2.0	2.5	1.5
10	43.3	21.6	47.3	20.9	24.2	15.0
12	55.7	26.0	62.9	25.3	29.4	18.1
15		33.3		32.1	37.6	23.0
20		46.4		44.6	53.8	30.1
30						47.4

Source: Adapted from information supplied by courtesy of Automation Sperry, a unit of Qualicorp.

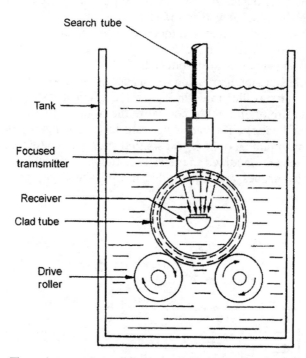

Figure 11.33 Through transmission of a clad tubular element using immersion testing technique. This test can be automated by rotating the element with rollers while scanning it with transducers.

squirter probe can be used for angle beam or reflection testing or two squirter probes can be used for through transmission testing. Squirter probes are used for high-speed scanning of plates, cylinders, and other regularly shaped parts. The bubbler probe is primarily used for angle beam testing. Some of these specialty probes are shown in Figure 11.34.

Immersion testing with wheel-type search units eliminates the need for a tank and provides portability. Ultrasonic wheel units can be used for straight beam, angle beam, or surface wave examinations. The wheel-type search unit has several unique features. It can be used for straight beam testing by either moving the wheel past the test material or the test material past the test wheel. Wheels (Figure 11.35) can incorporate the use of one or more crystals and sound can be projected into the test object at virtually any direction. Wheels can also be used to introduce Lamb waves into thin sheets and plates. The versatility of the wheel and its angle of sound propagation depend on the mounting angle of the transducer in the wheel and the angular position of the wheel mounting mechanism.

Resonance testing may be accomplished by contact or immersion techniques. When ultrasonic energy is coupled into an object, it tends to vibrate in the direction of wave propagation. This amplitude of vibration will increase and reach its maxima at the test object's natural resonant frequency. There may be several modes of vibration for a test object such as axial, longitudinal, flexural, harmonic, radial, and diametral. A node is a point or line in a vibrating body that does not vibrate. Nodes will be present in a test object and their location will be dependent on the mode of vibration, shape, and other physical properties of the object. Resonance testing is frequently used to make

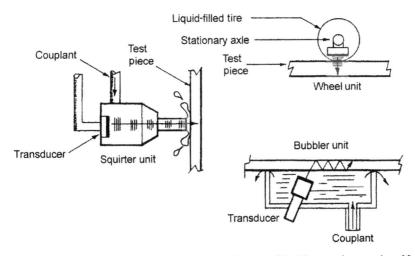

Figure 11.34 A number of specialty probes using modified immersion testing. Note that a liquid couplant is used in all cases.

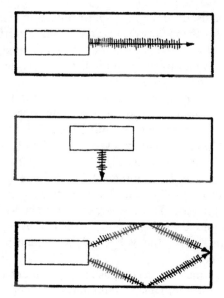

Figure 11.35 Some sound projection patterns available with wheel-type search units. Arrows represent the direction of wave propagation.

thickness measurements on an object from one side. The fundamental resonant frequency of a test object is given by Eq. (11.12):

$$f = v/2t \qquad (11.12)$$

where f = resonant frequency
v = longitudinal velocity
t = object thickness

Note: Resonant-testing equipment generates continuous longitudinal waves.

11.7 ULTRASONIC TESTING EQUIPMENT

Flaw detector operation is possible because of the difference in acoustical impedance of various materials. The acoustical impedance of air is about 150,000 times greater than most metals and therefore the velocity of sound travels about 18 times faster in steel than in air.

Pulse-echo instruments create a burst or pulse of sound when a piezoelectric crystal or transducer is struck with an electrical pulse. The sound wave thus produced travels through the transducer, its coupling agent, and into the material being tested. The sound pulse continues until it encounters a sudden

ULTRASONIC TESTING EQUIPMENT

increase in acoustical impedance such as when it hits an air interface. The air acts as a transmission barrier, causing the sound pulse to echo or return. For defect geometries other than flat, the echo amplitude will be reduced. Several hundred reflections can be received, processed by electronic circuitry, and displayed as a single precise reading.

Modern pulse-echo instruments consist of a shock pulse generator, variable attenuator, amplifier, filter, and spectrum analyzer or oscilloscope with marker, delay, and various gating circuits. Most pulse-echo instruments provide A-scan-type data presentation. The sensitivity of the ultrasonic test system depends on search unit, pulser, and amplifier characteristics.

The shock pulse generator produces a burst of energy for the sending transducer and may also produce a synchronous output pulse for the horizontal amplifier or sweep circuit of the readout. Sweep circuits provide time baselines for the visual display. Modern pulser circuits have adjustments for controlling the pulse repetition rate, pulse dampening, and pulse energy. Increasing the pulse length for crystal activation increases the ultrasonic strength and decreases the resolution power.

The pulse repetition rate is the number of pulses produced per unit time, such as pulses per second. The time from the start to end of the initial pulse determines the maximum feasible pulse repetition rate. In turn, the pulse repetition rate determines the maximum scanning speed. With contact testing, ghost indications may be produced with excessively high repetition rates.

Depending on the transducer backing material, pulse dampening may be required to optimize transducer response and prevent transducer damage. Maximum pulser voltage typically varies from -120 to $-330\,V$ depending on load impedance and design. Pulse repetition rates typically vary from 0.1 to 5 kHz. Risetime (the time required for the pulse to rise from 10 to 90% of its final peak value) varies from 3 to 10 ns. A clock circuit is used to coordinate the timing and operational sequence of the entire system.

A variable attenuator to increase the dynamic range of the instrument may precede the receiver or echo amplifier. One advantage of a frequency-independent attenuator over a continuously adjustable gain control is that signal amplitude measured is independent of frequency. Since the echo pulses must also exceed a threshold voltage, noise, scatter echoes, and small flaw echoes can be suppressed. The variable attenuator also prevents amplifier saturation that would result in a nonlinear output with respect to input voltage.

Receiver amplifier circuits amplify the return signals and modify them for the display. The amplifier's output is often filtered to reduce spurious noise pulses or "hash." The variable attenuator can provide 0 to 68 dB attenuation in 2 dB steps. Available amplifier gain is about 40 dB maximum. The ideal amplifier has uniform frequency response over its entire frequency range.

After amplification, the ultrasonic pulses are sent to a spectrum analyzer or gating and detector circuits for measurement and discrimination of pulse amplitudes. The gate circuits can delay the main bang or selected echo pulses 0.1 to 200 µs, provide display blanking delay of 3 to 100 µs, and provide gate widths of 50 ns to 200 μs, and provide RF markers. Radio frequency indications are above and below the baseline on an A-scan presentation. Marker circuits provide visible distance marks on the display screen. Electronic gates control what information is processed by an instrument; amplitude gates are necessary for automatic examinations. The peak detector can provide a linear output for peak amplitudes of 10 mV to 1.0 V, output reset, and adjustable decay rates.

With the A-scan presentation, the oscilloscope displays the initial pulse or front surface indication, back surface reflection or echo pulse, and reflected pulses from any discontinuities. Pulse amplitude is displayed on the vertical axis of the oscilloscope and elapsed time or distance is displayed on the horizontal axis. Strong, random pulses on an A-scan can be caused by electrical interference. With A-scan, the dead zone refers to the front surface pulse width and recovery time. Vertical indications represent the signal amplitude or amount of energy returning to the search unit. When a vertical indication has reached its maximum signal height, it has reached its "vertical level." Numerous, small-amplitude screen indications are known as *hash* or noise. Event markers and pulse gates may also be displayed on the time baseline. A separate baseline superimposed on the viewing screen is referred to as a marker. The A-scan display is commonly used with ultrasonic inspection of weld areas.

B-scan instruments are used where fast inspection rates are required. Ultrasonic data representative of the cross section of the part is shown on a B-scan presentation. Scanned information is presented on the vertical sweep of the oscilloscope and the mechanical scanning equipment controls the horizontal sweep. B-scan is very effective in scanning rotating tubing and pipe. A rotational signal is used to control the horizontal sweep. A crack penetrating the outer surface at an angle will move across the display screen and change its distance from the transducer. The flaw-to-transducer distance is displayed on the vertical sweep of the display. When the two sweeps are combined, the flaw appears as an inclined line and resembles a cross-sectional view of the flaw. A B-scan shows the relative distance the discontinuity is from the transducer and its length in the direction of transducer travel.

B-scans similar to the one shown in Figure 11.36 have been obtained by Sigma Research during their in-service inspection of tension legs of offshore oil platforms in the North Sea. Their one-of-a-kind system is capable of detecting fatigue cracks in any of the 16 legs, which have 4-in.-thick steel walls and are 500 ft long. Inspection requires that a large number of ultrasonic signals be processed because of the complex geometry of the additional 17-threaded couplings per leg.

A special 17-ft ultrasonic probe fits inside the 3-in. bore of each tension leg. The probe collects and transmits data topside where it is analyzed by a computer and displayed on an operator's console. Fully automatic inspection of straight-walled sections is accomplished by pulling the rotating probe head (Figure 11.37) through the legs with a winch, thus providing a helical scan of the inner surface. The probe head contains 10 specially angled 5 MHz lithium niobate transducers in an anodized aluminum housing.

ULTRASONIC TESTING EQUIPMENT

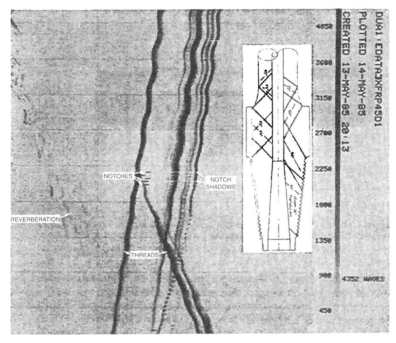

Figure 11.36 B-Scan presentation of threaded tension leg element of offshore drilling platform. Data shows electric discharged machine (EDM) notches used to verify test sensitivity. Courtesy of Sigma Research, Inc.

Figure 11.37 Multiple transducer element used for examining the legs of offshore oil drilling platforms. Courtesy of Sigma Research, Inc.

To provide a thorough inspection of the threaded couplings, the probe is locked to the bore wall, and the transducer head performs a precision scan by rotating and translating across the threaded area. A number of inspection angles are required to provide full coverage inspection of all critical areas. Fatigue cracks at the thread roots are the most potentially dangerous defects and the computer must be able to separate these crack signals from background thread echoes.

With today's high-memory, high-speed computers, data gathering and analysis are primarily limited by the time required to physically scan the legs. In the threaded couplings, the ultrasound travels complex wave paths and, under these conditions, flaw detection has been equated to locating a grain of rice in 72 in. of steel.

With a C-scan presentation, the display shows the flaw area as if it were viewed from above. A C-scan presentation is most often used for high-speed automatic scanning. C-scan can produce a facsimile recording of flaws superimposed on a plan view of the test part. The technique shows the location and size of the flaw in the plan view, but does not give any information with regard to the depth of the flaw between the top and bottom surfaces of the test piece. For C-scan mapping, an X-Y scanner is used to translate the transducer in a plane parallel to the test surface. Figure 11.38 shows a typical C-scan display of a 0.5-in.-thick plate coated on the top surface with a 0.008-in.-thick layer of flame-sprayed aluminum. The plate contains two rows of four flat-bottomed drilled holes of 0.031, 0.062, 0.125, and 0.250 in. diameters. The first row of holes had been drilled to a depth of 0.125 in. and the second row of holes is 0.062 in. deep. The notches shown at the right are circular saw cut notches 0.125 in. wide × 1.2 in. long × 0.062 in. deep and 0.062 in. wide × 0.85 in. long × 0.062 in. deep from top to bottom respectively.

Probe manipulators enable the user to position the ultrasonic search unit normal to the material surface to be inspected or at an angle to detect obliquely oriented discontinuities. The manipulator adjusts and maintains the transducer angle; it is also used to set the proper water path length in immersion testing. Figure 11.39 shows a closeup of a manual probe manipulator. As shown, the probe manipulator can be raised or lowered, translated, and rotated

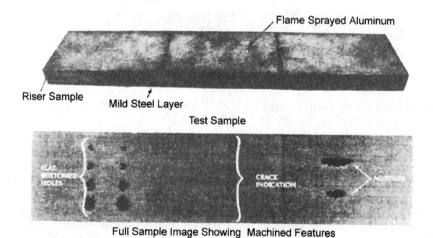

Figure 11.38 C-Scan presentation of composite plate standard. Courtesy of Sigma Research, Inc.

ULTRASONIC TESTING EQUIPMENT

Figure 11.39 Photograph showing partial view of immersion tank, probe manipulator, and search unit. Courtesy of TAC Technical Instrument Corp.

in two directions so that the transducer can be positioned at virtually any angle with respect to the test surface. Manual adjustments and clamp locking arrangements are clearly shown in the figure. This manipulator is mounted on an immersion inspection tank and is used for ultrasonic inspection of tubing and bar stock.

Figure 11.40 shows a probe manipulator, small immersion tank, and X-Y-Z (three-dimensional) probe *manipulator bridge* arrangement. The manipulator bridge is called a bridge because it spans the width of the immersion tank. In automatic scanning, the bridge supports the manipulator and moves the scanner tube. As shown, the probe can be moved across the tank with the screw drive arrangement, and raised or lowered and moved down the length of the tank with two rack and gear arrangements. The probe manipulator system may be driven manually or automatically by small electric motors to provide manual or automatic scanning of the test object. Manual scanning is often used to locate and pinpoint small defects.

Figure 11.40 Sealed immersion tank, probe bridge, and manipulator. Courtesy of Testech, Inc.

Figure 11.41 shows an automatic scanning arrangement using focused ultrasonic transducers for the on-mill inspection of welded pipe. Note that liquid coupling agent can be added through the clear plastic tubing arrangement. Note also that the wear surfaces of the transducers are contoured to the outside diameter of the pipe. Contoured shoes may be spring-loaded to accommodate small irregularities in the pipe surface.

Figure 11.42 shows a complete automatic scanning immersion-type system consisting of a bridge scanner, search tube and mount with probe manipulator, and turntable with drum recorder. An X-Y recorder (optional) can be used in place of the turntable and drum recorder, depending on the size and shape of the test object. The local console or a host computer can control the automatic scanning bridge. The unit shown can be benchtop mounted and is considered a small to midsize unit by the manufacturer.

A large-scale ultrasonic tubing and bar inspection system is shown in Figure 11.43. The control console also includes controls for material handling on the input and output conveyors. Since most tubing and bar stock is helically scanned, Eqs. 11.13–11.17 can be used to help determine running time or the time required for inspection:

$$\text{sv}_{(m/min)} = 0.06 \, \text{pps} \times \text{displacement}_{(mm)} \qquad (11.13)$$

$$\text{sv}_{(m/min)} = \text{rpm}/1000 \times \pi \times d_{(mm)} \qquad (11.14)$$

ULTRASONIC TESTING EQUIPMENT

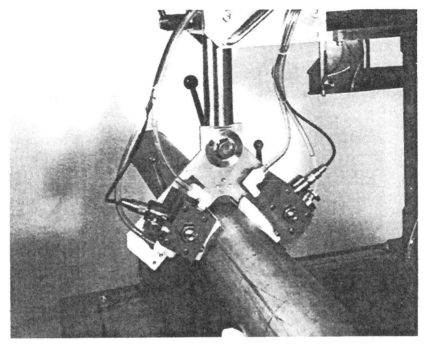

Figure 11.41 Squirter probes with contoured shoes for automatic high-speed testing of pipe and tubing. Courtesy of TAC Technical Instrument Corp.

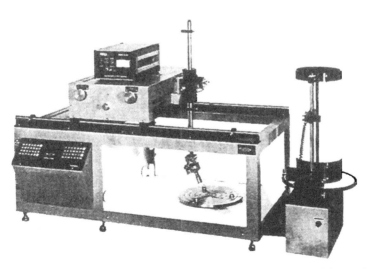

Figure 11.42 Medium-sized ultrasonic inspection system. Turntable and drum recorder can be removed and replaced with an X-Y recorder. Courtesy of Testech, Inc.

Figure 11.43 Large-scale ultrasonic testing station for high-speed inspection of tubes, bars, and pipes. Courtesy of TAC Technical Instrument Group.

$$lr_{(m/min)} = rpm/1000 \times pitch_{(mm)} \qquad (11.15)$$

$$pitch_{(max)} = \text{defect length} - \text{beam length} \qquad (11.16)$$

$$lr = sv \times pitch/(\pi \times d) \qquad (11.17)$$

where sv = surface velocity
pps = pulses per second
d = diameter
rpm = revolutions per minute
m/min = meters per minute
mm = millimeters

The preceding equations do not take into account setup time, loading and unloading time, or the time required for calibration, record keeping, and defect marking and segregation.

High technology has also been incorporated into ultrasonic inspection systems for the detection and measurement of intergranular stress corrosion cracking (IGSCC) in stainless-steel pipes that carry the circulation water of a

ULTRASONIC TESTING EQUIPMENT 507

boiling water reactor (BWR). Three conditions that contribute to the formation and growth of IGSCC are:

1. The formation of chromium carbide precipitates at the grain boundaries in the heat-affected zone (HAZ) about 0.25 in. on either side of a weld. Most cracking occurs in these regions.
2. The presence of tensile stress caused by external forces or internal forces during welding or machining. These forces can cause microscopic cracking at sensitized grain boundaries that allows oxygen to diffuse into minute cracks.
3. The presence of oxygen in cooling water when electrochemical potential is not kept below −350 mV. This is a difficult level to achieve because normal cooling water has an electrochemical potential of −150 mV.

One system, designed by AMDATA systems, Inc. and known as Intraspect/98, consists of a robotic ultrasonic scanning system, called AMAPS (AMdata Automatic Pipe Scanner), and a microprocessor-based data acquisition and imaging system. Figure 11.44 shows the compact automatic scanner and a typical ultrasonic image.

The Intraspect/98 system is composed of an HP computer and display. The processor provides system control and evaluates results on a real-time basis during the test. The scanning subsystem consists of a scanner, search unit, scan controller, and track assembly. The scanner may be used with other manufacturers' search units and other types of sensors, such as eddy current probes.

The scan controller provides timing signals and two-way communications with the computer. The controller and power supply measure 15 in. wide, 7 in. high, and 28 in. deep; the unit weighs about 30 lb.

The AMAPS scanner can be installed by one person in about 5 min. on a variety of surfaces; the contact-type transducer is gimbal mounted. Encoders that operate independently of the drive train are used to determine scanner position. The heavy stepper motors are chain-and-sprocket driven; their bearings are permanently lubricated.

The Y-axis stepping motor uses a belt-and-pulley-driven lead screw and nut arrangement to position the probe in the vertical direction as shown in the figure. The X-axis stepping motor positions the scanner in the circumferential direction around the pipe. It is chain driven from the stepping motor to two magnetic wheel assemblies. A flexible, mild steel track guide conforms to almost any pipe geometry and curvature. The magnetic coupling between the hardened track wheels and mild steel track provides for quick and easy assembly or removal.

11.7.1 Equipment Operation

In their most basic form, ultrasonic flaw detectors consist of a pulser unit and amplifier/receiver with built-in display. The StressTel FlawMike®, shown in

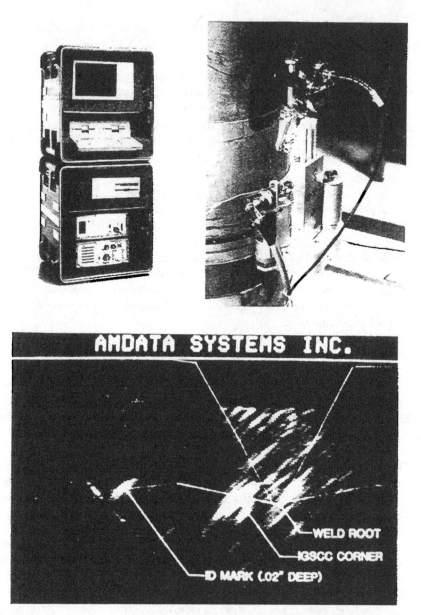

Figure 11.44 Automatic testing equipment for evaluating intergranular stress crack corrosion (IGSCC) in stainless-steel welds. Courtesy of AMDATA Systems, Inc.

ULTRASONIC TESTING EQUIPMENT

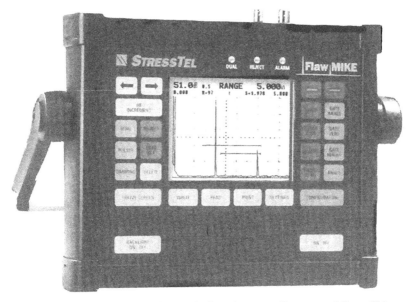

Figure 11.45 FlawMike ultrasonic flaw detector. Courtesy of StressTel.

Figure 11.45, is a tough, easy-to-use portable ultrasonic flaw detector with direct access keys to all functions. One key press changes gain, turns on the backlight, prints a report, and displays all instrument settings. The flaw detector features a large, high-resolution LCD scale that is easy to read from all angles. The readout has excellent daylight visibility and is updated at a 60 Hz rate. The FlawMike is housed in a metal enclosure, has a weather-resistant industrial keypad, and comes with a durable instrument case.

11.7.2 Flaw Transducers

Flaw detector transducers include single-element contact, single-element delay line, dual-element, and angle beam units designed for optimum performance with the StressTel FlawMike and other compatible flaw detectors. Figure 11.46 shows nine types of available transducers.

11.7.2.1 Instrument Features The main instrument features include:

- On-screen amplitude measurement
- Two gates for echo-to-echo digital thickness measurements
- Trig functions for calculation of sound path, projection distance, and depth
- Setup memory: 80 data sets for rapid instrument setup
- Long 8-hour battery life

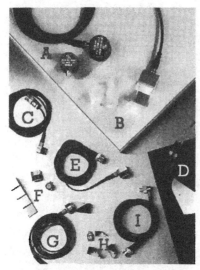

Flaw transducers in photo:
A. standard contact
B. AWS angle beam wedges
C. fingertip contact
D. dual element
E. fingertip contact
F. fingertip contact
G. mini angle beam w/wedge
H. mini angle beam w/wedges
I. delay line

Figure 11.46 A selection of ultrasonic flaw transducers. Courtesy of StessTel.

- Printer support: Epson, HP Deskjet, HP Laserjet, and Seiko DPU
- Languages: English, German, French, Spanish, Italian, and Russian
- Optional *distance amplitude correction (DAC)* to facilitate evaluation of flaws at various depths in the test material

11.7.2.2 Ultrasonic Specifications
Frequency range: 0.5 to 15 MHz @ −3 dB
Measuring range (steel): 0.14 in. to 120 in (3.6 mm to 3040 mm)

11.7.2.3 Physical Description and Power Supply
The FlawMike is 9.6 inches wide by 7.7 inches high and 2.1 inches deep. The unit weighs 5 pounds with batteries and is environmentally protected. The power supply uses 4 C alkaline or 4 NiCad rechargeable batteries. Battery life is rated as 8 hours without backlighting and 6 hours with backlighting. The instrument has a full two-year warranty excluding batteries.

With the proper probe and calibration the FlawMike serves as a lightweight, portable thickness gauge as well and can be calibrated for steel thickness ranges from 0.14 to 120 inches. With a dual-element probe operating on the pulse-echo principle, the ultrasonic signal travels through the material, strikes the opposite surface, and is reflected back to the probe. Material thickness is determined by the precise measurement of time required for the round trip.

A block diagram of a Panametrics Model 502UA ultrasonic analyzer is shown in Figure 11.47. The transducer under test is driven by shock pulse generator and the received signal is fed into a variable attenuator followed by

ULTRASONIC TESTING EQUIPMENT

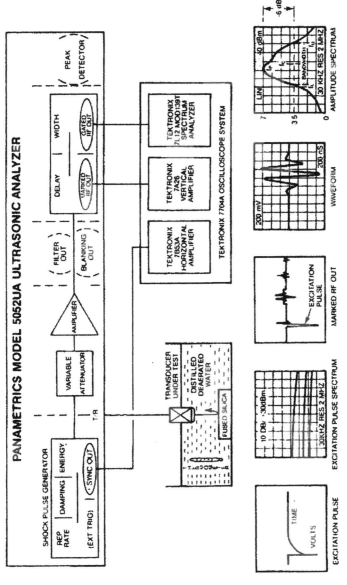

Figure 11.47 Block diagram of modern ultrasonic analyzer. Courtesy of Panametrics, Inc.

an amplifier. Signal delay and gating are provided. Key waveforms are shown at the bottom of the illustration in sequential order.

The equipment described in this section represents some of the current state-of-the-art ultrasonic testing equipment. Other manufacturers of similar equipment may incorporate many of the same and/or other design features. A comparison of design features by manufacturer is beyond the scope of this book.

Some considerations for the care of electronic equipment of this nature are as follows:

- Search units and electronic equipment should not be subjected to temperature extremes. Consult manufacturer's data sheets for operating and storage temperature ranges.
- Search units and electronic equipment should not be dropped or subjected to excessive shock or physical abuse.
- Search units designed for immersion testing should not be energized until they have been immersed unless it is known that they will work equally well in air.
- Sensitive electronic equipment should not be subjected to excessive high humidity or harsh chemical environments.

11.7.3 Testing Procedures

The selections of test parameters depend on what is to be tested, the expected discontinuities, availability of test equipment, and time and cost constraints that may exist. Many test parameters may be selected based on actual test trials. Table 11.8 lists some frequency considerations for contact testing.

Search unit size and type are important test parameters. Large-diameter transducers, rectangular-shaped paintbrush transducers, and mosaic (multiple-crystal) transducers can be used to provide rapid inspection of relatively large surface area with some corresponding decrease in sensitivity. Large, thin, high-

TABLE 11.8. Frequency Considerations

Test Parameter	Recommended Frequency or Range (MHz)
Cast iron and coarse-grained materials; high-carbon steel	0.5
Refined grained steels; small discontinuities (bursts, flaking, and pipe); large forgings	2.25–5.0
Small forgings	5.0–10
Microscopic defects; fatigue cracks and segregation; drawn and extruded products	10

frequency search units are more susceptible to breakage. Large, lower-frequency units permit testing at greater sound path lengths and depths. Smaller, focused-beam search units have higher sensitivity and provide greater accuracy in mapping defect areas.

With immersion testing, focused transducers are necessary for optimizing sensitivity when testing tubular elements. Convex acoustic lenses are required for testing inside surfaces and concave acoustic lenses are required for testing outer surfaces. The water distance, distance from the face of the search unit to front surface of the test piece, should be such that the second water reflection from the front surface is to the right of the first back surface reflection on the oscilloscope presentation. Because sound travels four times faster in aluminum and steel test objects than it does in water, the water distance should be a minimum of $\frac{1}{4}$ of the test object thickness plus $\frac{1}{4}$ in.; with a 1-in.-thick steel object, the minimum water distance would be $\frac{1}{2}$ in. Using sweep delay, the water distance can be blanked from the oscilloscope screen if desired. Test blocks with known discontinuities and geometries can be used to calibrate both contact- and immersion-type ultrasonic test equipment.

Many factors affect the maximum scanning and indexing speeds that can be reasonably obtained. Some of these factors are:

- The size and shape of the test object
- The type of presentation, A-, B-, or C-scan
- Pulse repetition rate
- Whether alarms or recordings are to be provided
- Whether automatic marking of defects is to be provided
- Transducer size and frequency

Higher automatic scanning speeds can usually be provided with immersion testing or modified immersion-testing techniques using bubbler, squirter, or wheel- or array-type transducers. Lightweight pipe and tubing can be helically scanned at relatively high rates of speed. Some typical inspection speeds for pipe, tubing, and barstock are listed in Table 11.9.

TABLE 11.9. Typical Inspection Speeds for Tubular Elements and Barstock

Type of Material	Rotational Speed (rpm)[a]	Linear Speed (fpm)[b]
Lightweight pipe and tubing	240	20
Normal testing of pipe and barstock	120	10
Critical inspection of all material	60	5

[a] rpm = revolution per minute.
[b] fpm = feet per minute.

Table 11.9 provides some guidelines for inspection, but Eqs. 11.13–11.17 should be used for more accurate determinations of scanning speed. Ideally, scanning speeds would be continuously adjustable over a wide range of settings.

Calibration of ultrasonic instruments is accomplished by comparison with known standard reference blocks usually containing flat-bottomed holes of various diameters. Reference standards are used for the estimation of defect size. Notches have proved to be useful reference standards with immersion testing of rotating tubing. Notches are frequently used for distance amplitude calibration with shear waves; they provide good simulation for root weld lack of penetration. A list of some operating parameters (controllable) and material parameters (fixed) affecting calibration are given in Table 11.10.

In order to obtain reproducible signals for known discontinuities, distance amplitude, area amplitude, resolution, and International Institute of Welding reference blocks have been developed. The ASTM distance amplitude blocks consist of 19 standard reference blocks each with the same-size, flat-bottomed test hole. The variable in distance amplitude blocks is the metal distance above the drilled hole. Available hole diameters are 3/64, 5/64, and 8/64 ($\frac{1}{4}$ in.). Metal path distances are 1/16, 1/8, $\frac{1}{4}$, 3/8, $\frac{1}{2}$, 5/8, $\frac{3}{4}$, 7/8, 1, $1\frac{1}{4}$, $1\frac{3}{4}$, $2\frac{1}{4}$, $2\frac{3}{4}$, $3\frac{1}{4}$, $3\frac{3}{4}$, $4\frac{1}{4}$, $4\frac{3}{4}$, $5\frac{1}{4}$, and $5\frac{3}{4}$ in. Transducer response to discontinuity depth variations is determined by the relative echo response to blocks within this group. When the nearest hole gives the poorest results, the problem could be inconsistent surface of the test block, near-field effects, or incorrect hole geometry.

The ASTM area amplitude blocks consist of 8 standard reference blocks having flat-bottomed test holes with diameters ranging from 1/64 to 8/64 in. in 1/64 in. steps. There is a 3 in. metal path distance to each of the flat-bottomed holes. Transducer response to discontinuity area variation is determined by relative echo response to blocks within the group.

TABLE 11.10. Calibration Parameters

Operating Parameters	Material Parameters
Coupling method	Attenuation
Distance–amplitude effect	Geometry
Equipment type	Impedance
Scanning method	Noise level
Search unit characteristics	Surface
beam shape	Velocity
frequency	Flaw characteristics
size	depth
type	impedance
	location
	orientation
	shape
	size

ASTM combination distance/area amplitude blocks consist of 10 standard reference blocks. Three of the blocks have one flat-bottomed hole each of 3/64, 5/64, and 8/64-in.-diameter holes with metal path distances of 1/8, $\frac{1}{4}$, $\frac{1}{2}$, $\frac{3}{4}$, $1\frac{1}{2}$, and 6 in. One block has an 8/64-in.-diameter hole with a metal path distance of 6 in.

A typical resolution block has 38 holes with metal path distances of 0.05 to 1.250 in. There are 10 holes each of 3/64, 5/64, and 8/64 in. in diameter. In addition to these blocks, angle beam blocks and welding reference blocks have been developed to determine the sensitivity and resolution of shear wave transducers. These blocks typically have reference discontinuities and engraved scales for surface path and defect depth measurements. When adjusting the flaw-locating rule for a shear wave weld inspection, the zero point of the rule must be at the sound beam exit point of the wedge. A side hole, drilled parallel to the test surface and perpendicular to the sound path, provides a reference discontinuity that is not dependent on beam angle. Most reference blocks are available in a clear anodized aluminum or nickel-plated steel construction. Figure 11.48 shows the distance amplitude block set, welding reference block, and miniature angle beam calibration block.

One manufacturer's procedure for testing tubing and pipe for nuclear and other critical applications is outlined below:

Procedure Contents

Scope
Type of test
Equipment
Comparative calibration standards
Choice of transducer
Longitudinal defect detection
Dynamic analysis
Static analysis
Transverse defect detection
Transducer focusing
Spherically focused transducers
Cylindrically focused transducers
Feed pitch adjustment
Speed adjustment
Static calibration
Longitudinal calibration notches
Transverse calibration standards
Dynamic calibration
Testing

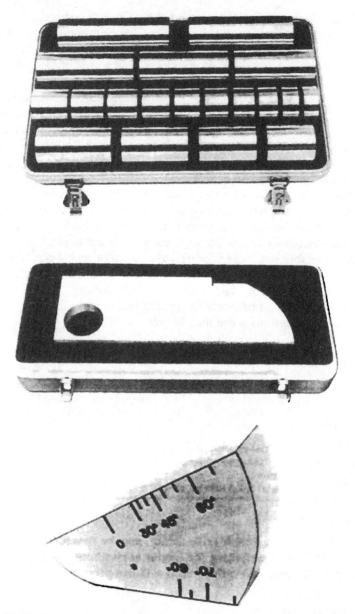

Figure 11.48 Standard amplitude-distance blocks. Courtesy of Panametrics, Inc.

Defect analysis and marking
Calibration checks
Packing
Specification form for specific inspection program

Other users of ultrasonic testing equipment with similar applications may develop procedures with similar table listings. The general specification form used with this procedure is shown in Table 11.11.

11.7.3.1 Variables Affecting Results It would be nice if all discontinuities were well-defined flat-bottomed holes or sharp-edged notches. However, in the real world, discontinuities are not usually that well defined. The ultrasonic impedance mismatch for most defects, such as nonmetallic inclusions, gas porosity, cracks, blowholes, and so on, makes it difficult if not impossible to make direct comparisons with the indications received from calibration standards. For example, gas discontinuities are reduced to flat disks parallel to the surface by rolling and discontinuities with concave surfaces cause the reflected beam to be focused at a point dependent on the curvature of the discontinuity. Therefore, there is less reflection from the actual discontinuity than might be expected and considerable interpretation of test results is required. Errors can also exist in ultrasonic measurements if the velocity of propagation varies significantly.

Perpendicular discontinuities produce the largest-amplitude display indications. Discontinuities parallel to the sound beam will produce small indications compared to their length. To minimize this decrease in sensitivity, transducers used for parallel flaw detection should have a large surface area

TABLE 11.11. Specification Form[a]

Project No. _____
Customer _____
Customer's address _____
Governing specifications _____
Remarks _____

Comparative Standard Notches
Longitudinal   o.d. _____   i.d. _____   length _____   depth _____
Transverse   o.d. _____   i.d. _____   length _____   depth _____
Instrumentation _____

Transducer beam size Longitudinal notch _____
                    Transverse notch _____
Prepared by _____ NDT level III certification
Date _____

[a] Courtesy of TAC Technical Instrument Corp., Trenton, N.J.

and high frequency. Differences in signals received from identical reflectors at different distances from a transducer may be caused by material attenuation, beam divergence, or near-field effects.

Small defects can produce indications with fluctuating amplitude when they are in the near field. Beam divergence occurs only in the far field. The energy reflected from a flaw depends on its size, orientation, and type. The maximum echo reflection is obtained when the discontinuity thickness is one-fourth wavelength. A smooth, flat discontinuity, not perpendicular to the direction of sound propagation, may be indicated by the loss of the back surface reflection or by echo amplitude larger or equal to the back surface indication.

Facilities with a metallurgical lab have a considerable advantage; they can destructively examine discontinuities to determine the severity of the defect and then compare these results with ultrasonic recordings.

Other techniques, such as changing search beam frequency and angle, can be employed to help determine the area, location, and severity of the defect. Some variables that complicate test analyses are:

Beam spread—caused by finite transducer size

Beam attenuation—caused by scattering or dampening

Beam nonuniformity—caused by near-field effects or physical properties

Defect impedance—dependent on defect size, shape, and surface condition

Geometry—false signals caused by irregular surfaces, fillets, or corners

Noise problems—caused by external RF interference, internal signal-to-noise characteristics of the equipment, loose transducer crystals, dirt, grease, air bubbles, poor transducer coupling, or grain size of the material

Transducer loading—caused by couplant characteristics or specimen characteristics

It should also be remembered that the presence of a discontinuity will not produce a specific indication with the through transmission method and that the depth of a discontinuity cannot be determined.

Inspection of castings is often impractical because of their coarse grain structure. Some casting discontinuities are blowholes produced by escaping gases during the solidification of metals. When grain size is one-tenth wavelength or larger, excessive scattering may occur.

Testing of coarse-grained materials is sometimes possible after grain refining heat treatment. Coarse-grained materials should be tested at lower frequencies because higher-frequency beams are more easily scattered. The random distribution of large crystal directions determines the acoustic noise level, selection of test frequency, and scattering of sound. Attenuation losses are higher at higher crystal frequencies. However, higher frequencies can distinguish better between large planar defects and stacked laminations in rolled plate.

In forgings and wrought products, discontinuities are usually aligned with the direction of grain flow or forging lines. The preferred method for testing a complex-shaped forging combines thorough inspection of the billet before forging with careful inspection of the finished part. With shrink fit materials, a tighter shrink fit will increase the back surface reflection while reducing the interface reflection.

Rough entry surface conditions cause loss of echo amplitude and an increase in the width of the front surface echo that can decrease resolution and conceal surface defects. Contoured surfaces produce broad-based false signal indications with immersion testing. With surface wave testing, oil, loose scale, dirt, sharp-edged corners, and contoured surfaces can produce nonrelevant display indications. Surface waves can also be highly attenuated by heavy couplants. When rough-surfaced parts are tested by techniques other than surface waves, lower frequencies and heavier couplants should be used.

11.8 TIME-OF-FLIGHT DIFFRACTION (TOFD)

The TOFD method is relatively new. It relies on the diffraction of ultrasonic energies from defects in the butt weld joints of heavy pipe and pressure vessels. Early applications include the periodic inspection of offshore platform piping. In most cases, TOFD can supplement conventional UT and RT methods—in some cases it may be able to replace them. See Chapter 13, Visual and Optical Testing, for additional information.

12

VIBRATION ANALYSIS METHOD

12.1 INTRODUCTION

Vibration analysis is widely used in predictive maintenance programs involving pumps, motors, gearboxes, turbines, fans, and compressors, as well as all types of vehicles, heavy machinery, bridges, and civil engineering structures. Excessive vibration of equipment under load can be caused by wear, corrosion, or even forces of nature, and is a major cause of equipment or structure failure. In many cases, excessive vibration causes excessive noise.

Perhaps one of our earliest encounters with vibration and sound was in an elementary school music class when the teacher showed us a set of tuning forks and struck them one at a time on a solid object. Each tuning fork produced a different tone as it vibrated at its natural or resonant frequency. By touching both tines of the fork, the teacher dampened the vibration and stopped the sound. We learned that sound and vibration are closely related.

Vibration testing is the shaking or shocking of a component or assembly to see how well it will stand up to everyday use and abuse. Vibration analysis is a broad subject involving many techniques. Applications include the testing of airplanes, appliances, bridges, buildings, circuit boards, computers, instruments, rotating equipment, spacecraft, satellites, and vehicles. Equipment used in vibration testing includes climate-controlled chambers, controllers, data analyzers, instruments, sensors, and impulse and vibration exciters.

In general, we must shake the structure and sense the amplitude and frequencies of vibration. With heavy rotating equipment like engines, motors, tur-

Introduction to Nondestructive Testing: A Training Guide, Second Edition, by Paul E. Mix
Copyright © 2005 John Wiley & Sons, Inc.

bines, and compressors, we want the main rotor to turn smoothly and quietly. When you listen to vibrations in bearing housings, you may hear periodic thumping sounds instead of a smooth whirling sound, indicating potential bearing problems.

Vibration analysis is an important consideration in many industries. Automobiles are protected against shock, vibration, wind, and rain by their structural design, pneumatic tires, gaskets, shock absorbers, motor mounts, springs, and protective finishes. Military equipment and components must be "hardened" or protected against shock, vibration, dust, sand storms, high temperatures, and moisture. Individual components and complete assemblies are often tested using "shakers" or "shaker tables" capable of generating variable but controlled shock and vibration loads.

Shakers can be mechanically, electrohydraulically, or electromagnetically driven over frequency ranges of 10 to 55 Hz, 0 to 500 Hz, and 5 to 2 KHz, respectively. Mechanical shakers are the easiest to maintain and the least expensive, but they cannot produce random vibrations. Electromagnetic shakers produce the purest sine waves, but they are more difficult to maintain and the most expensive. Mechanical shakers are limited to a stroke of about 0.1 in., whereas the other shakers have stroke ranges of several inches. Reaction masses are typically used with both mechanical and electrohydraulic shakers, but not electromagnetic shakers.

12.2 PRINCIPLES/THEORY

12.2.1 Modes of Vibration

Almost everything from relatively simple molecules to some of the world's largest buildings vibrates at what is known as fundamental or resonant frequencies. A string fixed at both ends, as shown in Figure 12.1, illustrates simple transverse modes of vibration. The fixed ends of the string are nodes and the wavelength and frequency of the string for the fundamental and first two harmonic modes of vibration are shown in Eqs. 12.1–12.3 below.

$$\lambda = 2L; f_1 = v/2L \text{ (fundamental mode)} \quad (12.1)$$

$$\lambda = L; f_2 = v/L \text{ (1st harmonic)} \quad (12.2)$$

$$\lambda = 2L/3; f_3 = 3v/2L \text{ (2nd harmonic)} \quad (12.3)$$

Notes:

- An infinite number of harmonics, odd and even, can be developed for a vibrating string.
- The nodes and antinodes of vibrating strings are opposite to those of air columns.

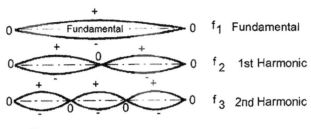

Figure 12.1 Modes of vibration showing frequency relationships.

- Tension and mass are additional factors affecting the velocity of string vibration.
- *Mode shape*—the relative position of all points on a structure at a given natural frequency. Mode shapes are frequently exaggerated for illustration purposes.

12.2.2 Resonance

Resonant frequency is a natural frequency of vibration determined by the physical parameters of a vibrating object. Objects resonate when the frequency of the driving force coincides with the natural frequency of the driven system. Physically determined natural frequencies apply universally to the fields of modern physics, mechanics, electricity, and magnetism. Increasing loads at resonant frequencies can increase stress and even cause catastrophic failures. Soldiers keeping cadence while marching across a bridge could be an example of this. Some characteristics of resonance are:

- Most objects have multiple resonant frequencies.
- Any object composed of an elastic material will vibrate at its own set of natural or resonant frequencies when disturbed.
- It is easy to get an object to vibrate at its resonant frequencies and extremely difficult to get them to vibrate at other frequencies.

- Objects will resonate at their resonant frequencies even when they are excited by a complex source of multiple frequencies. The resonant frequencies tend to filter out other frequencies.
- The greatest amplitude of vibration will take place when the frequency of the driving force coincides with the natural frequency of the driven system.

For building vibrations, the natural frequency for each mode of vibration can be calculated using Eq. 12.4:

$$f_n = 1/2\pi \times \sqrt{K/M} \tag{12.4}$$

where f_n = natural frequency in Hertz
K = stiffness factor associated with this mode
M = mass of the building associated with this mode

Therefore, buildings have lower natural frequencies when they have heavier masses or lower stiffness factors. As might be expected, tall buildings tend to have lower stiffness factors than short buildings.

With lighter and stronger machine designs today, mechanical resonance is a growing problem. Since resonance occurs at frequencies where damping is very low, every mechanical system requires some damping. There are only two things that can be done when a machine is vibrating due to resonance—remove the source or the response in the mechanical system.

12.2.3 Degrees of Freedom

If an object or system has one degree of freedom (DOF), it has one mode of vibration and one natural or resonant frequency. However, if the system has 2, 3, 4, or 5 degrees of freedom, it will have 2, 3, 4, or 5 modes of vibration and the same number of resonant frequencies. A continuous system has an infinite number of modes of vibration.

12.3 SOURCES OF VIBRATION

Rotating equipment, which is subject to excessive vibration and wear, are motors, gearboxes, pumps, blowers, fans, compressors, and turbines. Individual components within this type of equipment that are most subject to wear, vibration and noise are bearings, couplings, rotors, gears, and impellers. Correct alignment and balance of new rotating equipment is a key to preventing early component failures. Original equipment manufacturers often supply detailed information on installation, alignment, and balance of new, heavy industrial equipment.

12.4 NOISE ANALYSIS

Noise analysis is often used to assure environmental compliance with OSHA and local zoning regulations. Noise detectors can also be used to find and identify the sources of noise and paths of transmission, and identify radiating surfaces.

Sound level meters (SLMs) are frequently used for noise analysis. The sound level meter measures the decibel (dB) level of noise and is used to perform octave and $\frac{1}{3}$ octave band analysis. The output of the SLM can be fed into a spectrum analyzer to identify discrete frequency components in the noise spectrum.

Acoustic intensity testing with two microphones is a preferred noise analysis tool. The dual microphone system measures sound vectors whose calibration enables rapid determination of overall radiated sound power levels. Acoustic intensity is a useful tool for positive identification of individual noise sources.

12.5 STRESS ANALYSIS

Strain gauges are frequently used to determine material stress due to static and dynamic loads coming from internal and external sources, such as mechanical, thermal, and pressure. Strain gauges are typically bonded or welded to the test surface. A single strain gauge in a $\frac{1}{4}$-bridge arrangement can be used to measure strain along a single axis.

Mohr's circle is a graphical method for showing the possible results of a stress transformation. A rosette of three $\frac{1}{4}$-bridge circuits can be used to determine primary stresses for constructing a Mohr's circle of the stress state. As shown by Figure 12.2, the principal stresses σ_1 and σ_2 and the maximum shear stress τ_{max} are known as soon as the Mohr circle is drawn. Normal stresses are shown on the horizontal X-axis and shearing stresses are shown on the vertical Y-axis. Tensile normal stresses are positive and compressive normal stresses are negative. The pair of shearing stresses that cause a clockwise couple are plotted above the X-axis and the pair of shearing stresses that cause a counterclockwise couple are plotted below the X-axis. In the figure shown, there is a tensile normal stress of 20 MPa, a compressive normal stress of −5 MPa, and clockwise and counterclockwise shearing stresses of ±15 MPa. A number of interactive Mohr circle applets are available for use on the Internet.

Full bridge circuits (four $\frac{1}{4}$-bridges) can be used to measure bending, tension, and torsion. Full bridge circuit results can be transmitted by telemetry to indicate blade tension due to centrifugal force, bending due to lift or drag, and torsion in the driveshaft of an axial flow fan.

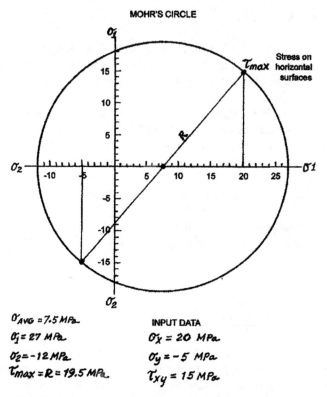

Figure 12.2 Mohr's circle—a simple aid for calculating principle stresses and shear stress λmax.

12.6 MODAL ANALYSIS

Modal analysis determines a structure's dynamic characteristics such as resonant frequencies, damping values, and patterns of structural deformation known as mode shapes. Modal analysis is performed when the test object is not running.

An instrumented impact hammer or electrodynamic shaker is used as an input force at various geometric locations on the test object. Responses to the input are measured in the X, Y, and Z direction with accelerometers. The *frequency response function FRF* of the object is the response per unit force over the frequency range of interest. The modes are used to visualize how the test object is moving and determine the number of modes present.

Modal analysis software is used to fit data into a matrix to determine the unique mode shape associated with each resonant frequency. Associated mode shapes are then displayed on a computer screen. Then additional test data are input to the structural modification software to study the effectiveness of

VIBRATION ANALYSIS/TROUBLESHOOTING

potential corrective action before making actual modifications. If vibration troubleshooting indicates a problem is due to a forced vibration rather than a resonant enhancement, an operating deflection shape analysis is performed.

Operating deflection shape (ODS) displays the peak and phase deflection measurements of an operating machine. A few operating forces that can excite natural frequencies are imbalance, vane pass, and turbulent flow. The machines vibrate when the inherent design properties of the system such as mass, stiffness, and dampening are overridden. This analysis is similar to the modal analysis and results can be also be displayed as an animated structural deformation.

12.7 VIBRATION ANALYSIS/TROUBLESHOOTING

12.7.1 Rotating Equipment Analysis

Rotating equipment analysis is used to assess the general mechanical health of rotating and reciprocating machinery. Waterfall plots and order tracking can be used to quantify operational characteristics and the critical speed of rotating machinery in both the frequency and time domains. A *waterfall plot* shows test object vibration frequency on the x-axis, amplitude of vibration on the y-axis, and time on the z-axis. Another popular presentation, used in conjunction with order tracking, is to show orders on the x-axis, amplitude on the y-axis, and rpm on the z-axis. A spectrogram is another method for showing 3D data. It is similar to viewing a waterfall plot from the top so that color or shades of gray accentuate the amplitude of vibration. Waterfall analysis is used to examine data trends versus machine shaft speed (rpm) and elapsed time. A typical plot is shown in Figure 12.3.

Order/s is an expression that relates a subsynchronous, synchronous, or nonsynchronous frequency to rotating shaft turning speed (TS). Order frequency can be calculated from Eq. 12.5:

$$\text{Order} = f/\text{TS} \qquad (12.5)$$

where f = frequency of shaft rotation
TS = normal turning speed of the shaft

12.7.2 Order Analysis

Order analysis refers to frequency analysis where the frequency axis of the spectrum is expressed in orders of rpm instead of hertz or rpm. *Order tracking* is usually associated with waterfall plots where running speed and its harmonics always occur at the same frequencies or orders in the spectrum regardless of machine speed. Line frequency effects show up as curves on the waterfall plot.

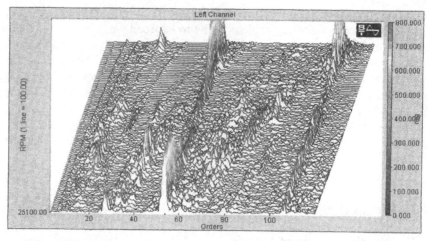

Figure 12.3 Waterfall plot showing high amplitudes of vibration at 53rd order. Courtesy of HEM Data Corporation.

12.8 TRANSFER FUNCTIONS

A transfer function is the output-to-input relationship of a structure. Mathematically, it is the Laplace transform of the output divided by the Laplace transform of the input. The six functions that transfer functions are used for with regard to vibration analysis are:

1. Inertance = Acceleration/Force (g's/lb)
2. Mobility = Velocity/Force (in./s/lb)
3. Compliance = Displacement/Force (mils/lb), and their inverse as follows
4. Mass = Force/Acceleration (lb/g's)
5. Impedance = Force/Velocity (lb/in./s), and
6. Stiffness = Force/displacement (lb/mils)

Transfer function plots can be used to help analyze rotor modes.

12.9 PREDICTIVE MAINTENANCE

In the case of rotating machinery, the first step is to create a custom database using vibration sensors and vibration analysis software. The initial database should include complete machine identification, running speeds of the motor and all driven components, vibration analysis data for every bearing, and every bearing measuring position that can be recorded.

The next step is to evaluate the individual spectra for each bearing in the machine to determine if there are any problems noted for a specific rotating unit. The numeric amplitude of vibration for each bearing can then be compared to ISO tolerances to determine when bearing replacement is justified. Absolute bearing vibrations using velocity transducers in addition to accelerometers can be an integral part of the program. Similar records should be kept for each new piece of rotating equipment.

12.10 FAILURE ANALYSIS

Failure analysis attempts to explain what happened to cause an unexpected, premature, or catastrophic structural failure. All contributing factors must be examined to determine the root cause.

Structures, large and small, fail because of fatigue damage caused by restrained thermal expansion, repetitively applied external loads or forces, and internal vibration sources. Failure analysis uses stress analysis and strain gauge testing to determine the stress state in the region of failure initiation. Based on the material's properties, it can be determined if the failure was initiated by low cycle failure, high cycle failure, or if a crack grew to critical size and catastrophically failed due to repetitive stress and other factors such as corrosion.

When applied to rotating and reciprocating equipment, failure analysis can be used to identify contributing premature component failures such as bearing breakage or reduced bearing life. When combined with strain gauge testing, failure analysis identifies the source of the fatigue damage that caused a pressure vessel to fail or a fan blade to break. Visual and microscopic examinations of failed components can also provide valuable insights as to the causes and modes of failure.

12.11 IMPACT TESTING AND FREQUENCY RESPONSE

An impact test involves hitting the test object with a loaded or automated impact hammer and studying the modal and dynamic stiffness characteristics. From the dynamic acceleration plot, it can be determined if there was a double hit or if the force of response was too high or too low. Acceptance criteria are shown at the top. The second plot shows the phase response and the third plot shows coherence. A coherence of 1.0 indicates perfect data, but the value will drop slightly as more averages are acquired. Figure 12.4 shows impact test responses.

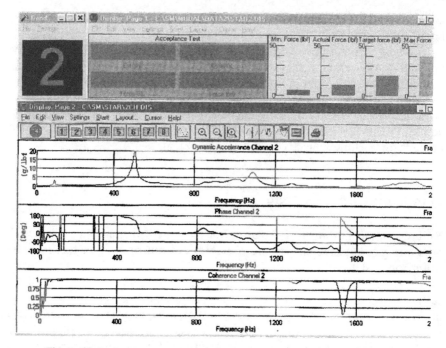

Figure 12.4 Impact test results. Courtesy of HEM Data Corporation.

12.12 PASS AND FAIL TESTING

The top waveform in Figure 12.5 shows an acceptable waveform for the frequency range of interest. The second waveform shows newly acquired data with three peaks at 8, 10, and 13 Hz that exceed the threshold value, thus failing the test.

12.13 CORRECTION METHODS

12.13.1 Alignment and Balance

Correcting the alignment and balance of new rotating equipment produces smooth-running equipment and extends its life. Figure 12.6 shows a precision shaft-to-shaft alignment system manufactured by Machine Dynamics, Inc. of Albuquerque, NM.

12.13.2 Beat Frequency

Beats are produced when two waves have frequencies that vary only slightly and they interfere with each other. When two waves are in phase with each

CORRECTION METHODS 531

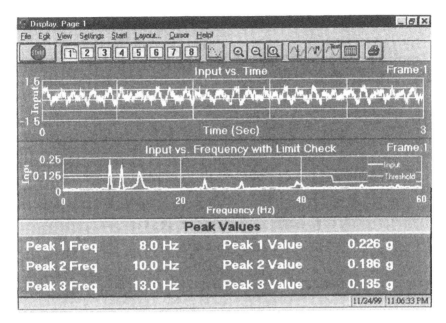

Figure 12.5 Pass/fail testing. Courtesy of HEM Data Corporation.

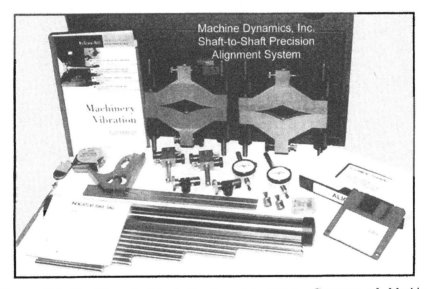

Figure 12.6 Precision shaft-to-shaft alignment system. Courtesy of Machine Dynamics.

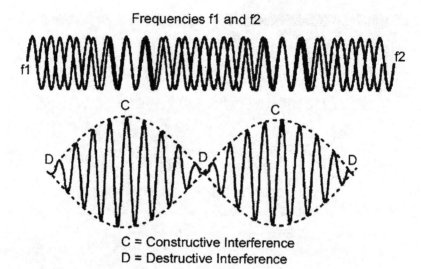

Figure 12.7 Alignment-beat frequency. Envelope of beat frequencies showing constructive and destructive interferences.

other, the interference is constructive or additive and the waves combine to produce a wave with larger amplitude of vibration, typically the sum of the two peaks. When the amplitude of vibration is the same, but the two waves are slightly out of phase, the interference is destructive. When the amplitudes of vibration oppose each other, they will cross at a null point of zero amplitude. Beat frequency is defined by Eq. 12.6:

$$f_{beat} = f_2 - f_1 \qquad (12.6)$$

Figure 12.7 shows two frequencies slightly out of phase and the resultant beat frequency.

12.13.3 Vibration Damping

Damping refers to anything that soaks up energy, reducing the amount of energy converted into vibratory movement. Motor mounts on an automobile engine would be an example. Critical damping refers to applying the smallest amount of damping that is required to return a system to vibration-free equilibrium. The *damping factor* or *damping ratio* (z) is the ratio of actual damping to critical damping.

Vibration damping is important in sports such as archery, tennis, cycling, skiing (snow and water), surfboarding and Lacrosse. It is equally important for all modes of transportation from bicycles and motorcycles to jet skis, boats, cars, trucks, farm tractors, and airplanes. There are literally thousands of industrial and military applications.

Materials for vibration damping are flexible and varied. Some materials used for vibration damping are neoprene o-rings, viscoelastic gels, closed-cell foam, and rubber. Wagner Manufacturing (a division of DiversiTech) considers cork and rubber the industry standard for economical vibration damping. Their standard pads consist of oil-resistant padding designed to withstand 50 psi. Standard pads consist of a cork center laminated between two ribbed rubber sheets. The pads are laminated so that the ribbed rubber sheets are oriented 90° from each other and are ideal for vibration dampening of air conditioners, compressors, cooling towers, presses, machines, and other equipment. The Wagner rubber/cork vibration pad is shown in Figure 12.8.

The same company also manufactures natural-rubber Iso-Cubes. These cubes are designed to reduce vibration and noise emanating from refrigeration and air conditioning equipment. The cube design with circular openings in each pad provides a suction cup effect that eliminates the need to bolt pads down. A standard 18″ pad is divided into 80 2″ segments, each of which may be cut or torn to provide smaller pads. These 50 durometer pads will withstand a load of 180 psi with greater resiliency.

The latest-generation Extreme Vibration Attenuation (EVA) pad, shown in Figure 12.9, provides the most effective vibration-dampening pad available on the market today. The enhanced performance of this pad is made possible by a special composite foam center that is structurally sounder than cork. Oils

Figure 12.8 Antivibration mounting pad. Courtesy of Wagner Manufacturing (division of DiversiTech).

Figure 12.9 Wagner Extreme Vibration Attenuation (EVA) pad. Courtesy of Wagner Manufacturing (division of DiversiTech).

and chemicals that can sometimes break down cork material do not affect the foam center. The EVA pads are also much more effective at vibration dampening than solid rubber. Wagner antivibration performance is compared with other antivibration pads in Figure 12.10.

All vibration attenuation pads are available in surface sizes from 2" × 2" to 18" × 18". Rubber pads are available in 3/8 in. thickness, composite rubber and cork pads are available in 7/8" thickness, and EVA pads are available in 7/8" thickness.

12.13.4 Dynamic Absorber/Increasing Mass

Adding mass tends to both dampen vibrations and lower vibration frequencies. A dynamic absorber has been successfully used to reduce the peaks of transient vibrations of a new Japanese high-speed, high-quality printing press (shock-streakless at 16,000 sheets per hour). To accomplish this performance, a roll (blanket cylinder) with a built-in dynamic absorber was used.

The Japanese have also employed a dynamic absorber to minimize the swing of ropeway gondola lift carriers. The swing of ropeway carriers is caused by the wind, which limits their operation to wind velocities below 15 m/s (33 mph). Matsushita showed that the swing of a carrier could be reduced by a dynamic absorber if it was located far above or below from the center of oscillation. Based on this finding a dynamic absorber composed of a moving mass on an arc-shaped track was designed for practical use. It was installed in chairlift-type carriers and gondola-type carriers in Japanese resorts for the first time in 1995. A dynamic absorber with one-tenth the weight of the carrier can reduce the swing to half.

An undamped tuned dynamic absorber is shown in Figure 12.11 at the right. The initial mass-spring arrangement shown on the left has 1 degree of freedom

CORRECTION METHODS

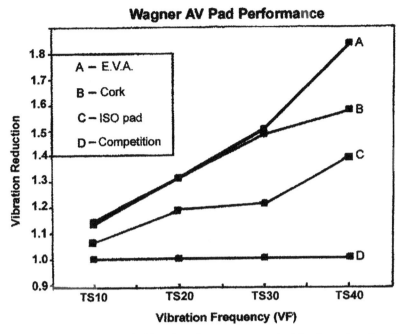

Figure 12.10 Comparison of AV pad performance vs. materials of construction. Courtesy of DiversiTech Corporation.

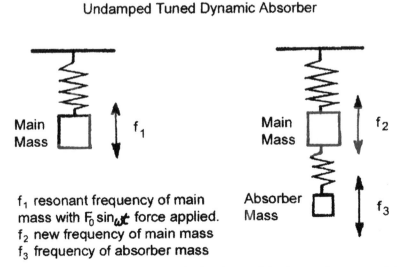

Figure 12.11 Undamped tuned dynamic absorber at right changes frequency of vibration of main mass.

(1DOF) and will resonate at its natural frequency when a sinusoidal force of $F_0 \sin \omega t$ acts on it. However, resonant frequencies can cause severe problems for vibrating or rotating systems.

When an absorbing mass-spring system is attached to the main mass and the absorber is tuned to match that of the main mass, the motion of the main mass can be reduced to zero at its resonant frequency (not shown). For this to happen, the vibration of the main mass must be totally absorbed by the tuned dynamic absorber without damping. However, the 1-DOF system has become a 2-DOF system as shown in the illustration at the right. The 2-DOF system has two resonant frequencies, both of which are different than original resonant frequency.

With no damping, the 2-DOF system can vibrate at its resonant frequencies, which can become a problem when the new system is stopped and started. Therefore, some amount of damping is required to prevent resonance at the new frequencies. However, when the additional damping is applied to either mass spring, the vibration response of the main mass will no longer be zero at its target frequency.

12.13.5 Looseness/Nonlinear Mechanical Systems

If a linear mechanical system has imbalance, a single vibration will be noted at its resonant frequency. However, if looseness also exists, the system becomes nonlinear and multiple harmonics of the resonant frequency will exist. When a nonlinear system exists, the looseness must be corrected first before the mechanical system can be balanced or rebalanced. In modal analysis, a nonlinear system produces erroneous results. A good example of a nonlinear system can be demonstrated by impacting a rotor with sleeve bearings. The rotor will move about the bearing causing looseness in the spectrum.

Journal bearings are typically cylindrical bearings with greater surface area than roller or ball bearings. They can be lubricated or self-lubricating and are relatively maintenance free. They are frequently used in high-load, slow-speed, linear-motion applications. Maintaining proper lubrication and clearance is the key to maximum bearing life.

High-speed tilting pad journal bearings are typically used in high-speed turbomachinery. This design provides direct lubrication using flooded oil or a thin film of cool oil. Adjusting the clearances on tilting pad journal bearings to design optimums can reduce pad friction, heating, power loss, and oil consumption, thereby increasing bearing life while minimizing oil consumption.

12.13.6 Isolation Treatments

There are a number of companies specializing in vibration isolation products. Load cells, which measure the weight of contents in chemical and petrochemical plants, can also serve as vibration isolation products. Vibration iso-

CORRECTION METHODS

lation products are also designed for benchtop applications, and as basic building blocks for vibration isolation platforms and floor platforms.

Applications for vibration isolation products include:

- *Microscopy* to lower vibrations and provide images with better resolution and greater detail
- *Micro-hardness testing* to provide finer details and more repeatable results
- *Aerospace/spacecraft systems* to simulate zero gravity and provide lower frequencies for dynamic testing and reducing responses to building vibrations
- *Metrology* for providing lower vibrations to provide more repeatable and accurate results
- *Microelectronics fabrication* to lower vibrations while simultaneously increasing production and yields
- *Medical and industrial laser optical systems* to protect fragile lasers from vibration while providing greater beam accuracy and stability
- *Biology patch clamping and cell manipulation* to protect against cell damage and provide more accurate manipulation and more sensitive measurements
- *Neuroscience—fluorescent dye imaging* to lower vibrations and improve image resolution

Vibration Isolation Theory. Nano-K™ vibration isolators by Minus K Technology provide vertical motion isolation by using a stiff spring that supports the weight load, combined with a *Negative-stiffness mechanism NSM*. The net vertical stiffness is made very low without affecting the static load capacity of the spring. Beam columns in series with the vertical motion isolator provide horizontal motion isolation. The horizontal stiffness of the beam columns is reduced by the "beam-column" effect, whereby the column behaves like a spring combined with an NSM. The result is a compact passive isolator capable of very low vertical and horizontal natural frequencies and high internal structural frequencies. Figure 12.12 shows the Model BM-4 Biscuit benchtop isolator. Manual vertical stiffness and load adjustments are available to the operator.

Nano-K™ isolators typically use three isolators stacked in series. A tilt motion isolator is placed on top of a horizontal motion isolator that is placed on top of a vertical motion isolator. Schematics of vertical and horizontal motion isolators are shown in Figure 12.13. The vertical isolator uses a conventional spring connected to an NSM. Note the two bars hinged at the center and supported by the outer ends on pivots that are loaded by forces P. The spring is compressed by the weight W to the operating position of the isolator as shown in the figure. The stiffness of the isolator is $K = K_S - K_N$,

Figure 12.12 Very low vertical and horizontal natural frequency. Courtesy of Minus K Technology.

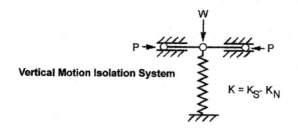

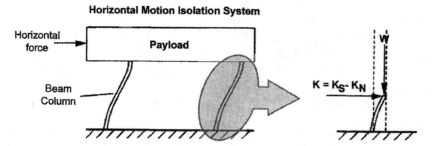

Figure 12.13 Horizontal and vertical motion isolation systems. Courtesy of Minus K Technology.

CORRECTION METHODS

where K_S is the spring stiffness and K_N is the magnitude of negative stiffness. Negative stiffness is a function of the length of the bars and the load P. Isolator stiffness can be made to approach zero (ideal) while the spring supports the weight W. Flexures are used instead of hinged bars in the actual isolators in order to avoid friction.

The horizontal motion isolation system shown in the same figure consists of two beam-column isolators. Each beam-column isolator behaves like two fixed-free beam columns loaded axially by a weight load W. Without the weight load the beam columns have horizontal stiffness K_S. With the weight load the lateral bending stiffness is reduced by the "beam-column" effect. This behavior is equivalent to a horizontal spring combined with an NSM such that the horizontal stiffness is $K = K_S - K_N$, where K_N is the magnitude of the beam-column effect. Horizontal stiffness can be made to approach zero (ideal) by loading the beam columns to approach their critical buckling load.

Figure 12.14 shows a schematic of a Series SP-1 vibration isolation platform consisting of a weighted platform supported by a Series SM-1 vibration isolator that incorporates both vertical and horizontal motion isolation systems. Flexures are used in the place of the hinged bars and a tilt flexure at the top serves as a tilt motion isolation system. A vertical stiffness adjustment screw is used to adjust the compression force on the negative stiffness flexures thereby changing the vertical stiffness. A vertical load adjustment screw is used to adjust for varying weight loads by raising or lowering the base of the support spring to keep the flexures in their straight, unbent operating position. This

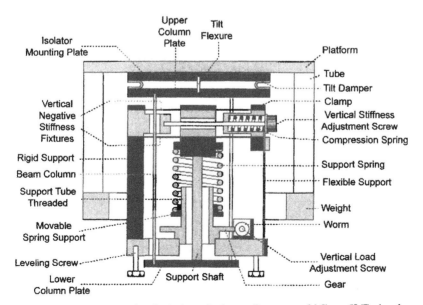

Figure 12.14 SP-1 vibration isolation platform. Courtesy of Minus K Technology.

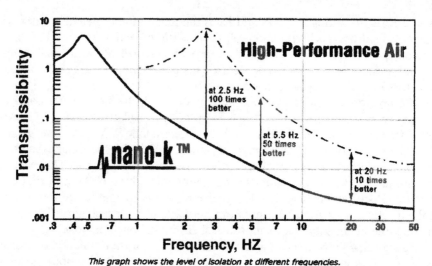

Figure 12.15 Comparison of Minus K isolators to top-performing air tables. Courtesy of Minus K Technology.

feature is automated in some single isolator systems and to achieve automatic leveling in some multiple isolator systems.

Structural damping in Nano-K™ isolators limits the resonant responses at the natural frequencies and this damping is magnified by the use of NSMs. In some applications additional damping is provided by the use of elastomeric damping elements in parallel with the isolators described above.

Figure 12.15 shows the transmissibility curves that compare Minus K isolators to top-performing air tables. Minus K isolators typically achieve 90% isolation efficiency at 2 Hz, 99% at 5 Hz, and 99.7% at 10 Hz. Their performance is typically 10 to 100 times better than high-performance air tables, depending on the frequency of vibration.

12.13.7 Speed Change

In cases where heavy industrial equipment uses adjustable speed controls, avoiding high vibrations at natural frequencies may be as simple as avoiding setting the speed control at those speeds that induce harmonic vibrations or limiting the speed control setting physically or procedurally.

Nonsynchronous (a.k.a. asynchronous frequencies) in a vibration spectrum can also exceed shaft turning speed and be potentially harmful, but they are not integer or harmonic multiples of turning speed.

12.13.8 Stiffening

Stiffening is an undesirable characteristic for both tall buildings as shown in Eq. 12.4 and as described in Vibration Isolation Theory above. Stiffening increases natural frequencies and has the opposite effect on both equipment and buildings when compared to the effect of adding mass.

12.14 MACHINE DIAGNOSIS

Vibration measurement and analysis provides a large amount of quality information for a low capital investment. Analytical diagnosis involves fault recognition using frequency analysis and dynamic behavior analysis of the equipment. Dynamic behavior analysis can be accomplished using self-excitation or external excitation.

Overall vibration and sound measurements are used to determine absolute bearing vibration, general bearing condition, and relative shaft vibration. Displacement measurements can be dependent on temperature, speed, and process variables such as pressure, flow, and load. Displacement measurements of concern are relative shaft displacement, relative shaft expansion, and absolute housing expansion. Typical shaft and bearing vibration measurements to determine the overall condition of a journal bearing are shown in Figure 12.16.

Absolute bearing vibration measurements are used to determine the overall bearing condition for rotating machinery. Absolute bearing vibration measurements are typically made at the six locations shown in Figure 12.17. Points 1 and 4 measure absolute bearing vibrations in the horizontal direction. Points 2 and 5 measure absolute bearing vibrations in the vertical direction. Points 3 and 6 measure absolute bearing vibrations in the axial direction. The largest amplitude of vibration noted during testing determines the maximum *vibration severity* for the machine being tested. Limit values for mechanical vibrations are determined by VDI Guidance 2056, which is based on machine group and motor rating. Figure 12.18 shows the limit values for evaluation of mechanical vibrations according to VDI Guidance 2056.

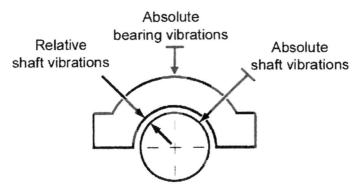

Figure 12.16 Types of vibrations used to evaluate journal bearing condition. Courtesy of Schenck Trebel Corporation.

542   VIBRATION ANALYSIS METHOD

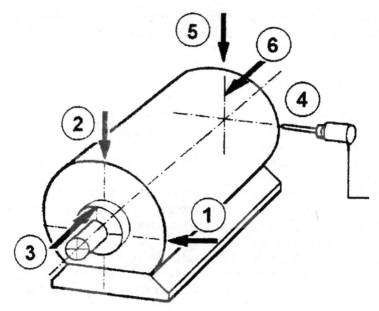

Figure 12.17 Six typical locations for measuring bearing vibrations. Courtesy of Schenck Trebel Corporation.

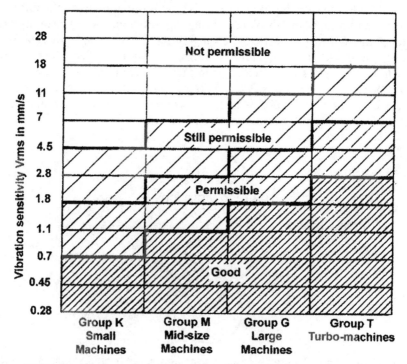

Figure 12.18 Limit values for evaluating mechanical vibrations per VD 2056. Courtesy of Schenck Trebel Corporation.

Mechanical vibrations above the severity levels shown below are not permissible:

Group K—small machines with motors up to 15 KW = 4.5 V_{rms} in mm/s
Group M—medium 1 sized machines to 75 KW and rigidly mounted engines up to 300 KW = 7.0 V_{rms} in mm/s
Group G—large machines with rigid and heavy foundations = 11.0 V_{rms} in mm/s
Group T—large machines mounted on soft foundations, such as turbo machines = 18.0 V_{rms} in mm/s

Mechanical vibrations are measurable vibrations at the surface of machines, construction elements, and foundations. Mechanical vibrations are a source of structure-borne sound. By proper filtering and signal conditioning these sounds may be heard and/or displayed on analog or digital frequency analyzers. Defective bearings may produce a clicking or thumping sound as compared to a smooth whirling sound.

12.15 SENSORS

12.15.1 Strain Gauges

Strain gauges are frequently used for failure analysis studies that may be run concurrently with vibration analysis studies. A resistance strain gauge consists of a thin wire that is strain sensitive and changes dimensions elastically. The gauge is mounted on a backing material that insulates the wire from the test structure. Strain gauges are calibrated based on a manufacturer's specified gauge factor F, which relates strain to resistance as shown in Eq. 12.7.

$$F = \frac{\Delta R/R}{\Delta L/L} \tag{12.7}$$

where R = resistance
L = length of wire

Strain gauge signals can be temperature compensated, digitized, amplified and displayed by data acquisition instruments.

Disadvantages of strain gauges include low sensitivity at low strain levels, fragility, and the requirement for temperature compensation in high-accuracy work.

12.15.2 Accelerometers

Accelerometers are transducers whose output is an electrical/mechanical signal that is directly proportional to the received acceleration forces (g). When shock or vibration is transmitted to the piezoelectric crystals, the accelerometer generates a current pulse or series of pulses proportional to the applied forces. The current is usually converted to a voltage, amplified, and displayed as a time waveform (acceleration force vs. time) or processed by a fast Fourier transform (FFT) to produce a frequency (voltage vs. frequency) display.

The accelerometer has internal damping so that it provides a linear response to acceleration over its useful range. Its natural frequency is very high and it is designed to operate at frequencies below its natural frequency. Figure 12.19 shows a compression-type sensor with piezoelectric ceramic discs preloaded with a seismic mass. Using this configuration, the ceramic discs act as a spring in the spring-mass system. When subjected to vibration, the seismic mass produces an alternating force on the discs, which in turn produces an alternating current proportional to the acceleration of the vibration. A charge amplifier converts the current to a voltage.

Disadvantages of the accelerometer are that it requires an external power source and has low-sensitivity at low-frequency measurements.

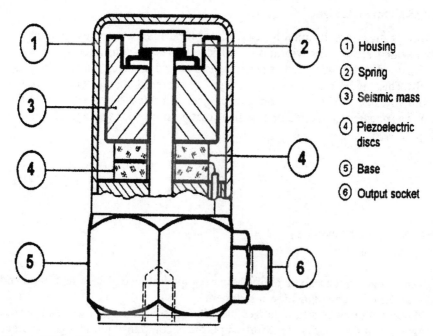

Figure 12.19 Piezoelectric accelerometer. Courtesy of Schenck Trebel Corporation.

12.15.3 Velocity Sensors

Velocity sensors are electrodynamic transducers that produce a voltage proportional to the vibratory velocity induced in a moving coil by a permanent magnet arrangement. As shown in Figure 12.20, the coil is suspended on two leaf springs and forms a frictionless spring-mass system. When the transducer is attached to a vibrating component, its coil remains stationary in space at frequencies above the transducer's natural frequency while the permanent magnet vibrates with the object, thus producing an induced voltage in the coil. The velocity sensor requires no additional power and is therefore considered an *active sensor*.

The natural frequency varies based on the manufacturer's model. Below the natural frequency of the transducer, the electronic frequency response is linearized. Instrument readout is calibrated to provide a high-accuracy measurement of absolute vibration. Absolute bearing vibrations can be measured with velocity transducers covering the range of 1.0 to 2000 Hz.

Disadvantages of the velocity sensor are its upper limit frequency of 2000 Hz and its susceptibility to error in the presence of strong magnetic fields.

12.15.4 Displacement Sensors

Noncontacting displacement sensors can be permanently mounted on large machines to determine:

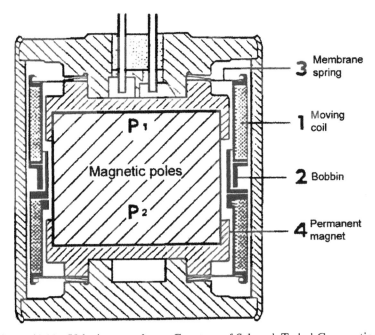

Figure 12.20 Velocity transducer. Courtesy of Schenck Trebel Corporation.

- Relative shaft vibration
- Relative shaft displacement
- Relative expansion between shaft and housing
- Rotational speed, direction, standstill
- Eccentricity

The installation of a pair of eddy current displacement sensors is shown in Figure 12.21. Eddy current sensor output can be conditioned to produce a linear voltage that is proportional to sensor/shaft spacing.

The advantages of eddy current sensors are that they can be used with any electrical conducting material and dielectric materials such as oil and water do not affect them. The disadvantage is that they can be adversely affected by shaft runout (clearance between the shaft and bore).

Noncontacting, eddy-current displacement sensors with built-in oscillators feature increased reliability, simplified installation, and lower cost through

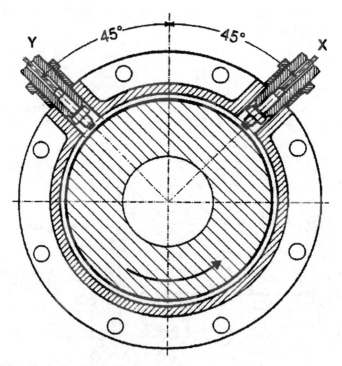

Figure 12.21 Typical eddy current displacement sensor installation. Courtesy of Schenck Trebel Corporation.

state-of-the-art technology. All components necessary for the eddy current displacement measurements are integrated into the basic sensor. Separate extension cables, oscillators, and oscillator housings are eliminated.

12.16 ROLLING ELEMENT BEARING FAILURES

In the manufacturing process, nonhomogeneous material or incorrect bearing tolerances can produce bearing defects. Hopefully quality control inspectors will find and discard these bearings before they are shipped to customers, but many customers also inspect incoming materials on a routine basis. Bearings can also be damaged by improper storage or shipping. In these cases, the damaged bearings should also be detected by routine inspection before being installed.

Considerable bearing damage can be done during the initial mounting and installation operations or later during actual machine operation. Excessive strain, incorrect preloading, incorrect bearing tolerances or clearances, and misalignment are the cause of breakdowns during initial mounting and installation operations. During normal operations, the causes of bearing breakdowns are excessive loading, lubrication failure, and foreign particle intrusion, along with rust and corrosion, caused by harsh atmospheres, thermal stress, and humidity. The major causes of roller element bearing failure are wearing and overload as shown in Figure 12.22.

Cause	Relative Importance, %
Wear	51
Abrasion	25
Fatigue	18
Corrosion	8
Overload	49
Deformation	14
Breakage	12
Cracking	12
Hot running	11

Figure 12.22 Causes of roller element bearing failures and relative importance. Courtesy of Schenck Trebel Corporation.

12.17 BEARING VIBRATION/NOISE

Rolling element bearing damage is caused by fatigue cracks on the contact surfaces of the bearings or breaking or cracking of the bearing elements themselves. Defective bearings produce relatively sharp fast-rising vibrations called "shock impulses." These higher-frequency shock vibrations are added to machine noise and the harmonic content of other machine vibrations that may be present. The amplitudes of shock vibrations continue to increase with time until the bearings are replaced or sudden failure of the system occurs. Sudden bearing failure must be avoided to prevent extensive machine damage as well as potential injury or death to operating personnel.

From rolling-element bearing dimensions such as the contact angle, ball diameter, pitch diameter, number of balls, and rotational speed of the shaft, it is possible to determine the frequencies of the outer race damage, inner race damage, rolling element damage, and cage damage. See Figure 12.23. By determining these frequencies, it is possible to locate the Bearcon (*bearing condition*) envelope for further Bearcon signature analysis.

Bearcon signal conditioning looks at the amplitude of the bearing vibrations over the frequency range of 15 to 60 kHz. The FFT transform converts this frequency envelope into a series of pulses. The filtered shock pulses are then passed through a peak detector and displayed as BCU (Bearcon units) representing the energy content of the peaks as shown in Figure 12.24.

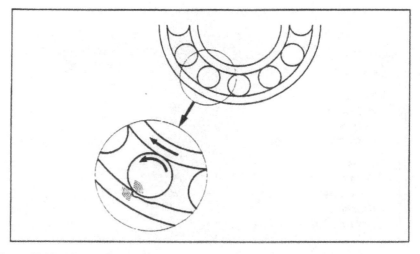

Figure 12.23 Illustration of pitting on contact surface of the inner race of a damaged bearing. Courtesy of Schenck Trebel Corporation.

BEARING VIBRATION/NOISE

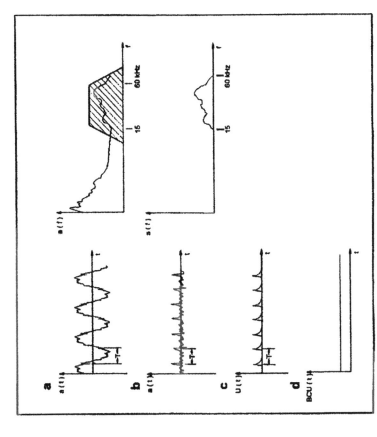

Figure 12.24 Bearcon units representing energy content of vibration peaks. Courtesy of Schenck Trebel Corporation.

12.18 BLOWERS AND FANS

Electric motors, compressors, and pumps are examples of machinery that produce a vibration spectrum with almost constant amplitude over years of operation. However, blowers and fans are rotating machinery that is prone to vibration. Over time their mechanical vibrations tend to worsen to a degree that can become dangerous. Mechanical vibrations need to be periodically monitored at bearing positions, on the housing, and at the foundations of the machines.

Because of impeller construction, locked-in welding stresses can relieve themselves due to centrifugal force. Hot-air blower impellers can relieve their inherent stresses through thermal effects, which can result in deformation and unbalance. Unbalances result in excessive machine vibrations.

The impeller life and performance for exhaust-gas fans, sintering blowers, and air separators can be deteriorated by fly ash deposits, stone or solid particle impact, and corrosion. Aerodynamic effects during operation can also excite mechanical vibrations at the blower housing and bearing positions, thereby further reducing impeller life expectancy. These factors along with other resonant vibrations can cause residual shaft bending, bearing pedestal breaking, foundation damages, and in extreme cases, building fractures.

Preventive and predictive maintenance (PPM) programs must therefore be established to assure that mechanical vibrations do not exceed permissible VDI Guideline 2056 and other applicable standards such as ISO Standard 1940.

12.19 VIBROTEST 60 VERSION 4

The Schenck Vibrotest 60 Vibration Analyzer is a lightweight hand-held instrument, which is ideally suited for accurately identifying machine condition and faults. Figure 12.25 shows an operator checking rolling element bearings using a contact probe and Vibrotest 60 analyzer. The new version 4 instrument offers manual entry to input machine process values, band-pass measurement to evaluate bearing condition on low-speed machines, and overall vibration versus speed and overall vibration versus time functions for enhanced analysis. New features for accurately diagnosing the cause of faults and damages include *selective envelope detection (SED)* to diagnose bearing problems on low-speed machines and *cepstrum function evaluation.*

A new *constant percentage bandwidth (CPB)* module helps simplify previous time-consuming FFT fault detection and signal analysis. The CPB analysis feature guarantees maximum spectral analysis in broadband fault detection so ongoing damage can be identified earlier. A listing of Vibrotest 60 version 4 modules and features follows:

Figure 12.25 Testing rolling element bearings using a Schenck Vibrotest 60 with contact probe. Courtesy of Schenck Trebel Corporation.

Module 1.1

- Absolute bearing vibrations
- Relative shaft vibration
- Bearing condition unit (BCU)
- Process values
- Speed measurement
- Band-pass measurement

Module 1.2

- Overall vibration vs. speed (Figure 12.26)
- Overall vibration vs. time (Figure 12.27)

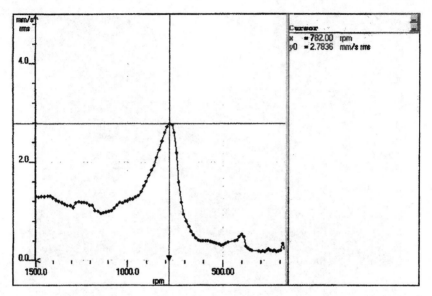

Figure 12.26 Module 1.2 indication of overall vibration vs. speed. Courtesy of Schenck Trebel Corporation.

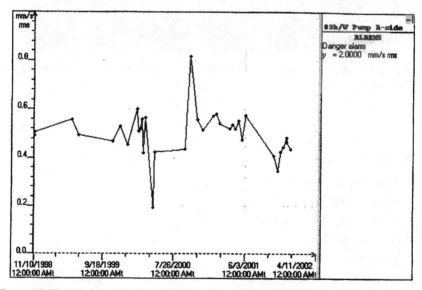

Figure 12.27 Module 1.2: Overall vibration vs. time (trend). Courtesy of Schenck Trebel Corporation.

Module 2.1

- FFT spectrum

Module 2.2

- Bearcon signature (BCS)
- Selective envelope detection (SED), shown in Figure 12.28
- Cepstrum function

Module 3

- Tracking. Order tracking analysis (amplitude and phase vs. speed) is shown in Figure 12.29

Module 5

- Dual channel function

Module 7

- Balancing expert

Module 8

- Constant percentage bandwidth, shown in Figure 12.30

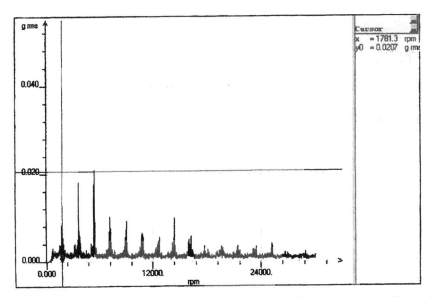

Figure 12.28 Module 2.2: Selection envelope detection (SED). Courtesy of Schenck Trebel Corporation.

554   VIBRATION ANALYSIS METHOD

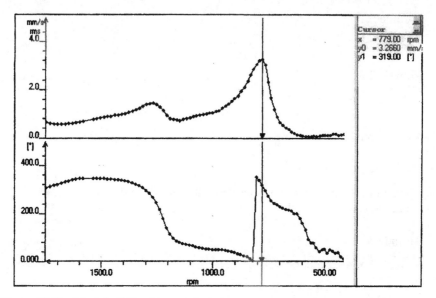

Figure 12.29 Module 3: Tracking analysis (amplitude and phase vs. speed). Courtesy of Schenck Trebel Corporation.

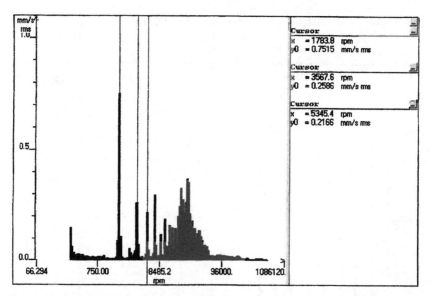

Figure 12.30 Module 8: CPB spectrum. Courtesy of Schenck Trebel Corporation.

DATA PRESENTATION 555

In addition to the Vibrotest 60 Vibration Analyzer for field balancing, the Schenck Trebel Corporation can provide a complete line of horizontal and vertical vibration balancing machines for manufacturers or heavy users of rotating equipment. They can also provide custom online vibration monitoring and diagnostic systems for cost-effective monitoring by large-scale users of industrial rotating equipment. The company also conducts educational training seminars on machine diagnosis, field balancing, and shaft alignment.

12.20 SIGNAL CONDITIONING

12.20.1 Acoustic Filters

There are three types of acoustic filters—low pass, high pass, and band pass. The low-pass filter blocks the high-frequency components of a wave. The high-pass filter blocks the low-frequency components of a wave. The band-pass filter, which is a combination of high- and low-pass filters, has two cut-off frequencies—one for the low-frequency end and one for the high-frequency end of the wave. The band-pass section of a wave would pass through this filter set. High-, low-, and band-pass filters are frequently built into vibration analyzers. The Butterworth filter is one form of band-pass filter.

12.21 EQUIPMENT RESPONSE TO ENVIRONMENTAL FACTORS

12.21.1 Temperature/Humidity

Climate control is provided to control one or more of the following variables—temperature, humidity, light, and/or pollutants. The physical size of climate-controlled chambers can range from relatively small chambers to large walk-in rooms. Climate-controlled chambers are frequently used to verify that circuit components and final equipment designs operate satisfactorily over their design temperature and humidity ranges. Climate-controlled chambers are fully instrumented to monitor the variables being controlled over their environmental range for the time required for thorough testing.

12.22 DATA PRESENTATION

12.22.1 Acceleration, Velocity, and Displacement

Once a force is applied to a surface, the unit of vibration is called *acceleration*, which is proportional to the applied force. In Figure 12.31 acceleration is the measured waveform. By integrating acceleration, *velocity* is derived. The amount of distance the impacted surface travels in one second is its velocity. The velocity peak is a measure of fatigue. By double integrating acceleration,

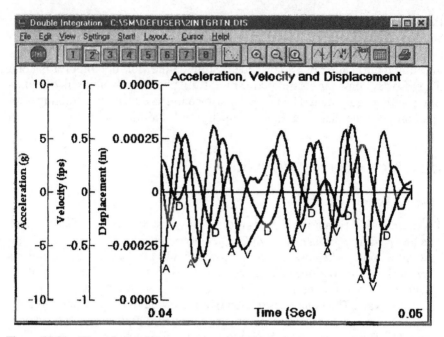

Figure 12.31 The relationship among acceleration, velocity, and displacement. Courtesy of HEM Data Corporation.

displacement can be calculated. The peak-to-peak displacement is the peak-to-peak movement of the impacted surface. As might be expected, displacement is 180° out of phase with acceleration. As acceleration increases, displacement decreases, and as acceleration decreases, displacement increases. Velocity is 90° out of phase with acceleration and has a zero crossing when acceleration peaks. The vibration cycle will continue until all of the energy is dissipated.

12.22.2 Fast Fourier Transform (FFT)/Time Waveform

The fast Fourier transform is the algorithm that converts a digital signal from a time to frequency domain that allows a computer to calculate the discrete Fourier transform (DFT) very quickly. The *discrete Fourier transform* is a complex mathematical calculation that converts the sampled and digitized time waveform into a sampled frequency spectrum.

The FFT analyzer is a device that uses the FFT algorithm to calculate a spectrum from a time to frequency domain signal. It is the most common type of spectrum analyzer today and the heart of most predictive maintenance programs for machinery.

By generating an infinite summing of sine and cosine waves, an infinite number of fundamental and harmonic excitation waves can be generated to

DATA PRESENTATION 557

excite a test object. For an interactive example of the FFT, see the FFT Laboratory website.

12.22.3 Cepstrum Analysis

The cepstrum is the forward Fourier transform of a spectrum or the spectrum of a spectrum. One of its most powerful characteristics is that repetitive patterns in a spectrum will be sensed as one or two specific components in the cepstrum. Spectrums containing several sets of sidebands or harmonic series can be confusing because of overlap. Cepstrum analysis identifies and separates harmonic families. With cepstrum analysis overlapping bands are separated in a manner similar to the way the initial spectrum separates repetitive time patterns in the waveform. Therefore, the worst of the harmonic series of vibrations are separated from the complex FFT spectrum and identified.

Gearboxes and rolling bearing vibrations are ideal cepstrum analysis applications. Since gear boxes have a prime number of teeth, wear is spread out more evenly on the teeth and different harmonic families do not usually overlap, but there can often be several harmonic families which are difficult to separate in a spectrum. Cepstrum is a practical tool for identifying different harmonic families. These families can be monitored to determine when something goes wrong. *Comb liftering* is a term used to describe a spectrum analyzer's ability to remove one or two harmonic families from the cepstrum. Advanced cepstrum analysis allows cepstra to be edited and permits editing to be done on a frequency spectrum before the cepstrum is calculated in order to analyze harmonic or side-band families.

In Figure 12.32, acceleration versus time in the revolution or angle domain is shown at the top right. Performing a real-time FFT produces the lower plots. The lower left plot shows the entire order range and the bottom right is zoomed around the 53rd order to show the modulation effect of the operating speed. Traditional cepstrum analysis is shown on the top right.

Some instruments can replay data previously recorded (left plot) and analyze their spectrum as shown in the (right plot). The user defines the region of interest, then marks the starting and stopping points, and analyses the data to get the FFT plot on the right. Figure 12.33 shows the retrieved and analyzed data.

12.22.4 Nyquist Frequency/Plot

Digital signal processing requires analog-to-digital (A/D) conversion of the input signal. The first step in this conversion is to sample the instantaneous amplitudes of the input signal at specific times determined by the selected sampling rate. For greatest accuracy, the input frequency must not exceed $\frac{1}{2}$ the sampling rate frequency because higher input frequencies contain spurious components, producing what is known as *aliasing*. This theoretical maximum input frequency is called the *Nyquist frequency*.

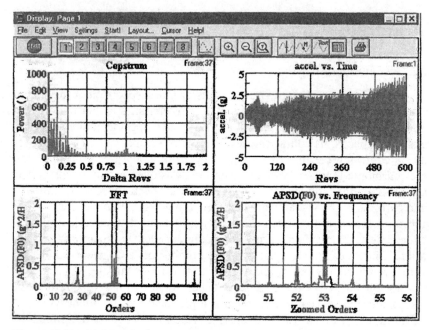

Figure 12.32 Multiple waveform display. Courtesy of HEM Data Corporation.

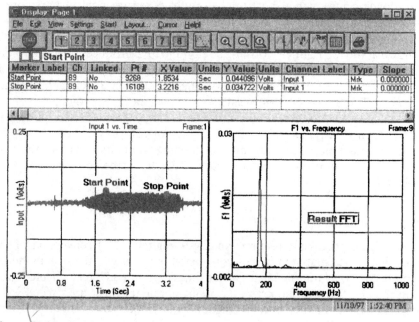

Figure 12.33 Retrieved (L) and analyzed data (R) data. Courtesy of HEM Data Corporation.

DATA PRESENTATION 559

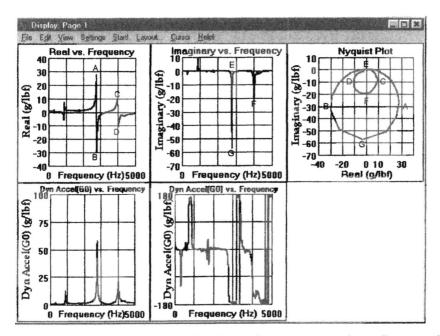

Figure 12.34 Nyquist plot with acceleration vs. frequency comparisons. Courtesy of HEM Data Corporation.

The *Nyquist plot* is a plot of the real and imaginary parts of the frequency response function. For a 1-DOF system, the Nyquist plot is a circle. The Nyquist plot represents the frequency function by graphing the real part versus the imaginary part. Resonance shows up as a circle, but there is no indication of frequency. A Nyquist plot is shown in Figure 12.34 at the top right. The real and imaginary plots that are used to make the Nyquist plots are shown on the top left. The bottom plots show amplitude and phase for comparison.

12.22.5 Orbit, Lissajous, X-Y, and Hysteresis Plots

Orbit plots are plots with changes in amplitude and phase between two measurement points that are physically located 90° apart. In Figure 12.35, the orbit plot on the right is an ellipse because the amplitudes are not identical. The ellipse is not horizontal because the phase angle between the two inputs is not 90° out of phase as might be expected from the transducer positions. They are actually 150° out of phase.

An X-Y plot or Lissajous pattern are general plots of one measurement versus another. A hysteresis plot has two similar sensors with no initial phase shift except what is measured.

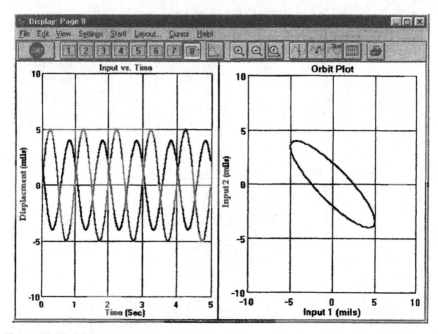

Figure 12.35 Orbit plot (R) is elliptical because amplitudes of displacement (L) are not the same. Courtesy of HEM Data Corporation.

12.23 ONLINE MONITORING

12.23.1 Trend Analysis

Trend analysis takes place over a relatively long period of time, say days or perhaps even a week or more. The purpose of a trend analysis is to study the amplitude of acceleration over relatively long enough periods of time. Accelerations can be observed for the established time period to see if amplitudes exceed the established upper or lower thresholds. The subject of a trend analysis might be a newly constructed factory floor with many pieces of machinery installed. If the lower threshold of allowable vibration is exceeded, FFT spectra can be plotted for shorter time periods, such as an hour, to determine the frequencies of vibration that exceed the lower threshold limit. Then a corrective measure such as vibration damping or vibration isolation can be employed to eliminate the problem.

12.24 PORTABLE NOISE AND VIBRATION ANALYSIS SYSTEM

NOVA™ is a portable PC-based noise and vibration analysis system (Figure 12.36) designed for use with a Windows-based operating system, which com-

Figure 12.36 NOVA-MD6 analyzer with computer. Courtesy of HEM Data Corporation.

bines signal conditioning, data acquisition, and specially designed (Snap-Master) noise and vibration analysis software.

HEM Data has designed the user interface for use by noise and vibration analysis novices. The analyzer's software features a series of built-in wizards to perform critical tests with a few mouse clicks. A wide variety of noise and vibration tests have been preconfigured by experts and are included with the system. When possible, the system also provides automatic calibration and ranging routines to help match hardware settings with the data being analyzed.

NOVA combines the capabilities of a dynamic signal analyzer, spectrum analyzer, digital storage oscilloscope, stripcart recorder, and data logger. By integrating these components in a single system, the user needs to learn the characteristics of only one user interface to perform a variety of tests. This reduces operator training time and the need for expert personnel to perform most of the testing.

Snap-Master software permits administrative users to add new tests for specialized applications. Once new programs are implemented and tested, the end user can be familiarized with the new test methods, parameters, and requirements, then perform the actual testing with an interface he or she is already familiar with.

DAWN (Data Acquisition With a Network) is a complete in-vehicle data acquisition and analysis software program that simultaneously acquires data from both in-vehicle networks and analog data sources. This powerful software tool allows engineers to display and analyze live data directly from the vehicle network in real-time without wasting time on complicated setup and configuration procedures. DAWN is based on HEM Data's Snap-Master software.

12.24.1 Typical Applications

- Order tracking to analyze rotating machinery
- Modal analysis to identify resonant frequencies
- Frequency response (transfer) functions
- Power spectrum and power spectrum density
- Predictive maintenance and diagnosis
- Production line (pass/fail) testing
- Product development
- Identifying noise sources and material noise absorption properties
- Field and in-vehicle testing

12.24.2 System Requirements

NOVA requires a 486 CPU or more powerful computer with 8 MB RAM, 10 MB hard disk space, a mouse, and Windows®95/98 operating system. NOVA-DB is a 2–4-channel analyzer, NOVA-MD is an 8-channel analyzer, and NOVA-WB is an 8–64-channel system that handles various sensor types.

12.25 LASER METHODS

High-tech entries in the field of stress and vibration analysis involve laser-testing methods. Ettemeyer's 3D-Vibro-ESPI-System Q-500 is a three-dimensional (3D) electronic speckle pattern interferometry (ESPI) system that provides a highly sensitive noncontacting full-field measurement of harmonic vibration analysis for almost any component of interest. The compact head, shown in Figure 12.37, is flexible and portable, as well as capable of providing both static deformation analysis and dynamic measurements.

The Q-600 system is a double-pulsed ruby laser system using three cameras to evaluate large-scale acoustic events such as brake squeal, car body vibrations, engine vibrations, and other dynamic events such as shock waves, explosion tests, etc. New ESPI techniques instantly record the full-field dynamic deformation of brake components.

LASER METHODS

Figure 12.37 3D-Vibro-ESPI-System. Courtesy of Ettemeyer Corporation.

12.25.1 Theory of Operation

The Q-500 uses a powerful external laser optically coupled to the test object via a flexible glass fiber, thereby allowing objects up to $1\,m^2$ to be analyzed. An integrated optical light modulator and trigger system produces stroboscopic laser illumination, which is synchronized to the vibration frequency of the test object. Using this configuration, the system is capable of showing the vibration amplitude at any frequency and mode without contacting the test piece.

The phase of the laser illumination pulses can be adjusted with respect to object vibration to analyze the phase distribution at every object point. A built-in zoom lens provides flexibility for testing objects of different sizes and geometries.

The Q-600 uses three cameras and a double pulsed ruby laser system, shown in Figure 12.38, to capture all dynamic deformations within microsecond time ranges and can provide a subsequent display and animation of quantitative results. In order to accomplish this, the object under investigation is illumi-

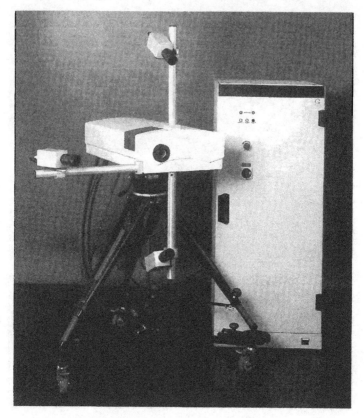

Figure 12.38 3D-PulseESPI-System. Courtesy of Ettemeyer Corporation.

nated with short light pulses from a pulsed laser. Illumination time is several nanoseconds (ns) and is simultaneously observed from three different directions with three cameras.

The measured results from the three cameras represent the deformation field in the three directions established by the optical setup. The optical image distortion from the three cameras is automatically compensated for and the 3D-deformation vector is calculated for every point on the test surface. The use of a double-pulsed laser provides two sets of data in a very short duration time, based on the separation time of the two laser pulses. This provides sufficient data for analysis of object deformation, even during highly dynamic processes. The pulsed three-dimensional ESPI technique provides separation of the in-plane and out-of-plane components of deformations (vibrations) without environmental disturbances. Amplitude resolution is in the submicrometer range and independent of the measuring field. The system can measure surface quality on areas up to several square meters.

12.25.2 Applications

The Q-500 is best suited for the development and testing of complex components and structures with regard to acoustics and stress/strain analysis in the electronics, automotive, aerospace, machining, and research industries. Typical applications include testing the vibration modes of a flexible impeller blade at different frequencies and vibration analysis of catalytic converters.

Typical applications for the Q-600 include brake squeal analysis (Figure 12.39), car body and engine vibration analysis, and the evaluation of other dynamic events such as shock waves and explosion tests. The fully automatic

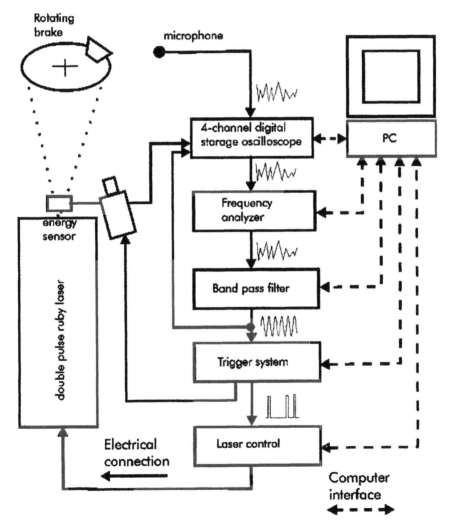

Figure 12.39 Q-600 break squeal analysis. Courtesy of Ettemeyer Corporation.

brake squeal inspection system captures the squealing signal, automatically fires the laser, and provides a complete deformation map of the part under test.

12.25.3 Specifications

Technical specifications applicable to the Q-500 are:

- Measuring sensitivity: 0.031–20 μm adjustable
- Measuring range: static 1–20 μm per measuring step, with any serial measurement, dynamic 0.03–3 m amplitude
- Measuring area: static up to $1 m^2$, dynamic up to 400×600 mm
- Measuring frequency: 25–40,000 Hz
- Working distance: variable, 0.1 m to >2.5 m
- Operation modes: automatic, manual, static, dynamic, 1D, 2D, 3D analysis
- Data acquisition speed: 2.5 s for 3D analysis
- Data analysis: automatic or semiautomatic
- Dimensions of sensor head without arms: $80 \times 80 \times 120$ mm
- Weight: 2.5 kg, total sensor
- Processor: Pentium PC
- Operating system: Windows®98/2000/NT

Technical specifications for the Q-600:

- Measuring sensitivity: 0.06 μm
- Measuring range: 1–10 μm per image
- Measuring area: up to 2×1.5 m ($80'' \times 60''$)
- Working distance: adjustable, depending on glass fiber (typ. 1–3 m)
- Operating modes: single pulse, double pulse
- Pulse separation: 2–800 μsec
- CCD resolution: 1300×1030 pixels
- Data interface: TIFF, ASCII, Windows metafile
- Data analysis: automatic phase calculation (patent pending)
- Laser: 1 joule ruby laser, computer controlled
- Processor: Pentium PC
- Operating system: Windows®98/NT

12.26 TEC'S AVIATION PRODUCTS

TEC's Material Testing Division deals directly with vibration analysis products sold under the registered trademark ACES Systems for use in engine test cells and overhaul facilities. Aviation products utilize vibration and acoustic analysis for balance and engine performance trend monitoring. This equipment is used on small and large jets, propeller-driven aircraft, and helicopters for a variety of aviation markets such as corporate, commercial, and military. TEC periodically recalibrates customer instruments at its Knoxville, TN factory.

12.26.1 Analyzer Plus Model 1700

The ACES Analyzer Plus Model 1700, shown in Figure 12.40, is a digital system with adjustable backlit screen. It measures and analyzes acoustic noise and is capable of locating problems within any rotating equipment. The ACES graphics display with built-in software stores and displays noise surveys on user PCs as:

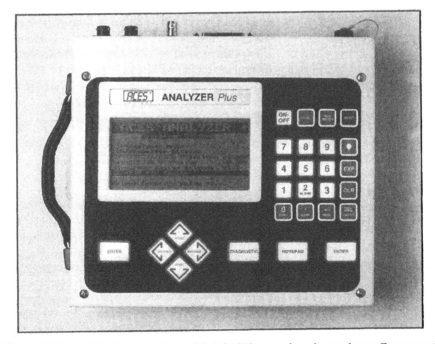

Figure 12.40 ACES Analyzer Plus—Model 1700 acoustic noise analyzer. Courtesy of TEC Aviation Group.

- Single spectrum
- Waterfall spectra
- Multiple spectra
- Comparison spectra

The unit is used as a "controller" for ACES Systems Jet Engine Data Acquisition Module (JEDA). The system is easy to use—even novice technicians can get expert results.

12.26.1.1 Flexible System

The heart of the unit is a Procedure Card, which is a credit-card-size memory device that plugs into a slot on the Analyzer Plus. Each card contains all the information and technical data needed to perform a vibration survey and analysis of specific engines. This flexible approach allows the Analyzer Plus to become your "expert" on the TFE731 engine, JT15D engine, or any other engine by simply inserting a different Procedure Card.

12.26.1.2 User Friendly

All Procedure Card programs are based on current maintenance manual procedures. This allows new Analyzer Plus users, even those with little experience, to quickly feel at home with the equipment. During each test, the analyzer leads the user, step by step, through the process of collecting the required survey data. The collected data are then graphically displayed alongside the manufacturer's limits and markers for N1, N2 and other major components for quick comparison and evaluation. The Analyzer Plus is designed to allow the technician to get all the information needed, in the right order, the first time.

12.26.1.3 Expandability

The Analyzer Plus also serves as the foundation for expanding user engine testing and analysis capabilities. For example, when combined with the ACES JEDA module, it becomes capable of performing certain engine performance tests and data logging with accuracy rivaling that of a test cell. This "building block" approach protects the customer's initial investment by upgrading the capability of the unit without replacing the entire system.

12.26.1.4 Quality Commitment

ACES Systems products are manufactured at the TEC facility in Knoxville, TN. Aviation products are manufactured by the same people and to the same standards as those built and supplied to the nuclear industry through TEC's nuclear products division. Products are designed and manufactured to provide years of trouble-free service. This employee-owned company also provides quick customer response for service and technical needs.

 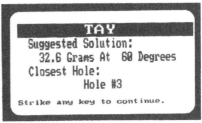

Figure 12.41 Data screens for ACE Analyzer Plus 1700 instrument for fan balancing application. Analysis Data are shown on the screen at the left. Screen at right shows suggested solution. Courtesy of TEC Aviation Group.

12.26.1.5 Engine Fan Balancing Application Step-by-step instructions take the technician through the proper data collection procedure. Then the collected information is analyzed and displaced on the Analyzer Plus screen as a graphic display of the fan layout and suggested corrective trim balance weights are recommended. See Figure 12.41. Subsequent runs allow the analyzer to evaluate the corrective actions taken to insure that the fan is correctly balanced.

12.26.1.6 Technical Specifications

Power Supply

Camcorder-type battery: VDO-PAK Model RB 85 or equivalent
Operation time: 7–8 hours (24 hours typical)
Voltage: 12 VDC
Charging time: 2 hours
Capacity: 2.2 ampere-hour

Environmental Conditions

Enclosure: Type NEMA-2
Humidity: 0 to 95% relative humidity
Temperature: –22 to 122°F
Shock: MIL-STD-810C, Procedure II (48-inch drop test)

Physical Specifications

Height: 3.371 inches
Width: 11.802 inches, including handle
Depth: 9.375 inches
Weight: 6 pounds

12.26.2 ProBalancer Analyzer 2020

This unit combines the Model 1000 ProBalancer with spectrum analysis capabilities of top-of-the-line ACES analyzers. It features true two-channel simultaneous inputs, which makes performing twin-engine propeller balancing a snap, as well as acquiring both vertical and lateral helicopter main rotor vibration measurements without switching channels.

The analyzer, shown in Figure 12.42, provides step-by-step on-screen instructions for performing propellar balancing, or rotor track and balance measurements used in vibration surveys. The user may also extend the basic capabilities of the system by defining and saving setup information for common procedures. Saved settings can be recalled later to provide on-screen setup information, consistent procedures, and one-shot balance solutions without the difficulty of configuring the unit and reentering data each time it is used.

Complete information for balance jobs is stored in memory, eliminating the need for pen and paper. For helicopter rotors, track information may also be entered and stored with the conditions under which the data were collected.

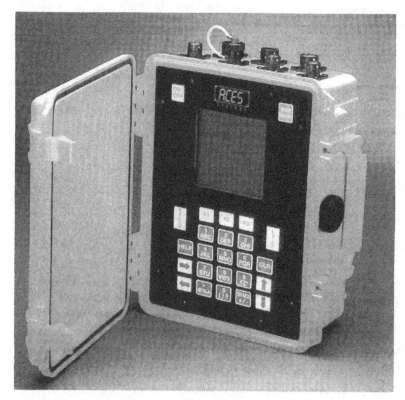

Figure 12.42 ACES Model 2020 ProBalancer Analyzer for propeller or rotor track balancing. Courtesy of TEC Aviation Group.

TEC'S AVIATION PRODUCTS

Complete Propeller Balance Kit

- Model 2020 ProBalancer Analyzer
- 991 Vibration Sensor
- 25' Vibration Sensor Cable
- Phototach Speed Sensor
- 25' Tachometer Sensor Cable
- Phototach Mount Assembly
- 1/4 x 28 Vibration Sensor Mount
- Eight piece case-bolt adapter set
- Battery Charger
- Pocket Pro Tackle Box
- Communications / Printer Cable
- Training Video
- Carrying Case
- User's Manual
- FAA-Approved *ACES Guide to Propeller Balancing*
- Digital Gram Scale

Figure 12.43 Complete Propeller Balance Kit. Courtesy of TEC Aviation Group.

When a job is completed, it is easy to print a summary of the job suitable for logbook entry. The high-capacity rechargeable battery or ship's power input provide extended operation for changing or recharging the battery, thus increasing productivity. A complete propeller balancing kit is shown in Figure 12.43.

12.26.2.1 Software Features
- Simple menu screens for stepping logically through function
- Graphical displayed data aid in quick interpretation
- Configuration can be saved as setups, eliminating repetitive input
- Data can be entered manually or automatically
- Data can be transferred to a PC for storage
- Stored setups provide speedy solutions
- Data and menu functions are readily accessible
- Multiple sensors can be configured and saved

12.26.2.2 Technical Specifications

Accuracy

Vibration amplitude: ±5%, 0 to 10 IPS
Frequency range: 0 to 10 kHz
Tachometer inputs: ±0.3%, 100 to 10,000 rpm

Power Supply

Camcorder-type battery Model RB 85 or equivalent (12 VDC, 2.3 Ah internal lead acid battery)
Operation time: 10–12 hours
Voltage: 12 VDC battery or 14–28 VDC ship's power
Charging time: 2 hours

Physical Specifications

Height: 9.3"
Width: 7.5"
Depth: 4.4"
Weight: 4.8 lb

12.26.3 Viper 4040

The Viper 4040 is a versatile compact instrument that combines all the technologies required for high-end engine vibration analysis, rotor track and balance, fan trim balancing, and acoustic analysis. All these aviation analysis functions can be performed on virtually any airframe and engine type using the Viper 4040.

12.26.3.1 Automated Track and Balancing

The compact Viper 4040, shown in Figure 12.44, provides accurate solution in the minimum number of runs, saving costly run times and fuels. Setups are entered into the analyzer, which can be customized by the user to accommodate a number of engine and airframe types.

Potentially hundreds of setups can be stored in the Viper 4040's 32 MB of memory and recalled for quick reuse. By using the Viper's ChartBuilder™ function, influence and adjustment criteria from polar charts can be entered into setups so that raw data acquired are automatically translated into easy-to-follow adjustment solutions such as "inboard tab up 2 degrees." Sample charts are shown in Figure 12.45.

With four channels available for simultaneous data acquisition, the Viper 4040 gathers data quickly. With each run, the analyzer adjusts and refines the solution based on the data gathered during the previous runs, taking into account the unique properties of each job, automatically shortening the

TEC'S AVIATION PRODUCTS 573

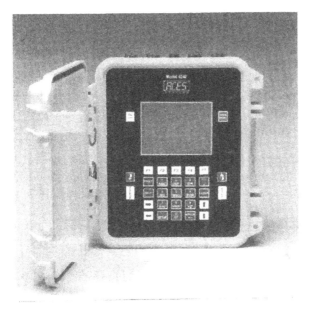

Figure 12.44 Compact ACES Viper 4040 Analyzer. Courtesy of TEC Aviation Group.

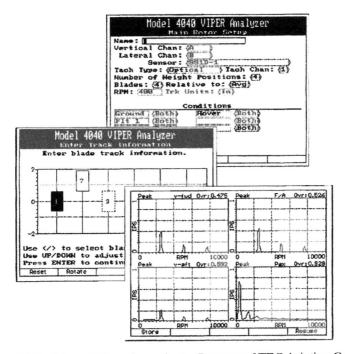

Figure 12.45 Viper 4040 analyzer charts. Courtesy of TEC Aviation Group.

process until vibration is reduced to the lowest possible level. The flexible Viper 4040 can be used with any tachometer, detecting speeds at up to 32,000 rpm. Tach input is used to synchronously sample and average data and report phase to ±3 degrees, reproducible to 1 degree. For additional flexibility, the Viper 4040's SmartTach™ speed and phase signal processing uses single or double interrupter, high tooth, low tooth, offset tooth, or standard TTL inputs. The 4040 will acquire transient, tracked vibration surveys that it can then replay or plot in various formats for review and analysis.

12.26.3.2 Vibration Analysis The compact Viper 4040 weighs 7.3 pounds, is powered by an internal battery with minimal cabling, and delivers vibration analysis at speed and accuracy levels typically available only in manufacturer's test cells.

Vibration data acquired from up to four channels simultaneously at the rate of 10 spectra per second are recorded at frequency levels to 30 kHz per channel. Antialiasing filters are used with FFT to convert data from time to frequency at 100, 200, 400, 800, 1600, and 3200 lines. The massive amounts of data acquired in a transient are easily stored in the Viper's 32 MB of memory. Flexible architecture provides for the use of virtually any sensor type.

12.26.3.3 Acoustic Analysis The Viper 4040 can also interface with microphones for measuring sound levels and frequencies in order to diagnose airframe noise problems.

12.26.3.4 Technical Specifications

AC input: +3.8 volts, peak to peak

Tachometer inputs: ±0.01%, 100–32,000 rpm

Sensor types: accepts any vibration sensor (accelerometer, velocity, and displacement)—any voltage-generating sensor

Autoranging sensor inputs: adjusts gains by factors of 2 (1–512) independently for all channels

Vibration amplitude: ±5%, 0 to 190 ips with 20 mV per ips sensor

Frequency range: 0–30 kHz

Microprocessors: 5

Memory: 32 MB

Display: LCD, 4.7 in., 320 × 240 pixels, backlit

Power: rechargeable nickel cadmium battery

Dimensions: 10.5" wide, 9.75" long, 5" deep

Weight: 7.3 lb (3.31 kilos)

13

VISUAL AND OPTICAL TESTING

13.1 FUNDAMENTALS

Visual inspection continues to develop as an American Society for Nondestructive Testing (ASNT) evaluation method. In the past, visual inspection was considered highly subjective in nature and provided little or no hardcopy documentation. Successful results depended on trained operators, cleanliness and condition of the test object, quality of the optical instrument, and proper illumination of the test part. These factors are still important today, but equipment has become much more sophisticated.

Visual examination is one of the most basic nondestructive evaluation (NDE) methods. Quality control inspectors follow procedures that range from simply looking at a part to see surface imperfections to performing various gauging operations, which assure compliance with acceptable physical standards. Today's optical systems may include special probes, spectrometers, and real-time imaging and analysis using notebook computers.

High-tech pipe weld inspection systems often combine multiple NDT methods. For example, remote magnetic pipe crawlers can be easily equipped with compact video cameras and eddy current or ultrasonic probes to visually inspect the weld and determine the extent of defects.

Introduction to Nondestructive Testing: A Training Guide, Second Edition, by Paul E. Mix
Copyright © 2005 John Wiley & Sons, Inc.

13.2 PRINCIPLES AND THEORY OF VISUAL TESTING

Visual inspection aided or unaided, direct or remote, is a valuable NDE tool. Visual inspection with good lighting is usually one of the first methods employed for locating suspected defect areas in large structures and heavy equipment. Once located, the areas of interest can be more thoroughly examined and evaluated in detail. There is much to be learned by being able to look directly or remotely at the end of a heat exchanger or inside large vessels via an open manhole or inspection port.

Physical measurements using micrometers and spring-loaded depth gauges can often provide inspectors with positive indications that confirm eddy current and ultrasonic test results. The visual appearance and color of a corroded pipe or vessel area can also provide useful insights as to the cause and extent of corrosion, and help determine the remaining useful life of the vessel, pipe, weld, or elbow.

Each NDT technique is unique with its own advantages and disadvantages. Usually, visual inspection cannot be used to verify test results, nor can it be used to replace other NDE methods. However, when direct visual inspections cannot be made, sophisticated optical instruments can often be used to provide remote viewing of critical areas. Some of these instruments are discussed in this chapter.

13.3 SELECTION OF CORRECT VISUAL TECHNIQUE

For castings, forgings, extrusions, and machined parts a thorough visual inspection is usually made after cleaning. For surface and subsurface defects, ultrasonic testing or a combination of ultrasonic and eddy current testing can be used to test pipes and tubing. For long tubing, fully automated tests may be cost effective. For small-diameter tubing or hidden areas, careful manual inspection with flexible fiberscopes and bright LED lights might be appropriate. When industrial pipes and equipment handle hazardous or radioactive materials, remote viewing cameras with real-time imaging, analysis, and recording features can be used. For large welded pipes and vessels, magnetic crawlers with camera, lights, and TOFD transducers could be a likely choice for toe weld crack inspections.

In many cases, the most cost effective and efficient method for visual testing of specific items may have been already determined by company or agency policy. However, when new parts are to be inspected or manufacturing procedures have changed, existing test methods may need to be reevaluated to assure that potential defects are not being overlooked. Some important questions to consider are:

- Based on the manufacturing processes, what defects could be present?
- What would the nature of expected defects be? Would they be visible or hidden?

SELECTION OF CORRECT VISUAL TECHNIQUE 577

- What should be used as a basis for determining pass/reject test criteria?
- Can artificial standards be made and used effectively?
- What is the smallest defect that can be detected?
- What is the smallest defect that could possibly cause part/equipment failure?

Recently, there has been a great deal of emphasis placed on the probability of detection (POD). Obviously, any defect or discontinuity that could cause part or equipment failure must be detected. However, we don't live in a perfect world and we are often blind-sided by the unexpected or unforeseen. Therefore, we must often look beyond our own discipline and expertise in order to enlist the help of others to make the best, most informed decisions. Multiphase engineering, math, science, chemistry, metallurgy, and thermodynamics may all be factors affecting part integrity and life. The goal of NDT practitioners continues to be to create a safer world through personal integrity and knowledge.

The following items/pieces of equipment are frequently inspected using visual testing methods:

Pumps. Stainless steel pump impellers or pistons are subject to distortion and binding through erosion, corrosion wear, and fatigue. Cast stainless-steel castings are subject to corrosion, erosion, and chloride stress corrosion cracking. Corrosion products usually can be seen by visual inspection.

Pump casings, bearings, seals, and gaskets are all sources of potential leaks. Maintenance inspectors that are familiar with the equipment can usually tell when installed pumps and compressors are running properly. Periodic visual inspections can determine when observed noise and vibration are increasing and when small leaks occur. Sometimes small leaks can be stopped by adjusting the pump packing. However, these adjustments are limited by physical restraints. Scheduled plant shutdowns can often provide a window of opportunity for making any necessary pump adjustments and repairs before failure occurs.

Valves. Stainless steel valves are subject to wall thinning from corrosion and erosion. Internal stainless steel valve parts may be subject to distortion as well. Stress corrosion cracking is a problem when chlorine gas or salt air environments are present. Valve seats, seals, and gaskets may be subject to wear, creep, or hardening, which can cause leakage. Routine visual inspections can detect early signs of corrosion and leakage.

Bolting. Improper torque, the wrong materials of construction, and corrosion are the prime causes of bolt failure. Torque wrenches can minimize the use of improper torque. Alloy analyzers can be used to detect the wrong materials of construction and visual inspection can determine when early signs of corrosion exist. In many cases, the wrong materials of construction can cause sudden and catastrophic failure that may endanger personnel as well as equipment. It is much safer to check all materials of construction prior to plant or process fabrication.

Castings. Most ASTM specifications require that the surface of a casting be examined visually and be free of adhering sand, scale, cracks, and hot tears.

Visual method MSS-SP-55 may be used to define acceptable surface discontinuities. This standard contains a series of photographs that define acceptable and unacceptable discontinuities. Other visual standards may also be used as long as the supplier and user agree on the criteria.

Forgings. Visual inspection of critical aerospace forgings usually follows a dye penetrant inspection under ultraviolet light. A 100% visual inspection is used to assure that all functional and dimensional requirements are met.

Extrusions. Quality control departments typically require a 100% visual inspection of precision metal, thermoplastic, and multiple thermoplastic extrusions regardless of use. Again, photo guidelines may also be used as long as they are agreeable to all concerned parties.

Microcircuits. Optical comparators, microscopes, and laser profilometers are frequently used to inspect microcircuit chips to determine overall quality, connection reliability, and solder integrity, and to determine if any physical abnormalities are present.

Environmental Factors. Heat and humidity generally accelerate corrosion and the formation of corrosion products.

13.4 EQUIPMENT

13.4.1 Borescopes

Borescopes have long been used for the inspection of small-diameter pipe and tubing. They can get into tight places and provide reliable visual indications of internal surface conditions. Borescopes are optical instruments designed for remote viewing of objects. In some cases, they are necessary because it is impossible to get close to the objects we want to inspect, such as the internal parts of jet engines. In other cases, it may be too dangerous to get close to the objects we want to inspect because of heat or radiation. Because of the variety of applications and multitude of inspection needs, borescopes are manufactured in rigid, extended, flexible, and micro designs. In general, the diameter of the borescope determines the size of the minimum opening into which it can be inserted. Figure 13.1 shows typical rigid and microstyle borescopes.

The borescope's three main optical components are the objective lens system, the relay lens system, and the eyepiece, which determine its magnification. The objective lens is located at the end of the borescope and it acts similar to a camera lens. It forms the primary image of an object on the back of the lens. Relay lenses reform or relay the primary image every few inches along the length of the borescope. In long borescopes, several relay lenses are used. The last set of relay lenses produce the final image at the eyepiece of the borescope. The eyepiece lens enables the human eye to see the final image formed. Each of these three components produces a

EQUIPMENT 579

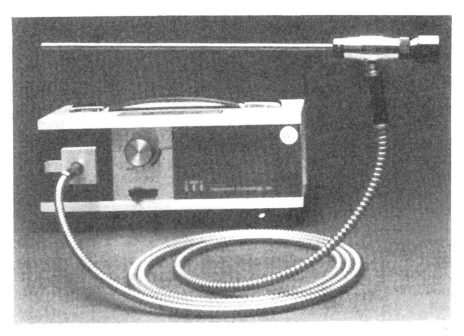

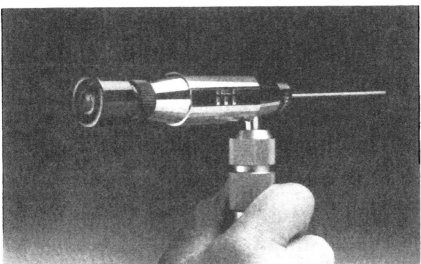

Figure 13.1 Rigid and micro-style borescopes for jet engine inspection. Courtesy of ITI/Instrument Technology, Inc.

magnification—the total magnification is the product of each magnification in accordance with Eq. 13.1:

$$M_b = (M_o)(M_r)(M_e) \qquad (13.1)$$

where M_b = total borescope magnification
M_o = magnification of objective lens
M_r = magnification of relay lens, usually 1
M_e = magnification of eyepiece

The manufacturer determines relay lens and eyepiece magnification. The magnification of the objective lens is complicated by the fact that it varies with distance from the object to the objective lens. One trick of the trade in determining the effective magnification of the objective lens is to place a wire of known length near the unknown defect while positioning the borescope a known distance from the object. When this can be done, the magnification of the objective lens can be calculated.

Therefore, in specifying the magnification of a borescope, it is necessary to specify the magnification at a specific distance, such as 2× magnification at 1 in. Visual magnification is a logarithmic function; magnification is highest at very close distances, decreases with increasing object distances, changes quickly at short distances, and then changes only slightly with increasing object distance. The accuracy of defect image measurement depends on the probe-to-object distance and the operator's ability to focus the instrument. Under ideal conditions, the measurement of image defects can be made to an accuracy within 0.001 in. using a special measuring filar eyepiece (graduated reticle mounted on a transversing slide attached to a micrometer) and jet engine borescope.

Rigid borescope models come in diameters ranging from 4 to 19 mm. A number of light sources, UV, VIS, high intensity, and LED, are available to aid viewing. Extendable model diameters range from 9.5 to 44 mm. They feature a variety of extender lengths and interchangeable viewing heads. The diameters of flexible models range from 3 to 10 mm and they are also available in a number of different lengths. Micro models come in diameters from 1 to 3.5 mm with lengths ranging from 2 to 6 in. Other models are available for underwater use (periscopes) and high or low pressures and temperatures.

Many companies now offer a complete, standalone video information processing system. Features of these closed-circuit television (CCTV) systems include defect image measuring capability, computerized video enhancement, alphanumeric keyboard for data identification, high-resolution display and recording, hardcopy documentation, and audio recording. The high-resolution video camera used with this system can be attached to virtually any borescope, fiberscope, periscope, or telescope. This system is said to be ideal for applications involving long inspection periods, multiple operator viewing, or cases

EQUIPMENT   581

Figure 13.2 Portable CCTV system mounted on an internal dolly. Courtesy of ITI/ Instrument Technology, Inc.

where complete documentation is required. Figure 13.2 shows an older portable CCTV system mounted on an integral dolly.

13.4.2 Jet Engine Inspection

Straight in-line and double swivel eyepiece borescopes, designed for inspecting jet turbine engines, are shown in Figure 13.3. These high-resolution borescopes, with 2× zoom capacity, were designed and developed in conjunction with commercial and military jet engine users for inspecting all major sec-

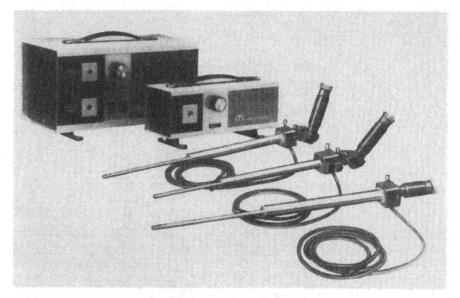

Figure 13.3 Straight in-line and double swivel eyepiece borescopes for jet engine inspection. Courtesy of ITI/Instrument Technology, Inc.

tions of the engine. The borescopes have a 35° field of view and an 80° scan (50° forward and 30° retro) capability. Complete inspection of the burner section, along with compressor and turbine blade inspection, from root to tip, can be achieved. Optional equipment, such as binocular eyepieces and cameras, can be used to reduce operator fatigue and photograph defects, respectively. In some cases, borescopes can save up to $150,000 in repairs on a single aircraft engine by eliminating the need for costly takedown and inspection.

13.4.3 Nuclear Applications

Custom periscope systems are finding increasing use in nuclear inspection systems. Large periscopes, 3 to 5 in. in diameter and 50 ft in length, have been designed for continuous use in sodium breeder reactors at 700°F. These units operate in an argon atmosphere in the presence of radiation and sodium vapors. Different viewing heads, one with 90° scan capability, have been provided.

Three identical special periscopes, capable of inspecting a fusion reactor during sustained operation, have been developed for the Tokamak fusion test reactor (TFTR) at Princeton University. These periscopes have to operate in high electromagnetic fields, at high temperatures and high radiation levels. In addition, high-resolution video recording and photography capabilities had to be provided. A Princeton computer automatically controls rotation, azimuth

EQUIPMENT

scan, elevation scan, power range, camera selection, focus, derotation of image, aperture control, and filter selection.

Other nuclear periscope inspection systems include:

- An underwater periscope for the inspection and photography of fuel control rods in loss of flow tests
- A 4-in.-diameter periscope system, with scan and power shift capability, to inspect and photograph underwater reactor components
- A 5-in.-diameter periscope with offset viewing head for inspecting and photographing reactor vessel welds under 90 ft of water
- Floor and wall periscopes for inspecting and photographing the processing of fuel elements
- The development of high-intensity illuminated periscopes for photography of waste storage tanks at depths of 30 ft
- Wall and ceiling periscopes for plant-wide monitoring of fueling at some reactor locations
- Complete remote viewing and inspection of advanced gas-cooled reactors using periscopes with remote scan, azimuth, drive, and focus controls

Underwater periscopes are primarily used for the inspection of spent fuel tube bundles. Normal operating lengths are 10 to 15 m, but units up to 30 m in length have been built for use in reactor vessels. Standard dual power units can resolve 0.001 in. at 2 ft. Magnification range is $1\frac{1}{2}\times$, $3\times$, $6\times$, and $12\times$. Options include azimuth and elevation scan, illumination, manual or motorized pool mounts, photo, and CCTV camera adapters.

A modular 3-in.-diameter periscope can be assembled in 6 ft lengths up to a maximum length of 100 ft for general observation and inspections. The instrument can be configured differently using optical elbows and viewing heads. Models are available for ambient, underwater, and high-temperature use. Magnification range is $1\frac{1}{2}\times$, $3\times$, $6\times$, and $12\times$. Options include illumination, photo, and CCTV camera adapters.

Wall periscopes provide 180° viewing of hot cells, reactor rooms, processing rooms, and other remote locations. These units provide high-resolution photography systems with built-in power change, lead shielding, and derotational features. They are available with wall rotation, lock, and counterweights. Their magnification range is $2\times$, $4\times$, $10\times$, and $20\times$. A less expensive, general-purpose 90°-wall viewer is also available with $1\frac{1}{2}\times$ magnification.

Underwater, $1\frac{1}{2}$-in.-diameter periscopes are designed for in-service inspection of pumps, fuel, and other items located in the reactor or spent fuel pool. Lengths of 100 ft are possible with 5 and 10 ft relay extenders. Several viewing heads and camera adapters are available.

Underwater telescopes are used to aid operators during fuel transfer operations. The telescope allows the operator to see fuel bundles from a service bridge. By placing the lower end of the telescope in the water, the operator can get a clear view of all underwater objects. The telescope has a rotating power shift control and a magnification range of 2×, 4×, 10×, and 20×. Camera adapters are optional.

13.4.4 Other Applications

For the Statue of Liberty restoration project, the Olympus Corporation donated flexible fiberscopes, rigid borescopes, halogen light sources, and photographic recording accessories. This equipment was used by National Park Service engineers to examine the statue's internal iron skeleton. In-depth observations revealed a hazardous assortment of wraps, sags, leaks, and failed joints.

When the Statue of Liberty was first built, an insulating material was placed between her outer copper skin and inner iron skeleton to minimize electrolysis. Later, moisture and decaying insulation created severe corrosion problems. Borescopes and fiberscopes were used to look into these hidden crevices to assess damages. They were also used to inspect the inside of the fingers holding the torch because they were key stress points.

Damaged copper skin was replaced and holes were repaired. Old insulation was removed and corroded iron pieces were replaced with 316L low-carbon stainless-steel and Teflon insulation. It should be several hundred more years before the Statue of Liberty needs any additional restoration work.

A partial listing of other applications for borescopes includes:

- Inspecting large gasoline and diesel engines, eliminating the need to tear them down
- Looking inside combustion boilers when "loss of flame" alarms occur to make sure there is no excess fuel that could cause an explosion on reignition
- Inspecting the inside of molding rams after cleaning, to assure that they will function properly when used
- Inspecting steam generators in power plants as preventive maintenance
- Examining the interior of heavy equipment hydraulic cylinders for flaws
- Pinpointing weld imperfections on inner pipe surfaces
- Inspecting helicopter spars internally
- Inspecting the bores of turbine shafts
- Inspecting the bores of large gun barrels

13.5 FIBERSCOPES AND VIDEOSCOPES

Modern fiberscopes and videoscopes eliminate or minimize many of the deficiencies of traditional borescopes. Their small size and flexibility provide

FIBERSCOPES AND VIDEOSCOPES

access to internal areas inaccessible to rigid borescopes. Digital images can be captured and processed in real time. With the aid of laser lights, the area and depth of many surface defects can also be determined.

AEI North America, Inc. provides alternative visual inspection systems. Equipment and services include borescopes, fiberscopes, inspection services, fiber-optic lighting, light sources, and repairs of most borescope/fiberscope/videoscope models.

The Econoscope™ provides a low-cost, high-resolution expendable inspection system that outperforms many more expensive models. The unique rod lens optical system provides high-quality images comparable, and in some cases, superior to conventional borescopes. The fiber-optic connector and eyepiece are compatible with current illumination and CCTV accessories.

Main Econoscope features include:

- Body—molded high-impact plastic
- Light guide fittings—stainless steel
- Optical components—glass and plastic
- Insertion tube—stainless steel; insertion diameter—10 mm; insertion length—285 mm
- View of direction—forward 0° with optional 60° forward oblique, 90° side-view, and 120° retrograde (with mirror tube)
- Field of view—75° with fixed focus from 6 mm to infinity
- Magnification factors—1:1 at 30 mm, 5:1 at 6 mm.

Light guide fittings are industry standards.

The Econoscope is fully compatible with integrated technology, which includes a range of battery light sources for complete portability. Standard adapters are used for total compatibility with CCD cameras, providing videotape and printer use for virtually any application. In addition, the wide optical focal range eliminates the need for focusing. The Econoscope is shown in Figure 13.4.

13.5.1 Applications

Some applications for the Econoscope include:

- Automotive repair
- Aircraft maintenance
- Casting industry inspections
- Pumping and power stations
- Locksmithing and safesmithing
- Sea wall/bridge inspections for civil engineers
- HVAC inspections

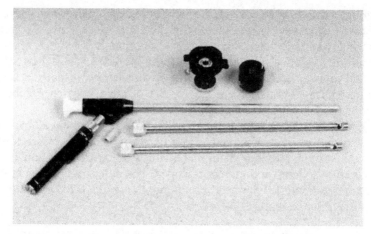

Figure 13.4 Econoscope. Courtesy of AEI North America, Inc.

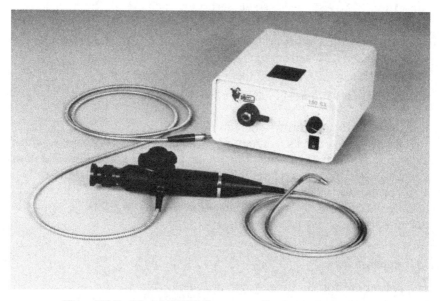

Figure 13.5 Model FS236 fiberscope. Courtesy of AEI NA, Inc.

The Econoscope can be used for most internal structural inspections within its size and design limitations.

Figure 13.5 shows a Model FS236 two-way articulating fiberscope. Specifications for this unit are shown in Figure 13.6. AEI North America, Inc. has also put together a special automotive inspection kit, known as the Car

Model # FS236 Fiberscope

Diameter	.236in (6.0mm)	Operating Temperature	32°F to 140°F (0°C to 60°C)
Length	40in. (102cm) 60in. (153cm) 80in. (203cm)	Light guide length Storage Temperature	60in (152cm) -70°F to 160°F (-57°C to 71°C)
Articulation	120° up/down	Shaft construction	Braided stainless steel
Depth of Field	0.4in. - 4.0in. 10mm - 100mm	Environmental - shaft immersible in:	Water JP4, JP5, Jet A, Gasoline
Field of View	60°	Eyepiece Diopter range	DIN 32mm +4 to -8
Bend Radius (shaft)	2.0in (50mm)	Right Angle	Prismatic

Figure 13.6 FS236 fiberscope specifications. Courtesy of AEI North America.

InspectaKit™ (not shown). The kit saves time in problem diagnosis, ensures more accurate estimates, and saves money on unbillable labor time. The kit features a 5/16″-diameter flexible fiberscope with 24″ length, 90° viewing tip, portable 12 V light source, battery adapter cable, durable foam-fitted carrying case, and light source powered by auto battery. Contact the manufacturer for other options and prices.

Suggested automotive applications include:

- Check combustion chamber for cracks, valves for damage, or carbon buildup without pulling the engine head.
- Check disk brakes without removing wheels.
- Check differential.
- Check for rust in wheel wells and body panels.
- Inspect radiator block.
- Inspect locking mechanisms without removing doors.
- Inspect wiring under dash and in other tight places.

13.6 SNAKEEYE™ DIAGNOSTIC TOOL

The SnakeEye™ diagnostic tool is a portable, lightweight, easy-to-use remote video inspection device with a high-resolution (320 × 234 pixels) flat panel 5″ color thin film transistor/liquid crystal display (TFT/LCD). Figure 13.7 shows the display. The basic kit for the unit, shown in Figure 13.8, consists of the high-resolution display, 4′ flexible cable, rigid wand, camera head, battery charger, rechargeable battery pack, AC adapter, 12-volt auto adapter, and ring adapter for ring-finger control of the camera.

Several uses for the SnakeEye system are industrial, building, aviation, vehicle maintenance, border patrol security, and public safety. In industry the

588 VISUAL AND OPTICAL TESTING

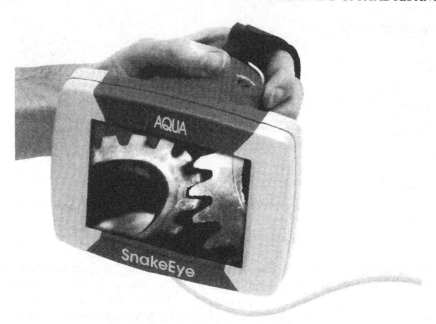

Figure 13.7 AEI NA, Inc. SnakeEye™ monitor. Courtesy of AEI NA, Inc.

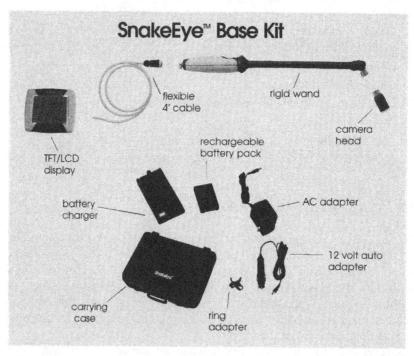

Figure 13.8 AEI NA, Inc. SnakeEye™ base kit. Courtesy of AEI NA, Inc.

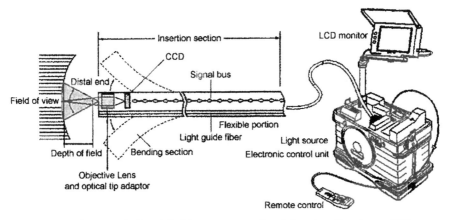

Figure 13.9 Olympus industrial videoscope structural drawing. Courtesy of Olympus Industrial America, Inc.

unit is used to inspect machinery, pipes, tanks, valves, and welds. In building maintenance work, it can be used to check wiring and plumbing behind walls and in ceilings. For aviation, the unit can be used to inspect airframes, engines, and instrument panels. Automotive uses include inspections under dashboards and in car doors for contraband. For public safety work, the unit can be used in search-and-rescue missions and for inspection of vehicles and crime scenes.

13.7 INDUSTRIAL VIDEOSCOPES

Olympus is a leader in the design and fabrication of industrial videoscope systems. The IPLEX™ industrial videoscope, shown structurally in Figure 13.9, combines advanced optodigital technology for intelligent imaging. Micron-scaled components have been incorporated into a scope with a diameter of just a few millimeters to provide superior performance for capturing and reproducing images as accurately as possible. All necessary components have been combined in an integrated portable IPLEX package that is easy to use, handle, and transport to the inspection site. Inspection results can be integrated with a PC to record, store, manage, and display data and provide inspection reports.

13.7.1 Equipment and Features

There are three basic IPLEX models: IPLEX SA (stereo measurement), IPLEX SA (standard model), and IPLEX (limited-function model).

Features and advantages include:

- Detachable 5.6″ high-resolution LCD monitor. The bright, high-resolution monitor is designed to minimize fatigue during extended inspections.
- Insertion tube diameters of 4.4 mm or 6.0 mm.

- Compact, lightweight multifunction remote control. The slim remote control fits easily in the palm of the hand and weighs only 160 grams. The remote control provides two joysticks and total control for all functions required to operate the IPLEX. Functions include Brightness/mark button, zoom lever, Live/gain button, Menu/exit button, Index button, Power button, Angulation control with lock, Center button, Measurement/enter joystick, Record button, and Freeze button.
- Recording card slot. Inspection result images can be recorded directly on the dedicated recording card and transferred to a PC. When the card is full, image files can be erased or a new card can be used.
- USB connector port. The USB port can be used for direct transfer of inspection data to a USB-compatible PC for storage, data processing, and analysis. Remote operation of IPLEX using the mouse of the connected PC is also possible.
- Angulation driving motor provides smooth high-precision tip angulation that can be controlled with the joystick on the remote control.

Other features include:

- Video output terminal
- S-video input/output terminal
- Multifunction CCU
- Shock-resistant case
- 50 W metal halide lamp
- Insertion tube winding handle and drum

Figure 13.10 shows a selection of available accessories.

13.7.2 Instrument Setup

Instrument setup consists of the following:

1. *Adjusting the monitor position.* Raise the integral monopod with the attached LCD monitor. Adjust the position for convenient viewing.
2. *Connecting unit to the power supply.* Pull out the power cable stored in the main unit and plug it into the AC power outlet.
3. *Holding the remote control.* The compact, lightweight remote control fits snugly in the palm of the hand. It is also provided with a mount for the insertion tube so the operator can hold the remote control and insertion tube simultaneously.
4. *Pulling out the insertion tube.* The insertion tube is wound around the built-in drum so it can be pulled out smoothly. As soon as it is pulled out, the built-in illumination system is automatically activated.

INDUSTRIAL VIDEOSCOPES

ACCESSORIES

Optical Adapters

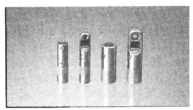

Stereo Optical Adapters

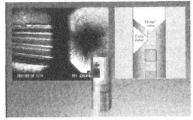

Forward/Side-Viewing Adapter

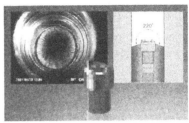

Ultra Wide-Angle Adapter

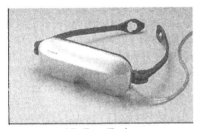

3D Eye-Trek

Accessory Bag

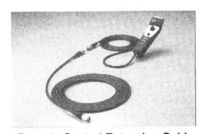

Remote Control Extension Cable

Rigid Sleeve

Figure 13.10 IPLEX industrial videoscope accessories. Courtesy of Olympus Industrial America, Inc.

13.7.3 3D Viewing

In addition to looking at flat 2D images on monitoring screens, operators can now virtually put themselves inside the inspection object and examine interiors from every angle and explore surface textures for defects that might be difficult or impossible to detect in a 2D image. With stereo measurement and the optional 3D Eye-Trek face-mounted display, the operator can utilize the power of 3D imaging to accurately observe and bring a new level of confidence to his or her inspection work.

13.7.4 Applications

A versatile 6.2-mm-diameter videoscope, which includes a working channel, has been designed to meet the specific requirements of gas turbine inspections. The working channel can be mounted on the remote control unit for further convenience. This IPLEX SX unit provides improved access to intricate engine interiors, enhanced hooking and retrieval operations for exceptional ease of use, advanced observation, and proven stereo measurement performance.

High-precision stereo measurements may be made from any angle. The six measurement modes available are *distance, point-to-line, depth, lines, area*, and *profile*. Point-to-line, lines, and area measurements are illustrated in Figure 13.11. When both ends of a crack are plotted, distance measures the straight line distance between the ends. Depth measures the depth/height from a hypothetical plane inside three designated positions to the required point. This measurement is effective for the measurement of corrosion, protrusion, and indentation. Finally, for profile, the cross section of two designated points is represented with computer graphics. An irregularity that is not comprehensible on the monitor is now easy to view and understand.

A 4.4-mm-diameter scope (without a channel) facilitates inspection of minute parts. Interchangeability of the 4.4 and 6.2 mm scopes provides a simple cost-effective system.

The rigid sleeve is useful as an auxiliary insertion tube. It also makes the scope easier to handle. The sleeve is simply fit and locked to the tip of the insertion tube.

Olympus's innovative Tapered Flex technology is known for its unequaled insertion capability. The IPLEX SX's Tough Tapered Flex tube has been designed to provide optimum flexibility in aircraft engine inspection and excellent resistance to crushing.

13.7.5 Working Tools

A number of working tools are available to meet a wide range of inspection requirements. These tools typically assist the operator in the retrieval of foreign objects and dropped objects. Available retrieval tools currently include a snare basket, three-prong grasper, magnets, and alligator forceps. Please contact the manufacturer for additional information.

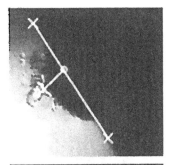

Point-to-Line

Measures the distance between a hypothetical line between two designated positions and the required point. This is effective for the detection of the dimensions of a chipped section in a blade.

Lines

The focal length of an abnormal section can be measured in accordance with its actual shape. This is useful for estimating the total length of a crack with complicated contours.

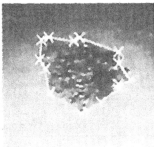

Area

The circumference of the abnormal section can be plotted, allowing its area to be measured. This mode allows the dimensions of a burn or surface coating defect to be qualified.

Figure 13.11 Stereo measurement modes showing point-to-line, lines, and area methods. Courtesy of Olympus Industrial America, Inc.

13.8 PROJECTION MICROSCOPES

The standard microscope has been the workhorse of the medical industry and analytical chemistry laboratories for many years. Operating microscopes, with swivel arms and floor stands or ceiling supports, have taken the microscope out of the laboratory and put it in the operating room where microsurgery on ears and eyes is commonplace. Likewise, the development of projection microscopes has led to the incorporation of the common microscope into the inspection lines of many industries. The optical projector reduces eyestrain and enables the operator to visually inspect large numbers of small parts at rea-

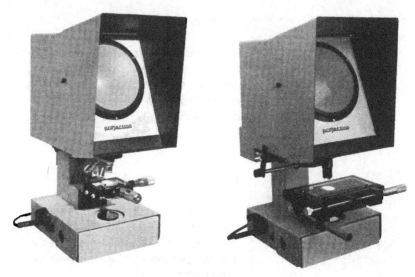

Figure 13.12 Older-style micro- (L) and macro- (R) projection microscopes.

TABLE 13.1. Projection Microscope Applications

Profile projection diascopy, episcopy	Particle size inspections
Textile studies	Paper surface and ink print studies
Metallurgical studies	Wood fiber and cellulose studies
Examination of electronics and cables	Measuring thicknesses to 1 μm
Inspection of precision mechanics	Examination of masks, integrated
Food industry studies	circuits (thick, thin, and hybrid)
Plastic and packaging studies	Visual melting point determinations

sonable production rates. Thirty five-millimeter cameras can be attached to virtually any projection microscope and many of them can project their images on the walls of darkened rooms.

So-called micro-macro projection microscopes (Figure 13.12) have magnification ranges of 3× to 500× or 3× to 3000×, depending on make and model. The lower-power units provide up to 150 mm free space clearance between the carrier and objective lens. Various lamps are provided for transmitted, inclined, or vertical viewing. Macro viewing is done at magnifications of 3 to 50× and micro viewing is done at magnifications above 50×. Several viewing screen materials are available and most projection microscopes feature wide-angle viewing so that several people can simultaneously view the screen. Table 13.1 is a partial listing of applications.

Stereomicroscopes and stereomacroscopes are used in many similar applications. They provide good working distances, typically 55 to 112 mm, depending on magnification. The stereomacroscope provides excellent depth of field, high resolution, good image brightness, and comfortable workstation viewing. Some defects are easier to detect and evaluate when viewed in three dimensions.

Comparison projection microscopes and macroscopes (Figure 13.13) permit side-by-side, and in some cases, overlap comparison of apparently identical objects. These comparison-type techniques can show small differences or similarities in "identical" objects. They are especially useful in overlap comparison studies for quality control in the electronics industry.

Perhaps the most interesting use of comparison microscopes is in the field of police science and criminology. Television shows dealing with police work and crime scene investigations (CSIs) have illustrated how the criminal can be brought to justice as the result of comparison microscope studies dealing with fired cartridges, rifling marks on spent bullets, hairs, typewriter impressions, and fibers. Recent sniper attacks in the Washington, DC area provide a grim reminder of the importance of projection microscopes in police work. In this case, a number of spent bullets from multiple locations were tied to one of the sniper's weapons. Figure 13.14 shows the differences between hair and textile fiber samples. Such differences can sometimes prove the innocence or guilt of a suspect.

Figure 13.13 Stereo comparison microscope and macroscope.

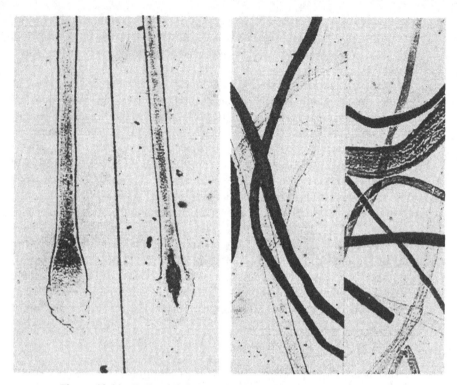

Figure 13.14 Police lab hair and textile fiber sample comparisons.

13.8.1 Leica FS4000 Forensic Comparison Microscope

The Leica FS4000 (Figure 13.15) is a modern, state-of-the-art system for trace evidence examination in crime lab work. It combines current optical and mechanical microscope features in an easy-to-use fully integrated bridge design. This ergonomic workstation allows the user to concentrate on his or her work and achieve the highest accuracies possible with a minimum of effort.

The Leica FS4000 provides precise comparison of two objects at magnifications up to 100× and is sensitive to minute differences in microstructure, texture, and color. All common contrasting techniques such as fluorescence, polarization, and brightfield are available and can be selected in a fraction of a second. Side-by-side, superimposed image and mix image viewing modes can be achieved by pressing a single button.

Main instrument features include:

- *Automated comparison bridge* with one-button control of all functions, built-in tilting ergo tube, color neutral beam splitter prism in bridge, separate control unit for bridge functions.

PROJECTION MICROSCOPES

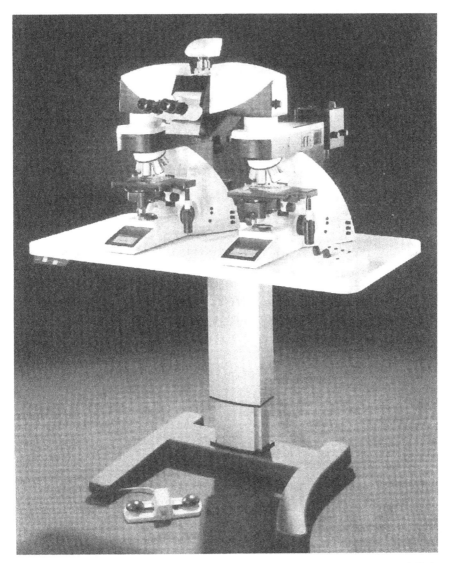

Figure 13.15 Leica FS4000 forensic comparison microscope. Courtesy of Leica Microsystems, Inc.

- Based on two modified Leica digital microscope stands equipped with Variolux, the Leica system for perfect color balance in both light paths. Light intensity is twice that of earlier systems.
- One-button control and selection of contrast method on each microscope stand.

- Illumination manager for Auto Kohler illumination (even illumination) from 12.5× to 1500× total magnification.
- Two status displays for exact match of illumination conditions. The new bridge allows ±4% zoom adjustment on the right side for comparing deformed specimens or temperature-sensitive materials. Calibrated matched configurations are easily reproduced and indicated by LEDs.
- *Constant color intensity control (CCIC)* for color neutral intensity adjustments in both transmitted light paths.

Figure 13.16 shows a highly magnified grayscale image of a multicolored textile fiber.

Observation modes include:

- Full-screen left, full-screen right.
- Superimposed image comparison. Complimentary color filters render unmatched sample portions in color. Overlying details appear in their original color only in those places where there is no structural deviation.
- Split-image with adjustable strip for partial superimposed image. The width and position of the dividing line can be adjusted by the user or set

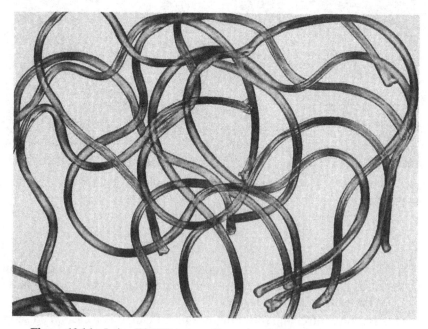

Figure 13.16 Leica FS4000 image. Courtesy of Leica Microsystems, Inc.

PROJECTION MICROSCOPES 599

as a wide strip, where both objects can be overlaid in the accordance with the above.
- Split image with adjustable dividing line.

Leica also manufactures the FS C Macroscope for simultaneous macroscopic comparisons. This instrument has a working distance of up to 60mm (2.36″), with new 1×, 2×, and 4× telecentric objectives that provide exact magnifications of 4× to 60× regardless of what the z position of the objectives are relative to the sample. Built-in iris diaphragms in each objective are used to increase contrast and depth of field for optimization of each sample.

Remote-controlled cold-light sources are applied in three illumination techniques—axial incident light, transverse incident light, and transmitted light. Motorized stages and focused drives can be synchronized at the touch of a button. Figure 13.17 shows a split-image comparison of breech face markings on shell casings.

Documentation options include photo systems and cameras with dedicated software, high-output image recording, processing, and archiving.

Other applications for comparison microscopes include comparisons of industrial filter materials, microorganisms and bacteria, metal heat treatment, grain size and surface finish, rare stamps and forgeries, paint, ink, and pigments, diamond quality, and chemical crystal structures.

A special optical comparator, known as an interferometer, is used for testing flatness and parallelism. In electronic work, the interferometer is used to check the flatness of semiconductor substrates. In mechanics, it is used for testing the flatness of lapped surfaces and the parallelism of slip gages relative to a reference surface. Finally, in optics, the interferometer is used for testing the flatness and parallelism of graticules, photographic plates, and prism faces. Figure 13.18 shows several common interference patterns.

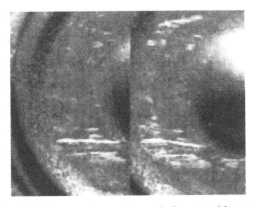

Figure 13.17 Leica FS C comparison of breech face markings. Courtesy of Leica Microsystems, Inc.

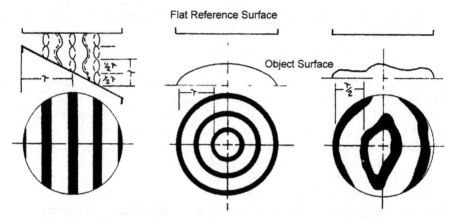

Figure 13.18 Several interference microscope patterns. Light bands are produced when transmitted and reflected light are in phase. Dark bands are produced by out-of-phase light waves.

13.9 THE LONG-DISTANCE MICROSCOPE

The Infinity Photo-Optical Company is the inventor and recognized world leader of long-distance and continuously focusable microscope technology. Infinity manufactures long-distance microscopes (LDMs), continuously focusable microscopes, macro systems, internal-focusing devices, and other lenses for industrial inspection, process/product monitoring, machine vision, QC, advanced imaging, noncontact gauging/inspection, and laser/biomedical research.

13.9.1 New Developments

Four exciting new developments were announced by Infinity from Stuttgart, Germany (Vision 2003), October 21, 2003. These were:

1. The new and improved HDF-2, which focuses directly and internally from infinity down to 30 mm and can be used with all camera formats from $\frac{1}{4}$ in. to 8×10 in. sheet film cameras (with accessories). It is therefore considered a *universal format lens (UFL)*. With its pronounced macro depth-of-field and excellent imaging characteristics, system optics can be further supplemented with motor-focus and motor-zoom capabilities.

2. The new and improved InfiniVar CFM-2 further refines the microscope front-end assembly to allow greater close-focus convenience and the ability to lock the internal focus at any point in its continuous infinity to 25 mm range of operation. Consequently, with $\frac{1}{2}$ in. sensor cameras, InfiniVar CFM-2 provides 0–250× on 13 in. monitors; 0–500×, when sup-

plemented by a 2× DL tube. Motor-focus and motor-zoom options make the CMF-2 a state-of-the-art imaging system.

Note: The new model has an objective front that provides a range of 0.2× to 6× when zoomed down to 25 mm working distance.

3. Infinity's new concept in machine vision systems—the Accordian™ configuration. For the first time, a series of motorized machine vision lenses can be combined in any number, up to 10 each, and locked in a row. This configuration allows end users to build vision machines taking into account the combined system. Accordian provides a compact answer to growing demands for standardization combined with the versatility in the machine vision field. This permits vision machines to be built around predetermined dimensions—mixing and matching a variety of different lenses—instead of trying to retrofit lenses to predetermined machines.
4. New upgraded InFocus™ Insert Series, designed to fit directly into trinocular observation tubes or into side/video ports of virtually all infinity-corrected microscopes. This insert can be used with virtually all brands of microscopes. The InFocus Insert is a simple "plug-in" which can be moved from one microscope to another (with proper connecting adapter). At present, the InFocus Insert is the world's only universal active photo/video adapter. The insert allows microscopes to be focused without external movement—controlling and correcting spherical aberration, while increasing contrast and resolution.

Note: InFocus Inserts are strictly photo/video adapters; the visual binocular image remains unaffected.

13.9.2 Model K-2 Long-Distance Microscope

The Model K-2 for large-format video and photo (Figure 13.19) is designed to operate from infinity down to 51 mm, depending on its optical system setup. This LDM can be used in visual, photographic, or video modes. Accessories can be used to multiply, halve, or zoom the K-2's magnification or adapt it to a range of video/photo cameras. Figure 13.20 shows the working distance, magnification, and field of view for available K-2 lenses and the relatively linear relationship between magnification and working distance on a log/log scale.

The K-2 is composed of a main body that contains its patented *Afocal Variation System™ (AVS)* internal focusing system, activated by rotation of the focusing ring. A clamp is provided to allow positioning and mounting capabilities. An optional focusing ring lock can be provided for special requirements. A dovetail flange at the front links various supplemental objectives. Each objective, except the CF-4, has a built-in holder with M49 threads, allowing standard photographic polarizers and color filters to be used.

The K-2 clamp can be exchanged with stereomicroscope stand adapters if required. Behind the standard clamp various optical amplifiers such as a 2×

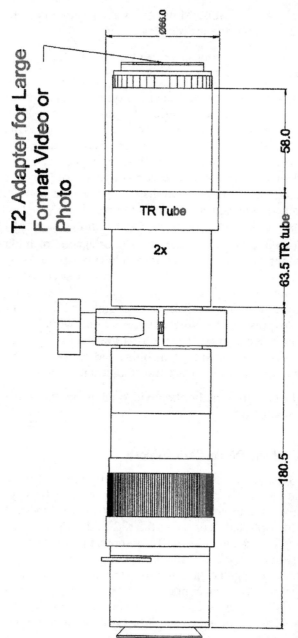

Figure 13.19 K2 Direct Mode LDM with 2× TR tube and T2 adapter for large-format video or photo applications. Courtesy of Infinity Photo-Optical Company.

THE LONG-DISTANCE MICROSCOPE

K2	STD		CF-1		CF-1/B		CF-2		CF-3		CF-4	
	Near	Far	Near	Far	Near	Far	Near	Far	Near	Far	Near	Far
WD mm	440	1350	272	660	225	457	140	214	90	125	51	63
MAG	1.5	0.45	2.3	0.9	2.8	1.5	4.0	2.5	5.8	4.0	10.7	8.0
FOV mm	4.4	14.1	2.8	7.0	2.3	4.2	1.6	2.6	1.1	1.6	0.6	0.8

*FOV based on 1/2" video format (6.4mm horizontally). See Video Format Page.

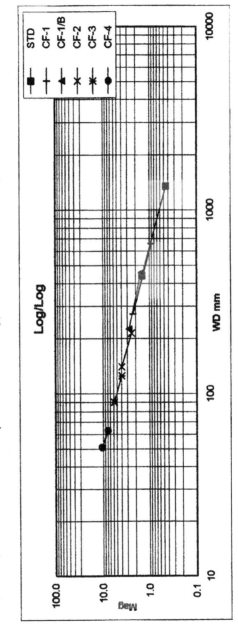

Figure 13.20 Working distance (WD), magnification (MAG), and field of view (FOV) for various K2 lenses (Top). Percent magnification as a function of WD on a log/log scale (Bottom). Courtesy of Infinity Photo-Optical Company.

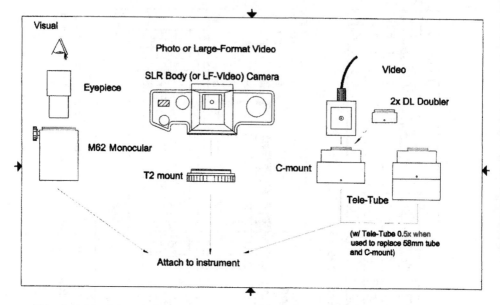

Figure 13.21 Direct-imaging options with K-2 long-distance microscope. Courtesy of Infinity Photo-Optical Company.

TR tube, zoom module, and/or 2× DL tube doubler can be attached. Figure 13.21 shows direct imaging options for the K-2 LDM.

13.9.2.1 Numerical Aperture (NA) Infinity instruments typically operate at exceptionally long working distances, yet provide equal or better resolution than instruments that function at much closer distances. The following information can be used to determine the instrument's theoretical resolving power at any front of lens to object distance.

1. Divide the radius of the objective by the working distance in mm. This provides the tangent. Consulting mathematical tables or calculators for the sine of this tangent will provide the "exact" NA.
2. Multiply NA by 3000 (Rayleigh's formula) to get the theoretical resolution in lines/mm (LPM).
3. Dividing 1000 by LPM provides the theoretical resolution in µm (microns).

For example, with the CF series of objectives for Model K-2 and CFV series objectives for Model KV, the objectives have apertures of 38 mm or a radius of 19 mm (with the exception of CF-4 and CFV-4).

Using the procedure described above, the following results are obtained:

1. With a CF-2 objective working distance of 166 mm, the NA is 19/166 = 0.114.
2. Then, the theoretical resolution in lines/mm = 0.114 × 3000 = 342 LPM.
3. The theoretical resolution in microns = 1000/342 = 2.92 µm.

13.9.2.2 Care and Cleaning The K-2 and all other Infinity instruments discussed in this section should be treated as fine optical instruments. Keep dust and dirt off external lens surfaces. Clean metal parts only with a soft cloth moistened with alcohol. External optical surfaces should be cleaned only when necessary, and then only with a soft cotton swab moistened with an approved optical glass cleaner. For additional information, please contact the Infinity Photo-Optical Company.

13.9.3 InfiniVar CFM-2 Video Inspection Microscope

The InfiniVar CFM-2 (Figure 13.22) refines the front end of the LDM to provide greater close-focus convenience and the ability to lock the internal focus at any point in its continuous infinity to 9 mm range of operation. It is the only video-dedicated microscope that focuses continually without "blackout" from infinity down to 1.0" (25 mm) by a 180° twist of its focus control. There are a large variety of accessories. Consequently, with $\frac{1}{2}$ in. sensor cameras, InfiniVar CFM-2 provides 0–250× on 13 in. monitors, 0–500× when supplemented by a 2× DL tube. Motor-focus and motor-zoom options make the InfiniVar CFM-2 into a state-of-the-art imaging system.

InfiniVar CFM-2 has no objective lens changes and does not require any additional optics to operate throughout its more than 30:1 range. As long as the minimum distance is not exceeded, it is impossible not to get an image. You see everything that happens as you use the microscope.

The InfiniVar CFM-2 is composed of a unitized optics module, which contains all of its lens elements. Spacer tubes have international "T" thread—M42 × 0.75—a mounting clamp and C-mount adapter. All C-mount cameras can be connected to this assembly and once connected, it is ready for use.

Note: The CFM-2 optics module is a sealed unit. Operators should not attempt to take the optics apart or enter any part of the optical assembly unit.

On the front optics module is a section that looks like a standard microscope objective. This part is topped with a 28-mm-diameter flange, permitting fiber-optic ring illuminators of that diameter to be attached. This flange also has a mounting depression that mates with the three setscrews of the 66 mm *ring light adapter*. Once mounted, 66 mm ring lights can be used. The 66 mm ring light adapter can be left on the module during normal use. InfiniVar CFM-2 optical data are shown in Figure 13.23.

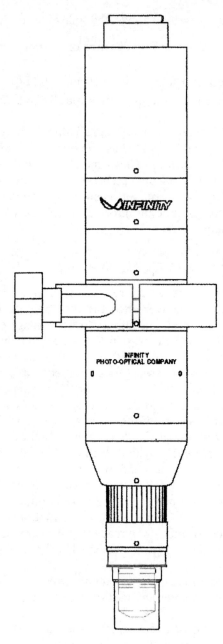

Figure 13.22 InfiniVar CFM-2 video inspection microscope. Courtesy of Infinity Photo-Optical Company.

INFINIVAR CFM-2 OPTICAL DATA (0.5 RANGE)

WD mm	9	12	16	25	35	50	100	200
Mag	4.3	2.4	1.5	0.9	0.5	0.4	0.2	0.1
FOV	1.5	2.7	4.2	7.5	12	17	36	74

*FOV based on 1/2" video format (6.4mm horizontally). See Video Format Page for all conversions. For 1/3" sensors multiply FOV in chart by 0.75 for new 1/3" FOV.

WD mm	9	12	16	25	35	50	100	200
Mag	8.0	3.4	2.1	1.1	0.7	0.5	0.2	0.1
FOV	0.8	1.9	3	5.7	8.7	13.9	28	57

Figure 13.23 InfiniVar CFM-2 optical data charts (0.5 range). Courtesy of Infinity Photo-Optical Company.

Care and cleaning requirements are the same as previously discussed for the Model K-2.

13.9.4 Accordion™ Machine Vision

Infinity has introduced a series of motorized machine vision lenses that can be combined in any number, up to 10 each, and locked in a row. This configuration allows end-users to build vision machines taking into account the combined system. With known dimensions and standardized components, Accordion™ is a compact answer to the ever-growing demands of standardization and versatility in the machine vision field. Now, vision machines can be built around predetermined dimensions, mixing and matching a variety of different lenses, instead of trying to retrofit lenses to existing machines. Accordion will also mate with many of Infinity's other products. Contact Infinity for compatibility questions.

13.9.5 InFocus Microscope Enhancement System

InFocus is the world's-first optical focusing system for microscopes. InFocus translates focus above, through, and below the object. Its action is identical to that of the mechanical focus. For the first time in 400 years, microscopes can be focused without external movement—controlling and correcting spherical aberration while increasing contrast and resolution.

13.9.5.1 Spherical Aberrations What is *spherical aberration* and how serious is it? Spherical aberration occurs when an image cannot be brought into complete focus at one nearly approximate point. Instead, light focuses in many places, none of which are entirely suited for imaging. Because every microscope slide is also an optical system, the thickness of the cover glass and the depth of the mounting material must be within specified precalculated tolerances. The cover glass must not deviate from a thickness of about 0.17 mm; the object must be positioned in close proximity. Even oil immersion objectives and the newer water immersion objectives are susceptible to image deterioration caused by spherical errors. The best solution prior to the development of the InFocus insert was to incorporate movable lens elements within the objective, activated by a correction collar. However, setting the correction collar is difficult and specific to only one working distance at a time. Moving the objective upsets the correction and the process must be repeated every time it is moved.

13.9.5.2 InFocus Corrections The InFocus insert corrects cover glass induced spherical aberration in high-apertured microscopic objectives. When InFocus is set to its central marking, it is corrected for cover glasses of 0.17 mm thickness. Turning the InFocus control ring right or left translates focus through a specimen while the objective remains at its initial working distance. All this is accomplished with virtually no magnification changes during translation.

If the cover and specimen preparation are correct to begin with, focal translations can be ranged with no further consideration. However, few preparations are perfect. Therefore, if the cover/preparation shows less contrast than expected, InFocus can be turned clockwise for covers that are too thin, and counterclockwise for covers that are too thick. For routine work, it is usually possible to select one of three positions on the InFocus dial. Although that setting may not be the theoretically exact one, it will surpass the results obtained without the InFocus insert in the system.

InFocus's focal sweep or translation range depends on the focal length of the objective. Unlike depth of field, which is interdependent on magnification and NA, InFocus's translation range is not limited by high-NA objectives. Therefore, significant translation can be accomplished by high-NA 40× (generally 4 mm f.l.) or 60× (generally 2.5 mm) objectives of up to 1.40 NA. When used with a 40×/1.3 NA/4 mm f.l. objective, InFocus's sweep is better than 60 μm, conservatively rated. In addition, if preparations are of known depth, the InFocus marker ring can be set to translate focus primarily in that direction. Most often, even the thickest preparations can be accommodated. Consequently, InFocus is ideal for correcting and scanning 3D images.

13.9.5.3 Applications Using conventional microscopes, it is relatively easy to damage slides, the microscope, or sensitive semiconductor wafers by accidentally contacting the slide or wafer while focusing the instrument. The InFocus insert eliminates this problem by providing the sharpest focus possible without moving the instrument's objective.

Drs. Z. Kam, D. A. Agard, and J. W. Sedat have successfully implemented InFocus, with some modifications, for three-dimensional biological microscopy solving sample motion, perturbed by conventional focus and spherical aberration problems at high resolution in thick live biological specimens. See the Infinity website to download this paper.

One unique application for the Accordian mounting system uses dual-ganged Infinimini w/CFM attachments as shown in Figure 13.24. The CFM

THE LONG-DISTANCE MICROSCOPE

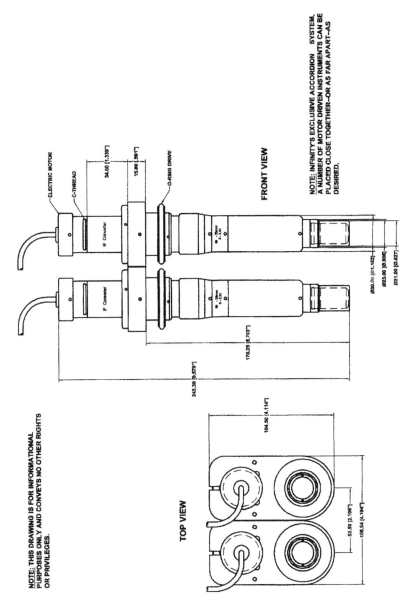

Figure 13.24 According mounting of two motorized Infini-Mini lenses with CFM attachments. Courtesy of Infinity Photo-Optical Company.

attachment provides continuously focusable microscope technology from infinity to 25 mm as shown in Figure 13.25 with a wide range of magnification based on lens options. The compact system shown has been found to be useful for making forensic comparisons.

INFINIMINI w/CFM ATTACHMENT
(used with IF Converter and SpheraCone inserted)

WD mm	25	35	40	50	68	100	200
Mag	3.2	1.6	1.0	0.7	0.5	0.3	0.1
FOV	2	4.1	6.4	9.2	12.8	23	50

Use DL Tube to amplify all above magnifications by 2x. The FOV is consequently reduced by 0.5x.
*FOV based on 1/2" video format (6.4mm horizontally). See Video Format Page for all conversions.
For 1/3" sensors multiply FOV in chart by 0.75 for new 1/3" FOV.
The SpheraCone™ fits into the IF Converter and is supplied with all CFM Attachments.

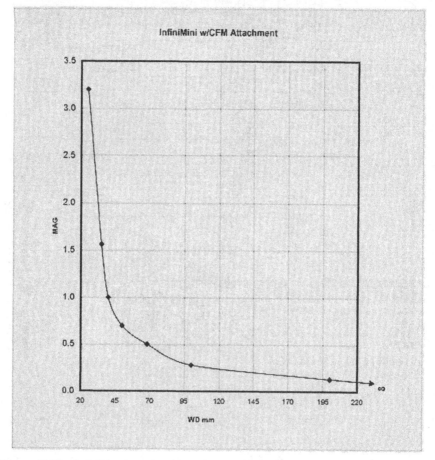

Figure 13.25 InfiniVar CFM-2 optical data charts. Courtesy of Infinity Photo-Optical Company.

REMOTE VISUAL INSPECTION 611

13.10 INFINIMAX™ LONG-DISTANCE MICROSCOPE

The InfiniMax LDM has working distance of 381 mm to 65 mm (15.0″ to 2.56″) and an OD of 48 mm. The InfiniMax-SD™ has similar optics with a smaller OD of 33 mm. The InfiniMax can be used with photo formats up to 36 × 43 mm (35 mm single lens reflexes SLRs, digital cameras, or very large video formats up to 2/3 in. The InfiniMax LDM is shown in Figure 13.26 and available objective lens optical characteristics are shown in Figure 13.27.

13.11 REMOTE VISUAL INSPECTION

The Everest VideoProbe® XL PRO™ is an advanced, portable, and modular remote visual inspection system consisting of a hand-held controller/display (hand-piece), one of two processor/light source units, one of five probes featuring all-way probe tip articulation, a portable shipping/operating case or optional backpack for added portability and battery operation. Figure 13.28 shows some basic system components. For a complete list of modular system accessories, see manufacturer's Videoprobe brochure or website information listed in Appendix 2 under Company Contributors.

Figure 13.26 Infinity InfiniMax long-distance microscope. Courtesy of Infinity Photo-Optical Company.

InfiniMax Objective	CW	Mid	CCW	
MX-1	10.1	10.2	11.1	FOV mm
	315	335	381	WD mm
	0.64x	0.63x	0.58x	Mag
MX-2	7.9	7.9	8.2	FOV mm
	250	255	285	WD mm
	0.81x	0.81x	0.78x	Mag
MX-3	6.4	6.4	6.4	FOV mm
	174	178	194	WD mm
	1.00x	1.00x	1.00x	Mag
MX-4	5.5	5.5	5.5	FOV mm
	169	175	186	WD mm
	1.16x	1.16x	1.16x	Mag
MX-5	4.5	4.5	4.5	FOV mm
	134	140	146	WD mm
	1.42x	1.42x	1.42x	Mag
MX-6	2.2	2.2	2.2	FOV mm
	65	66	68	WD mm
	2.9x	2.9x	2.9x	Mag

Field of View for 1/2-in. CCD Format camera.
Note: For 1/3-in. camera format reduce Field of View by 25%.

Figure 13.27 InfiniMax optical characteristics. Courtesy of Infinity Photo-Optical Company.

XL PRO VideoProbe features include:

- 3.5 in. floppy drive
- 16 MB internal flash memory
- Integrated temperature warning system
- Probe interchangeability with five different diameter probes
- Complete portability with AC, DC, or battery operation
- Three measurement modes—ShadowProbe®, StereoProbe®, and comparison
- iView™ image management

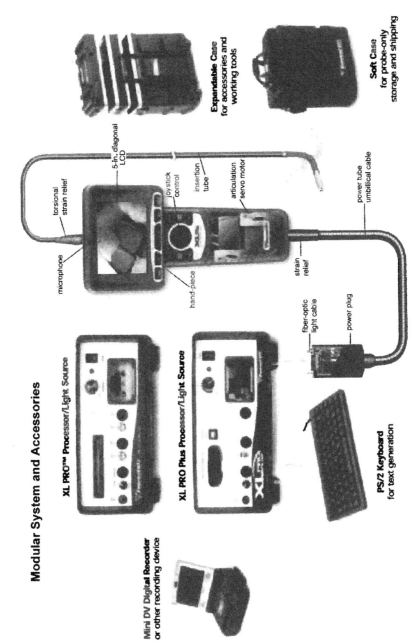

Figure 13.28 Videoprobe modular system and accessories. Courtesy of Everest VIT, Inc.

Additional XL PRO Plus features include:

- Two hours of DVD-format, MPEG2 full-motion video
- CompactFlash® digital storage media
- USB streaming video port
- 32 MB internal flash memory

The VideoProbe XL PRO and VideoProbe XL PRO Plus systems consist of the following major components:

System. The main system components are the shipping/operating case and AC and DC power supplies.

Camera. Probes with diameters of 5.0 mm, 6.1 mm, 7.3 mm, and 8.4 mm use a 1/6 in. SUPER HAD™ CCD image sensor with 380,000 pixels NTSC; 440,000 pixels PAL in a titanium housing (8.4 mm in stainless-steel housing). The 3.9 mm probe uses a 1/10 in. SUPER HAD CCD image sensor with 250,000 NTSC; 290,000 pixels PAL in a titanium housing.

Hand-Piece. The lightweight hand-piece has overall dimensions of $12.0 \times 5.8 \times 5.0$ inches and weighs in at 5.1 pounds with insertion tube, power plug power tube, and hand-piece. The hand-piece is housed in ABS housing with integrated Santoprene® bumpers. The LCD monitor is a 5.0 in. diagonal integrated TFT color LCD. The unit enables 360° All-Way® servo-motor tip articulation. Other controls include a joystick and complete function button set and an integrated high-sensitivity microphone.

Insertion Tube. Figure 13.29 shows a typical camera head, bending neck, and insertion tube arrangement with construction details.

Processor/Light Source. A 12 pound, $11 \times 11.5 \times 5.5$ in. aluminum chassis with polyurethane bumpers; 32-bit Pentium®-class imbedded computer; 50 W metal-halide lamps with automatic and variable brightness control; PS/2 style input connector; built-in front panel speaker and volume control; video input and output and audio output jack. The unit also has a RS-232 remote control connector.

Note: See manufacturer's data for complete specifications.

13.11.1 Industries—Applications

Aviation—commercial, military, OEM, MRO, and business.

Power—nuclear, fossil, combustion turbine, combined-cycle, hydro, and wind.

Process—offshore, refining, chemical, distribution/transportation, food, ultrapure, and water treatment. Figure 13.30 shows Videoprobe inspection of a refinery fin fan heat exchanger.

Other—manufacturing, shipping, law enforcement.

REMOTE VISUAL INSPECTION

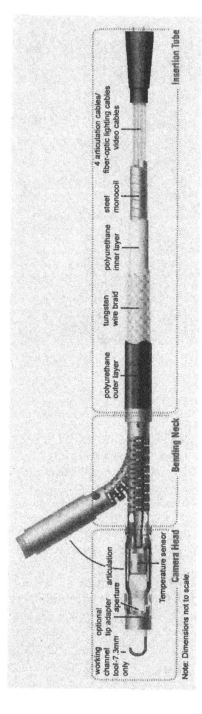

Figure 13.29 Insertion tube operation and construction details. Courtesy of Everest VIT, Inc.

Figure 13.30 Videoprobe inspection of a refinery fin-fan heat exchanger. Courtesy of Everest VIT, Inc.

The Everest VIT Ca-Zoom® PTZ 6.0 advanced pan-tilt-zoom camera system, shown in Figure 13.31, is ideal for inspecting large, hazardous, or inaccessible areas, such as those typically found in the chemical, petrochemical, and nuclear industries.

The system features include interchangeable camera heads, with wide-angle and tele-zoom viewing options, high-resolution video and still-image capture, and remote operation with a hand-held controller. The controller has a large color-VGA LCD monitor and controls zoom, lighting, image, and measurement.

Operating and display features include on-screen control of a miniDV recorder, an accessory for parallel laser measurement, on-screen indications of temperature, pressure, PTZ position, time/date, and character generation display.

13.11.2 Camera Head Options

There are two camera heads for the Ca-Zoom PTZ 6.0 system as shown in Figure 13.32. Both camera heads are enclosed in rugged metal housings, pressurized with an inert gas to protect sensitive electronics and sealed to protect the systems when submerged underwater.

REMOTE VISUAL INSPECTION

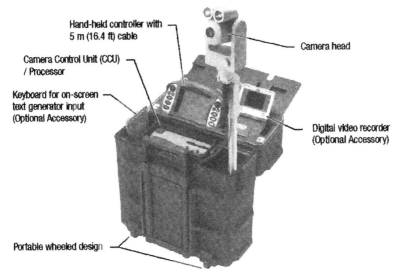

Figure 13.31 Ca-Zoom PTZ-6.0 system. Courtesy of Everest VIT, Inc.

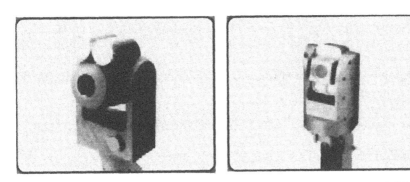

Figure 13.32 PTZ 140 and PTZ 100 camera heads. Courtesy of Everest VIT, Inc.

The PTZ140 camera provides a 25× optical and 12× digital zoom for a 300× total zoom capability. The camera is equipped with two high-power 35 W lights with a range of narrow and widebeam spreads.

The PTZ100 camera provides a 10× optical with a 4× digital zoom for a 40× total zoom capability. The camera is equipped with four 5 W LEDs offered with wide- and narrow-beam spreads.

13.11.3 Camera Pan and Tilt Features

Figure 13.33 illustrates the pan and tilt range of the PTZ140 camera. As shown the pan range is 360° (±180°) and the tilt range is 234° (+105°/−129°). Pan and tilt speed is 0–9°/s in both directions.

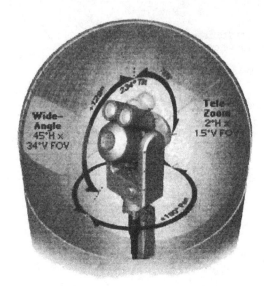

Figure 13.33 Ca-Zoom camera pan, tilt, and zoom ranges. Courtesy of Everest VIT, Inc.

By contrast, the PTZ 100 camera has a pan range of 354° (±177°) and a tilt range of 280° (±140°). Pan speed is 0–9°/s in both directions and tilt speed is 0–11°/s in both directions.

Available pan and tilt torque for both cameras is 8 in. lb max for both cameras in both directions.

Note: For a complete list of camera specifications, please contact Everest VIT, Inc.

Other camera options include:

- Tool head cameras for robotic equipment installations
- Vibration-resistant cameras for aircraft engine test cells
- PTZ cameras for robotic crawlers

Figure 13.34 shows a 4-light PTZ camera with wide-angle lens adapter and a tool head zoom camera.

13.11.4 Hand-Held Controller

The hand-held controller operates at distances up to 50 feet from the camera control unit. The controller has a 6.4-inch diagonal color VGA LCD monitor, dual joysticks for camera pan and tilt, and control buttons for zoom, lighting, and image control, and extensive menu functions letting you capture, file, compare, annotate, measure, and image review.

REMOTE VISUAL INSPECTION

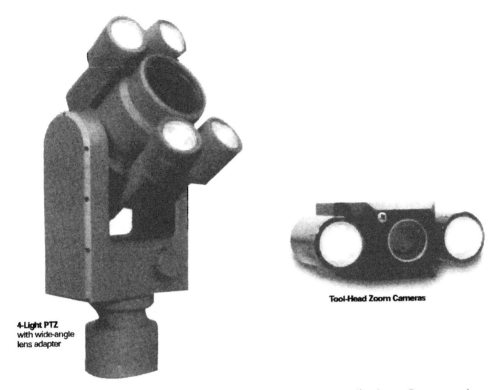

Figure 13.34 Special PTZ camera heads for customer applications. Courtesy of Everest VIT, Inc.

The miniDV recorder can be operated by the hand-held controller. Its recording format delivers 3× the color bandwidth of a VHS tape and includes a Memory Stick Media® terminal, MPEG movie capture capabilities from tapes, and USB streaming video port.

13.11.5 Camera Control Unit

The camera control unit (CCU) is portable and can be case-mounted for operation/shipping or rack-mounted for operation/shipping or storage. The camera control unit has a Pentium® class processor that runs a real-time, multitasking software operating system that supports file storage on internal flash memory or floppy disks.

Other accessories include:

- Parallel laser measurement accessory
- iView™ Remote PTZ software for controlling the Ca-Zoom® camera from a PC

- Second video hand-held controller and 15 m (50 ft) controller cables
- Rack-mount consoles
- PS/2 keyboard
- Telescoping and interconnecting poles; swivel ball or standard tripod
- Extension cables for operating up to 500 m (1640 ft) distance
- Slip-ring cable reels
- MiniDV and super VHS videotape recorders
- High-resolution video monitors

Use of the case minimizes setup and teardown time by incorporating all components into a portable, wheeled shipping case. The system can be rolled to the inspection site and operated from the case.

13.11.6 Hand-Held Controller Details

Monitor: 6.4″ full-resolution color VGA 640 × 480 pixels.

Recording options: The miniDV video recorder can be controlled by the hand-held controller. The hand-held controller is shown with main features identified in Figure 13.35.

Controls: Backlit buttons and joystick menu navigation.
- Menu: Enable on-screen menus button.
- Enter: Selects highlighted on-screen menu selections.
- Exit: Exits on-screen menu selections.
- Eight soft buttons: Context-sensitive buttons using on-screen labels.
- Zoom: Activate optical zoom and control digital zoom with soft buttons.
- Lights: Select flood and/or spotlights and control brightness with soft buttons.
- Image brightness: Controls camera gain, iris, and electronic shutter.
- Volume: Raise or lower playback volume.

iView™ Image Management. Software features include freeze, capture, store, and/or recall images. In addition, areas of inspection can be measured and there is a full-featured embedded file management system with thumbnail-based image recall system.

Multifunction Joystick. Dual joystick controls for camera pan and tilt, menu navigation, and on-screen character generation and access to advanced features.

Intuitive Drop-Down Menu System. Access the advanced feature set via a drop-down menu system.

Advanced Camera Setup. Menu-driven setup features provide options for enabling or disabling the digital zoom range, brightness (shutter, iris, and gain), invert positioning control of pan and tilt, and camera home positioning.

REMOTE VISUAL INSPECTION

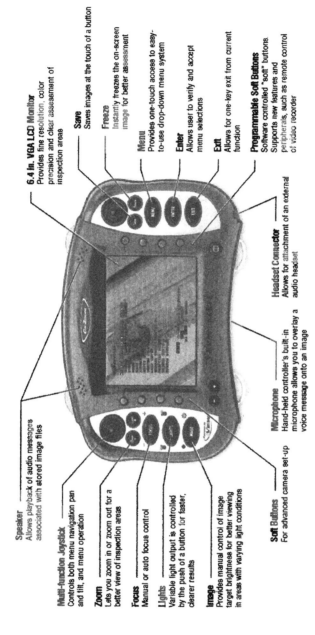

Figure 13.35 Video hand-held controller for Ca-Zoom PTZ-6.0 advanced pan-tilt-zoom camera. Courtesy of Everest VIT, Inc.

Laser Measurement. Parallel laser accessory mounted on the camera head provides a reference for measurement on a perpendicular surface.

Position Settings. Set camera home and zero positions by moving the camera to the desired location and saving this setting. Set up to 10 additional preset locations, for pan-tilt-zoom.

Image Freeze and Storage. A single press of a button activates the freeze frame. A separate button provides a freeze-only function.

File Manager. A full-feature imbedded file management system with a thumbnail-based image recall system. Create and frame file folders, and move and store files between folders on internal flash memory and floppy discs.

Integrated Text Generator with Arrow Annotation. Text annotation and graphics callout arrows can be placed onto the live display or any frozen or recalled image. Text can be created using an external keyboard or the pendant's joystick. Create preset text messages in advance and quickly recall them during inspection operations. Company logos and other bitmap files can be stored with captured images and displayed on-screen.

Digital Video Recording. On-screen control of Sony® miniDV digital video recorder.

Multilanguage Interface. Select from English, German, Spanish, Italian, French, Portuguese, Japanese, or Swedish for on-screen text annotation. Custom language capability is also available.

13.11.7 Applications

iView Remote PTZ allows full capability remote inspections from a field control room, central control room, or a control room across the world via a high-speed Internet connection. Once the Ca-Zoom PTZ 6.0 is connected to a PC at the inspection site, a remote user has full access to all menu functions, and can control camera movement and utilize all advanced camera capabilities. iView Remote supports MPEG 4-based video capture and allows still capture to laptop or PC.

Typical applications for visual inspection and defect detection include:

- Inspection of chemical plant vessels
- Detection of broken agitator blade bolts
- Steam generator bowl inspections
- Pharmaceutical tank weld inspections
- Oil refinery vessel inspections
- Glass-lined vessels and equipment inspection
- Tower tray inspections for corrosion
- Nuclear reactor vessel inspection
- Identification of nuclear fuel bundle serial numbers
- Examination of damage to vessel diffuser plates

Fuel Bundle Serial Number Identification

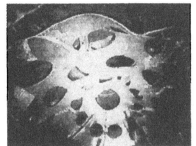
Damaged Diffuser Plate

Figure 13.36 Two applications for the Ca-Zoom PTZ-6.0 camera. Courtesy of Everest VIT, Inc.

The last two applications are illustrated in Figure 13.36.

13.12 ROBOTIC CRAWLER UNITS

Everest VIT manufactures modular Rovver™ systems featuring camera, control unit, cable reel, and lighting that are interchangeable on three crawler models. Figure 13.37 shows a compact Rovver® 400 robotic crawler.

13.12.1 Control Unit

The AC-powered control unit, which can also be rack-mounted, controls camera pan, tilt, and lighting as well as crawler direction and speed. A lightweight pendant control can also be used with the unit to control steering, speed, and light intensity. Control unit options include:

- 110 AC, 60 Hz, or 220 AC
- 19" rack-mounting option
- Outputs for Sony digital recorder and pendant control

13.12.2 Cable Reels

A standard cable reel for high-strength multiconductor cable with slip-ring cable drum for various length cables or automatic motorized cable reel with level wind feature can be provided (optional).

Figure 13.37 Rovver 400 advanced camera, lighting crawler system. Courtesy of Everest VIT, Inc.

13.12.3 Crawler and Camera Options

The series 400, 600, and 900 crawlers are shown in Figure 13.38. Note that both axial and pan and tilt cameras are available and crawler designs may be selected based on expected customer use. System features include:

- Waterproof construction for damp or underwater environments
- Low maintenance quick disconnects for time-saving operation
- Color forward pan and tilt cameras with remote focus
- Remote viewing up to 660 feet from operator
- Lightweight portable design

13.12.4 Applications

The Rovver systems are designed for underground use, environmental inspections, maneuvering through wreckage and steam in process lines, and keeping workers safe during search-and-rescue missions. The Rovver systems can be

ROBOTIC CRAWLER UNITS

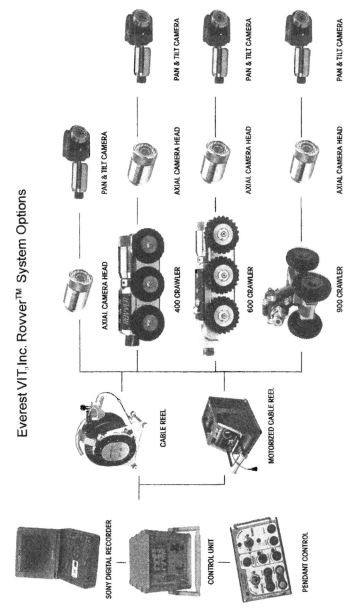

Figure 13.38 Rovver system control, cable, camera, and transport options. Courtesy of Everest VIT, Inc.

used in pipes having internal diameters of 4 to 24 inches. This system is usable in pipes having:

- Restricted piping
- Large offsets
- Protruding pipe taps
- Pipe with many bends

13.13 PIPE AND VESSEL INSPECTIONS/ METAL JOINING PROCESSES

Welding is one of the most common metal joining processes that is inspected by a number of NDT techniques including visual, radiography, ultrasonic, and other. Weld inspection for thick-walled pipe and pressure vessels is of critical importance for off-shore, nuclear, chemical, and petrochemical industries. The TOFD method has gained favor in many of these applications as a supplemental if not final inspection method.

ScanTech Instruments makes compact, rugged four-wheel magnetic pipe crawling units that can be equipped with numerous ultrasonic and eddy current probes and arrays. They also make the CVm remote visual inspection system (Figure 13.39) with miniature pan and tilt, integrated illumination, and high-resolution video. The C1 Spider drive train is the proven mechanical heart for several scanning systems. By using the joystick controls, the operator can concurrently drive the scanner and position the camera to points of interest. Optional equipment for the CVm includes the advanced recirculating bubbler attachment, time of flight diffraction (TOFD) assembly, and sealed optical encoder assembly (options not shown). UT and ET system designs are application specific, based on customer requirements.

Figure 13.40 shows the CVm remote pipe crawler system equipped with TOFD UT transducers. In practice, the magnetically coupled pipe crawler would be positioned to straddle large pipe butt welds, while visually inspecting them. The TOFD method is rapidly becoming the preferred inspection method for testing circumferential butt welds on heavy pipe and reactor sections. In practice, the entire pipe weld can be checked for pipe flaws using the TOFD method. The TOFD method can locate and in most cases determine the extent of defects at any point within the weld. Computer programs are used to calculate the location and size of pipe weld defects and display TOFD indications.

The use of noncontacting electromagnetic acoustic transducers (EMATs) has been found advantageous in some TOFD applications because they minimize problems associated with mode conversion, beam skewing, and distortion. For the same reason, perfect air/gas matched noncontacting transducers should prove useful. Linear sensor arrays are also used for TOFD applications because they substantially reduce test time.

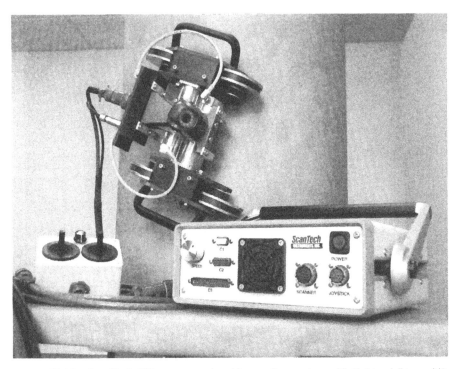

Figure 13.39 ScanTech CVm remote visual inspection system with lightweight pan/tilt camera and proven C1 Spider drive. Courtesy of ScanTech Instruments, Inc.

The geometric and mathematic TOFD model is shown in Figure 13.41. The time, T, for the ultrasonic energy to interact with the flaw tip, D, and return to the specimen is given by Equation 13.2:

$$CT = \left[d^2 + (S-X)^2\right]^{1/2} + \left[d^2 + (S+X)^2\right]^{1/2} \quad (13.2)$$

where C = the ultrasonic velocity
T = time of flight
d = the depth of flaw tip D below the specimen surface
X = the displacement of the diffractor from the center plane between the probes.

Note: If X = 0, ideal condition, the flaw is vertical. Transmitting transducer is at the left and receiving transducer is at the right.

The TOFD method uses both a transmitter and receiver transducer operating in the shear horizontal (SH) mode. There are four UT wave modes generated during TOFD inspections. The transmitting sensor generates a longitudinal wave that is partially transformed to a spherical wave when the

Figure 13.40 CVm scanner with TOFD UT probes. Courtesy of ScanTech Instruments, Inc.

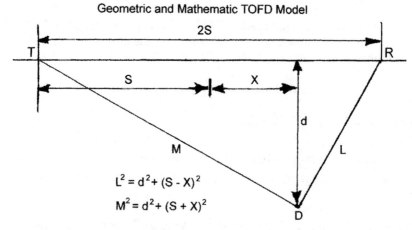

Figure 13.41 TOFD mathematical model for computerized calculations of TOFD.

SPECIFICATIONS	
CAMERA	**PAN & TILT**
Type: Sony CCD chipset	High Torque
Focus: Manual	Rare Earth Motors
Iris: Auto	All Metal Gears
Sensitivity: .0003 lux	Range: 360° x 180°
Resolution: 420 lines	Ht: 3 5/8"
Lens: 3.6mm micro	Dia.: 1.97"
ILLUMINATION	Wt: 1.25 lbs
Variable Intensity	**CONTROLLER**
Mega Bright White LEDs	110 / 120 VAC
33,600mcd	12.5" x 12.25" x 5"
High Efficiency	8 lbs.
Low EMI	

Figure 13.42 ScanTech CVm remote visual system specifications. Courtesy of ScanTech Instruments, Inc.

UT beam crosses the tip of the defect. A lateral wave follows the surface between the transmitter and receiver transducers. A longitudinal wave is reflected back to the receiver from the back wall of the pipe or vessel. Finally, shear waves are generated by mode conversion at the interface with discontinuities.

So-called "D-scans" are typically used for testing weldments. With D-scan, or multiple B-scans as it is sometimes called, the propagation of the wavefronts is perpendicular to the direction of the weldments—it is the most accurate scanning method for determining the size and location of weld defects.

ScanTech Instruments CVm remote visual system specifications are shown in Figure 13.42.

13.14 OCEAN OPTICS PHOTOMETERS

Ocean Optics is a diversified electro-optics company and global leader in optical sensing, display optics, and biophotonics technologies.

The coupling of linear CCD-array detector spectrometers with computers has resulted in high-speed, high-resolution spectral data that are much more complex than data received from single-wavelength devices. The development of low-cost, high-performance detectors and the availability of lower-cost, higher-performing fiber optics have greatly advanced optical spectrometer development. In addition, the personal computer has substantially lowered the cost of spectral data acquisition, processing, storage, and analysis.

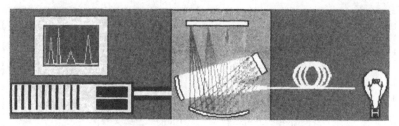

Figure 13.43 Basic operating principles of the Ocean Optics S2000 spectrometer. Courtesy of Ocean Optics, Inc.

As computers continue to become smaller and smarter, in-situ real-time imaging and analysis applications are expanding rapidly. Today an entire spectrometer system consisting of spectrometer, light source, fiber optics, accessories, and notebook PC can easily fit into a small briefcase. Currently hand-held PCs are popular in process control and nondestructive testing applications. In addition, universal serial bus (USB) ports have greatly simplified PC device interconnection.

Ocean Optics is a major supplier of spectrometers, optical sensors, associated software, and data acquisition units. They are also a supplier of sampling accessories, light sources, optical fibers and probes, thin films, and optics.

Figure 13.43 shows the operating principle of the S2000 spectrometer. The S2000 series spectrometer was used by amateur astronomers Nick Glumac and Joseph Sivo to record the spectra of the Hale-Bopp comet on the outskirts of New York City. At the time, the night skies were clear with moderate light pollution.

As shown in the figure, light enters the optical fiber and is transmitted to the spectrometer. Once inside the spectrometer, the light diverges and is collimated by a spherical mirror. The collimated light is then diffracted by a planar grating and focused by a second spherical mirror. The spectrum image is projected onto a single-dimensional linear CCD array.

The reversed-biased photodiodes of the linear CCD array discharge a capacitor at a rate proportional to the photon flux. When the integration period of the detector is reached, a series of switches close and transfer the charge to a shift register. After the transfer to the shift register is complete, the switches open and the capacitors attached to the photodiodes are recharged and a new integration period begins. At the same time the light energy is being integrated, the data is read out on the shift register by an A/D converter and the digitized spectrum data is displayed on a computer.

13.14.1 Optical Resolution

Optical resolution measured as full width half maximum (FWHM) of a monochromatic source depends on the groove density (lines/nm) of the grating and the diameter of the entrance optics (optical fiber or slit). In configuring a spectrometer, there are two important tradeoffs:

1. Resolution increases with increases in groove density of the grating, but at the expense of spectral range.
2. Resolution also increases as the slit width or fiber optic diameter decreases, but at the expense of signal strength.

The approximate optical resolution in nm (FWHM) can be calculated as follows:

1. Dispersion (nm/pixel) = spectral range of the grating ÷ number of detector elements (2048 for the S2000 series).
2. Typical resolution (in pixels) can be determined from slit size/fiber diameter as shown below:
   - 5 micron slit = ~3.0 pixels
   - 10 micron slit = 3.2 pixels
   - 25 micron slit = 4.2 pixels
   - 50 micron slit = 6.5 pixels
   - 100 micron slit = 12.0 pixels
   - 200 micron slit = 24.0 pixels
3. Optical resolution (in nm) = dispersion (nm/pixel value) from number 1 × resolution (pixel value) from number 2.

For example, the S2000 Spectrometer with Grating #3 and 10 micron slit has an optical resolution of 650 nm/2048 = 0.32 nm/pixel × 3.2 pixels or 1.02 nm (FWHM). The intended use for Grating #3 is for the measurement of the VIS/color spectrum. The *spectrometer efficiency curve* for Grating #3 is shown in Figure 13.44. The *blaze wavelength* is the peak wavelength in the typical efficiency curve for a ruled grating or for a holographic grating, the most efficient wavelength.

For comparison purposes, the Grating #13 spectrometer efficiency curve is shown in Figure 13.45. The blaze wavelength is the same, but there are substantial differences. The intended use for Grating #13 is UV/VIS/NIR spectrum. This grating has half the groove density of Grating #3 and a very broad spectral range, 300–2000 nm. However, the S2000 spectrometer itself has a spectral range of only 200 to 1100 nm. Optical resolution for an S2000 spectrometer with Grating #13 and 10 micron slit is 1700 nm/2048 = 0.83 nm/pixel × 3.2 pixels = 2.65 nm (FWHM). In this case, the broad spectral response limits optical resolution and makes it much more difficult to eliminate undesirable second-order effects even when using order-sorting filters and other techniques.

632 VISUAL AND OPTICAL TESTING

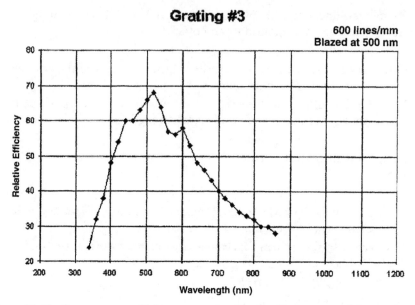

Figure 13.44 Spectrometer efficiency curve for Grating #3. Courtesy of Ocean Optics, Inc.

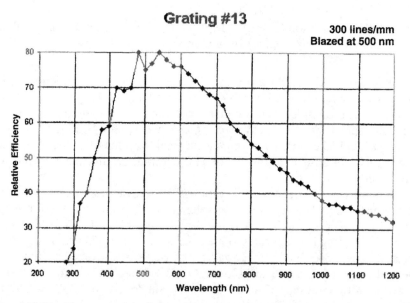

Figure 13.45 Spectrometer efficiency curve for Grating #13. Courtesy of Ocean Optics, Inc.

13.14.2 System Sensitivity

The subject of system sensitivity comes up when someone wants to convert the amplitude of the raw data spectra (scope mode presentation) into meaningful energy spectra. Unfortunately, it is not practical to apply correction factors to the various phenomena that affect amplitude. Ocean Optics, Inc. provides a more useful alternative, namely a NIST-traceable radiant standard (the LS-1-CAL), which can be used to normalize the spectra to energy terms. In their OOIBase32 operating software, these normalized data can be processed in the "I" (Irradiance) mode as relative energy (scaled 0 to 1). In OOIIrrad Irradiance Measurement Software, the data can be processed in absolute terms (calculated in $\mu W/cm^2/nm$ or in lumens or lux per unit area). For experiments investigating transmission or reflection, the data are normalized to the spectra of a physical standard such as transmission in air or reflection of a diffuse white standard.

Some of the factors affecting spectrometer system amplitude are:

- *CCD detector response.* Check manufacturer for specific detector response curve.
- *Fiber attenuation.* It is typically very flat in VIS region, but increases dramatically in UV region. In NIR region there are water absorption bands at 750 nm and 900 nm.
- *Grating efficiency.* All ruled or holographically etched gratings optimize first-order spectra at certain wavelengths depending on blaze wavelength and other factors.
- *Collection optics.* Sampling optics can have spectral signatures such as chromatic aberrations, which vary with focus.
- *Sources and samples.* Light sources and samples also have their own spectral response. If the light is the sample, its spectral response is what is being measured.
- *Other factors.* Detector dark current signal and amplifier zero set point identified as "Dark." This value varies from pixel to pixel so it must be subtracted from each CCD element. Variations in pixel response (fixed pattern noise) mean that normalization on a pixel to pixel basis must also be done.

The S2000 Miniature Fiber Optic Spectrometer is shown in Figure 13.46. The spectrometer with its PC interface is flexible and portable; optical fiber allows the operator to bring the spectrometer to the sample. The modular configuration also allows the user to specify components that are built into the optical bench and choose from hundreds of spectroscopic accessories for optimizing the system for specific applications. Figure 13.47 shows the S2000 optical bench and how light moves through the bench.

Spectrometers can also be multiplexed. The modular stackable units permit connection of up to eight spectrometers for expanded wavelength range, multipoint sampling, or reference monitoring.

Figure 13.46 S2000 miniature fiber-optic spectrometer. Courtesy of Ocean Optics, Inc.

13.14.3 Specifications

Dimensions—141.6 mm × 104.9 mm × 40.9 mm (with enclosure, master channel only)

Weight—390 g (with enclosure, master channel only)

Power consumption—110 mA @ 5 VDC (master); 60 mA @ 5 VDC (slave)

Detector—2048 element linear silicon CCD array

Gratings—14 gratings; UV through shortwave NIR

Entrance aperture—5, 10, 25, 50, 100, or 200 um wide slits

Order-sorting filters—installed band-pass and long-pass filters

Focal length—42 mm (input) 68 mm (output)

Optical resolution—~0.3 to 10 nm FWHM (depending on grating and size of entrance aperture)

Stray light—<0.05% at 600 nm; <0.10% at 435 nm; <0.10% at 250 nm

Dynamic range—2×10^8 (system); 2000:1 for a single scan

Sensitivity (estimated)—86 photons/count; 2.9×10^{-17} joules/count; 2.9×10^{-17} watts/count (for 1-second integration)

Fiber-optic converter—SMA 905 to single-strand optical fiber (0.22 NA)

Integration time—3 ms to 60 s with 1 MHz A/D card; 2 ms to 60 s with 2 MHz A/D card

Note: For complete details and specifications on the S2000, visit the website at OceanOptics.com/Products/S2000.asp.

OCEAN OPTICS PHOTOMETERS

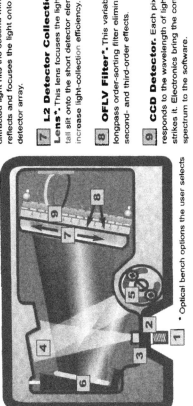

1. **SMA Connector.** Light from a fiber enters the optical bench through the SMA connector.

2. **Installed Filter*.** Light passes through a filter installed in the SMA connector.

3. **Slit*.** Light passes through the installed slit, which acts as the entrance aperture. (A fiber can also act as the entrance aperture.)

4. **Collimating Mirror.** Light coming from the entrance aperture reflects from this mirror, as a collimated beam, toward the diffration grating.

5. **Grating*.** Light is diffracted by the fixed grating and directed to the focusing mirror.

6. **Focusing Mirror.** When the diffracted light hits the second mirror, it reflects and focuses the light onto the detector array.

7. **L2 Detector Collection Lens*.** This lens focuses the light from the tall slit onto the short detector elements to increase light-collection efficiency.

8. **OFLV Filter*.** This variable-longpass order-sorting filter eliminates second- and third-order effects.

9. **CCD Detector.** Each pixel responds to the wavelength of light that strikes it. Electronics bring the complete spectrum to the software.

* Optical bench options the user selects

Figure 13.47 "S" Series optical bench showing how the light moves through the bench. There are no moving parts that can wear or break. (*) items are specified by the user. Courtesy of Ocean Optics, Inc.

Figure 13.48 Palm-SPEC spectrometer 152 × 89 mm with hand-held PC. Courtesy of Ocean Optics, Inc.

The Palm-SPEC™ Spectrophotometer, shown in Figure 13.48, features a 390–950 nm wavelength range for measuring concentration, absorbance, and percent transmittance (%T) values and full spectra data in about 1 second. The overall dimensions of the spectrometer are 152 mm × 89 mm × 203 mm.

This PC-based system comes loaded with touch-screen software and Windows CE 3.0. This easy-to-use software has method saving QC-watch and security features. QC-watch monitors system performance, logs outcomes, and presents a diagnosis to ensure the integrity of the data. The security feature tracks users. For data transfer, the Palm-SPEC comes with a serial cable that connects to RS-232 ports on desktops or portable PCs.

Optional add-on software can be used to calculate tristimulus and color-space values. Tristimulus methods generate three values (x, y, and z) to describe the color of an object and provide a more descriptive and precise measurement based on the complete spectral content of the transmitted radiation.

An optional Windows CE driver can be added for writing custom software for the Palm-SPEC for interfacing the hand-held PC to another Ocean Optics spectrometer. Please refer to the latest Ocean Optic, Inc. product catalog for a complete listing of products, specifications, and options.

13.14.4 Applications

Ocean Optics' photonics technologies have enabled exciting applications everywhere from research vessels in the Gulf of Mexico to Australia's Great Barrier Reef.

The Great Barrier Reef (GBR) of Australia is one of the world's greatest natural resources. Dr. Justin Marshall of the Sensory Ecology Laboratory and his research team are studying color vision in tropical marine habits. Their current research projects include:

- Color Vision in the Marine Environment
- Communication Between Cleaner Fish and Their Hosts: The Role of Color
- Vision in Pelagic Fish (Billfish Group)
- Neuroethology of Color and UV on the Great Barrier Reef
- Neural Processing of the World's Most Complex Color Vision System: Vision in Stomatopods (Mantis Shrimp)
- Adaptive Coloration in the Marine Environment
- A Comparative Survey of Retinal Neuorophysiology in relation to Visual Ecology in Australian Birds
- Color Communication in Parrots and Birds of Paradise
- Development of Vision in Elasmobranchs
- Ultraviolet Cues and Mate Choice in Reef Fish
- Evolution and Anatomy of Lateral lines of Deep-Sea Fishes
- Color Communication in Cephalopods and Fish

Wide Range of Optical Sensing Applications

- **Color measurement**
  - Inks and dyes
  - Foods
  - Cosmetics
  - Paper and pulp
  - CRT displays
  - Injection molding
- **Semiconductor and materials process monitoring**
  - Deposition thickness
  - Plasma chemistry monitoring
- **Light source characterization (power & wavelength)**
  - LEDs, Lasers, Fluorescent
- **Packaging**
  - Modified Air (foods & pharmaceuticals)

- **Water and air quality monitoring**
  - Fence line monitors for air pollutants
  - Cl_2, Br_2, O_3, organics and other water quality parameters
- **Life sciences**
  - Fluorescence (DNA, Quantum dot labels, immunological assays)
  - Direct blood spectroscopy
  - Direct tissue spectroscopy
- **Homeland security**
  - Detection of contraband
  - Detection of biological or chemical agents, e.g., anthrax
  - Sterilization verification
  - Early warning sensors

Figure 13.49 Optical sensing applications. Courtesy of Ocean Optics, Inc.

According to Dr. Marshall, animals living in the GBR environments have well-developed color vision greatly superior to our own, for reasons that are not clear. Therefore, his team plans to concentrate on the extraordinary visual system of mantis shrimps and examine a variety of reef fish with reference to color and polarization vision. He hopes to answer the question, What do marine animals see when they look at sources of food, rivalry and predators?

Ocean Optics spectrometers are suitable for a wide range of optical-sensing applications as shown in Figure 13.49.

14

OVERVIEW OF RECOMMENDED PRACTICE NO. SNT-TC-1A, 2001 EDITION

14.1 PURPOSE

The purpose of Recommended Practice No. SNT-TC-1A is to provide the general framework for the qualification and certification program for personnel working with various nondestructive test methods. Recommended education, training, and experience for each level of certification and each test method is proposed. Typical test questions and answers for each level of certification and each test method are also included in the recommended practice. Employers are responsible for addressing their specific needs and modifying the practice guidelines as appropriate to comply with their written procedures. Employers are encouraged to use the verb "shall" in written procedures to emphasize their needs.

14.1.1 Personnel Qualification and Certification in Nondestructive Testing

This section of the recommended practice covers scope, definitions, a list of recognized nondestructive testing methods, the levels of qualification, namely levels I through III, importance of a written practice by employers, a review of education, training, and experience requirements for initial qualification, training program goals, and examination guidelines.

Introduction to Nondestructive Testing: A Training Guide, Second Edition, by Paul E. Mix
Copyright © 2005 John Wiley & Sons, Inc.

14.2 NDT LEVELS OF QUALIFICATION

An individual is considered a trainee while being trained, qualified, and certified for a specific test method. He or she must work under the tutorage of a certified individual.

NDT Level I. Level I individuals should be qualified to set up and calibrate test instruments. They are expected to determine the acceptance or rejection of test objects based on approved written procedures, and accurately record test results. A certified NDT Level II or III individual should provide instruction and supervision for NDT Level I personnel.

NDT Level II. Level II individuals should be qualified to set up and calibrate test equipment. They should also be able to interpret and evaluate test results in accordance with applicable codes, standards, and specifications. They are expected to know the scope and limitations of qualified NDT methods, and assist in the training and guidance of Level I personnel. NDT Level II personnel should be able to organize and summarize test results.

NDT Level III. Some responsibilities of Level III individuals are as follows:

- Developing, qualifying, and approving procedures
- Establishing and approving techniques
- Interpreting codes, standards, specifications, and procedures
- Designating the specific NDT methods, techniques, and procedures to be used
- Assisting in establishing acceptance criteria when none are available (based on their background knowledge and experience)
- Helping to train and certify Level I and II personnel in qualified methods

14.3 RECOMMENDED NDT LEVEL III EDUCATION, TRAINING, AND EXPERIENCE

- Graduation from a minimum four-year college or university with a degree in engineering or science, plus one additional year of experience beyond Level II requirements in NDT in an assignment comparable to that of an NDT Level II in applicable NDT method(s), or:
- Completion with passing grades of at least two years of engineering or science study at a university, college, or technical school, plus two additional years of experience beyond the Level II requirements in NDT in an assignment at least comparable to that of NDT Level II in the application of NDT method(s), or:
- Four years of experience beyond the Level II requirements in NDT in an assignment at least comparable to that of an NDT Level II in the applicable NDT method(s).

The above Level III requirements may be partially replaced by experience as a certified NDT Level II or by assignments at least comparable to NDT Level III as defined in the employer's written practice.

14.4 WRITTEN PRACTICE

It is the employer's responsibility to establish a written practice for the control and administration of NDT personnel training, examination, and certification. The employer's written practice shall be reviewed and approved by a Level III individual. Employers can obtain an up-to-date version of Recommended Practice No. SNT-TC-1A by contacting:

The American Society for Nondestructive Testing

1711 Arlingate Lane

PO Box 28518

Columbus, OH 43228-0518

14.5 CHARTS

Charts are presented, showing the recommended number of written questions to be given based on certification level and test method. A table of recommended booklets for the various NDT methods is also provided. Examination criteria are discussed in detail.

Certification, Technical Performance Evaluation, Interrupted Service, Recertification, Termination, and Reinstatement requirements are also outlined and briefly discussed. Large charts are presented showing Recommended Initial Training and Experience Levels and Alternate Initial Training and Experience Levels. The goal is to provide a harmonious program that fulfills the employers' needs as well as the ASNT certification and qualification goals.

14.6 RECOMMENDED TRAINING COURSES

14.6.1 Acoustic Emissions Testing Method (TC-8)

The recommended hours of instruction for each training level varies based on the educational level of the student as follows, with a designation of **A** designating a high school graduate or equivalent and a designation of **B** indicating completion with passing grades of at least two years of engineering or technical school.

Basic Acoustic Emission Physics Course:

	A	B
Recommended Training for Level I		
Recommended Hours of Instruction	12	10

1. Principles of Acoustic Emission Testing
2. Sensing the Acoustic Emission Wave

Basic Acoustic Emission Technique Course:

	A	B
Recommended Training for Level I		
Recommended Hours of Instruction	28	22

1. Instrumentation and Signal Processing
2. Acoustic Emission Test Techniques
3. Codes, Standards, and Procedures
4. Applications of Acoustic Emission Testing (covers 3 out of 12 categories of Laboratory Studies and 4 out of 14 categories of Structural Applications).

Acoustic Emission Physics Course:

	A	B
Recommended Training for Level II		
Recommended Hours of Instruction	12	12

1. Principles of Acoustic Emission Testing
2. Sensing the Acoustic Emission Wave

Acoustic Emission Technique Course:

	A	B
Recommended Training for Level II		
Recommended Hours of Instruction	28	28

1. Instrument and Signal Processing
2. Acoustic Emission Test Techniques
3. Codes, Standards, Procedures, and Societies
4. Application of Acoustic Emission Testing (same as Level I)

Note: Consideration as Level II is based on satisfactory completion of both Level I and Level II certification in all training courses.

Level III Topical Outline

1. Principles and Theory
2. Equipment and Materials

3. Techniques
4. Interpretation and Evaluation
5. Procedures
6. Safety and Health
7. Applications

14.6.2 Electromagnetic Testing Method (TC-5)

Basic Electromagnetic Physics Course:

	A	B
Recommended Training for Level I		
Recommended Hours of Instruction	24	12

1. Introduction to Electromagnetic (Eddy Current/Flux Leakage)
2. Electromagnetic Theory

Electromagnetic Technique Course:

	A	B
Recommended Training for Level I		
Recommended Hours of Instruction	16	12

1. Readout Mechanism
2. Types of Eddy Current Sensing Elements
3. Types of Flux Leakage Sensing Elements

Electromagnetic Evaluation Course:

	A	B
Recommended Training for Level II		
Recommended Hours of Instruction	40	40

1. Review of Electromagnetic Theory
2. Factors That Affect Coil Impedance
3. Factors That Affect Flux Leakage Fields
4. Signal-to-Noise Ratio
5. Selection of Test Frequency
6. Selection of Method of Magnetism for Flux Leakage Testing
7. Coupling
8. Field Strength and its Selection
9. Field Orientation for Flux Leakage Testing
10. Instrument Design Considerations
11. Applications
12. User Standards and Operating Procedures

Level III Topical Outline

1. Principles/Theory
2. Equipment/Materials
3. Techniques/Calibrations
4. Interpretation/Evaluation
5. Procedures

14.6.3 Laser Testing Methods—Holography/Shearography (TC-14)

Basic Holography/Shearography Physics Course:

	A	B
Recommended Training for Level I		
Recommended Hours of Instruction	10	8

1. Introduction
2. Basic Principles of Light and SNDT
3. Lasers
4. Laser Safety
5. Basic Holography/Shearography System

Basic Operating Course:

	A	B
Recommended Training for Level I		
Recommended Hours of Instruction	10	10

1. Introduction
2. Laser Safety Review
3. Laser Systems
4. Holography/Shearography Camera
5. Holography/Shearography Image Processor
6. Operation of the Image Processor
7. Primary Test Methods
8. Documentation
9. Holography/Shearography Systems

Basic Application Course:

	A	B
Recommended Training for Level I		
Recommended Hours of Instruction	20	20

1. Introduction
2. Basic Fringe Interpretation
3. Mechanical Loading

RECOMMENDED TRAINING COURSES

4. Thermal Stress
5. Vacuum Stress
6. Pressurization Stress
7. Vibration Excitation, Mechanical
8. Vibration Excitation, Acoustical
9. Complex Structures

Intermediate Physics Course:

	A	B
Recommended Training for Level II		
Recommended Hours of Instruction	10	8

1. Physics of Light
2. Physics of Lasers
3. Laser Safety Officer
4. Physics of Materials

Intermediate Operating Course:

	A	B
Recommended Training for Level II		
Recommended Hours of Instruction	10	8

1. Holography and Shearography Systems
2. Sources of Noises and Solutions
3. Fixturing
4. Speckle Interferometry Camera
5. Speckle Interferometry Image Processor
6. Stressing Systems Setup and Operation
7. Method Development
8. Documentation

Intermediate Applications Course:

	A	B
Recommended Training for Level II		
Recommended Hours of Instruction	20	20

1. Materials and Applications
2. Fringe Interpretation
3. Mechanical Loading
4. Thermal Stress
5. Vacuum Stress
6. Pressurization Stress
7. Vibration Excitation, Mechanical

8. Vibration Excitation, Acoustical
9. Other Stressing Methods

14.6.4 Laser Testing Methods—Profilometry (TC-14)

	A	B
Recommended Training for Level I		
Recommended Hours of Instruction	8	8

1. Introduction
2. Lasers and Laser Safety
3. Theory of Laser Profilometry Testing
4. Laser Profilometry Testing
5. Introduction to data processing and analysis

	A	B
Recommended Training for Level II		
Recommended Hours of Instruction	24	12

1. Introduction
2. Laser Safety
3. Intermediate Theory of Profilometry Testing
4. Conducting Laser Profilometry Inspection
5. Evaluation of Indications
6. Inspection Procedures and Standards

Recommended Training for Level III

1. Introduction
2. Knowledge of Other Basic NDT Methods
3. Laser Safety
4. Codes, standards, specifications, and procedures
5. Advanced Theory of Profilometry Testing
6. Evaluation of Indications
7. Reporting Inspection Results
8. Training Level I and II Personnel for Certification

14.6.5 Leak Testing Methods (TC-7)

Fundamentals in Leak Testing Course:

	A	B
Recommended Training for Level I		
Recommended Hours of Instruction	14	8

1. Introduction
2. Leak Testing Fundamentals

Safety in Leak Testing Course:

Recommended Training for Level I	**A**	**B**
Recommended Hours of Instruction	14	8

1. Safety Considerations
2. Safety Precautions
3. Pressure Precautions
4. Safety Devices
5. Hazardous and Tracer Gas Safety
6. Types of Monitoring Equipment
7. Safety Regulations

Leak Testing Methods Course (not outlined):

Recommended Training for Level I	**A**	**B**
Recommended Hours of Instruction	14	8

Principles of Leak Testing Course:

Recommended Training for Level II	**A**	**B**
Recommended Hours of Instruction	24	12

1. Introduction
2. Physical Principles in Leak Testing
3. Principles of Gas Flow

Pressure and Vacuum Technology Course:

Recommended Training for Level II	**A**	**B**
Recommended Hours of Instruction	12	6

1. Pressure Technology
2. Vacuum Technology

Leak Test Selection Course:

Recommended Training for Level II	**A**	**B**
Recommended Hours of Instruction	12	6

1. Choice of Leak Testing Procedure

Level III Topical Outline

1. Principles/Theory
2. Equipment/Material
3. Techniques/Calibration

4. Interpretation/Evaluation
5. Procedures
6. Safety and Health

14.6.6 Liquid Penetrant Testing Methods (TC-4)

	A	B
Recommended Training for Level I		
Recommended Hours of Instruction	4	4

1. Introduction
2. Liquid Penetrant Processing
3. Various Penetrant Testing Methods
4. Liquid Penetrant Testing Equipment

	A	B
Recommended Training for Level II		
Recommended Hours of Instruction	8	4

1. Review
2. Selection of the Appropriate Penetrant Testing Method
3. Inspection and Evaluation of Indications
4. Inspection Procedures and Standards
5. Basic Methods of Interpretation

Level III Topical Outline

1. Principles/Theory
2. Equipment/Materials
3. Interpretation/Evaluation
4. Procedures
5. Safety and Health

14.6.7 Magnetic Particle Testing Method (TC-2)

	A	B
Recommended Training for Level I		
Recommended Hours of Instruction	12	8

1. Principles of Magnets and Magnetic Fields
2. Characteristic of Magnetic Fields
3. Effects of Discontinuities of Materials
4. Magnetization by Means of Electric Current
5. Selecting the Proper Method of Magnetization
6. Inspection Materials

RECOMMENDED TRAINING COURSES

7. Principles of Demagnetization
8. Magnetic Particle Testing Equipment
9. Types of Discontinuities Detected by Magnetic Particle Testing
10. Magnetic Particle Test Indications and Interpretations

	A	**B**
Recommended Training for Level II		
Recommended Hours of Instruction	8	4

1. Principles
2. Flux Fields
3. Effects of Discontinuities on Materials
4. Magnetization by Means of Electric Current
5. Selecting the Proper Method of Magnetization
6. Demagnetization Procedures
7. Equipment
8. Types of Discontinuities
9. Evaluation Techniques
10. Quality Control of Equipment and Processes

Level III Topical Outline

1. Principles/Theory
2. Equipment/Materials
3. Techniques/Calibrations
4. Interpretation/Evaluation
5. Procedures
6. Safety and Health

14.6.8 Neutron Radiographic Testing Method (TC-6)

Neutron Radiographic Equipment Operating and Emergency Instructions Course:

	A	**B**
Recommended Training for Level 1		
Recommended Hours of Instruction	8	8

1. Personnel Monitoring
2. Radiation Survey Instruments
3. Radiation-Area Surveys
4. Radioactivity
5. Radiation-Area Work Practices/Safety

6. Explosive-Device Safety*
7. State and Federal Regulations

Basic Neutron Radiography Physics Course:

Recommended Training for Level I	A	B
Recommended Hours of Instruction	7	4

1. Introduction
2. Physical Principles
3. Radiation Sources for Neutrons (specific descriptions)
4. Personnel Safety and Radiation Protection

Basic Neutron Radiographic Technique Course:

Recommended Training for Level I	A	B
Recommended Hours of Instruction	13	8

1. Radiation Detection Imaging
2. Neutron Radiographic Process: Basic Imaging Considerations
3. Test Result Interpretation

Neutron Radiographic Physics Course:

Recommended Training for Level II	A	B
Recommended Hours of Instruction	14	14

1. Introduction
2. Review of Physical Principles
3. Radiation Sources for Neutrons
4. Radiation Detection
5. Personnel Safety and Radiation Protection

Neutron Radiographic Technique Course:

Recommended Training for Level II	A	B
Recommended Hours of Instruction	26	26

1. Neutron Radiographic Process
2. Test Result Interpretation

---

* Required only for personnel involved in neutron radiography of explosive devices.

Level III Topical Outline

1. Principles/Theory
2. Equipment/Materials
3. Techniques/Calibrations
4. Interpretation/Evaluation
5. Procedures
6. Safety and Health

14.6.9 Radiographic Testing Method (TC-1)

Radiographic Equipment Operating and Emergency Instructions Course:

	A	B
Recommended Training for Level I		
Recommended Hours of Instruction	5	5

1. Personnel Monitoring
2. Survey Instruments
3. Leak Testing of Sealed Radiographic Sources
4. Radiation Survey Reports
5. Radiographic Work Practices
6. Exposure Devices
7. Emergency Procedures
8. Storage and Shipment of Exposed Devices and Sources
9. State and Federal Regulations

Basic Radiographic Physics Course:

	A	B
Recommended Training for Level I		
Recommended Hours of Instruction	20	15

1. Introduction
2. Fundamental Properties of Nature
3. Radioactive Materials
4. Types of Radiation
5. Interaction of Radiation with Matter
6. Biological Effects of Radiation
7. Radiation Detection
8. Exposure Devices and Radiation Sources
9. Special Radiographic Sources and Techniques

Radiographic Technique Course:

Recommended Training for Level I	A	B
Recommended Hours of Training	15	10

1. Introduction
2. Basic Principles of Radiography
3. Radiographs
4. Radiographic Image Quality
5. Film Handling, Loading and Processing
6. Exposure Techniques—Radiography
7. Fluoroscopic Techniques

Film Quality and Manufacturing Processes Course:

Recommended Training for Level II	A	B
Recommended Hours of Instruction	20	15

1. Review of Basic Radiographic Principles
2. Darkroom Facilities
3. Indications, Discontinuities, and Defects
4. Manufacturing Processes and Associated Discontinuities
5. Radiographic Safety Principles Review

Radiographic Evaluation and Interpretation Course:

Recommended Training for Level II	A	B
Recommended Hours of Instruction	20	20

1. Radiographic Viewing
2. Application Techniques
3. Evaluation of Castings
4. Evaluation of Weldments
5. Standards, Codes and Procedures for Radiography

Level III Topical Outline

1. Principles/Theory
2. Equipment/Materials
3. Techniques/Calibration
4. Interpretation/Evaluation
5. Procedures
6. Safety and Health

RECOMMENDED TRAINING COURSES

14.6.10 Thermal/Infrared Testing Method (TC-10)

Basic Thermal/Infrared Physics Course:

	A	B
Recommended Training for Level I		
Recommended Hours of Instruction	12	10

1. Nature of Heat (what is it and how is it measured/expressed?)
2. Temperature (what is it and how is it measured/expressed?)
3. Heat Transfer Modes Familiarization
4. Radiosity Concepts Familiarization

Basic Thermal/Infrared Operating Course:

	A	B
Recommended Training for Level I		
Recommended Hours of Instruction	10	10

1. Introduction
2. Checking Equipment Calibration with Blackbody References
3. Infrared Image and Documentation Quality
4. Support Data Collection

Basic Thermal/Infrared Applications Course:

	A	B
Recommended Training for Level I		
Recommended Hours of Instruction	10	10

1. Detecting Thermal Anomalies Resulting from Differences in Thermal Resistance (Quasi-Steady-State Heat Flow)
2. Detecting Thermal Anomalies Resulting from Differences in Thermal Capacitance, Using System or Environmental Heat Cycles
3. Detecting Thermal Anomalies Resulting from Differences in Physical State
4. Detecting Thermal Anomalies Resulting from Fluid Flow Problems
5. Detecting Thermal Anomalies Resulting from Friction
6. Detecting Thermal Anomalies Resulting from Nonhomogeneous Exothermic or Endothermic Conditions
7. Field Quantification of Point Temperatures

Intermediate Thermal/Infrared Physics Course:

	A	B
Recommended Training for Level II		
Recommended Hours of Instruction	12	10

1. Basic Calculations in the Three Modes of Heat Transfer
2. Infrared Spectrum

3. Radiosity Problems
4. Resolution Tests and Calibration

Intermediate Thermal/Infrared Operating Course:

Recommended Training for Level II	A	B
Recommended Hours of Instruction	10	10

1. Operating for Infrared Measurements (Quantification)
2. Operating for High-Speed Data Collection
3. Operating for Special Equipment for "Active" Techniques
4. Reports and Documentation

Intermediate Thermal/Infrared Applications Course:

Recommended Training for Level II	A	B
Recommended Hours of Instruction	12	12

1. Temperature Measurement Applications
2. Energy Loss Analysis Applications
3. "Active" Applications
4. Filtered Applications
5. Transient Applications

Level III Topical Outline

1. Principles/Theory
2. Equipment/Materials
3. Techniques
4. Interpretation/Evaluation
5. Procedures
6. Safety and Health

14.6.11 Ultrasonic Testing Method (TC-3)

Basic Ultrasonic Course:

Recommended Training for Level I	A	B
Recommended Hours of Training	20	15

1. Introduction
2. Basic Principles of Acoustics
3. Equipment
4. Basic Testing Methods

Ultrasonic Technique Course:

Recommended Training for Level I	A	B
Recommended Hours of Training	20	15

1. Testing Methods
2. Calibration (Electronic and Functional)
3. Straight Beam Examination to Specific Procedures
4. Angle Beam Examination to Specific Procedures

Ultrasonic Evaluation Course:

Recommended Training for Level II	A	B
Recommended Hours of Training	40	40

1. Review of Ultrasonic Technique Course
2. Evaluation of Base Material Product Forms
3. Evaluation of Weldments
4. Evaluation of Bonded Structures
5. Discontinuity Detection
6. Evaluation

Level III Topical Outline

1. Principles/Theory
2. Equipment/Materials
3. Techniques/Calibrations
4. Interpretations/Evaluations
5. Procedures

14.6.12 Vibration Analysis Method (TC-11)

Basic Vibration Analysis Physics Course:

Recommended Training for Level I	A	B
Recommended Hours of Instruction	24	24

1. Introduction
2. Transducers
3. Instrumentation

Basic Vibration Analysis Operating Course:

1. Machinery Basics
2. Data Collection Procedures
3. Safety and Health

Intermediate Vibration Analysis Physics Course:

	A	B
Recommended Training for Level II		
Recommended Hours of Instruction	72	48

1. Review
2. Additional Terminology
3. Diagnostic Tools

Intermediate Vibration Analysis Techniques Course:

1. Data Acquisition
2. Signal Processing
3. Data Presentation
4. Problem Identification
5. Reporting Methodology
6. Safety and Health

Level III Topical Outline

1. Principles/Theory
2. Equipment
3. Techniques/Calibration
4. Analysis/Evaluation
5. Procedures
6. Safety and Health

14.6.13 Visual Testing Method (TC-9)

	A	B
Recommended Training for Level I		
Recommended Hours of Instruction	8	4

1. Introduction
2. Definitions
3. Fundamentals
4. Equipment (as applicable)
5. Employer-Defined Applications (includes a description of inherent, processing, and service discontinuities)
6. Visual Testing to Specific Procedures

	A	B
Recommended Training for Level II		
Recommended Hours of Training	16	8

RECOMMENDED TRAINING COURSES

1. Review of Level I
2. Vision
3. Lighting
4. Material Attributes
5. Environmental and Physiological Factors
6. Visual Perception
7. Equipment
8. Employer-Defined Applications
9. Acceptance/Rejection Criteria
10. Recording and Reports

Level III Topical Outline

1. Principles/Theory
2. Equipment Accessories
3. Techniques/Calibration
4. Interpretation/Evaluation
5. Procedures and Documentation
6. Safety and Health

Basic Examination

- General Level III Requirements
- Basics of Common NDT Methods
- Basic Materials, Fabrications and Product Technology
- Predictive Maintenance (PdM) Basic Examination

14.6.14 Appendix

14.6.14.1 Example Questions Example questions are limited to the following 11 methods:

1. Acoustic Emission Testing Method
2. Electromagnetic Testing Method (includes Eddy Current Testing Method and Flux Leakage Testing Method)
3. Leak Testing Method (includes Bubble Leak Testing Method, Halogen Diode Leak Testing Method, Mass Spectrometer Leak Testing Method, and Pressure Change Measurement Leak Testing Method)
4. Liquid Penetrant Testing Method
5. Magnetic Particle Testing Method
6. Neutron Radiography Testing Method
7. Radiographic Testing Method

8. Thermal/Infrared Testing Method
9. Ultrasonic Testing Method
10. Visual Testing Method

Note: Questions for No. 11, Laser Testing Methods, including Holography, Shearography, and Profilometry, are not included at this time. Neither are questions for the Vibration Analysis Method.

14.6.14.2 Answers to Example Questions Answers to all questions are given at the end of the current edition Appendix of Recommended Practice No. SNT-TC-1A.

14.6.15 A Dynamic Document

Recommended Practice No. SNT-TC-1A is a dynamic document. It is also changing at a rapid rate. High-tech instruments have modernized inspection and analysis equipment used with all NDE and NDT methods. Old techniques are constantly being improved and new techniques are constantly being developed. Two relative new techniques include time-of-flight diffraction (TOFD), which is an ultrasonic technique used extensively for the evaluation of welds and butt-welds in vessels and heavy piping. This technique provides a reliable method for the detection of cracks and sharp grooves at the inner wall of welded vessels or pipes. In most cases it can supplement conventional UT and RT methods, and in some cases it may be able to replace them.

A second relatively new method is alternating current field measurement (ACFM), which provides reliable crack detection and depth sizing without the need to remove paint and other coatings. This saves a considerable amount of time normally related to building scaffolds, removing old coating, and repainting pipes and vessels. This electromagnetic (ET) technique was originally developed for offshore rig inspections and is currently used for most outdoor applications involving painted steel structures. It can be used for all conductive materials, and it is also highly effective for weld inspections.

Whenever a new technique or method is introduced, Recommended Practice No. SNT-TC-1A must be eventually revised to include the recommended Level I, II, and III certification training hours to be used with these new methods. Therefore, the recommended practice itself must be periodically updated to keep up with the technology. In some respects, it seems like a never-ending story. The same can be said for ANSI, ASTM, and ISO standards that apply to the various NDT methods.

The ASNT has hopes of releasing the 2005 edition of Recommended Practice No. SNT-TC-1A sometime in 2005. Revision D or E of this document is currently in the discussion stage. The recognition of new methods is of prime concern because it involves new certification and training for all ASNT certification levels, which also impacts company costs.

14.6.16 Special Disclaimer

This chapter provides a simple overview and discussion of Recommended Practice No. SNT-TC-1A, 2001 Edition, and its highlights. The recommended practice is an $8\frac{1}{2} \times 11$-inch soft-cover book containing over 165 pages. Readers who are actively engaged in technician certification and training programs should purchase the most current edition of the recommended practice.

APPENDIX 1

BIBLIOGRAPHY OF CREDITS

PAPERS/ARTICLES LISTED ALPHABETICALLY BY NDT METHOD

Acoustic Emissions

Brunner, Andreas and Bohse, Jurgen, "Acoustic Emission Standards and Guidelines 2002: A Comparative Assessment and Perspectives," NDT.net, Vol. 7, No. 9, September 2002.

Dunegan, H. L., "Prediction of Earthquake with AE/MS? Why Not?" *Sixth Conference on AE/MS Activity in Geological Structures and Materials*, May 1996.

Dunegan, H. L., "Considerations for Selection of Advanced AE Transducers," DECI Report, May 2003.

"Field Instruments for Nondestructive Evaluation of Concrete and Masonry," Impact-Echo Instruments, LLC, 2002.

DOT/FAA/AR-97/9, "An Acoustic Emission Test for Aircraft Halon 1301 Fire Extinguisher Bottles," Final Report, NTIS Public Information Document, April 1998.

Prosser, W. H. and Gorman, M. R., "Propagation of Flexural Mode AE Signals in GR/EP Composite Plates," Department of Aeronautics, Naval Post Graduate School, Monterey, CA, 1992.

Vallen, Hartmut, Dipl. Ing., "AE Testing—Fundamentals, Equipment, Applications," Vallen-Systeme GmbH, Munich, Germany, April 2002.

Introduction to Nondestructive Testing: A Training Guide, Second Edition, by Paul E. Mix
Copyright © 2005 John Wiley & Sons, Inc.

ET Method

Topp, Dr. David A., "Quantitative In-Service Inspection Using Alternating Current Field Measurement (ACFM) Method," NDT.net, Vol. 5, No. 3, March 2003.

Zhou, Jiannong, Wong, Brian Stephen (Assoc. Prof.), Seet Gim Lee, Gerald (Assoc. Prof.), Robotics Research Center, Nanyang Technological University, Singapore, "An ACFM Automated Crack Detection System Deployed by an Underwater Roving Vehicle," NDT.net, Vol. 9, No. 9, September 2003.

General

Grills, Robert H., "Probability of Detection," NDT Solution, July 2001.

Laser Methods

"Advanced Measuring Tool for Precision Stamping Analysis," TQS Application Note No. 2008, Trilion Optical Test Systems, date unknown.

Andersson, Joakim and van den Bos, Barend, "NDI on Bonded Sandwich Structures with Foam Cores and Stiff Skins—Shearography the Answer?" Ettemeyer AG Application Report No. 02/01, CMS Materialteknik AB, Reprint, 2001.

"Deformation Analysis of Car Components", TQS Application Note No. 2009, Trilion Optical NDT Systems, date unknown.

Dennis, John and Doyle, James, "Robotic Laser Scanner Being Developed to Improve the Safety of Future Space Shuttles," TechLink Press Release, August 29, 2003.

Gustafsson, Jeanette, "Nondestructive Large Area Testing Using Shearography," Ettemeyer AG Application Report No. 01/02, CMS Materialteknik AB, Reprint, 2002.

Honlet, Michel, "Nondestructive Testing of Complex Composite Materials and Structures Using Optical Techniques," Honlet Optical Systems GmbH, date unknown.

Reichard, Karl M., "DSPI Techniques (Digital Shearography and Holography)," PennState Applied Research Laboratory, 2000.

Roberts, Richard D., NDT Solution, "Laser Profilometry as an Inspection Method for Reformer Catalyst Tubes," American Society for Nondestructive Testing, April 1999.

Trost, Thomas, "Optical Measuring Methods," PACKFORSK Report No. 192, June 2000.

Tyson, John and Schmidt, Timothy, "Advanced Photogammetry for Robust Deformation and Strain Measurements," Trilion Quality Systems, LLC, *SEM 2002 Proceedings*, June 2002.

Waltz, Thomas and Dr. Ettemeyer, "Automatic Shearography Inspection System for Helicopter Rotor Blades," Dr. A. Ettemeyer Application Report No. 03/97, Dr. Ettemeyer GmbH & Co., Elchingen, Germany 1997.

Leak Detection Methods

Varian, Inc., "Leak Detection," September 2003, authors unknown.

MT Method

Chedister, William C., "Evaluation of Magnetic Gradients for Magnetic Particle Testing," *Materials Evaluation*, June 2002.

Chedister, William C., "Technical Advances Simplify Your Remote Magnetic Particle Inspections," NDT Solution, April 1997.

Chedister, William C. and Long, Jeffery, "MT Inspects Fans for Power Utilities," *Welding and Design Fabrication*, 1995.

Erlanson, R. Mark, "New Pigment Bonding Approach to Magnetic Particle Testing," *Quality*, May 1987.

Morris, John, "Induced Current Magnetization Made Simple," NTD Solution, June 1999.

Potter, Bob, "NDT of Precipitation Hardened Steels Using Fluorescent Magnetic Particle Testing," NDT Solution, December 2003.

Skeie, K. and Hagemaier, D. J., "Quantifying Magnetic Particle Inspection," *Materials Evaluation*, May 1987.

Stanley, Roderic K., "Magnetic Particle Inspection: Some Unsolved Problems," *Materials Evaluation*, 1994.

Stanley, Roderic K., "Simple Explanation of the Theory of the Total Magnetic Flux Method for the Measurement of Ferromagnetic Cross Sections," *Materials Evaluation*, January 1995.

Stanley, Roderic K., "Use of Inductive Ammeters to Ensure Saturation of Tubular Magnetic Materials Using the Internal Conductor Method," *Materials Evaluation*, June 1987.

Neutron Radiography

Berger, H. and Iddings, F., "Neutron Radiography—A State of the Art Report," Report NTIAC-SR-98-01, Nondestructive Testing Information Analysis Center, Austin, TX, October 1998.

Berger, H. and Dance, W. E., "Accelerator-Based Neutron Radioscopic Systems," *AIP Proceedings 475*, pp. 1084–1087, AIP, Woodbury, NY, 1999.

Brenizer, J. S., Berger, H., et al., "Development of a New Electronic Neutron Imaging System, "Nuclear Methods and Instruments in Physics, A, Vol. 424, No. 1, pp. 9–14, 1999.

Radiography

Bavendiek, Klaus, "Automatic Test of X-ray Imaging Quality," presented by Juergen Klicker and Frank Wiencek, YXLON International, 2002.

Berger, H. and Schulte, R. L., "Volumetric X-Ray Testing," *Materials Evaluation*, Vol. 60, No. 9, pp. 1028–1031, September 2002.

Brant, Frank, "The Use of X-ray Inspection Techniques to Improve Quality and Reduce Costs," YXLON International, 1998.

Jones, T. S. and Berger, H., "Performance Characteristics of an Electronic Three-Dimensional Radiographic Technique," *ASNT Spring Conference*, 2000.

Muenker, Martin, "Industrial Computed Tomography," YXLON International, 2004.

Pontefract, Rita, "Safety in X-Ray Inspection," YXLON International, 1998.

Schulte, R. L., Jones, T. S., and Berger, H., "A Digital X-Ray Inspection System for Full Volumetric Imaging," *ASNT Topical Meeting Digital Imaging 4*, 3 pages, August 6–8, 2001.

Schulte, R. L., "Volumetric X-Ray Inspection of Welds," *Inspection Trends* (AWS), Vol. 6, No. 3, pp. 20–23, Summer 2003.

Strong, Clark and Pontefract, Rita, "The New Image of Automatic Defect Recognition," YXLON International, 2002

Thermal/Infrared Testing Method

ASTM E977-84, "Standard Practice for Thermoelectric Sorting of Electrically Conductive Materials," May 1984.

Campbell, M. B. and Heilweil, E. J., "Non-invasive Detection of Weapons of Mass Destruction Using THz Radiation," *Proceedings of SPIE*, Vol. 5070, *Terahertz for Military and Security Applications*, edited by R. Jennifer Hwu, Dwight Woolard (SPIE, Bellingham, WA, 2003), in press.

Derby, Roger W., "Identifying Saw Blades: A Good Application for Thermoelectric Sorting," *Materials Evaluation*, July 1986.

Everett Infrared, "Physics of Detectors," 1998–2001.

Kim, Quiesup and Kayali, Sammy A., "Radiative Measurement of Temperature in a MESFET Channel," NASA Jet Propulsion Laboratory, Pasadena, CA, March 1998.

Raytek Corp., "Two Papers—Developing an Inspection Program and The Basics of Predictive/Preventive Maintenance," 2003.

Ross, Steven C., "Infrared Imaging Guide," FLIR Systems, Inc., date unknown.

Wilming, John and Rirozzola, Paul, LECO Corporation, "Advance Analysis," *World Coal*, June 2002.

Ultrasonic Testing Method

Berke, Michael, "Nondestructive Material Testing with Ultrasonics—Introduction to the Basic Principles," Krautkramer GmbH, 1990–1992.

Betti, F., Guidi, A., Raffarta, B. et al., "TOFD—The Emerging Ultrasonic Computerized Technique for Heavy Wall Pressure Vessel Welds Examination," 2002.

Bhardwaj, M. C., "High Efficiency Non-Contact Transducers and a Very High Coupling Piezoelectric Composite," WCNDT, 2004.

Bhardwaj, M. C., "Evolution of Piezoelectric Transducers to Full Scale Non-Contact Ultrasonic Analysis Mode," WCNDT, 2004.

Bhardwaj, Mahesh C., "Non-Contact Ultrasound: The Final Frontier in Nondestructive Analysis," Publication #SW302, March 2002.

Frielinghaus, Rainer, "Examples of Ultrasonic Application for Nondestructive Testing of Plastics," Krautkramer GmbH, 1990.

Hoover, Kelli et al., "Destruction of Bacterial Spore by Phenomenally High Efficiency Noncontact Ultrasonic Transducers," November 2002.

Krautkramer, "The Krautkramer Ultrasonic Book," Krautkramer GmbH & Co., 1998.

Kobe Steel, Ltd., "New Non-destructive Examination (TOFD System)," date and author unknown.

NDT.net, "Nondestructive Testing Encyclopedia—Ultrasonic Testing—TOFD Principles," date unknown.

Rao, B. P. C., Jayakumar, T., Kalyansundaram, P., et al., "TOFD for Sizing of Defects Using Shear Horizontal Waves Generated by Electromagnetic Acoustic Transducers (EMATs)," *Proceedings of NDE-98*, Trivendrum, 1998.

Sansalone, Mary J. and Street, William B., "Impact-Echo Nondestructive Evaluation of Concrete and Masonry," 1997.

Sommer, Jörg, "The Possibilities of Mobil Hardness Testing," Krautkramer, GmbH date unknown.

Splitt, G., "Ultrasonic Probes for Special Tasks—The Optimum Probe for Each Application," Krautkramer GmbH, date unknown.

Splitt, G., "Piezocomposite Transducers—A Milestone for Ultrasonic Testing, Krautkramer GmbH, date unknown.

Vibration Analysis

Cabanilla, Norman, et al., "Determining the Dampening Factor for the First Four Frequencies of Vibration of a Fixed-Free Beam and Locating Nodes of Vibration," date unknown.

Edney S. and Mellinger, F., "Advances in Tilting Pad Journal Bearing Design," Dresser-Rand, date unknown.

Fry, Johnson and Rossing, "Lecture Note 5, Modes of Vibration and Resonance," 1997.

Krupta, R., Walz, T., and Ettemeyer, A., Dr. Ettemeyer Application Report, No. 04-00, "New Techniques and Applications for 3D-Brake Analysis," *SAE Brake Colloquim*, San Diego, October 1–4, 2000.

Marshall Space Flight Center, "Software for Selecting Modes of Vibration of a Structure," October 2001.

Modeshape Technologies, Inc. Online "Vocabulary" at www.modeshape.com.

"Optomechanical Engineering and Vibration Control," *Proceedings of SPIE*, Vol. 3786, Reprint, July 20–23, 1999.

Wang, Zhiguo and Ettemeyer, A., Dr. Ettemeyer Application Report No. 03-98, "Some Applications of Pulsed ESPI to Brake Squeal Analysis," Germany 1998.

"Why the Nano-K™ Platform Outperforms Conventional Pneumatic Isolators by 10 to 100 Times," Minus K Technical Bulletin #2, date unknown.

Wismer, Johan N., "Gearbox Analysis Using Cepstrum Analysis and Comb Liftering," Bruel and Kjaer, Denmark, date unknown.

Visual Test Methods

Marshal, Dr. Justin, "Sensory Ecology Laboratory—Current Research," 2004.

Mendonsa, Ruth A., "Spectrometer Makes Astronomical Debut," *Photonic Spectra*, October 1997.

APPENDIX 2

COMPANY CONTRIBUTORS

The author wishes to thank the following company contributors, listed alphabetically by last name first, and their companies for reviewing and commenting on much of the rough draft material that was used in the preparation of this second edition of *Introduction to Nondestructive Testing: A Training Guide, Second Edition.*

Angelo, Bob, Mkg., Gigahertz-Optik, Newburyport, MA, www.gigahertz-optic.com

Ambramo, Dr. Kimberly Harris, Shimadzu Scientific Insts., Inc., Columbia, MD, www.shimadzu.com

Beehler, Don, Kistler Instrument Corp., Amherst, NY, www.kistler.com

Berger, Harold P. E., President, Industrial Quality, Inc./Digitome Corp., Montgomery Village, MD Digitome Corp., www.propylon.com

Bhardwaj, Mahesh C., CEO, Ultran Labs, Boalsburg, PA, www.ultrangroup.com

Brunner, Andreas and Bohse, Jurgen, Authors, AE Standards and Guidelines, NDT.net

Campbell, Fred, Gen. Mgr., Varian, Inc. Lexington, MA, www.varianinc.com

Carlos, Mark, V. P., Mistras Holding Grp., Princeton Junction, NJ, www.mistrasholding.com

---

Introduction to Nondestructive Testing: A Training Guide, Second Edition, by Paul E. Mix
Copyright © 2005 John Wiley & Sons, Inc.

Chedister, William C., V. P., Circle Systems, Inc., Hinckley, IL, www.circlesafe.com.

Creager, Brian, DiversiTech Corp., Decatur, GA, www.diversitech.com

Cynoweth, Jon, V. P., Mikron Infrared Technologies, Louisville, KY, www.mikroninfrared.com

Davidson, Paul, President, WIS, Inc., 117 Quincy St., NE Albuquerque, NM, emats@sprintmail.com

Dayagi, Stefan Y., V. P., Ettemeyer, AG, www.ettemeyer.com

Dorsey, Alan, President, ScanTech Instruments, Inc., Longview, TX, www.scantechinstr.com

Doyle, James, President, Laser Techniques, Bellevue, WA, www.laser-ndt.com

Dyer, Catherine W., President, Industrial Instruments, Inc., www.industrialinstruments.com

Dunner, Stephanie, President, AEI North America, Inc., www.aeinorthamerica.com.

Ettemeyer, Dr. Andreas, President, Ettemeyer AG, Elchingen, Germany, www.ettemeyer.com

Fay, Charles, Vice Provost, Cornell University, Ithaca, NY, www.cornell.edu

Fix, Robert M., President, American Stress Technologies, Pittsburgh, PA, www.astresstech.com

Gilardoni, Marco T., Mkg. Dir., Gilardoni Research Lab, Lario, Italy, www.gilardoni.it

Gilbrich, Sebastian, Ad. Consultant, Leica Mircosystems, Wetzler, Germany, www.light-microscopy.com

Goranson, Burt, UniWest, Pasco, WA, www.uniwest.com

Herbsthofer, Frau Dr. Elke, Mng. Dir., Castell Verlag, GmbH, Munich, Germany, www.castell-verlag.de

Idtensohn, Richard R., V. P., Schenck Trebel Corp., Deer Park, NY, www.medibix.com

Jenkins, Patrick R., President, Magwerks Corp., Indianapolis, IN, www.magwerks.com

Joshua, Peter, Solarius Development, Sunnyvale, CA, www.solarius-inc.com

Justice, Jerry, Prod. Mgr., ACES Systems TEC Aviation Division, Knoxville, TN, www.acessystems.com

Kratzert, John R., Sr. Mkg. Specialist, JSR Ultrasonics, Pittsford, NY, www.jsrultrasonics.com

Lee, William, Owner, Alpha Labs, Inc., Salt Lake City, UT, www.trifield.com

Lopez, Joe, Everest VIT, Inc., Flanders, NJ, www.everectvit.com

Lugg, Martin, TSC Inspection Systems, Milton Keynes, UK, www.tscinspectionsystems.com

APPENDIX 2: COMPANY CONTRIBUTORS **669**

Maciejowski, Cyndi, Mkg., Cardinal Health, Cleveland, OH, www.cardinal.com

Menne, Dave, V. P., Olympus Industrial America, Orangeburg, NY, www.olypusindustrial.com

McGuire, Kate, Raytek Corp. P. R., Santa Cruz, CA, www.raytek.com

Margolis, H. Jay, President, Infinity Photo-Optical, Boulder, CO, www.infinity-usa.com

Menne, Dave, Olympus Industrial Inc., Orangeburg, NY 10962, www.olympusindustrial.com

Morris, Rob, Dir. Mkg., Ocean Optics, Inc., Dunedin, FL, www.oceanoptics.com

Moss, Clive, Mkg., Specac Ltd., Orpington, Kent, UK, www.specac.com

Nguyen, Julie, Mkg. Com., Thermo Electron Corp., Houston, TX, www.thermo.com

Peck, P. Michael, President, Dynamold, Inc., Fort Worth, TX, www.dynamold.com

Plasek, Terry L., Western Reg. Mgr., Fuji N.D.T. Systems, Roseville, IL www.fujimed.com/ntd/

Platus, David L., President, Minus K Technology, Inc., Inglewood, CA, www.minusk.com

Pontefract, Rita, Mkg. Specialist, YXLON International, Akron, OH, www.yxlon.com

Rogers, B. J., Mkg., CMC Electronics, Cincinnati, Mason, OH, www.cinele.com

Sapienza, Rocky, Mkg., Liquid Crystal Resources, Chicago, IL, www.lcr-usa.com

Scheib, Randy, Mkg., PAC/Mistras Holdings Grp., Princeton Junction, NJ, www.pacndt.com

Schmidlein, Ken, LD Mkg. Specialist, Varian, Inc., Lexington, MA, www.varianinc.com

Shields, Susan, Sr. Prod. Mgr., Everest VIT, Inc., Flanders, NJ, www.everestvit.com

Skeie, Kermit, Uresco/Ardox (retired), Whitefrog@uia.net

Smith, Rosanne, Editor, *World Coal* Magazine, www.worldcoal.com

Sowers, Pat, Mkg. Dir., Ansonics Inc., Halfway, OR, www.ansonics.com

Stanley, Rod K., NDE Information Consultants, Houston, TX, www.ndeic.com

Street, William B., Impact-Echo Instruments, Ithaca, NY, www.impact-echo.com

Sy, Dagmar, Institut Dr. Foerster, Reuttlingen, Germany, www.foerstergroup.de

Thomas, Benoy George, Asst. Editor, *Voice & Data* Magazine

Tyson, John, President, Trilion Quality Systems, West Conshohocken, PA, www.trilion.com

Vallen, Harmut, Dipl. Ing., Mkg. Dir.,Vallen-Systeme GmbH, Munich, Germany, www.vallen.de

Wakabayashi, Jiro, Mkg., Fuji Ceramics Corp., Tokyo, Japan, www.fujicera.co.jp

Wallace, Robert, Mkg. Com., Vacuum Instrument Corp., Ronkonkoma, NY 11779, www.vacuuminst.co

Walter, Rick, President, HEM Data Corp., Southfield, MI, www.hemdata.com

Whitacre, Nikki, G., Mkg., Pyromation, Ft. Wayne, IN, www.pyromation.com

Willming, John, Copywriter, Leco Corp., St. Joeseph, MI, www.leco.com

Wowh, Victor, President, Machine Dynamics, Inc., Albuquerque, NM

Wong, Brian S., Assoc. Prof., Nanyang Tech. University, Singapore

Wuelfing, Troy, Mgr. Magwerks Corp., Indianapolis, IN, www.magwerks.com

Zwigart, Dennis A., Mgr., StressTel, State College, PA, www.stresstel.com

INDEX

A-scan, 470, 499, 500
Abrasion-resistance thermocouples, 418
Absolute bearing vibrations, 529, 541
Absolute shaft vibrations, 541
Absorptivity, 421
Absorbed dose, 326, 327
Accelerators, 303, 309, 323, 324, 337, 345, 359, 387, 397
Acceleration, 555, 556
Accelerometers, 544
Accordion machine vision, 607
Accumulated dose, 332
Accumulation testing method, 194
Acoustic filters, 555
Acoustic focusing, 485
Acoustic impedance, 477–479
Acoustic pressure vs. target distance, 461, 462
AE burst, 18–20, 25, 34
AE Coupler, 36–38
AE parameters, 18, 19
AE process chain, 17
AE measurement chain, 20
AE preamplifier, 23

AE sensor features, 36
AE sensors, 21, 35, 36, 39
AE theory, 39
Adhesive bonding, 54, 323
Adoptive reference subtraction, 391
Advanced mercury analyzer, 448–450
Advanced pan-tilt zoom camera system, 616–623
Afocal variation system (AVS), 601
Agreement states, 338
Air/gas propagation transducers, 460–463
Airport security, 9
Alignment, 114, 170, 390, 444, 524, 530, 555
Alignment and balance, 530
Alkali metals and water reactions, 408
Alloy analyzer, 10, 577
Alloy sorting, 450–456
Alpha particles, 313, 343–346
Alternating current field measurement, (ACFM), 124–130, 658
Aluminum corrosion, 321, 323
Ambient radiance, 423, 424

Introduction to Nondestructive Testing: A Training Guide, Second Edition, by Paul E. Mix
Copyright © 2005 John Wiley & Sons, Inc.

Amdata automatic pipe scanner (AMAPS), 507, 508
American Society for Testing and Materials (ASTM), 51–53, 56, 106, 238, 246, 268, 269,
Amplitude-distance blocks, 515, 516
Analyzer Plus Model 1700, 567–569
Angio-genesis therapy, AGT, 6
Angular resolving power, 428
Angulation, 495, 590
Annual limit on intake (ALI), 326
Anthrax spores, 466
Antialiasing filters, 574
Ariane 5 launch vehicle, 166
ARAMIS optical system, 168–170, 172
Arterial research, 434
Artificial intelligence (AI), 390
Artificial standards, 239, 514, 515, 577
As low as reasonably achievable (ALARA), 326
Attenuated total reflection (ATR), 446
Attenuation domain transducers, 458
Attenuation losses, 482, 518
Autoclave ink, 413
Autofocus principle, 142
Automated comparison bridge, 596
Automatic defect recognition (ADR), 387–393
Automatic film processing, 375
Automatic marking, 513
Automotive applications, 181, 434, 587
Autoranging sensor, 574
Aviation analysis functions, 572, 573

B-scan, 500, 629
Backscatter radiation, 316
Bandwidth (BW), 11, 42, 46, 361, 362, 413, 442, 458, 461, 468, 485, 487, 550, 553, 619
Bandwidth center frequency (BCF), 458
Basic modes of heat transfer, 407
Bearcon signature (BCS), 553
Bearing condition (Bearcon), 548, 549
Bearing vibration/noise, 548, 549
Beat frequency, 530, 532
Beam collimation, 314, 315
Beam divergence, 311, 481, 482, 489, 518

Beam purity indicator, 316
Becquerel, Henry, 338
Beta particles, 345, 346, 354
Betatrons, 345
Bimetal dial thermometers, 416
Biological damage, 134, 349
Biological effects (of radiation), 348–351
Bioterrorism, 466
Blackbody, 421, 427, 428, 439, 653
Bluetooth technology, 1
Blower and fans, 550
Boiling water reactor (BWR), 507
Bolting, 577
Bone density, 400
Bottom-blown furnaces, 409
Borescopes, 578–584
Break squeal analysis, 565
Breeder reactor, 340, 582
Broad-band optical pyrometers, 424–426
Bubble leak testing, 183, 199–202, 657
Building vibration, 524
Bulk polymerization, 409
Bunsen-Roscoe reciprocity law, 368
Butterfly plot, 127

C-scan, 122, 458, 502, 513
Cable reels, 623, 625
Calculated dose rate, 333
Calibrated leaks, 198
Camera control unit, 619, 620
Camera pan and tilt features, 617, 618
Carbon composite/s, 53
CFRP, carbon fiber reinforced plastic, 164, 165
Castings, 226, 239, 263, 289, 338, 356, 387, 393, 576, 577, 652
Casting defects, 398
Celsius temperature, 410
Cepstrum analysis, 557, 558
Characteristic curve, 315, 366–368
CCD camera, 132, 141, 152, 154, 160, 161, 163, 165, 168, 170, 172, 322, 387, 585
CCD detector, 131, 146, 149, 566, 614, 629, 633, 634
Cineradiographic work, 361
Climate controlled chambers, 521, 555
Closed circuit TV (CCTV), 580, 581, 583, 585

INDEX   673

Clustering, 27, 29, 30
Coercive force, 91, 92, 249, 278, 279
Cold neutrons, 316
Cold-vapor atomic absorption
   spectroscopy (CVAAS), 450
Collection phase, 450
Color change crayons, 413
Color change paints, 412
Color change thermometry, 411–415
Color contrast, 224, 227, 232, 237–239,
   264, 412
Comb liftering, 557
Compliance, 525, 528, 575
Composite material, 39, 53, 138, 161, 162,
   164, 165, 483
Compressed fiber, 3, 460
Comptom effect., 346
Constant color intensity control (CCIC),
   598
Crawler and camera options, 624, 625
Crime scene investigations (CSI),
   595–599
CT scanners, 383
CT system components, 384–387
Comparison microscopes, 593–600
Computer interface dosimeter (CID),
   352, 353
Condensation polymerization, 409
Conduction, 407, 408, 427, 429, 482
Confocal measurement, 144
Confocal microscopy, 145–147
Confocal point sensor, 142
Conoscopic holography, 143
Constructive interference, 532
Contact probe, 131, 180–182, 491, 550
Contact testing, 473, 478, 485, 489, 494,
   499, 512
Contamination, 179, 187, 188, 196, 228,
   238, 290, 293, 328, 329, 331, 350, 375,
   458, 466
Continuously focusable microscope, 600
Contoured focusing, 487
Contoured shoes, 504, 505
Contraflow, 187–188, 196
Control rods, 340, 583
Convection, 407, 408, 427, 429, 434
Conventional flow, 188, 189
Conventional ultrasound, 471–488
Cooled mid-wave IR detection, 434

Corrosion, 10, 24, 33, 41, 42, 52, 108, 110,
   121, 128, 130, 147, 152, 224, 228, 317,
   321–324, 430, 451, 472, 506, 521, 529,
   547, 550, 576–578, 584, 590, 622
Counts per minute (CPM), 328, 355
Coupling agents, 21, 56, 459, 474, 494, 498,
Crack propagation, 38, 41
Crystal mosaic, 488
Crayons, 410, 411, 413
Curie, Marie and Pierre, 338
Custom periscope, 582
Cylindrical focusing, 487

D-scan, 629
Damping factor, 532
Damping ratio (z), 532
Data acquisition with a network
   (DAWN), 562
Data analyzer, 521
Decomposition phase, 450
Defect identification, 391
Deformation, 15, 16, 32, 35, 38, 120, 121,
   137, 141, 154–156, 158, 159, 161–163,
   168, 170–172, 175, 177, 526, 527, 550,
   562–564, 566
Degradation of hybrid rocket motor
   insulation, 465
Degrees of freedom (DOF), 464, 524
Destructive interference, 532
Detection phase, 450
Detecting speeds, 574
Diffraction, dispersion and attenuation,
   481, 482
Digital speckle pattern interferometry
   (DSPI), 155
Digital technology, 1, 406
Digital video recording, 622
Digitome X-ray imaging, 393–396
Discrete Fourier transform (DFT), 556
Displacement measurements, 541, 555,
   556
Displacement sensors, 545–547
Distance amplitude correction (DAC),
   510
DLATGS Pyroelectric detector, 447
Dose rates, 309, 310, 327, 331, 332, 333,
   335, 341, 342, 345, 350, 354, 356
Double pulsed ruby laser system,
   563–565

Dry developer, 226, 228, 238
Dry pumps, 196
Dye penetrant, 172, 183, 222, 223, 227–230, 232, 236, 238, 578
Dye penetrant leak testing, 183
Dynamic absorber, 534–536
Dynamic behavior analysis, 541
Dynamic Tritop system, 177

Econoscope, 585, 586
Eddy current/s, 2, 3, 65–73, 79–82, 88–90, 96–108, 110, 111, 114, 121–123, 125, 150, 232, 507, 546, 547, 575, 576, 625, 643, 657
Edge detection, 391
Elastic energy, 16, 17
Electric discharged machined (EDM), 501
Electromagnetic radiation, 133, 343
Electromagnetic spectrum, 132, 134, 233, 234, 345, 421
Electron capture detection, 397
Emissivity, 421–429, 433, 436, 438, 440
Endothermic reaction, 408, 409
Engine fan balancing, 569
Engine vibration analysis, 565
Enhanced external counterpulsation, EECP, 6
Environmental factors, 328, 329, 426, 434, 440, 510, 525, 555, 569, 578, 624, 653, 657
Epithermal neutrons, 303, 312, 317, 321
Equivalent penetrameter sensitivity (EPS), 374
Error potential in radiant measurements, 429
ESPI interferometry, 135, 136, 154, 155, 157–159, 168, 169, 171, 172, 562–565
Examination of damage to vessel diffuser plates, 622, 623
Exothermic reaction, 375, 408, 409
Explosive devices, 9, 321, 650
Exposure factor, 362
Exposure time, 136, 172, 308, 310–314, 327, 332, 356, 360–370, 381, 396, 466
Extreme vibration attenuation (EVA) pad, 533, 534
Extrusions, 576, 578

Failure analysis, 170, 529, 543
Fan trim balancing, 572
Fast Fourier transform (FFT), 48, 544, 548, 550, 553, 556, 557, 560, 574
Fatigue cracks, 42, 128, 229, 253, 256, 263, 288, 500, 501, 548
Fault recognition, 541
Feature extraction, 25, 43, 44, 48
Fiberscopes and Videoscopes, 584–587
Fiberglass reinforced plastic (FRP), 165
Field of view (FOV), 171, 428, 433, 436, 438, 440
Film badge, 310, 326, 327, 330, 352–354
Film graininess, 316, 371
Film thickness on a silicon wafer, 448
Finite element method (FEM), 158, 168, 170, 269
Fission fragments, 345, 346
Fission reactor, 340
Flash radiography, 360
Flat panel digital imaging, 378–381
Flaw detection, 3, 39, 76, 79, 82, 84, 94, 111, 122, 123, 459, 471–473, 502, 512
Flaw image contrast, 371
Flaw transducers, 509, 510
Flexible fiberscopes, 576, 584–589
Fluorescent penetrant, 152, 153 223–225, 227, 229, 233, 236–239, 242, 245
Fluorescent penetrant inspection, laser-scanned (LSPI), 152, 153
Fluororesin film coatings, 447, 448
Fluoroscopy, 377, 378
Focal distance (FD), 487, 488
Focal plane array (FPA), 435, 436, 438, 440
Forensic comparison, 596, 599, 610
Forgings, 289, 519, 576, 578
Forming limit curve (FLC), 143
Fourier-transform infrared spectrometer, 444–448
Fraunhofer zone, 482
Frequency analysis, 541
Frequency considerations, 512
Frequency domain transducers, 458
Frequency response, 529
Frequency response function (FRF), 526, 599
Fresnel and Fraunhofer fields, 482, 483
Fresnel zone, 482, 487

INDEX   **675**

Fringe line/s, 140
Frisker stations, 329
Fuel rods, 340
Fuji Dynamix computed radiography, 379–381
Full field view, 132, 159
Full width half maximum (FWHM), 630
Fundamental resonant frequency, 475, 498
Fundamentals of radiation safety, 338

Gamma density measurement, 402, 404
Gas matrix piezo (GMP) transducers, 461
Gamma level gauging, 402, 403–405
General Atomics, 308
Generation of ultrasonic waves, 483, 484
G-M survey meter, 327–329
Geometric unsharpness, 314, 324, 363, 364, 371
Geometric and mathematical TOFD model, 627, 628
GLARE, fiberglass and aluminum, 165
Glare, 171, 227, 377
Golden image technique, 390
Graininess, 314, 316, 371
Gray (Gy) Si unit of absorbed dose, 326
Graybody, 422, 428
Greenberg, E. H., 451
Ground individual field of view (GIFOV), 428

H and D curve, 367
Half-life, 304, 342
Half-value layer, 335, 348, 360
Halogen diode, 209, 210–213, 657
Hall effect, 67, 83–87, 92, 95, 275, 276
Hall coefficient, 85
Hand-held controller, 620–622
Headphones, 180–182
Heat filters, 377
Heat flow, 407, 408, 653
Heat reactive ink, 415
Heat transfer, 210, 407, 408, 653
Helicopter main rotor vibration measurements, 570
Helium mass spectrometer leak testing, 184
High-end engine vibration analysis, 572
High-intensity illuminators, 376

High macroscopic cross section, 303
High-speed, high-resolution spectral data, 629
High-speed IR line camera, 442–444
High-speed temperature measurement of tires, 442–444
Holographic image, 136, 137, 141
Holographic interference, 138
Holographic sensor, 142
Holography, 132, 136, 137, 140, 141, 143, 644, 645, 658
Homeland security, 637
Horizontal ATR operation, 446, 447
Hydrogenous material, 303, 317, 321, 323
Hysteresis curve, 90, 277, 278

Identification of materials (IM), 305, 316, 448, 450–456
Identification of nuclear fuel bundle serial numbers, 622, 623
Illumination manager, 598
Image domain transducers, 458
Image freeze and storage, 622
Image quality indicators (IQI), 315, 372, 374
Image unsharpness, 371
Immersion transducers, 482, 489
Impact-echo, 54–57, 59–61, 63, 64
Impact testing, 529
Improper film handling, 374
Improper film processing, 376
Increasing mass, 534–536
Index of refraction, 481
Individual field of view (IFOV), 428, 429, 433, 436, 440
Industrial computed tomography, 382–387
Industrial polymerization, 409
Industrial radiography, 325, 337, 338
Industrial videoscopes, 589–593
Inertance, 528
InFocus corrections, 608
InFocus microscope enhancement system, 607
Infrared, 2, 10, 407, 408, 420, 421, 427–429, 435, 444, 458, 653, 654, 658
Infrared imaging, 408, 420, 427–429
Inspection speed, 76, 95, 168, 264, 266, 513

Inspection speeds for tubular elements, 513
IC packaging, 147
Intergranular stress corrosion cracking (IGSCC), 506, 507
Intensifying screens, 364, 365
Interference patterns, 138, 139, 599
Interferogram, 161
Instant picture analysis system (iPass), 463
Inverse square law, 333, 348, 366
Ion chamber, 217, 309, 310, 317, 327, 346, 352, 354, 399, 401, 402, 587
Ionizing radiation, 309, 312, 326, 327, 339, 346, 352, 359
Iso cubes, 533
Isolation treatments, 536–540
Isotope production, 338, 341
Irrelevant moving bands of color (heat), 436
Irreversible color change indicators, 411

JCAMP standard, 445
Jet engine inspection, 581, 582
Jet engine data acquisition module (JEDA), 568
Journal bearings, 536
Joystick menu navigation, 620

Kaiser effect, 16
Kelvin scale, 410
Kirchoff's Law, 420

Labels, 236, 312, 313, 411–415, 620
Lacquers, 411
Lamb waves, 476, 477, 493, 494, 497
Laser Methods, 562–566
Laser profilometer, 142, 578
Laser safety, 133, 644–646
Laser thermometer, 423–425
Laser-scanned penetrant inspection (LSPI), 152–154
LP-2000 laser profilometer, 150, 151
Leak rate/s, 180, 183, 184, 190, 192, 197–199, 207–209, 211, 214, 215, 217
Leak testing, bubble, 183, 199–201, 657
Leak testing, dye penetrant, 183
Leak testing, MSLD, 184, 186, 196, 198, 213, 216
Leak testing, pressure change, 183, 203, 205–209, 657
Leak testing, ultrasonic, 180, 183
L/D ratio, 307, 308, 311, 314, 316, 323, 324
Liquid crystal inks, 414
Liquid in glass thermometers, 415
Liquid penetrant testing, 221–224, 228, 231, 232, 239, 244, 245, 262, 648, 657
Linear diode array (LDA), 385–390
Locating leaks, 180, 191, 192
Long distance microscope (LDM), 600–607
Longitudinal wave, 35, 476–478, 489–492, 498, 627, 629
Looseness, 536
Lucite shoes, 487
Lucite wedges, 487

Machine diagnosis, 541–543
Machine vision systems, 601
Macroscopic cross section, 303
Magnetic crawlers, 575
Magnetism, 90, 112–115, 248, 251, 263, 273, 274, 277–279, 289, 523, 643
Man-made radiation, 326
Manipulator bridge, 503, 504
Mass, 36, 67, 171, 522–524, 527, 528, 534, 536, 541, 544, 545
Mass spectrometer leak detector, 184, 186, 198, 216
Material parameters, 514
Measured individual field of view (MIFOV), 428
Measuring area, 156, 158, 160, 164, 170, 172, 566
Measuring leaks, 180, 192, 193
Measuring sensitivity, 156, 158, 164, 165, 556, 566
Measurement range, 142, 146, 244, 401, 432, 438
Mechanical deformation, 137
Mechanical shakers, 522
Mechanical stress, 17, 113, 138, 140
Medical marvels, 5, 6
Medical radiography, 338, 345
Melting point indicators, 410, 411
Mercury in solids and solutions, 449
Meson particles, 345

INDEX 677

Micro-electromechanical systems (MEMS), 3, 5, 144,
Microcircuits, 578
Midstage flow, 188
Millirems per hour (mR/h), 327
Miniature fiber optic spectrometer, 633–636
Mixing of solutions, 409
Modal analysis, 526, 527, 537, 562
Mode conversion, 475, 478, 493, 495, 496, 626, 629
Modes of vibration, 477, 497, 522–524
Moderator, 303, 310
Modulation transfer function (MTF), 322
Mohr's circle, 525, 526
Moisture gauging, 397
Molybdenum processing, 408
Monitoring structures, 41
Multi-change paints, 412
Multilayer ultrasonic thickness gauge, 470, 471
Multiple flow, 188

Nanobots, 4
Nanosurf confocal microscopy, 145
Nanotechnology, 3, 4
Narrow-band optical pyrometers, 423, 424
National voluntary laboratory accreditation program (NLAP), 417
Natural background radiation, 348, 349
Natural focus, 482
Near field distances, 483
Near IR, 421, 444
Negative stiffness mechanism (NSM), 537–540
Neutral particles, 345
Neutron activation analysis (NAA), 304, 307, 309, 316, 317
Neutron cross section, 301, 303
Neutron energy, 303, 311–315
Neutron imaging, 312, 321–324
Neutron level gauges, 320
Neutron porosity logging, 319
Neutron reflections, 314
Neutrons, 301–317, 320, 321, 326, 327, 338–341
Noise, 3, 10, 16, 18–20

Noise analysis, 525
Noise vibration harshness (NVH), 159
Nominal frequency (F), 458
Non-agreement states, 337
Noncontacting, 3, 136, 142, 162, 422, 458, 460, 463, 489, 545, 626
Noncontacting pitch-catch mode, 463, 464
Noncontacting transducers (NCU), 460–463
Noncontacting ultrasonic testing, 458–466
Nonlinear mechanical system, 536
Nonrelevant indications, 71, 88, 263, 274, 297, 299
Nuclear applications, 582–584
Nuclear bomb fallout, 349
Nuclear fission, 326, 340
Nuclear reactor, 9, 41, 208, 303, 316, 321, 326, 340, 343, 346, 622
Nuclear Regulatory Commission (NRC), 52, 128, 308, 309, 325, 330, 331, 335, 337, 338, 349, 352
Numerical aperture (NA), 604
Nyquist frequency plot, 557, 559

Oak Ridge National Lab (ORNL), 316
Objective lens, 142, 144, 146, 578, 580, 594, 605, 611
Occupational dose limit, 349
Ocean optics photometers, 629–638
Operating deflection shape (ODS), 527
Optical inspection, 168
Optical pyrometers, 423–427
Optimum magnification (fluoroscopy), 377
Orbit plots, 559
Order analysis, 527, 528
Order frequency, 527
Order tracking, 527

Paint brush transducer, 488, 512
Pair production, 347
Pancake probe, 327, 329
Paper mill process, 434
Particle radiation, 303, 343
Particle vibration, 474, 476, 477, 490
Pass and fail testing, 531, 536
Peak frequency (PF), 458

Pellets, 313, 340, 410
Penetrameter, 289, 315, 370, 372–374, 378, 491
Penetrant materials, 224, 236–238
Penetrant testing, 221–224, 232, 239, 244, 245, 262, 648, 657
Percent transmittance (%T), 636
Permeability, 72, 73, 76, 79, 84, 88–90, 96, 108, 110, 125, 126, 249, 251, 253, 262, 264, 275, 305
Personnel monitoring, 310, 325, 337, 338, 649, 651
PCI bus, 10, 11, 34, 39, 42–49, 444
Phase-stepping, 162
Photoelectric effect, 346
Photofluorescent, 365
Photographic density, 315, 366, 367
Photons, 133, 341–346, 354, 355, 402, 408, 421, 428, 436, 436, 634
Piezoelectric sensor, 21, 23, 34
Pipe and vessel inspections, 626–629
Pipe weld inspection, 575
Planck's law, 423
Pocket dosimeter, 310, 326, 327, 330, 352
Polychromic inks, 415
Portable accelerators, 359
Portable noise and vibration analyzer, 560–562
Portable ultrasonic flaw detector, 507, 509
Position-sensing device (PSD), 149
Post emulsification, 226, 229
Predictive maintenance (PdM), 429–431, 521, 528, 550, 556, 562, 657
Pressure change leak testing, 183, 202, 205
Pressurization stress, 645
Preventive and predictive maintenance program (PPM), 429, 431, 550
Primary reaction control system (PRCS), 6, 7
Probalancer Analyzer 2020, 570–572
Probability of detection (POD), 577
Probe manipulator, 489, 502–504
Process Control, 1, 9, 10, 41, 101, 285, 291, 294, 393, 410, 439, 440, 455, 630
Profilometry, 134, 135, 142, 147, 646, 658
Projection microscopes, 593–600
Propellar balancing kit, 571

Proportional counter, 310, 313, 397
Pulse dampening, 499
Pulse generation, 475, 476
Pulse-echo method, 55, 465, 467, 469, 473, 490, 510
Pulse repetition rate, 463, 466, 467–469, 499, 513
Pulse width (PW), 458, 467, 485, 500
Pulsed 3D ESPI system, 563–566
Pulser, 50, 360, 463, 464, 466–468
Pulser/receiver, 468–470
Pumps, 4, 6, 10, 16, 186, 187, 195, 196, 216, 218, 239, 289, 402, 521, 524, 550, 577, 583
PXI bus, 10, 11
Pyrometers, 10, 422–427

Quality estimation, 391

Radiation area, 325, 332, 335
Radiation dose, 9, 310, 311, 326, 327, 331–336, 349, 350, 353, 354, 356
Radiation exposure calculator, 370
Radiation safety officer, 198, 311, 330, 336
Radiation shielding, 347, 378
Radiation survey, 310, 325, 327, 331, 335, 336, 337, 346, 649, 651
Radio frequency (RF), 475, 484, 500
Radiographic camera, 332, 335, 336
Radiographic contrast, 314, 315, 371
Radiographic sensitivity, 314, 370–372
Radiographic shadow formation, 363
Radioisotope, 198, 339, 340, 345, 358, 361, 366
Rare earth element/s, 301, 303
Rayleigh waves, 36, 122, 476, 493
Real-time holography, 137, 140
Refinery alloy comments, 453
Reflection and refraction, 478–481
Reflectivity, 421, 422, 427, 428
Region of interest (ROI), 387, 391, 441, 557
Relative shaft vibration, 541
Relay lenses, 578
Remote-controlled cold-light source, 599
Remote pulsers and preamplifiers, 467–470

INDEX   679

Remote magnetic pipe crawlers, 575
Remote visual inspection, 611–626
Repeatability, 143, 387, 418, 420, 432, 471
Research and development, 25, 39, 434, 470
Resistance temperature devices (RTDs), 204, 417–420, 428
Resistance temperature elements (RTEs), 420
Resolving power, 219, 428, 485, 604
Resonance testing, 497
Resonant frequency, 21, 473, 475, 485, 497, 498, 521–524, 526, 536
Restricted area, 331, 332, 335
Retentivity, 87, 264, 274, 275, 278, 279
Reverse flow, 186, 187
Reverse fluorescent, 227
Rigid borescope, 580, 584, 585
Ring light adapter, 605
Robotic crawler units, 623–626
Robotic inspection, 164, 168
Roentgen, William, 338, 347
Rolling element bearing failures, 547
Rotating equipment analysis, 527
Rotor track and balance measurements, 570
Rule base/specific algorithm method, 392

Scanning laser profilometry, 147–149
Self-adhesive strips and indicators, 411
Scintillation detector, 355, 402
Scintillation screen, 385, 389
Search unit construction, 484–489
Seebeck effect, 416, 451, 452
Selective envelope detection (SED), 550, 553
Semiautomatic defect recognition (SADR), 392
Sensor coupling, 21, 23, 26, 47
Shear waves, 57, 122, 123, 476, 478, 487, 490–495, 514, 515, 629
Shearography, 132, 161–168, 644, 645, 658
Shearography camera, 164, 644
Shielding, 80, 88, 289, 317, 332–335, 345–348, 378, 426, 583
Shock pulse generator, 499, 510
Signal-to-noise ratio (SNR), 79, 80, 84, 88, 111, 162, 460, 461, 643

SilverHawk, 5
Single color change paints, 412
Small-angle neutron scattering instrument (ANS), 316
SnakeEye diagnostic tool, 587, 588
Snell's law, 480
Sniffer, 184, 192, 193, 195, 212–214, 219
Sound concentrator, 182
Sound level meter (SLM), 525
Source locking mechanism, 336
Source strength, 341, 342, 366, 369
Sources of vibration, 524
Space elevator, 60,000-mile-high, 11, 12
Spatial resolution, 156, 177, 324, 384, 385, 436, 440, 443
Speckle interferogram, 161
Spectral range, 432, 443, 447, 631, 632
Spectrometer efficiency curve, 631, 632
Spectrometer tube, 185–188, 197, 214, 216–218
Spectrum analyzer, 76, 111, 499, 525, 556, 557, 561
Speed and phase signal processing, 574
Speed change, 540
Spherical aberration, 607
SpyGlass lens, 439, 440
Squirter unit, 489, 495, 497, 513
Stability and Decay, 341
Standard radiographic technique, 361
Standard reference blocks, 514, 515
Statue of Liberty, 584
Stefan-Boltzmann constant, 422
Stefan-Boltzmann law, 423
Stereomicroscopes, 594–598
Stereomacroscopes, 594, 595, 599
Sterling-cycle cooling engine, 435, 438
Stiffening, 540, 541
Stiffness, 524, 527–529, 537, 539
Strain gauges, 119, 155, 525, 543
Stress analysis, 120, 525, 529
Stress/strain analysis, 565
Structural integrity, 39, 40, 50, 53, 54, 124, 208
Surface waves, 36, 57, 457, 476, 477, 492–494, 497, 519
Superimposed image comparison, 598
Survey instrument/s, 327, 331, 342, 354, 649
Survey reports, 331, 651

INDEX

Tachometer inputs, 574
Target rods, 340
TEC's Aviation Products, 567–574
Technique charts, 261, 368–370
Temperature measurement, 207, 409, 410, 420, 424, 426, 427, 438, 442, 654
Temperature sensors with readouts, 416–420
TV camera, 131, 356, 583
TV holography, 136
Tele-zoom viewing, 616
Tenth-value layer (TVL), 348
Testing equipment (UT), 498–518
Test methods (UT), 489–498
TN alloy pro, 357–359
Thermal imager and visible light camera, 440, 441, 442
Thermal imaging, 432, 434–436, 442
Thermal neutron/s, 301, 303, 303, 310–314, 316, 317, 321–324
Thermal radiation, 407, 408, 421–423, 425, 427–429, 439
Thermal stress, 113, 138, 139, 547, 645
Thermal video imaging, 433
Thermochromic flex ink, 415
Thermochromic liquid crystal indicators, 413
Thermocouple sensors, 416, 417
Thermoelectric alloy sorting, 450–456
Thermoluminescent dosimeter, 326, 352–354
Thorium reduction, 408
3D confocal microscopy, 146
3D deformation vector, 564
3D ESPI, 154, 157, 158, 159, 171
3D profiler, 149
Through transmission, 467, 469, 490, 494, 495, 497, 518
Time domain transducers, 457
Time of flight, 457, 458, 463, 473, 627
Time-of-flight diffraction (TOFD), 519, 576, 626, 627, 658
Tool-head camera, 618
Tokamak fusion test reactor (TFTR), 582
Total borescope magnification, 580
Transducers (TOFD), 576
Track-etch neutron imaging method, 312
Tracking analysis, 554

Trained operators, 575
Transducer beam spread, 481
Transfer functions, 528
Translation device, 464
Transmissivity, 421, 422, 427
Trend analysis, 560
Tribology, 147
TRIGA reactor, 307, 308, 323
Tristimulus method, 636
Twin engine propellar balancing, 570

Ultran focused transducer, 463
Ultrasonic analyzer, 510, 511
Ultrasonic leak detector, 180, 183
Ultrasonic noise generator, 182
Ultrasonic wave generation, 476, 477
UV-A, 233, 234, 243, 244
Ultraviolet, 152, 222, 224, 378, 578, 637
Uncooled IR linear array, 442
Underwater periscope, 583
Universal format lens (UFL), 600
Unsharpness, 314, 324, 361, 363, 366, 371, 377

Vacuum stress, 645
Vacuum system, 179, 185–188, 190, 191, 194, 196, 197, 216–218, 220
Valves, 16, 41, 179, 180, 183, 190, 196, 200, 208, 217–220, 577, 587, 589
Variable attenuator, 136, 499, 510
Variables affecting results, 517–519
Velocity, 555, 556
Velocity sensors, 545
VDI Guidance 2056, 541–543
Vibration balancing machines, 555
Vibration damping, 532–534
Vibration excitation, 155, 162, 166, 645
Vibration isolation platform, 537–540
Vibration isolation theory, 537–540
Vibration survey and analysis, 568
Vibrational modes, 137, 141
Vibrotest 60 Version 4, 550–555
Viper 4040, 572–574
Visible dye penetrant processes, 227–229
Visual appearance, 576
Visual testing, 244, 519, 576, 577, 656, 658

Wall thickness measurement, 472
Wall thinning, 472, 577

Wall periscopes, 583
Ward Center, 307–309
Water washable, 221, 224, 226–229, 237, 238
Waterfall analysis, 527
Waterfall plots, 527, 528
Weapons of mass destruction (WMD), 3
Wear-resistance thermocouples, 418
Welding defects, 398–400
Wet developer, 225, 229, 230, 238
Wheel unit (UT), 489, 495, 497, 513
White light interferometry, 135

Wireless technology, 1
Working distance, 146, 147, 156, 158, 566, 594, 599, 601, 604, 605, 608, 611
Working tools, 592

X-ray fluorescence, 356, 358
X-ray generator, 345, 356

Yellow cake, 340
Young's modulus (stress/strain), 472

Zero-order Bessel function, 140

CPSIA information can be obtained
at www.ICGtesting.com
Printed in the USA
BVHW042054150719
553533BV00014B/139/P